Mössbauer Spectroscopy in Materials Science

NATO Science Series

A Series presenting the results of activities sponsored by the NATO Science Committee. The Series is published by IOS Press and Kluwer Academic Publishers, in conjunction with the NATO Scientific Affairs Division.

General Sub-Series

A. Life Sciences	IOS Press
B. Physics	Kluwer Academic Publishers
C. Mathematical and Physical Sciences	Kluwer Academic Publishers
D. Behavioural and Social Sciences	Kluwer Academic Publishers
E. Applied Sciences	Kluwer Academic Publishers
F. Computer and Systems Sciences	IOS Press

Partnership Sub-Series

1. Disarmament Technologies	Kluwer Academic Publishers
2. Environmental Security	Kluwer Academic Publishers
3. High Technology	Kluwer Academic Publishers
4. Science and Technology Policy	IOS Press
5. Computer Networking	IOS Press

The Partnership Sub-Series incorporates activities undertaken in collaboration with NATO's Partners in the Euro-Atlantic Partnership Council – countries of the CIS and Central and Eastern Europe – in Priority Areas of concern to those countries.

NATO-PCO-DATA BASE

The NATO Science Series continues the series of books published formerly in the NATO ASI Series. An electronic index to the NATO ASI Series provides full bibliographical references (with keywords and/or abstracts) to more than 50000 contributions from international scientists published in all sections of the NATO ASI Series.
Access to the NATO-PCO-DATA BASE is possible via CD-ROM "NATO-PCO-DATA BASE" with user-friendly retrieval software in English, French and German (© WTV GmbH and DATAWARE Technologies Inc. 1989).

The CD-ROM of the NATO ASI Series can be ordered from: PCO, Overijse, Belgium.

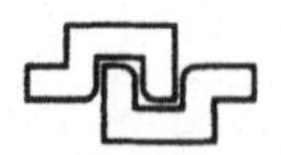

3. High Technology – Vol. 66

Mössbauer Spectroscopy in Materials Science

edited by

Marcel Miglierini
Department of Nuclear Physics and Technology,
Slovak University of Technology,
Bratislava, Slovakia

and

Dimitris Petridis
National Center for Scientific Research 'Demokritos",
Institute of Materials Science,
Athens, Greece

Kluwer Academic Publishers
Dordrecht / Boston / London

Published in cooperation with NATO Scientific Affairs Division

Proceedings of the NATO Advanced Research Workshop on
Mössbauer Spectroscopy in Materials Science
Senec, Slovakia
6–11 September 1998

A C.I.P. Catalogue record for this book is available from the Library of Congress.

ISBN 0-7923-5640-3 (HB)
ISBN 0-7923-5641-1 (PB)

Published by Kluwer Academic Publishers,
P.O. Box 17, 3300 AA Dordrecht, The Netherlands.

Sold and distributed in North, Central and South America
by Kluwer Academic Publishers,
101 Philip Drive, Norwell, MA 02061, U.S.A.

In all other countries, sold and distributed
by Kluwer Academic Publishers,
P.O. Box 322, 3300 AH Dordrecht, The Netherlands.

Printed on acid-free paper

Printed in the Netherlands

CONTENTS

IV. Experimental Techniques and Data Processing

*) corresponding author

PREFACE

Material science is one of the most evolving fields of human activities. Invention and consequent introduction of new materials for practical and/or technological purposes requires as complete knowledge of the physical, chemical, and structural properties as possible to ensure proper and optimal usage of their new features. In order to understand the macroscopic behaviour, one has to search for their origin on a microscopic level. A good deal of microscopic information can be obtained through hyperfine interactions.

Mössbauer spectroscopy offers a unique possibility for hyperfine interaction studies via probing the nearest order of resonant atoms. Materials which contain the respective isotope as one of the constituent elements (e.g., iron, tin, ...) but also those which even do not contain them can be investigated. In the latter case, the probe atoms are incorporated into the material of interest in minor quantities (ca. 0.1 at. %) to act as probes on a nuclear level.

This Workshop has covered the most evolving topics in the field of Mössbauer spectroscopy applied to materials science. During four working days, 50 participants from 19 countries discussed the following areas: *Chemistry, Mineralogy and Metallurgy, Artificially Structured Materials, Nanosized Materials and Quasicrystals,* and *Experimental Techniques and Data Processing.* A total of 42 contributions (30 keynote talks) reviewed the current state of art of the method, its applications for technical purposes, as well as trends and perspectives.

A total of 39 papers are included in the present volume. Applications in *Chemistry, Mineralogy, and Metallurgy* deal with practical questions and problems regarding technological and/or technical applications. The usage of different techniques of Mössbauer spectroscopy (transmission geometry, source experiments, detection of conversion electrons, backscattering geometry, different probe atoms, ...) are thoroughly discussed. Studies of technologically important materials comprising oxides, alloys, phosphates, fluorides, nitrides, steels, diamonds, minerals (terrestrial and extraterrestrial), zeolites, multilayers, amorphous solids, etc. are presented.

In the section *Artificially Structured Materials*, a variety of technological procedures like mechanical ball milling, ion beam-mixing, laser-nitriding and gas-nitriding, spark erosion, electrodeposition, plasma immersion are reported. Results from complementary techniques extend the interpretations and correlate the results obtained with the findings of Mössbauer spectroscopy.

Recently, nanostructured materials which show outstanding features for practical applications have emerged. Investigations dealing with their magnetic and structural properties are discussed in the section *Nanosized Materials and Quasicrystals*. Different approaches taken including different probing atoms provide invaluable information on structural arrangements including (nano)crystalline grains, the amorphous residual phase, and the interface zone.

Finally, methodological aspects and new trends are presented in the section *Experimental Techniques and Data Processing.* Experimental procedures comprising measurements *in situ*, under controlled atmosphere, at high pressure, diffusion studies using source experiments, as well as new instrumentation set-up (including a miniaturized spectrometer for space missions) and data evaluation software are discussed. The main interest, however, is focused on the usage of synchrotron radiation which, by enabling time domain experiments, opens another window intò the materials science investigations. A special round table discussion was organised on this topic on Wednesday evening. The scientists who already have an experience in using synchrotron for Mössbauer effect measurements shared their knowledge with the others. Both scientific and administrative questions were thoroughly assessed, the latter being even more important especially for those who would like to join the existing facilities and perform real experiments.

Mössbauer Spectroscopy in Materials Science (**msms**) is actually a name for a new series of scientific meetings held so far in Kočovce (Slovakia) and Lednice (Czech Republic). Starting with 33 scientists (12 - Slovakia, 10 - Czech Republic, 11 - from abroad) back in 1994, a tradition of biannual colloquia was established aiming to discuss the recent developments in Mössbauer spectroscopy mainly by scientists from the Central and Eastern Europe. However, it turned out that this sort of rather small meetings, which allow intensive „mutual interactions", became so popular, that except of Australia, all continents were already represented in Lednice two years later. This year, our meeting has acquired another dimension, being a workshop attended by distinguished scientists in the field of Mössbauer spectroscopy. In this respect, we are acknowledging financial support from the NATO Scientific Committee which made such an event possible. Our thanks are also due to Slovenské Elektrárne, a.s. (Slovak Power Plants) for a possibility to organise the workshop in their educational centre. Located at the banks of Sunny Lakes in Senec, the workshop site provided pleasant and relaxed atmosphere for numerous scientific discussions and helped with the initiation of new working contacts.

November 1998

Marcel Miglierini
Dimitris Petridis

STRUCTURE AND PROPERTIES OF TIN-DOPED METAL OXIDES

F.J. BERRY[1], C. GREAVES[2], Ö. HELGASON[3], K. JÓNSSON[3],
J. McMANUS[1] AND S.J. SKINNER[1]
[1)] Department of Chemistry, The Open University, Walton Hall, Milton Keynes, MK7 6AA, United Kingdom
[2)] School of Chemistry, University of Birmingham, Edgbaston, Birmingham, B15 2TT, United Kingdom
[3)] Science Institute, University of Iceland, Dunhagi 3, IS-107 Reykjavik, Iceland

Abstract

Tin-doped Fe_3O_4 prepared by solid state reactions is shown by X-ray powder diffraction and extended X-ray absorption fine structure to contain tin in the octahedral sites of the inverse spinel-related structure. *In situ* ^{57}Fe Mössbauer spectroscopy performed at elevated temperatures *in vacuo* shows the Curie temperature to decrease with increasing concentrations of tin and, after prolonged heating at elevated temperatures, the partial segregation of tin to form tin dioxide at the magnetite surface. Thermal treatment under oxidising conditions causes the oxidation of tin-doped Fe_3O_4 to tin-doped γ-Fe_2O_3 and tin-doped α-Fe_2O_3. Tin-doped γ-Fe_2O_3 prepared by precipitation techniques also contains tin in the octahedral sites of the inverse spinel-related structure. ^{57}Fe Mössbauer spectroscopy at elevated temperatures has been used to derive a Néel temperature for tin-doped γ-Fe_2O_3 and to examine the stabilising influence of tin on the conversion of γ-Fe_2O_3 to α-Fe_2O_3. Tin-doped α-Fe_2O_3 prepared by hydrothermal methods contains tin in both interstitial- and substitutional-octahedral sites in the corundum-related matrix. The defect structure is unusual and is rationalised by crystallographic- and chemical- arguments.

1. Introduction

The doping of tetravalent-titanium and -tin into Fe_3O_4 and γ-Fe_2O_3 [1-3] with the inverse spinel-related structure, and into α-Fe_2O_3 with the corundum-related structure [4-10] has been an area of activity for some time because of interest in the magnetic-, electrical- and other physical- properties of the systems. The doping of Fe_3O_4 has attracted attention because of the possible use of the materials in transformer cores, magnetic memories and heterogeneous catalysts. The tin-doped α-Fe_2O_3 system is currently attracting interest because of its sensing properties for gases such as methanol and carbon monoxide [8, 10-12].

M. Miglierini and D. Petridis (eds.), Mössbauer Spectroscopy in Materials Science, 1–12.

Assuming a similarity between titanium and tin, there are advantages in using tin since both iron and tin have Mössbauer isotopes such that phase transformations and magnetic properties at high temperatures are amenable to examination *in situ* by both ^{57}Fe- and ^{119}Sn- Mössbauer spectroscopy. We report here on the synthesis of tin-doped-Fe_3O_4, -γ-Fe_2O_3, and -α-Fe_2O_3 by different methods, their structural characterisation, and the use of Mössbauer spectroscopy to examine phase transformations and magnetic properties at high temperatures.

2. Experimental

Compounds of composition $Fe_{3-x}Sn_xO_4$ were prepared by heating stoichiometric quantities of powdered tin(IV) oxide, α-iron(III) oxide, and metallic iron in sealed evacuated quartz ampoules at 900°C for 24h and allowing the products to cool in the furnace. Compounds of the type γ-$Fe_{2-x}Sn_xO_3$ were prepared by adding aqueous ammonia to aqueous mixtures of iron(III) chloride hexahydrate, iron(II) chloride tetrahydrate, and tin(II) chloride until pH7. The mixtures were boiled under reflux (3h). The precipitates were removed by filtration, washed with 95% ethanol until no chloride ions could be detected in the washings by silver nitrate solution, and heated in air at 250°C (12h). Compounds of composition α-$Fe_{2-x}Sn_xO_3$ were prepared by precipitating aqueous mixtures of iron(III) chloride hexahydrate and tin(IV) chloride with aqueous ammonia and hydrothermally processing the precipitates in a Teflon-lined autoclave at 200°C and 15 atm pressure for 5h. The products were removed by filtration and washed with 95% ethanol until no chloride ions were detected by silver nitrate solution. The products were dried under an infrared lamp.

X-ray powder diffraction (XRD) data were recorded with a Siemens D5000 diffractometer in reflection mode using CuK_α radiation. The program FULLPROF [13] was used for Rietveld refinement and simulation of patterns for specific structural models.

The tin K-edge extended X-ray absorption fine structure (EXAFS) measurements were performed at the Synchrotron Radiation Source at Daresbury Laboratory U.K. with an average current of 200mA at 2GeV. The data were collected in transmission geometry on Station 9.2 at 298K. The raw data were background-subtracted using the Daresbury programme EXBACK and fitted using the non-linear least squares minimisation programme EXCURV92.

The ^{57}Fe Mössbauer spectra were recorded from powdered samples with a constant acceleration spectrometer and a *ca.* 400 MBq ^{57}Co/Rh source. The furnace designed for the *in situ* study of phase transformations has been described in detail elsewhere [14]. The sample thickness was 50-80 mg/cm^2. The line width (FWHM) of the calibration spectrum was 0.24 mms^{-1}. The chemical isomer shift data are quoted relative to the centroid of the metallic iron spectrum at room temperature. The ^{119}Sn Mössbauer spectra were recorded at 298K using a $Ca^{119}SnO_3$ source and a microprocessor controlled Mössbauer spectrometer. The chemical isomer shift data are quoted relative to tin(IV) oxide.

3. Results and Discussion

3.1. TIN-DOPED Fe_3O_4

X-ray powder diffraction data were collected for a sample with nominal composition $Fe_{2.7}Sn_{0.3}O_4$ [15]. The results clearly supported a structure containing octahedral Sn^{4+} ions whose charges are balanced by a corresponding excess of Fe^{2+} ions, and an excellent fit to the experimental data was obtained for this model. Tin K-edge EXAFS recorded from the compound [16] confirmed that the tin atoms adopt the octahedral-, as opposed to tetrahedral-, sites in the structure of Fe_3O_4. X-ray photoelectron spectroscopy showed the incorporation of tin to be accompanied by the partial reduction of Fe^{3+} to Fe^{2+} [16]. The ^{119}Sn Mössbauer spectrum recorded [17] from the material of composition $Fe_{2.9}Sn_{0.1}O_4$ (Figure 1) showed a sextet pattern with a hyperfine field of 20.9T at the Sn^{4+} ions which was interpreted in terms of the supertransfer of spin density via oxygen from the Fe^{3+} ions on the tetrahedral A sites to the Sn^{4+} ions on the octahedral B sites of Fe_3O_4.

The ^{57}Fe Mössbauer spectra recorded *in situ* from the material of composition $Fe_{2.9}Sn_{0.1}O_4$ [17] following heating over a range of temperatures *in vacuo* (pressure *ca.* 0.2Pa) are shown in Figure 2. The spectrum recorded at 300K showed a magnetically split sextet pattern which descreased in magnitude at 750K. On heating to 770K the broad line ^{57}Fe Mössbauer spectrum showed the onset of the collapse of the sextet patterns indicative of most of the tin-doped magnetite being in the paramagnetic state and within a few degrees of the Curie temperature. The ^{57}Fe Mössbauer spectra recorded at 780 and 790K showed the material to be completely paramagnetic. By fitting the magnetic fields at different temperatures to the relationship $B(T) = B_o (1 - T/T_c)\beta$, where B_o is the magnetic hyperfine field at $T = 0K$ and β is 0.3, the magnetic transition temperature was restricted to a temperature interval of *ca.* 10K and T_c calculated as 770 ± 5K which is 70K lower than the Curie temperature of pure magnetite. The ^{57}Fe Mössbauer spectra recorded from the compound $Fe_{2.8}Sn_{0.2}O_4$ showed the Curie temperature to be 680 ± 10K. The magnitudes of the hyperfine

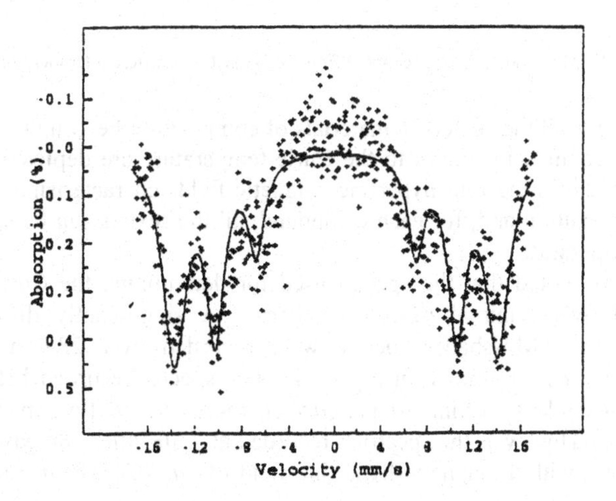

Figure 1. ^{119}Sn Mössbauer spectrum recorded from $Fe_{2.9}Sn_{0.1}O_4$ at 298K.

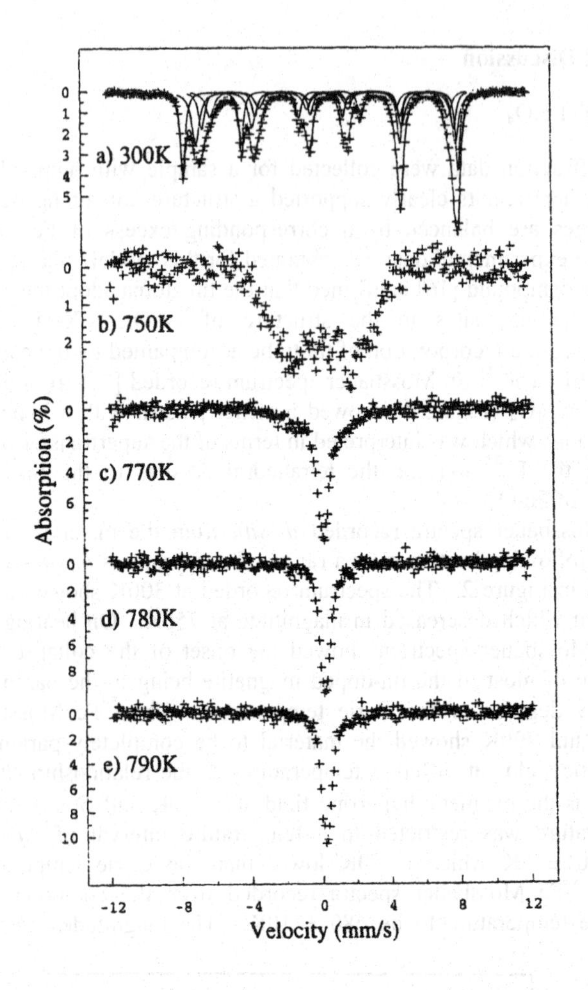

Figure 2. ^{57}Fe Mössbauer spectra recorded from $Fe_{2.9}Sn_{0.1}O_4$ at different temperature *in vacuo*.

magnetic fields for all the sextets in materials of composition $Fe_{3-x}Sn_xO_4$ (x=0.1 and 0.2) measured from room temperature to the Curie temperature are depicted graphically in Figure 3 which shows the four hyperfine magnetic fields characterising each sample to follow a similar pattern and, for each compound, all the sextets can be seen to collapse at the Curie temperature.

It should be noted that the spectra used for determining the Curie temperatures were recorded over minimal measuring times. A significantly different situation pertained when the Mössbauer spectra were recorded over 28-48h at the higher temperatures *in vacuo* as shown in Figure 4. The spectra recorded between 298 and 680K were amenable to fitting to patterns characteristic of tin-doped magnetite as described above. However, the spectrum recorded at 730K after 28h gave an additional sextet component with a magnetic hyperfine field of *ca.* 36T which can be associated

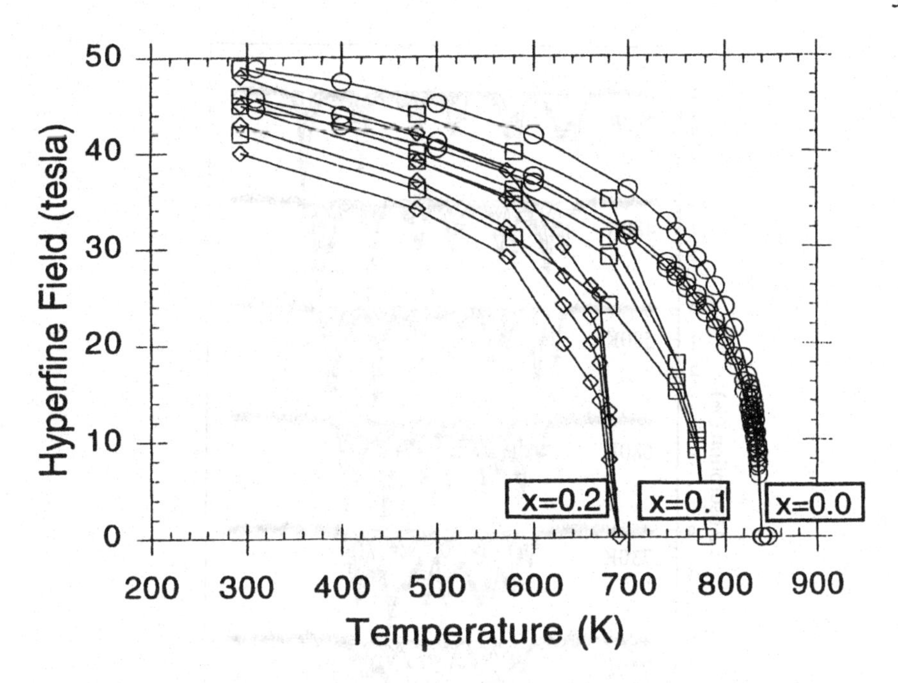

Figure 3. Variation of the magnitude of the magnetic hyperfine fields with temperature for $Fe_{3-x}Sn_xO_4$.

with Fe^{3+} occupying the tetrahedral A sites of pure Fe_3O_4. The spectrum recorded at 750K showed a component similar to that of magnetite together with a component characteristic of the tin-doped magnetite but in which the field had collapsed to *ca.* 12T. The X-ray powder diffraction pattern showed peaks attributable to tin dioxide. The results are consistent with the segregation of tin from tin-doped magnetite at elevated temperatures. The spectrum recorded at 775K, *i.e.* 5K above T_c, showed two distinct patterns, one corresponding to the tin-doped magnetite and the other to the double sextet pattern of pure magnetite.

A different situation was observed [18] when $Fe_{2.9}Sn_{0.1}O_4$ was heated to 680K under oxidising conditions with the furnace open to the air. The spectra are shown in Figure 5. The changes around the higher velocity peaks indicated rearrangements favouring the formation of components with stronger magnetic fields. The fitting of these spectra with two new sextet patterns characteristic of γ-Fe_2O_3 (maghemite) and α-Fe_2O_3 (hematite) allowed a transition to be followed which was interpreted in terms of the oxidation of magnetite to maghemite followed by a structural transition from maghemite to hematite.

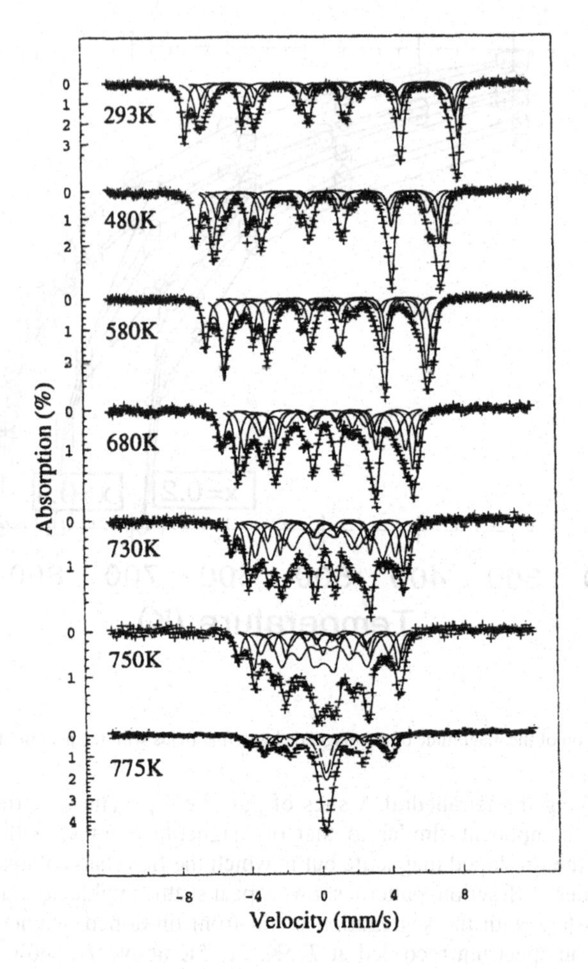

Figure 4. [57]Fe Môssbauer spectra recorded from $Fe_{2.9}Sn_{0.1}O_4$ at elevated temperatures *in vacuo*.

3.2. TIN-DOPED γ-Fe_2O_3

The tin K-edge EXAFS recorded from γ-$Fe_{1.9}Sn_{0.1}O_3$ showed the tin to be coordinated by six oxygen atoms at 2.03Å. The results demonstrate that the tin atoms in tin-doped γ-Fe_2O_3 adopt the octahedral, as opposed to tetrahedral, sites in the inverse spinel related structure.

The X-ray powder diffraction patterns recorded from γ-Fe_2O_3 following heating at various temperatures to 477°C showed the onset of conversion of γ-Fe_2O_3 to α-Fe_2O_3 at 377°C and its virtual completion at 427°C. In contrast, the X-ray powder diffraction patterns recorded from γ-$Fe_{2.9}Sn_{0.1}O_3$ demonstrated the stability of the tin-doped γ-Fe_2O_3 to conversion to tin-doped α-Fe_2O_3 up to 527°C.

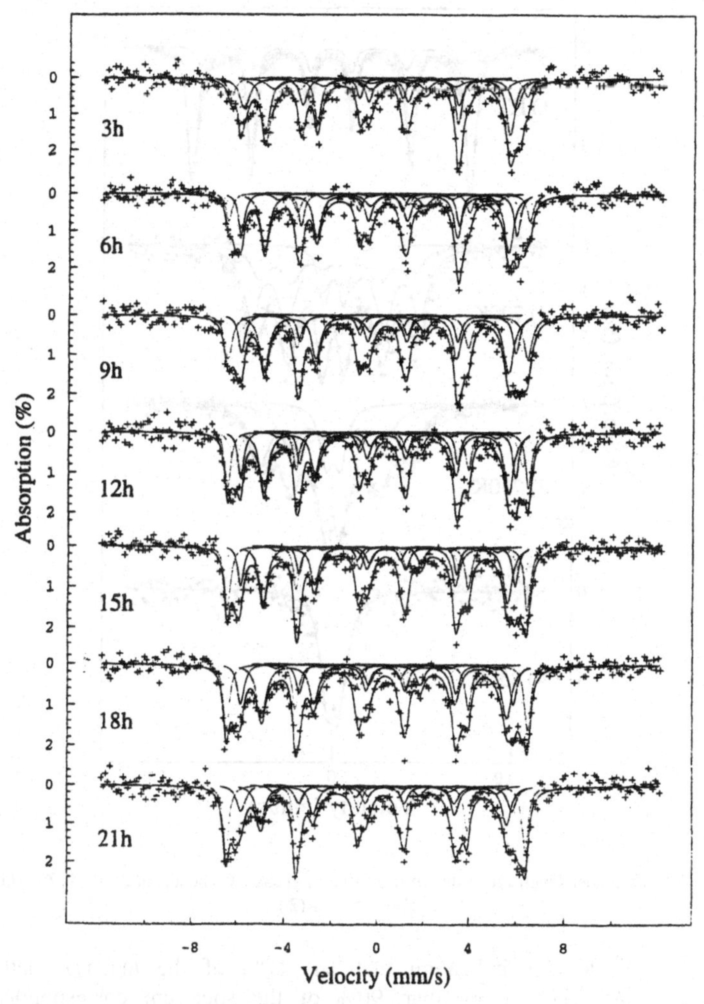

Figure 5. A sequence of ^{57}Fe Mössbauer spectra recorded from $Fe_{2.9}Sn_{0.1}O_4$ every 3 hours at 680K in oxidising conditions.

The ^{57}Fe Mössbauer spectra recorded [15] from γ-$Fe_{1.9}Sn_{0.1}O_3$ between 25 and 557°C (830K) (Figure 6) showed no evidenced for conversion of the γ-Fe_2O_3 to the α-Fe_2O_3 structure. The broad line spectrum recorded at 25°C was fitted to a distribution of sextet patterns with 80% of the spectrum being related to magnetic hyperfine fields between 48.3 and 49.2 Tesla. This distribution reflects the influence of the Sn^{4+} ions on the magnetic interactions within γ-Fe_2O_3. The isomer shift (0.32-0.35 mms^{-1}) confirms the presence of Fe^{3+} and the absence of quadrupole splitting is characteristic for γ-Fe_2O_3.

The spectrum recorded at 427°C (700K) remained dominated by a broad line sextet pattern with the magnetic field distributed between 26 and 30 T. The spectrum recorded at 477°C (750K) showed further reduction (and spreading) in the magnetic field (17 to

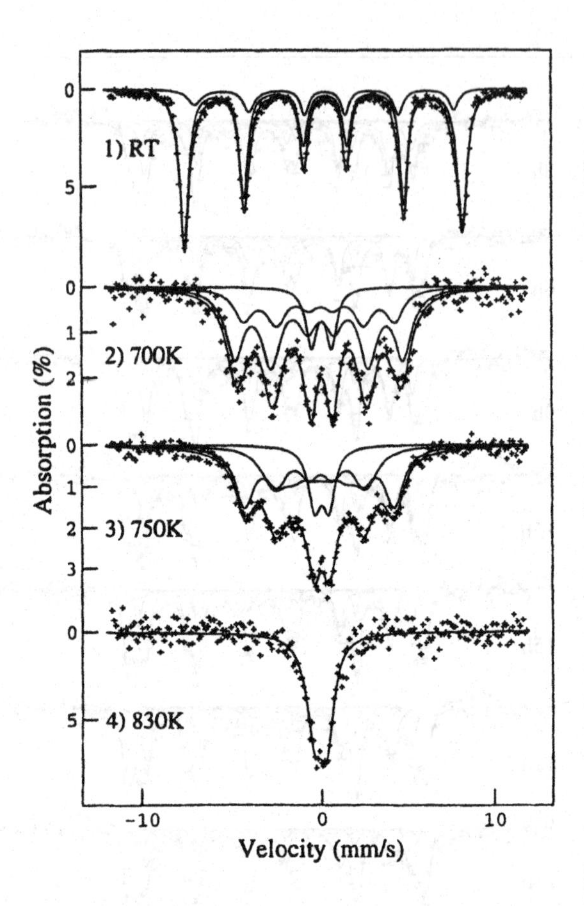

Figure 6. ^{57}Fe Mössbauer spectra recorded *in situ* from γ-$Fe_{1.9}Sn_{0.1}O_3$ heated at 25°C, 427°C (6h), 477°C (21h) and 557°C (2h).

26T) but also a doublet indicating that *ca.* 15% of the material had become paramagnetic. At 527°C more than 90% of the spectrum corresponded to the paramagnetic species whilst the spectrum recorded at 577°C (830K) showed the material to be completely paramagnetic. The results show that γ-$Fe_{1.9}Sn_{0.1}O_3$ has a Néel temperature ranging between 477 - 547°C (750 - 820K) compared to 840 - 860K for pure γ-Fe_2O_3 (an estimated value extrapolated from low temperature Mössbauer spectroscopy data due to the instability of γ-Fe_2O_3). Similar lowering was observed in tin-doped Fe_3O_4. The wide temperature range of the Néel temperature in the present case may be associated with the distribution in particle size (44 to 66 nm as shown by scanning electron microscopy).

The material was subsequently heated to 587°C (860K) and the ^{57}Fe Mössbauer spectra recorded every 4 hours for a total heating period of 24 hours. The first spectrum recorded between 4 and 8 hours (Figure 7) was fitted to a doublet characteristic of paramagnetic γ-Fe_2O_3.

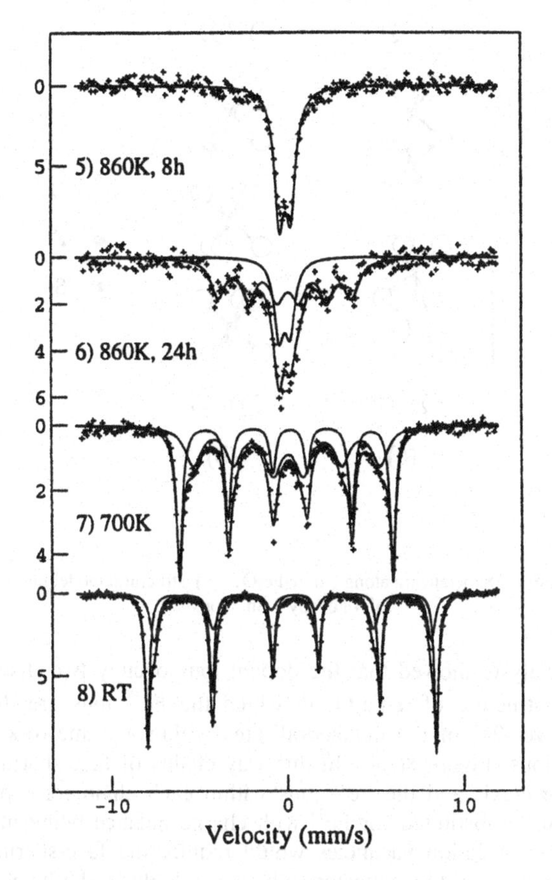

Figure 7. ^{57}Fe Mössbauer spectra recorded *in situ* from γ-$Fe_{1.9}Sn_{0.1}O_3$ heated at 587°C (4-8h), 587°C (20-24h), and on cooling to 427°C and 25°C.

After further heating at 587°C a magnetically split component began to emerge which was clearly observable in the spectrum recorded after 24 hours of heating. The hyperfine field of this sextet (23.5 T) and the quadrupole splitting (-0.1 mms^{-1}) are characteristic of α-Fe_2O_3. The two spectra recorded on cooling to 427°C (700K) and 25°C showed well defined patterns of two sextets. One of the sextet patterns can be related to the formation of tin-doped α–Fe_2O_3 with a magnetic field of 37.4T at 427°C (700K) and 51.0T at room temperature (in both cases the quadrupole splitting was -0.1 mms^{-1}). The second sextet pattern corresponded to a residual amount (38% by area at room temperature) of tin-doped γ-Fe_2O_3 with a magnetic hyperfine field of 33T at 427°C (700K) and 48.8T at room temperature.

3.3. TIN-DOPED α-Fe_2O_3

The structural characteristics of tin-doped α-Fe_2O_3 prepared by hydrothermal methods have been investigated by Rietveld structure refinement of the X-ray powder diffraction

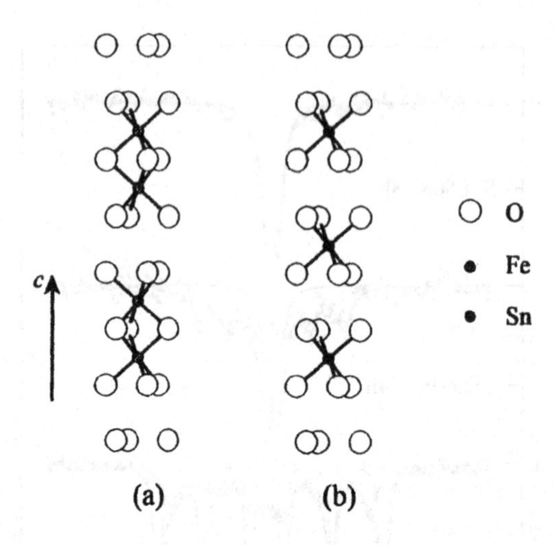

Figure 8. (a) Linking of FeO6 octahedra along *c* in α-Fe_2O_3; (b) structural model involving the substitution of 4 Fe^{3+} by 3 Sn^{4+} ions.

data [19]. The analysis showed that the dopant ions occupy two distinct sites in the corundum-related structure of α-Fe_2O_3 in which the Fe^{3+} ions are distributed in an ordered fashion over 2/3 of the octahedral sites within a framework of hexagonally close-packed O^{2-} ions (Figure 8(a)). In this way chains of face-sharing octahedra are directed along the *c*-axis and the Fe^{3+} ions within each chain form pairs. A simple model in which Sn^{4+} substituted for Fe^{3+} with charge balance being maintained by an appropriate number of cation vacancies would require the face-sharing of SnO_6 and FeO_6 octahedra which would be expected to be unstable due to high cation repulsion; in rutile-related SnO_2 only edge- and corner-shared octahedra occur. The best refinement to the X-ray powder diffraction data involved both interstitial *and* substitutional tin (Figure 8(b)). The presence of tin in the interstial sites would produce strong cation-cation repulsions from the two adjacent face-shared FeO_6 octahedra. Elimination of the cations from these two sites results in two additional octahedral sites (respectively above and below the cations removed) which do not involve face-sharing and which are therefore attractive for occupation by additional tin ions. In this way the structural model shown in Figure 8 was deduced: defect clusters are formed comprising a chain of three tin ions which all avoid face-sharing repulsions. In addition, the cluster is electrically neutral since $4Fe^{3+}$ ions are replaced by $3Sn^{4+}$ ions.

The ^{57}Fe Mössbauer spectrum recorded at 298K from 10% tin-doped α-Fe_2O_3 (Figure 9) was best fitted to two sextet patterns. One sextet accounting for *ca.* 60% of the spectrum was characteristic of α-Fe_2O_3 (δ = 0.37 mms^{-1} Δ = -0.10 mms^{-1}, H = 51.1T). The other sextet pattern (δ = 0.34 mms^{-1}, Δ = -0.10 mms^{-1}, H = 46.5T) may also

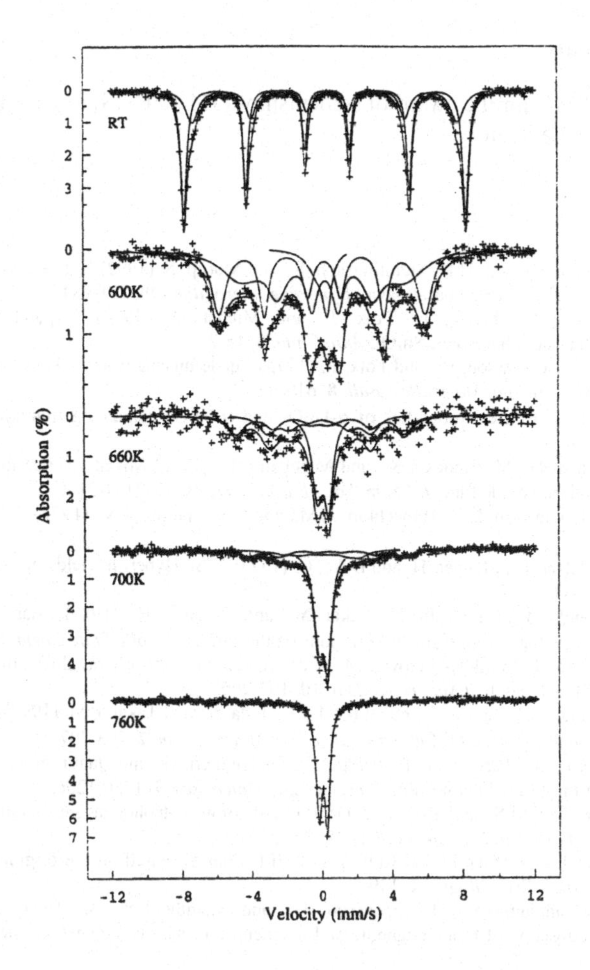

Figure 9. ^{57}Fe Mössbauer spectra recorded from α-$Fe_{1.80}Sn_{0.20}O_3$ at (a) 298K, (b) 600K, (c) 660K, (d) 760.

be associated with α-Fe_2O_3 with the smaller magnitude of the hyperfine magnetic field being ascribed to lower spin density at the Fe^{3+} ions in the vicinity of Sn^{4+} ions.

The spectrum recorded at 600K was also amenable to fitting to two sextet patterns but with smaller hyperfine magnetic fields (H = 36.5 and 30T) as expected at the higher temperature. The appearance of a doublet (δ = 0.15 mms^{-1}, Δ = 0.55 mms^{-1}) is indicative of some of the tin-doped α-Fe_2O_3 particles being sufficiently small to show paramagnetic behaviour at 600K. The growth of the doublet component in the spectra recorded at 660K and 700K is consistent with the progressive development of the paramagnetic phase with increasing temperature. It should be noted that the quadrupole splitting of all spectral components remained constant (~ -0.10 mms^{-1}) throughout the thermal treatment but, as the particles become paramagnetic, the quadrupole splitting increased to +0.55 mms^{-1}.

Acknowledgements

We thank the EPSRC for the award of studentships (JM and SJS) and for facilities at the SRS at Daresbury Laboratory.

References

1. Helgason, Ö., Gunnlaugsson, H.P., Steinthorsson, S., and Morup, S. (1992) High temperature stability of maghemite in partially oxidised basalt lava. *Hyperfine Interactions* **70**, 981-984.
2. Djega-Mariadassou, C:, Basile, F., Foix, P. and Michel, A. (1973) Preparation et proprieties crystallographiques des phases $Fe_{3-x}Sn_xO_4$, *Ann. Chim.* **8**, 15-20.
3. Basile, F., Djega-Mariadassou, C., and Foix, P. (1973) Sur le mechanisme de substitutions des cations Sn^{4+} et Fe^{2+} dans magnetite, *Mater. Res. Bull.* **8**, 619-626.
4. Morin, F.J. (1951) Electrical properties of α-Fe_2O_3 and α-Fe_2O_3 containing titanium, *Phys. Rev.* **83**, 1005-1010.
5. Uekawa, N., Watanabe, M., Kaneko, K., and Mizukami, F., (1995) Mixed valence formation in highly oriented Ti-doped iron oxide film, *J. Chem. Soc., Faraday Trans.* **91**, 2161-2166.
6. Fabritchnyi, P.B., Lamykin, E.V., Babechkin, A.M., and Nesmeianov, A.N., (1972) *Solid State Commun.* **11**, 343-348.
7. Schneider, F., Melzer, K., Mehner, H. and Deke, G. (1977) ^{119}Sn Hyperfine fields in α-Fe_2O_3, *Phys. Stat. Sol (a)* **39**, K115-118
8. Takano, M., Bando, Y., Nakanishi, N., Sakai, M., and Okinaka, H., (1987) Characterisation of fine particles of α-Fe_2O_3-SnO_2 with residual SO_4^{2-} ions on the surface *J. Solid State Chem.* **68**, 153-162.
9. Music, S., Popovic, S., Metikos-Hukovic, M., and Gvozdic, G. (1991) X-ray diffraction and Mössbauer spectroscopy of Fe_2O_3-SnO_2, *J.Mater. Sci. Lett.* **10**, 197-200.
10. Kanai, H., Mizutani, H., Tanaka, T., Funabiki, T., Yoshida, S., and Takano, M. (1992), X-ray absorption study of fine particles of α-Fe_2O_3-SnO_2 gas sensors, *J. Mater. Chem.* **2**, 703-707
11. Bonzi, P., Depero, L.E., Parmigiani, F., Perego, C., Sberveglieri, G., and Quattroni, G. (1994) Formation and structure of tin-iron oxide thin film CO sensors, *J. Mater. Res.* **9**, 1250-1256.
12. Liu, Y., Zhu, W., Tse, M.S., and Shen, S.Y. (1995) Study of new alcohol gas sensors made from ultrafine SnO_2-Fe_2O_3, *J. Mater. Sci. Lett.* **14**, 1185-1187.
13. Rodriquez-Carajal, J., (1994), FULLPROF V 261 (ILL France), based on the original code by Young, R.A., (1981) *J. Appl. Crystallogr.* **14**, 149.
14. Helgason, Ö., Gunnlaugsson, H.P., Jónsson, K., and Steinthorsson, S. (1994) High temperature Mössbauer spectroscopy of titanomagnetite and maghemite in basalts, *Hyperfine Interactions* **91**, 595-599.
15. Berry, F.J., Greaves, C., Helgason, Ö., and McManus, J. (1998) Synthesis and characterisation of tin-doped iron oxides, *J. Mater. Chem.*, in press.
16. Berry, F.J., Bilsborrow, R., Helgason, Ö., Marco, J.F., and Skinner, S.J. (1998) Location of tin and charge balance in materials of composition $Fe_{3-x}Sn_xO_4$ ($x<0.3$), *Polyhedron* **17**, 149-152.
17. Berry, F.J., Helgason, Ö., Jónsson, K., and Skinner, S.J. (1996) The high temperature properties of tin-doped magnetite, *J. Solid State Chem.* **122**, 353-357.
18. Helgason, Ö., Berry, F.J., Jónsson, K., and Skinner, S.J. (1996) A study of the oxidation of tin-doped magnetite by ^{57}Fe Môssbauer spectroscropy, *Proceedings of the International Conference on the Applications of the Môssbauer Effect*, Ed. I. Ortalli, S.I.F. Bologna, p.59-62.
19. Berry, F.J., Greaves, C., McManus, J., Mortimer, M., and Oates, G. (1997) The structural characterisation of tin- and titanium-doped α-Fe_2O_3 prepared by hydrothermal synthesis, *J. Solid State Chem*, **130**, 272-276.

THE KINETICS OF THE PHASE TRANSITION IN FERROSILICON SYSTEM

Studied by Mössbauer Spectroscopy and EXAFS technique

Ö. HELGASON[1], B. JOHANNESSON[2], B.PURSER[3] AND F. BERRY[3]
1) Science Institute, University of Iceland, Reykjavik, Iceland
2) Technological Institute of Iceland, Keldnaholti, 112-Reykjavik, Iceland
3) Department of Chemistry, The Open University, Milton Keynes, UK

Abstract

Ferrosilicon is an important additive in steel production. The phase dynamics of the silicon rich part of the binary Fe-Si system has been the subject of particular controversy and there remains uncertainty about the stability and the structure of the phases in this system. In particular there is confusion over the rate and mechanism of the transformation of the so-called zeta-alpha phase of iron disilicide to the zeta-beta phase. Iron-Mössbauer spectroscopy and EXAFS study have been used to investigate the phase transition in the silicon rich part of the ferrosilicon system in a temperature range 550-650°C. The purpose of this paper is twofold. Firstly, it will describe the use of these techniques as tool for "in situ" studies of phase transformations at elevated temperature. Secondly, the Debye temperature of the alpha, beta and epsilon phases is estimated from changes in the area ratio of Mössbauer spectra recorded at different temperatures. Then the ferrosilicon system in the region of 33 to 75% Si by weight will be discussed in some detail.

1. Introduction

Silicon rich ferrosilicon has been used in the production of cast iron and steel for more than a century. Throughout the history of the industry, disintegration of ferrosilicon has been a problem that has concerned both manufacturers and users of this product. The problem can arise in handling or crushing of the material, from processes taking place in the material such as phase transformations, especially due to differences between coefficients of thermal expansion of the phases involved. It has recently been pointed out, by Johannesson *et al.* [1], that phase transformations and thermal mismatch creates high tensile stresses in the alpha phase at room temperature, resulting in stress relaxation by cracking of the brittle alpha matrix. This causes unwanted generation of fines with the consequences of financial losses for the manufacturer and inconveniences for the user.

In order to shed light on the processes leading to disintegration, it is of crucial importance to understand how the microstructure develops during and after solidification and, in particular, the rate and extent of phase transformations taking place

M. Miglierini and D. Petridis (eds.), Mössbauer Spectroscopy in Materials Science, 13–24.

in the solid phase. Early investigations showed that these problems arose with alloys in the range 33-80% Si by weight, but the alloys in the range 53-57% Si were most susceptible. The ζ_α-phase or "Lebeauite" (now called alpha phase) has a composition of 53.5-56.5% Si and experimental work done in the 1950's and 1960's focused on this phase and how it is related to disintegration. In the 1950's it was realised that iron disilicide occurs in two forms, termed alpha (ζ_α) and ζ_β-phase (from now called beta phase), as shown in Figure 1. The alpha phase is formed on solidification and is a non-stoichiometric metallic phase. The structure is tetragonal and each unit cell has two Si sites and one Fe site, which is not always occupied [3,4]. The chemical composition is Fe_xSi_2 where x is in the range 0.77-0.87. The alpha phase is optically anisotropic and is stable above 937°C [5]. Below 937°C the phase decomposes into beta and Si by the reaction:

$$\text{alpha} \rightarrow \text{beta} + \text{Si} \qquad (1)$$

The beta phase is stoichiometric with the chemical formula $FeSi_2$ (66.7 atomic % Si, 50.14 wt % Si). The structure is orthorhombic with 48 atoms per unit cell [6]. Recently, the semiconducting beta phase (β-$FeSi_2$) has been studied extensively because it is considered to be one of the candidate materials for future Si optoelectronic devices, solar cells, photo detectors and thermoelectric devices [7, 8].

A volume increase is associated with the phase transformation in (1). Dilatometric studies by Holdhus [9] showed that upon cooling, the greatest fractional length increase (0.6%) is for an alloy with 56.6 wt% Si. Holdhus points out that the tendency of ferrosilicon to disintegrate is most pronounced for alloys with compositions in the range

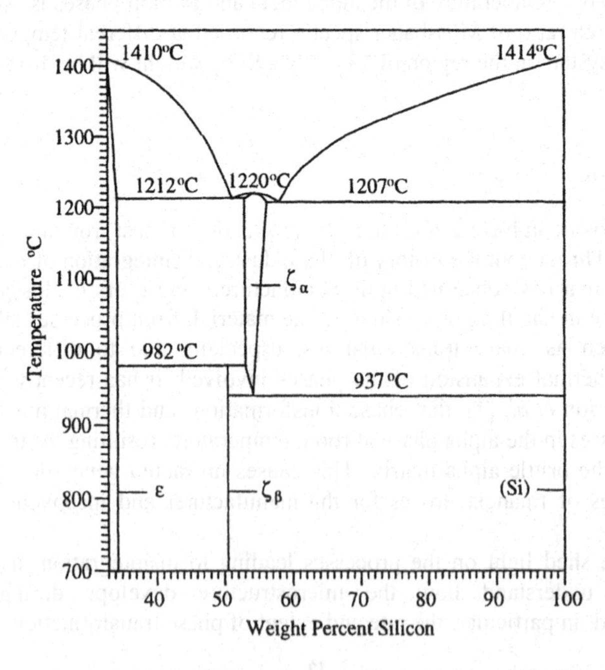

Figure 1. The Fe-Si phase diagram redrawn after Massalski, [2].

50-60% Si and that this is not surprising, considering the measured length increase associated with the phase transformation in (1). Calculations by Espelund [10] have shown that the expected length increase is about 1%, which is somewhat greater than that measured by Holdhus. The exact mechanism of the eutectoid reaction in (1) is not fully understood and Boomgaard [11] has suggested that the reaction takes place in two steps. In the first step, alpha transforms into beta(sat), which is supersaturated in Si. In the second step, silicon precipitates to form eutectoid particles in a beta matrix

i) alpha → beta(sat) (2)

ii) beta(sat) → beta + Si (3)

Mössbauer spectroscopy has been used to study the kinetics of phase transformations in ferrosilicon. Sigfusson and Helgason [12], used this technique at room temperature to determine the TTT-diagram (time-temperature-transformation) for pure ferrosilicon of alpha composition and found that the transformation proceeds fastest at about 720°C. In further work, [13], high temperature Mössbauer spectroscopy was used to study "in situ" the phase dynamics of the silicon rich part of the Fe-Si binary system and the influence of aluminium impurities.

This paper presents studies of the time-temperature transformation of the material of composition 50-57% Si by weight. The work will be confined to two parts of the TTT-diagram. At a transition rate comparable to the time domain of Mössbauer spectroscopy, series of spectra were recorded "in situ" at the relevant temperature. At still faster rate, high temperature EXAFS studies were performed. These two techniques span a wide time domain which means that it is possible to observe the phase transition over a temperature range which would not be accessible using one technique alone. This work demonstrates the way in which the EXAFS technique can be complimentary to the Mössbauer technique, enhancing the level of the results which can be obtained.

2. Experimental

2.1. SAMPLES

The ferrosilicon samples of high purity for the phase transition studies were provided by Metal Crystals & Oxides, Cambridge, UK. (+99,8% pure iron, Si +10ppm). The lump samples were of approx. 30 g of composition; 50.2%, 54.6% and 57% Si by weight. The first one can (before the heat treatment) be described as a mixture of epsilon (FeSi) and the alpha phase. This was verified by SEM (Scanning Electron Microscope) and XRD analysis and is in agreement with the Mössbauer spectroscopy, although the Mössbauer analysis alone cannot be conclusive in distinguishing between the beta and epsilon phases. The second one has the same composition as the high temperature alpha phase and the third one is still higher in Si content. For both these samples the SEM/XRD and MS verifies that only the alpha phase and some Si precipitation, but no beta phase, is present before the heat treatment.

The samples for the determination of the Debye temperature of the alpha, beta and epsilon phases have been described in earlier work, [14].

2.2. HIGH TEMPERATURE MÖSSBAUER SPECTROSCOPY

Mössbauer spectroscopy is useful in following phase transitions in the ferrosilicon system in two main time domains. At temperatures where the process is slow compared to the time needed for acquiring a spectrum with good resolution, spectra recorded at room temperature, before and after the process of annealing, are sufficient for determining the rate of transition. Then, at rates comparable to the time domain of Mössbauer spectroscopy, series of spectra recorded "in situ" at elevated temperatures become important. If, for example, two hours are needed for a reasonable spectrum, processes at temperatures where the phase transitions last for 12-60 hours can easily be studied, assuming the furnace can cope with this temperature range. At still higher temperatures (faster transition rate) EXAFS using synchrotron radiation can be used as described later.

The furnace used for the high temperature Mössbauer spectroscopy is shown schematically in Figure 2. The furnace is cut out of a block of aluminium with an elliptical inner shape. The sample holder is placed in one focal point and a halogen light bulb in the other. The light from the light bulb is by reflection at the inner surface collected at the quartz sample holder, which is then warmed up. The sample is placed between two BeO plates and a K-type thermocouple is in direct contact with the powdered sample. With a light bulb of 250 W the furnace can reach a temperature of 600°C in 4-5 minutes and the power regulator which flashes light on and off keeps the temperature on the samples fixed within +/- 0.5 °C for 24 hours. The pressure in the furnace is kept at less than 1 Pa during the phase transition. The aluminium block is water cooled to increase the lifetime of the light bulb. The furnace is described in more

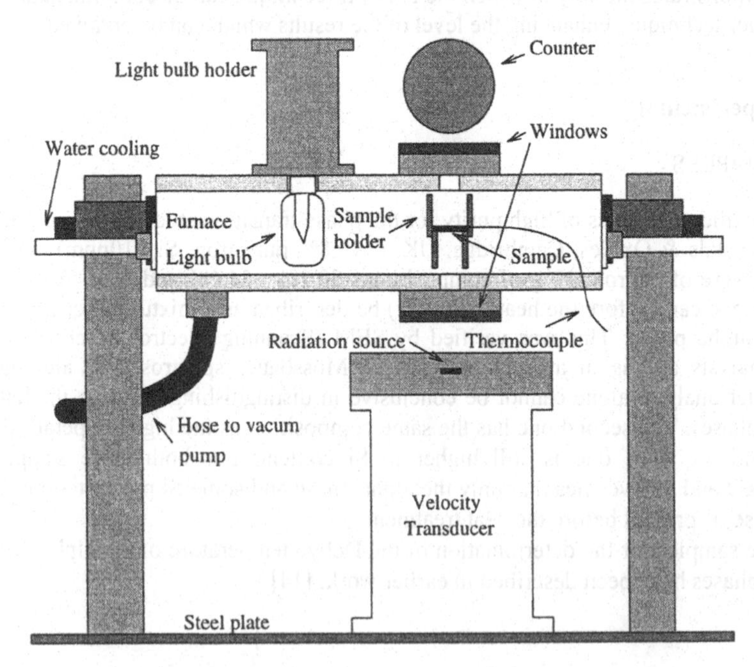

Figure 2. Set up for high temperature Mössbauer spectroscopy.

detail by Helgason *et al.* [15].

Samples of approximately 60 mg were used and Mössbauer spectra were recorded with a constant acceleration spectrometer with a line width (FWHM) of the calibration spectrum of 0.24 mm/s at room temperature.

2.3. COMBINED EXAFS AND XRD

Extended X-ray absorption fine structure (EXAFS), combined with simultaneous X-ray powder diffraction (XRD) measurements recorded at high temperature can be used to follow the phase transition in ferrosilicon at temperatures where the time domain is too short for successful investigation by Mössbauer Spectroscopy. Iron K-edge EXAFS of reasonable quality may be recorded in five minutes using synchrotron radiation, and hence the technique may be used where the time for the phase transition is as little as one hour.

The furnace used for the simultaneous high temperature iron K-edge EXAFS and XRD experiments can reach temperatures of up to 1300K at a rate of 20 K/min. A constant flow of nitrogen through the sample chamber reduces oxidation of both the sample and the platinum wires. The samples are ground to a fine powder (25 to 50μm) and mixed with dry boron nitride powder with a sample concentration of 25 - 50 wt %. Approximately 50 to 60 mg of this mixture is pressed into a pellet of 13 mm diameter under a pressure of 10 tonnes for about 3 minutes. The finished pellet must contain no cracks or pinholes and should absorb 10 - 15% of the incident radiation during the experiment. The pellet is held on a ceramic support in the centre of the furnace at an angle of 45° to the incident radiation.

The phase transition has been monitored by using this technique at 600°C and 650°C. At these temperatures the phase transition takes approximately 15 and 1 hour respectively. Alternate iron K-edge EXAFS and powder XRD data were recorded throughout the experiments. The EXAFS data were recorded over times between 5 and 30 minutes per scan, and XRD data were recorded over times between 5 and 10 minutes per scan, depending upon the time domain of the phase transition.

The monitoring EXAFS involves recording of the small fluctuations in the X-ray absorption spectrum of a material immediately after the atomic absorption edge. The Fourier transform of the background subtracted spectrum in K-space can be directly related to the environment of the absorbing atom. Hence a peak at 2.4Å in the Fourier transform represents a certain number of atoms surrounding the central atom in a shell at a radius of 2.4Å. The EXAFS data were analysed using the fast curved wave theory by the program EXCURV97 from Daresbury Laboratory, UK.

3. Results and Discussion

3.1 THE DEBYE TEMPERATURE OF THE α , β AND ε PHASE

When determining the relative amount of the different phases in a multi-phased sample by Mössbauer spectroscopy, it is necessary to know the respective f-factors or Debye temperatures. The area of the doublet in the spectrum is proportional to the number of iron atoms in the relevant phase, times the f-factor, which describes the possibility for the iron atom being detected. Then, the area of the phase (i), A_i, can be related to the number of iron atoms in the phase (i), N_i, and hence the amount of the phase by following equation:

$$A_i = \text{const} * N_i * f_{i,T} \tag{4}$$

where the constant describes the geometrical conditions in the set-up and $f_{i,T}$, is the f-factor of the phase (i) at the temperature T.

Assuming the Debye model gives a satisfactory description of the vibration modes of the phases involved, the temperature dependency of the f-factor is given by the following equation:

$$f = \exp\left(-\frac{6E_R}{k_B \Theta_D} \left\{ \frac{1}{4} + \left(\frac{T}{\Theta_D} \right)^2 \int_0^{\frac{\Theta_D}{T}} \frac{y dy}{e^y - 1} \right\} \right) \tag{5}$$

where Θ_D is the Debye temperature and E_R is the recoil energy in the Mössbauer gamma ray emission.

To keep track on the different phase transitions in ferrosilicon described in this work, it was necessary to determine the Debye temperature for the three phases involved. For this purpose Mössbauer spectra were recorded in the temperature range 300-850 K for three samples, where one had been characterised as pure alpha ferrosilicon, one as pure β-$FeSi_2$ and one as pure epsilon phase (FeSi) by SEM and Mössbauer spectroscopy. A comparison of the room temperature spectra before and after the temperature scans, gave no indication of a phase transition in any of the three cases. This is of special importance for the alpha sample since it is not stable in the higher temperature range.

Since all the spectra on the temperature scan for each phase are obtained at the

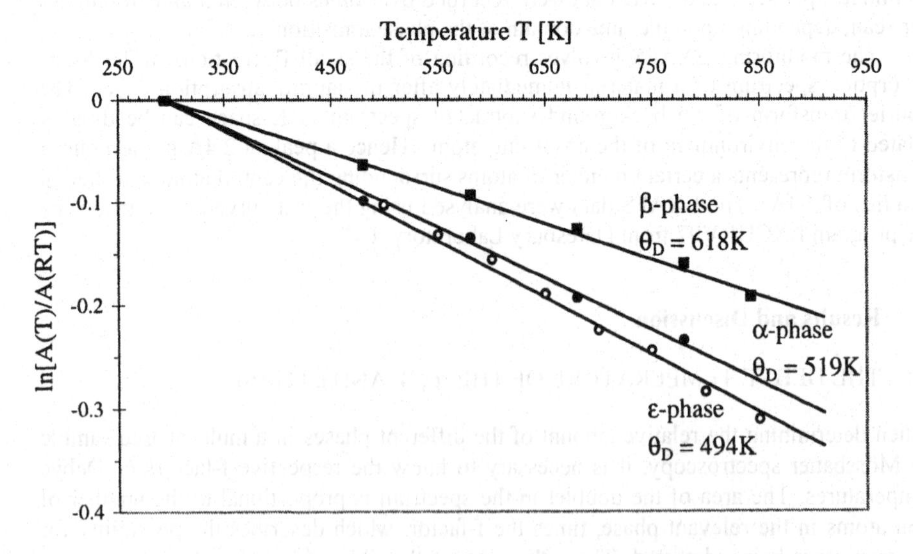

Figure 3. Determining the Debye temperature of the three phases.

same geometrical conditions, the changes in the total area of absorption, A_T, are only proportional to the changes of the f-factor. Then the set of data, area (A_i) versus temperature (T_i), obtained from the scan, can be related to the Debye temperature, Θ_D, by applying equation (4) and (5). A non-linear optimisation routine was used to fit the data by the least squares method and the results are shown in Figure 3. For comparison the area of all three samples are normalised to become 1 at 300K. The best fit gives the Debye temperature for the alpha ferrosilicon of 519 (+/-10) K, 618 (+/-10) for the β-$FeSi_2$ and 494 (+/-10) K for the epsilon phase.

TABLE 1. The f-factor and the hyperfine parameters at different temperatures.

T (K)	Debye temp / f-factor	1st. doublet (mm/s) IS	QS	Γ	2nd. doublet (mm/s) IS	QS	Γ
alpha phase	Θ= 519 K						
RT	0,85	0,20	0,44	0,37/0,41	0,27	0,76	0,30
500 K	0,77	0,06	0,42	0,36/0,41	0,14	0,70	0,29
850 K	0,65	-0,18	0,32	0,33	-0,13	0,60	0,27
beta phase	Θ= 618 K						
RT	0,89	0,15	0,41	0,28	0,02	0,43	0,28
500 K	0,83	0,01	0,40	0,28	-0,18	0,42	0,28
850 K	0,74	-0,23	0,38	0,26	-0,36	0,40	0,27
epsilon ph.	Θ= 494 K						
RT	0,84	0,26	0,51	0,33			
500 K	0,75	0,13	0,37	0,32			
850 K	0,62	-0,12	0,22	0,32			

All three phases have been studied in detail by Mössbauer spectroscopy as cited in [14 and 16]. The epsilon phase is characterised by one doublet but the alpha and beta phases can be characterised by two sets of quadrupole doublets at room temperature and in this work a set of two doublets were also applied at elevated temperatures for these phases. In the fitting procedure of the spectra from the temperature scans an additional constrain can be applied as the isomer shifts are expected to change with temperature according to the second order Doppler shift. The parameters at three different temperatures of the doublets (the isomer shift, (IS), the quadrupole splitting, (QS), and the line width, (Γ) used in the fitting procedure of the Mössbauer spectra, and the f-factor calculated from (5) are given in Table 1.

3.2. PHASE TRANSITION

The phase transitions for the 54.6% (Si by weight) sample, alpha composition, was studied at 582 and 606 °C followed by EXAFS at 600 and 650 °C as the transition rate becomes too fast for the Mössbauer studies. Mössbauer studies were also made at 600°C for the 50.2% and 57% Si samples .

3.2.1. Phase transition followed by EXAFS

The time scan of the XRD and EXAFS measurements at 600°C are depicted in Figure 4. The calculations of the time pattern in the phase transition is based on the EXAFS data, but XRD data are mostly used as control of the overall proceeding.

The iron K-edge EXAFS of ferrosilicon in the alpha phase is characterised by a shell of eight silicon atoms at a distance of 2.34Å and a shell of 3.4 iron atoms (an average due to the iron vacancies in the structure) at a distance of 2.68Å. Further shells cannot usually be resolved at high temperatures. The spectrum of ferrosilicon in the beta phase is characterised by a shell of eight silicon atoms at distances between 2.33 and 2.38Å and a shell of 2 iron atoms at a distance of 2.97Å. Hence the relative abundance of the two phases recorded by EXAFS during the phase transition may be calculated by fitting the fractional occupation numbers of the shells at 2.68 and 2.97Å. This is done by calculating a contour plot of fit indices for all combinations of occupation numbers, an example of which is shown in Figure 5a. Assuming that there is no transitional phase, the best fit must lie on the line characterised by following equation,

$$\frac{\text{average occupation no. of "}\alpha\text{" shell at 2.68Å}}{3.4} + \frac{\text{average occupation no. of "}\beta\text{" shell at 2.97Å}}{2} = 1 \qquad (6)$$

The EXAFS data corresponding to the contour plot shown in Figure 5a is therefore fitted to an average of 2.6 iron atoms in the "α" shell at 2.68Å and an average of 0.5 iron atoms in the "β" shell at 2.97Å, which implies approximately 80% ferrosilicon in the alpha phase is still present in the sample at that point in the experiment. In this way, the abundance of alpha phase ferrosilicon can be evaluated for each EXAFS data set recorded during the experiments. The results from the time scan at 650°C are shown in Figure 5b.

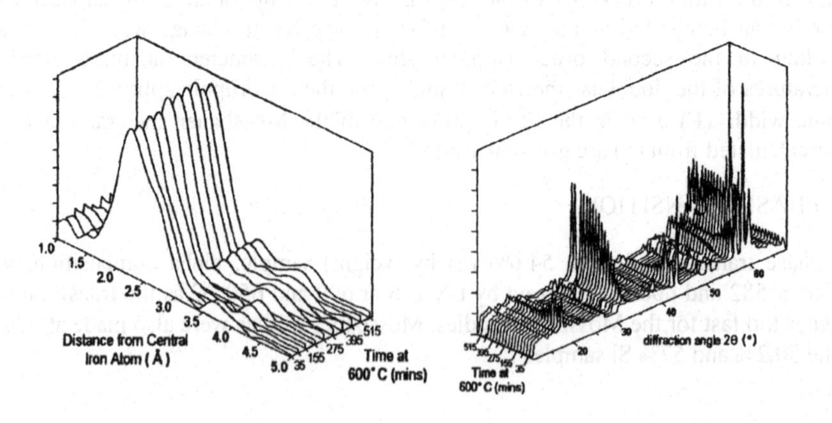

Figure 4. EXAFS and XRD spectra at 600°C as a function of time.

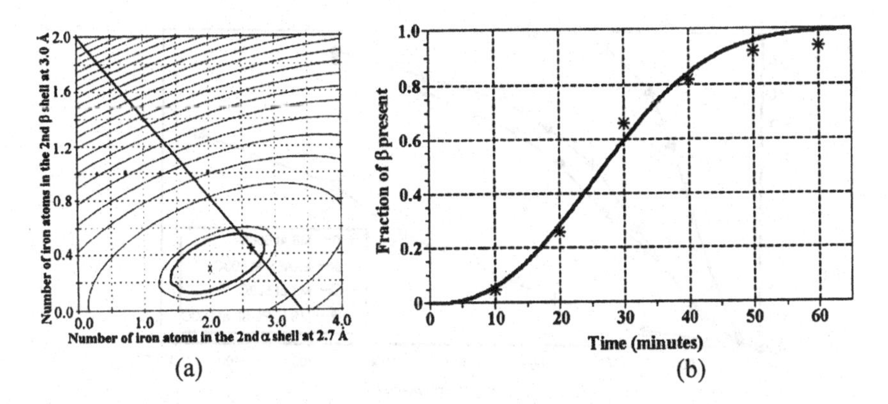

Figure 5. (a) A contour plot from the EXAFS used for determining the alpha/beta ratio. (b) The progress in the phase transition calculated from the time series of contour plots.

3.2.2 Phase transition followed by Mössbauer Spectroscopy

The Mössbauer spectra were collected every two hours at the elevated temperature. The relevant parameters used for the fitting procedure are similar to those listed in Table 1. Collected results from these measurements are displayed in Figure 6. The procedure of the work with the Mössbauer data is the same as described in detail in earlier work [13]. For comparison, data from the transition recorded with the EXAFS technique at 600°C are included in the figure. The increase in the transition rate as the temperature increases is obvious. The time needed to reach 50% transition decrease exponentially with increasing temperature in a good agreement with earlier work where only room temperature Mössbauer spectroscopy was applied [12]. There is a difference between the results from the EXAFS and Mössbauer spectroscopy as the transition seems to go somewhat faster in the EXAFS. On temperature scale, this is equivalent to a 5-8 degrees shift in the TTT-diagram. This could be due to differences in particle size, the particle size of the samples used in the Mössbauer spectroscopy was not thoroughly controlled like those of the EXAFS work. Also, the relatively slow heating of the EXAFS furnace (20K/min) might have some effect on the real starting point of the phase transition.

3.2.3. Studies of the 50,2% and 57% (Si by weight) samples

The purpose of this study was to investigate the effect of changes in the Si content, relative to the composition of the alpha phase, on the phase transition. The samples were of the same purity as the 54.6% sample. This is of importance, since it is known that even a small amount of impurity can change the rate considerably [13]. The increase of the Si/Fe ratio to 57% Si seems to have no detectable influence on the transition rate at 600°C, as can be seen from the results depicted in Figure 6.

The situation is somewhat different for the 50.2% sample. At the start, this sample is a mixture of alpha and epsilon phases. According to SEM studies, there was no sign of the beta phase before heat treatment, and from Mössbauer spectroscopy it was estimated that 34% of the sample was in the epsilon phase. In the case of samples containing the epsilon phase, it is of importance to have a cross check with the SEM-technique since Mössbauer spectroscopy cannot distinguish unambiguously between the beta phase and epsilon phase. After 1200 minutes at 597°C, about half of the sample had

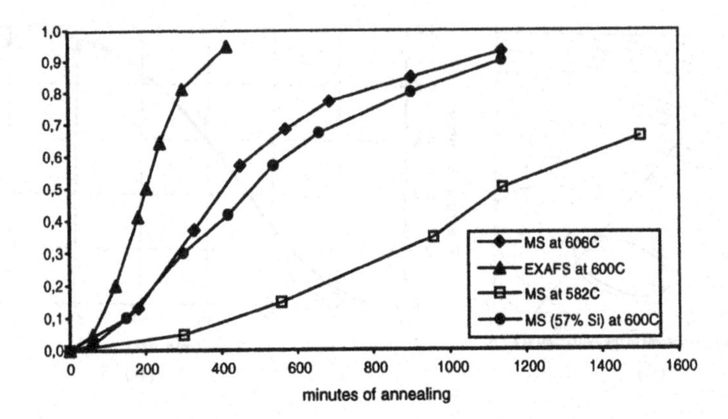

Figure 6. Phase transition at different temperatures as function of annealing time in minutes.

transformed into the beta phase and finally after 2500 minutes the sample consists of 83% of beta, 11% of epsilon and 7% of the alpha phase. Compared to the results in Figure 6 this sample transforms 2-3 times slower than the 54.6% and 57% Si samples. Even though the transition process is slower, when the epsilon phase is present, the result confirms that the beta phase is the most stable of the three phases involved. SEM pictures of the sample, before heat treatment and, after 18 hours annealing at 600°C are shown in Figure 7. The number and size of the "white" dots, which by microprobe were shown to have the composition of the epsilon phase (FeSi), has clearly decreased.

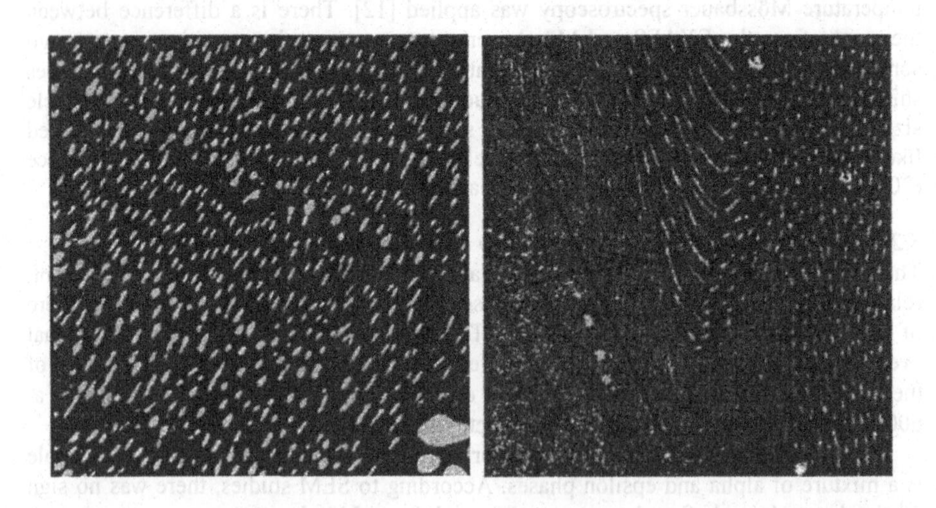

Figure 7. SEM-pictures (x400) in back scattering mode. To the left is the sample before heat treatment, to the right is after 18h at 600°C. The "white dots" are the epsilon phase.

3.3. THE KINETICS OF THE PHASE TRANSFORMATION

Another interesting aspect of the rate-time pattern, obtained from "in situ" measurement, is the possibility of studying the kinetics of the growth of a new phase. The rather complex theories on diffusional transformations and transitions controlled by interface migration are often presented by a simplified formula (6), describing the growth of the new phase,

$$x(t) = 1 - \exp(-kt^n) \quad (6)$$

where x is the fraction of the new phase formed at the time, t, and the "n " and "k" are constants related to geometrical conditions at the grain boundaries and rate of precipitation [17,18 and 19]. On the basis of the Mössbauer and EXAFS results, as depicted in Figure 5b and Figure 6, the value of k and the exponent "n" was estimated for the growth of the beta phase. The three temperature scans recorded by Mössbauer spectroscopy gave the same value for the exponent, n = 1.9 (+/- 0.2). The values obtained from the EXAFS at 600 and 650°C were 2.2 and 2.5, respectively. The reason for this difference is not clear. In both techniques, the calculations were mostly based on the time interval, where the beta fraction goes from 20% to 80%, as error in the ratio increases steeply outside of these values. The different n values might be due to difference in particle size. According to the theory, an increase of the n value is to be expected, when the growth front of the interface controlled transition, goes from a planar or cylindrical form, to a spherical one. This difference could also indicate, that at increasing temperature, the transition goes from more diffusional controlled transition to more interface controlled transition. Excess of Si, compared to the alpha composition, did not seem to affect the kinetics of the phase transition at 600°C. A similar results at 540°C was obtained in earlier work, [13]. Further conclusions will have to await a more systematic investigation and will not be attempted at this point.

4. Conclusion

High temperature "in *situ"* Mössbauer spectroscopy and EXAFS measurements have been used to follow in detail the transition of the alpha phase to beta phase for a time domain of less than an hour up to 50 hours in a ferrosilicon system.

The rate of transition increases exponentially with an increasing temperature in the range of this study. Excess in Si, compared to the nominal alpha composition seems to have no effect on the transition rate.

For a sample with lower Si content (50.2%) the third phase, the epsilon phase, is also involved. In this case the rate becomes 2-3 times slower, but the transformation shows that the beta phase is the most stable of the three phases.

The Debye temperatures of the three phases involved in these transitions were determined by Mössbauer spectroscopy.

Applying *"in situ"* methods for further studies on the kinetics of the phase transition looks promising.

Acknowledgements

We thank the EPSRC for granting beamtime at the Daresbury Laboratory, and acknowledge the use of the EPSRC's Chemical Database Service at Daresbury. Also, we express our thanks to Mr. K. Jonsson and B. Erlendsson, Science Institute, for valuable assistance in running the Mössbauer system.

References

1. Johannesson, B. and Sigfusson, Th. I. (1998) Effect of thermal history and internal stresses on disintegration of ferrosilicon, *8th International Ferroalloys Congress (INFACON 8)*, Beijing, China.
2. Massalski, T.B. (1990) *Binary Alloy Phase Diagrams*. ASM International.
3. Sidorenko, F. A. , Gel'd, P. V. and Dubrovskaya, L. B. (1959) *Fiz. Metall. Metalloved.* **8,** 735.
4. Aronsson, B. (1960) A note on the compositions and crystal structures of MnB_2, Mn_3Si, Mn_5Si_3, and $FeSi_2$. *Acta Chemica Scandinavica* **14**(6), 1414-1418.
5. Kubaschewski, O. (1982) *Iron- Binary Phase Diagrams,* Springer-Verlag.
6. Dusausoy, Y., Protas, J., Wandji, R. and Roques, B. (1971) Structure crystalline du disiliciure de fer, $FeSi_2\beta$, *Acta Crystallografia* **B27**, 1209-1218.
7. Shen, W. Z., Shen, S. C., Tang, W. G. and W, L. W. (1995) Optical and photoelectric properties of β-$FeSi_2$ thin films, *J. Appl. Phys.* **78**, 4793-4795.
8. Jin, S., Li, X. N., Zhang, Z., Dong, C., Gong, Z. X., Bender, H. and Ma, T. C. (1996) Ion beam syntheses on micro structure studies of a new $FeSi_2$ phase, *J. Appl. Phys.* **80**, 3306-3309.
9. Holdhus, H. (1962) The transformation of the zeta-phase in iron-silicon alloys, *Journal of the Iron and Steel Institute*, 200.
10. Espelund, A. (1964) Om desintergrering av teknisk ferrosilisium. *Tidskrift kjemi, bergvesen og metallurgi (Norwegian)* **1**, 13-20.
11. Boomgaard, J. van den. (1972) Stability of the high temperature phase $FeSi_2$. *Journal of the Iron and Steel Institute*, 276-279.
12. Sigfusson, Th. I. and Helgason, Ö. (1990) Rates of transformations in the ferrosilicon system, *Hyperfine Interactions* **54**, 861-867.
13. Helgason, Ö., Magnusson, Th. and Sigfusson, Th. I. (1998) High temperature Mössbauer spectroscopy on the phase dynamics in ferrosilicon system, *Hyperfine Interactions* **111**, 215-219.
14. Helgason, Ö. and Sigfusson, Th. I. (1989) Mössbauer spectroscopy for determining phase stability in the ferrosilicon system, *Hyperfine Interactions* **45**, 415-418.
15. Helgason, Ö., Gunnlaugsson, H.P., Jonsson, K. and Steinthorsson, S. (1994), High temperature Mössbauer spectroscopy of titanomagnetite and maghemite in basalts, *Hyperfine Interactions* **91,** 595-599.
16. Dobler, M., Reuther, H., Betzl, M., Mäder, M. and Möller, W. (1996) Investigations of ion implanted iron silicide layers after annealing and irradiation, *Nucl. Instr. and Meth. in Phys. Research* **B 17**, 117-122.
17. Doremus, R. H. (1985) *Rates of Phase Transformations*, Academic Press.
18. Porter, D. A. and Easterling, K. E. (1992) *Phase transformations in Metals and Alloys,* Chapman & Hall, London.
19. Magnusson, Th. (1996) *The rate of eutectoid transformation in silicon rich ferrosilicon*, Master's thesis, University of Iceland.

FLUORIDE-ION CONDUCTORS DERIVED FROM THE FLUORITE TYPE

Application of Mössbauer Spectroscopy to the Study of the Structure, Order/Disorder Phenomena and Conduction Mechanism

G. DÉNÈS, M.C. MADAMBA, A. MUNTASAR, A. PEROUTKA,
K. TAM AND Z. ZHU
*Concordia University, Department of Chemistry and Biochemistry,
Laboratory of Solid State Chemistry and Mössbauer Spectroscopy,
Laboratories for Inorganic Materials, Montréal, Québec, Canada*

Abstract

Mössbauer spectroscopy has seldom been used for the study of ionic conductors. When tin(II) is incorporated in the best binary fluoride ion conductors, namely in fluorite type MF_2, various new materials are formed, which have a structure related to the fluorite type, and their fluoride ion conductivity is enhanced by up to 10^3. Most of these new conducting materials show order/disorder phenomena. $PbSnF_4$, which is the highest performance fluoride ion conductor, can exist in several polymorphic forms, and has a very complex system of phase transitions. SnF_2 can be incorporated in BaClF to form a wide $Ba_{1-x}Sn_xCl_{1+y}F_{1-y}$ solid solution, which is fully disordered in terms of metals and of excess anion of one kind relative to the other. Tin-119 Mössbauer spectroscopy can be an invaluable technique for probing the tin sites. It has made possible the interpretation of the structure of the disordered phases. In addition, the information obtained about the valence electronic structure of tin(II) makes it very easy to distinguish between: (i) a stereoactive lone pair that is located on a tin hybrid orbital, and therefore cannot be a charge carrier, and (ii) a potentially mobile non-stereoactive lone pair located on the unhybridized *5s* orbital of tin.

1. Introduction

In the MF_2 ionic fluorides with M = Cd, Ca, Sr, Pb, Ba, Hg and Eu, the metal ion takes a cubic coordination, the fluoride ions are tetrahedrally coordinated, and the structure is called the CaF_2 *fluorite-type*. For weakly electropositive tin, bonding in the difluoride is highly covalent and the structure is not related to the fluorite-type. In the three phases of SnF_2, the tin(II) electron lone pair is located on a hybrid orbital [1-3]. Since it strongly distorts the tin coordination, such a lone pair is said to be *stereoactive*, and it reduces the tin coordination number. Lead, being more electropositive, can form ionic type structures, such as the fluorite-type β-PbF_2. Lone pair stereoactivity is the general rule when the ligand is small and highly electronegative, such as F, O and Cl. On the other

M. Miglierini and D. Petridis (eds.), Mössbauer Spectroscopy in Materials Science, 25–38.

hand, with larger and less electronegative ligands (Se, Te, I), a regular coordination with a non-stereoactive lone pair is observed [4]. If, in the fluorite-type MF_2, every second plane of fluoride ions is replaced by chloride ions, steric crowding and F/Cl order lead to tetragonal distortion, and the PbClF structural type is obtained [5].

In the MF_2 fluorite type structure, valence electrons are strongly held onto highly electronegative anions, and thus no electron mobility is present. However, in these materials, a significant electric charge transport is observed at high temperature, which is due to long distance fluoride ion motion. Electron transport is negligible in MF_2 fluorite type difluorides and of their derivatives, as shown by their very high fluoride ion transport number, since $\tau_i > 0.99$. When tin(II) fluoride is combined with a fluorite type MF_2, the fluoride ion conductivity is strongly enhanced, up to three orders of magnitude when half of the metal M is substituted by Sn(II) [6, 7]. Tin chloride fluorides have been found to be very poor electrical conductors with a low ionic transport number [8]. One might have expected the tin(II) lone pair of electrons to be mobile and give rise to considerable charge transport, since it is not engaged in bonding, however, this is not observed when the lone pair is stereoactive [6, 7]. On the other hand, when it is not stereoactive, i.e. when it is located on the *5s* non-hybridized orbital of tin, semiconduction is quite common, and in the case of $CsSnBr_3$, a transfer of the lone pair to the conduction band to give rise to metallic conduction has been observed at high temperature [9].

The two main purposes of this work were to (i) prepare new fluoride and chloride fluoride materials that might have high performance fluoride ion conductivity, and (ii) to use Mössbauer spectroscopy to study the structures and the possible conducting properties of the materials obtained.

2. Experimental

The materials were prepared either from aqueous solutions or by direct reactions of the fluorides/chlorides at high temperature. For reactions of SnF_2 with univalent chlorides, the MCl and SnF_2 solutions were mixed. No precipitation occurred and colorless needle shaped crystals were obtained in a few days. For the preparation of divalent metal/Sn(II) fluoride, an aqueous solution of metal(II) nitrate was added dropwise to an aqueous solution of SnF_2 (or vice versa) upon stirring. On contact between the two solutions, a white precipitate formed immediately, which was filtered, washed with distilled water, and allowed to dry at room temperature in air. Barium tin(II) chloride fluorides were prepared by the same method, using barium chloride dihydrate. In some cases, the partial decomposition of some metal/tin(II) fluorides by water, or the reaction of a slurry of α-PbF_2 with an aqueous solution of SnF_2 resulted in new materials being formed. The reactions between the divalent metal fluorides, or barium chloride, and SnF_2 were also carried out by heating the desired stoichiometric mixture of the anhydrous reactants under inert conditions, according to the procedure developed by Dénès [10]. Quenching to obtain metastable high temperature phases, or slow cooling or annealing at given temperatures for allowing sluggish phase transitions to take place, were also performed.

All the materials were characterized by X-ray powder diffraction using a Philips

PW 1050/25 instrument that has been automated by Sietronics. Data accumulation and processing was performed using the SIE112 software package from Sietronics. Phase identification was done by use of the JCPDS database and μ-PDSM Search Match program from Fein-Marquat. ^{119}Sn Mössbauer spectroscopy was carried out on appropriately selected samples, by use of an Elscint driving system and a Tracor Northern TN7200 multichannel analyzer. A nominally 10 mCi $CaSnO_3$ γ-ray source was used with a Harshaw (Tl)NaI detector. All chemical isomer shifts were referenced to a $CaSnO_3$ absorber as zero shift. Spectral splitting was performed by the GMFP5 software[11,12].

3. Results and Discussion

3.1. MATERIALS OBTAINED

3.1.1. $MSnF_4$ materials

$MSnF_4$ materials (M = Ba, Pb and Sr) under various degrees of strain, of crystallinity, and of preferred orientation were obtained by a variety of methods:

- highly strained and highly oriented tetragonal α-$PbSnF_{4(aq1)}$ is obtained by precipitation when an aqueous solution of $Pb(NO_3)_2$ is added to an aqueous solution of SnF_2 in the molar ratio lead(II) nitrate/tin(II) fluoride = 1:4;
- strained α-$PbSnF_{4(aq2)}$ is obtained when a slurry of solid α-PbF_2 in an aqueous solution of SnF_2 at a molar ratio α-$PbF_2/SnF_2 < 1$ is stirred for 1 hour;
- non-stressed and more moderately oriented α-$PbSnF_4$(ssr) was obtained by direct reaction of SnF_2 and PbF_2 at 250°C;
- non-oriented α-$PbSnF_4$ is obtained either by stirring μγ-$PbSnF_4$ in water or by annealing μγ-$PbSnF_4$ at 200°C under inert atmosphere;
- nanoparticles of α-$PbSnF_4$ are obtained by ball milling o-$PbSnF_4$ for 30 seconds;
- low stress mildly oriented α-$PbSnF_4$ is obtained when o-$PbSnF_4$ is annealed for a sufficiently long time between 50 and 270°C;
- orthorhombic o-$PbSnF_4$ is obtained when an aqueous solution of $Pb(NO_3)_2$ is added to an HF solution of SnF_2 in the molar ratio of $Pb(NO_3)_2/SnF_2$ = 1:4, and the magnitude of the orthorhombic distortion is a function of the amount of HF used;
- tetragonal β-$PbSnF_4$ is obtained when a 1:1 mixture of SnF_2 and PbF_2, or any phase of $PbSnF_4$ is heated above 290°C and cooled rapidly to ambient temperature;
- crystalline cubic γ-$PbSnF_4$ is obtained only in-situ above 390°C;
- microcrystalline μγ-$PbSnF_4$ is obtained at ambient temperature by ball milling α- or o-$PbSnF_4$ for 1 min. or longer, or by ball milling β-$PbSnF_4$ for 5 min. or longer;
- $SrSnF_4$ and $BaSnF_4$ are obtained by heating the stoichiometric mixture of the two fluorides at 500°C under inert conditions;
- $BaSnF_4$ has also been obtained by reaction of $BaCl_2$ and SnF_2 in aqueous solution for a molar ratio $X = BaCl_2/[SnF_2+BaCl_2] = 0.87$ when the solution of SnF_2 is added to the solution of $BaCl_2$.

It should be pointed out that $PbSnF_4$ is a much richer system than all the others.

3.1.2. Other materials in the MF_2/SnF_2 systems

Materials with stoichiometries other than M:Sn = 1:1 were obtained in the MF_2/SnF_2 systems, many of which are also excellent fluoride ion conductors:

- a cubic fluorite-type $Pb_{1-x}Sn_xF_2$ solid solution is obtained by reaction of PbF_2 and SnF_2 at high temperature (500°C) for $0 \leq x \leq 0.30$;
- a tetragonal β-$PbSnF_4$ type $Pb_{1-x}Sn_xF_2$ solid solution is obtained by reaction of PbF_2 and SnF_2 at 500°C for $0.30 \leq x \leq 0.50$, (x = 0.50 is stoichiometric β-$PbSnF_4$);
- cubic $PbSn_4F_{10}$ is obtained by heating a 1:4 PbF_2:SnF_2 mixture at ca. 250°C under inert condition, then quenching;
- orthorhombic $SrSn_2F_6{\cdot}H_2O$ precipitates when an aqueous solution of strontium nitrate is added to an aqueous solution of SnF_2 for a molar ratio X = $Sr(NO_3)_2/[Sr(NO_3)_2+SnF_2] < 0.5$;
- $BaSn_2F_6$ has been obtained by the reaction between aqueous solutions of barium chloride and tin(II) fluorides for a ratio $BaCl_2/[BaCl_2+SnF_2] = 0.33$;
- $CaSn_2F_6$ was obtained by adding a solution of calcium nitrate to a solution of tin(II) fluoride for a ratio $Ca(NO_3)_2/[Ca(NO_3)_2+SnF_2] < 0.23$;
- a cubic $Ca_{1-x}Sn_xF_2$ solid solution (x= 0.28 – 0.35) is formed when a solution of calcium nitrate is added to a solution of tin(II) fluoride for a ratio $Ca(NO_3)_2/[Ca(NO_3)_2+SnF_2] > 0.38$;
- the cubic $Ca_{1-x}Sn_xF_2$ solid solution is also obtained by water leaching SnF_2 from $CaSn_2F_6$.

3.1.3. Chloride fluoride phases

- $BaSn_2Cl_2F_4$ was obtained by addition of an aqueous solution of $BaCl_2{\cdot}2H_2O$ to an aqueous solution of SnF_2 in the ratio $BaCl_2/[BaCl_2+SnF_2] = 0.247$;
- $Ba_3Sn_3Cl_4F_8{\cdot}2.6H_2O$ was obtained by addition of an aqueous solution of $BaCl_2{\cdot}2H_2O$ to an aqueous solution of SnF_2 in the ratio $BaCl_2/[BaCl_2+SnF_2] = 0.363$;
- a BaClF type $BaCl_{1+y}F_{1-y}$ ($0 < y \leq 0.25$) solid solution was obtained by reaction of appropriate amounts of $BaCl_2$ and BaF_2 under inert conditions at high temperature (≥350°C);
- a BaClF type $Ba_{1-x}Sn_xCl_{1+y}F_{1-y}$ ($0 < x \leq 0.25$, $-0.15 \leq y < 0.25$) solid solution was obtained by reaction of appropriate amounts of $BaCl_2$, BaF_2 and SnF_2 under inert conditions at high temperature (≥ 350°C);
- a narrower BaClF type $Ba_{1-x}Sn_xCl_{1+y}F_{1-y}$ ($x \approx 0.1$, $y \approx 0.1$) solid solution was obtained by reaction between aqueous solutions of $BaCl_2{\cdot}2H_2O$ and of SnF_2 at $BaCl_2/(BaCl_2+SnF_2)$ ratios larger than 0.87;
- $M_3Sn_5Cl_3F_{10}$ and MSn_2ClF_4 (M = K and NH_4) were obtained in the form of indistinguishable needle shaped crystals by slow crystallization of an aqueous solution of MCl and SnF_2 in the ratio $MCl/SnF_2 = 1:1$.

3.2. ORDERED $MSnF_4$ STRUCTURES RELATED TO THE FLUORITE-TYPE

The X-ray diffraction patterns of α-$PbSnF_4$, β-$PbSnF_4$, o- $PbSnF_4$, $SrSnF_4$ and $BaSnF_4$ are clearly related to that of the MF_2 fluorite-type [fig.1]. All peaks of β-PbF_2 are present in the $MSnF_4$ materials, however most of them are split, due to symmetry break

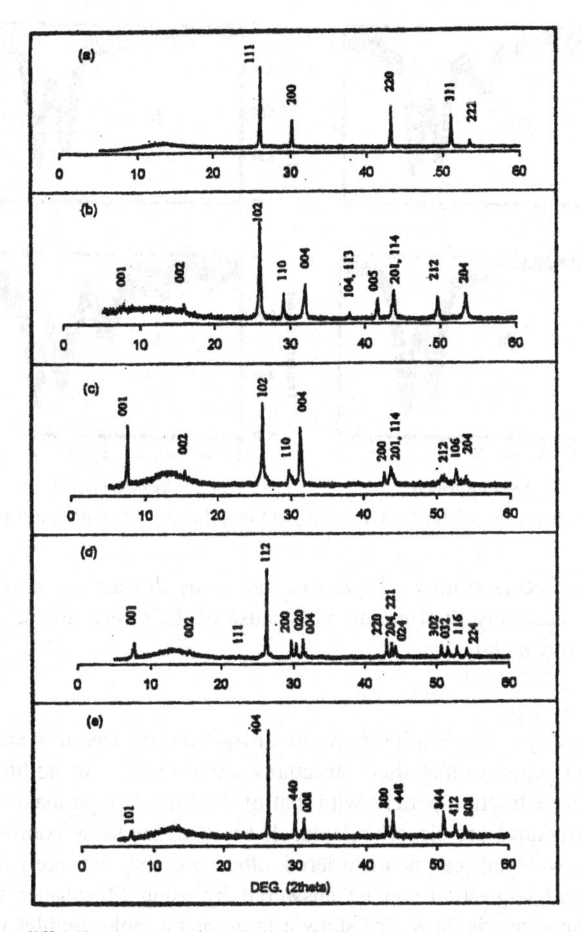

Figure 1. X-ray diffraction patterns of: (a) β-PbF_2, (b) $BaSnF_4$, (c) α-$PbSnF_4$, o- $PbSnF_4$, (e) β-$PbSnF_4$.

to tetragonal or orthorhombic. In all cases, peaks which are systematically absent in β-PbF_2 are observed in $MSnF_4$. This is due to a reduction of the number of lattice translations associated with a change of Bravais lattice from F in β-PbF_2 to P, except maybe in β-$PbSnF_4$. In all cases, new peaks are observed at very low angles, which are characteristic of a crystal periodicity much larger than in β-PbF_2 (superstructure). The *c* parameter is doubled in all ordered $MSnF_4$ except in β-$PbSnF_4$ which has **a** and **b** quadrupled relative to α-$PbSnF_4$, and *c* quadrupled relative to β-PbF_2. Peak shift shows that there is a contraction along *c* and an expansion in the *(a,b)* plane.

Solving the crystal structure of $MSnF_4$ required overcoming formidable challenges: (i) no suitable crystal could be obtained since reactions in aqueous solution give platelets too thin to be isolated. Direct reactions at high temperature produced only polycrystalline materials [13]; (ii) The high degree of preferred orientation makes it very difficult to use powder diffraction data for applying the Rietveld method (profile refinement). Use of powder data is also compounded by the broadening of some lines

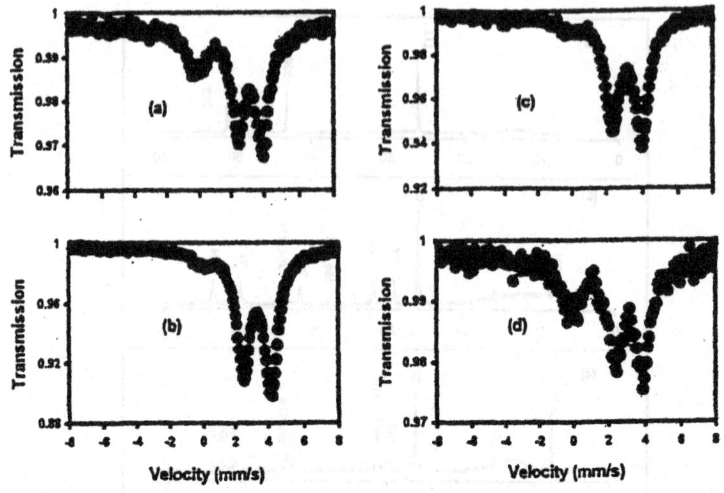

Figure 2. Mössbauer spectrum of: (a) α-$PbSnF_4$, (b) $BaSnF_4$, (c), o- $PbSnF_4$, (d) β-$PbSnF_4$.

due to stress or microcrystallinity. Moreover, by X-ray diffraction, it would be very difficult to locate accurately fluorine atoms because of their very low scattering factor compared to the heavy metals.

3.2.1. The α-$PbSnF_4$ type

Lone pair stereoactivity. The similarity of the diffraction patterns of $SrSnF_4$, α-$PbSnF_4$ and $BaSnF_4$ (fig.1) suggests that their structures are isotypic. In addition, the clear relationship of their diffraction pattern with that of the fluorite type makes it likely that the two types of structures are closely related. If tin takes the same coordination as the metal M, then the two metals can be disordered, otherwise, they are likely to be ordered. This very fundamental question can be answered by using Mössbauer spectroscopy. Their Mössbauer spectra (fig.2a & 2b) show a large quadrupole doublet in both cases. The isomer shifts (3.2 - 3.5 mm/s) are typical of divalent tin coordinated by fluorine [14]. In addition, in all cases, tin(II) in the $MSnF_4$ materials gives large a quadrupole splitting (ca. 1.5 mm/s) similar to the case of α-SnF_2 [14]. The large e.f.g. (electric field gradient) generating the quadrupole splitting is due to the presence of a strongly axial (and therefore stereoactive) lone pair on tin(II). Since the lone pair is on a hybrid orbital, it contains less s-electron density, than the non-hybridized $5s^2$ lone pair, and this is confirmed by the isomer shift (ca. 3.5 mm/s) smaller than for ionic Sn^{2+}. The small signal at ca. 0 mm/s is Sn(IV), most likely a very thin layer of amorphous SnO_2 at the air/material interface. We showed that, on SnF_2, it acts as a protective coating and passivate the material [15]. The amount of SnO_2 is less than. 10% of the tin(IV) signal ($f_a(SnO_2) << f_a(SnF_2)$) [10].

Sn/M order, and lone pair orientation, crystallite shape and preferred orientation. Since the tin(II) lone pair is strongly stereoactive, it is very likely that tin and lead are ordered. The superstructure obtained by doubling the *c* parameter suggests that the order of metal layers along *c* is as follows: .... Pb Pb Sn Sn Pb.... The most likely orientation of the tin-lone pair axis can be obtained from symmetry considerations and Mössbauer spectroscopy. The lone pair of each tin will be repeated by the 4-fold axis parallel to *c*, to give four lone pairs for each tin atom, unless the tin-lone pair axis is parallel to the 4-fold axis. It is assumed that tin is located on the 4-fold axis, thereby preserving the square shape of the two faces of the MF_8 cubes parallel to *(a,b)*, in accordance with the tetragonal distortion. The orientation of the tin-lone pair axis could also be obtained from the quadrupole doublet asymmetry if a sufficiently large single crystal were available. Unfortunately, the crystal growing habit of these materials did not make it possible. However, we were able, using a method we have developed, to prepare very highly oriented samples of α-$PbSnF_{4(aq1)}$, obtained by precipitation, and recrystallized in HNO_3 according to literature [16]. The stacks of very large and extremely thin crystals are filtered by suction, which further enhances the parallel stacking of the very thin paper-sheet like crystals. Samples quasi-perfectly oriented perpendicularly to *c*, i.e. normal to the 4-fold axis, are obtained. If, as suggested by symmetry, the tin-lone pair axis is parallel to *c*, and if the γ-ray beam travels in the same direction, then, the angle between the γ-ray beam and the tin-lone pair axis is zero. Since highly stereoactive lone pairs are the main contributors to the electric field gradient (e.f.g.) at tin, then V_{zz}, the main axis of the e.f.g, can be taken as being along the tin-lone pair axis, thus $\theta = (V_{zz}, \gamma) = 0$. At $\theta = 0$, the quadrupole doublet should be highly asymmetric, with $I_{\pm 1/2, \pm 1/2} / I_{\pm 3/2, \pm 1/2} = 2/6$. The very highly asymmetric spectra obtained and the decrease of asymmetry with increasing θ (fig.3) confirm that the tin-lone pair axis is most likely parallel to *c*.

Crystal structure, crystal shape and preferred orientation. Using the results from the X-ray diffraction patterns and Mössbauer spectroscopy, the structures of α-$PbSnF_4$ and

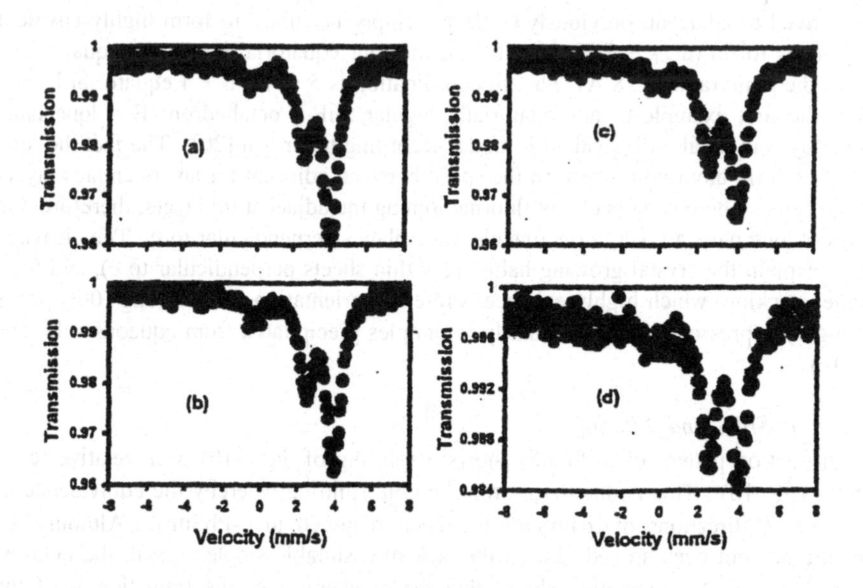

Figure 3. Mössbauer spectrum of highly oriented α-$PbSnF_4$ versus the angle between V_{zz} and the γ-ray beam direction: (a) 0°, (b) 15°, (c) 30°, (d) 45°.

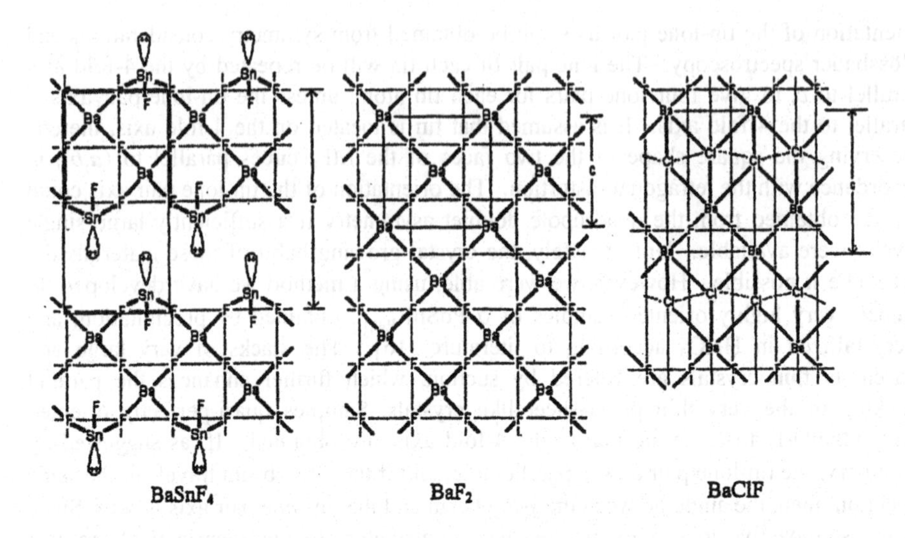

Figure 4. Projection of a slice of the structures of Ba SnF_4, BaF_2 and BaClF in the (a,b) plane.

$BaSnF_4$ were solved by neutron powder diffraction and EXAFS [17 & 18].
Figure 4 shows a projection of the structures of BaF_2 and $BaSnF_4$. Sn and Ba are ordered along *c* according to ...Ba Ba Sn Sn.Ba...,the tin lone pair is stereoactive, and the Sn-E axis is parallel to *c*. The fluoride ion sublattice is basically the same as in β-PbF_2, with the exception that every fourth layer of fluorine perpendicularly to *c*, that would be expected to bridge adjacent tin layers, has disappeared. These fluorine atoms have moved to adjacent, previously (in BaF_2) empty F_8 cubes, to form highly covalent axial bonds to tin ($d_{Sn\text{-}F}$ = 2.00 Å), whereas the four equatorial bonds are equal to the sum of the ionic radii (2.28 Å). The tin coordination is 5 (1 axial + 4 equatorial F) and this is the first example of an equatorially regular SnF_5E octahedron (E = lone pair) (model by Galy et al. [19]), called E model according to Brown [20]. The fact the lone pair of each tin atom points toward the space between adjacent tin layers creates layers of lone pairs. There is no bridging fluorine joining the adjacent tin layers, therefore, the planes of lone pairs are very effective cleavage planes perpendicular to *c*. The cleavage planes explain the crystal growing habit (very thin sheets perpendicular to *c*), and their parallel stacking, which highly enhances preferred orientation [very strong (00*l*) peaks and highly depressed (*hk*0)] particularly in samples precipitated from aqueous solutions (fig. 1c).

3.2.2. o-$PbSnF_4$ and β-$PbSnF_4$

The diffraction pattern of o-$PbSnF_4$ shows a splitting of the (110) peak relative to α-$PbSnF_4$ (fig. 1d). This is due to an orthorhombic splitting whereby the equivalence of the *a* and *b* directions of α-$PbSnF_4$ has been removed in o-$PbSnF_4$. Although its structure has not been solved, due to the lack of a suitable single crystal, the detailed study of the α → o transition shows that the mechanism of the transition is of the paraelastic→ferroelastic type, and that the structure of o-$PbSnF_4$ is very similar to that of α-$PbSnF_4$, with loss of the tetragonal symmetry. Scanning electron microscopy (SEM) of o-$PbSnF_4$ shows that the crystallites are also very thin plates, like in α-$PbSnF_4$, therefore the planes of lone pairs are conserved.

The unit-cell of β-$PbSnF_4$ is tetragonal, with a much larger superstructure than α-$PbSnF_4$ since $\mathbf{a}_\beta \approx 4\mathbf{a}_\alpha$ and $\mathbf{c}_\beta \approx 2\mathbf{c}_\alpha$. The diffraction pattern (fig. 1e) shows that it is closely related to the fluorite type. The superstructure reflections are not the same as for α-$PbSnF_4$, and they are so few that it is not possible to find the atomic positions from powder diffraction, especially considering the large size of the unit-cell. The $\beta \leftrightarrow \gamma$-$PbSnF_4$ transition is fully reversible without significant hysteresis [21, 22], whereas the $\alpha \rightarrow \beta$ transition gives metastable β-$PbSnF_4$. The $\beta \rightarrow \alpha$ transition is very sluggish, and γ-$PbSnF_4$ has the fully disordered fluorite type structure (see 3.3 below). Therefore, the structures of β-$PbSnF_4$ and of γ-$PbSnF_4$ must be very closely related, and the $\beta \leftrightarrow \gamma$ transition probably involves shifts of atoms rather than strong bond breaking. On the other hand, the $\beta \rightarrow \alpha$ transition is very sluggish and thus involves a more drastic structural reorganisation. Therefore, it is likely that the strongly covalent axial Sn-F bonds of α-$PbSnF_4$ disappear at the $\alpha \rightarrow \beta$ transition, giving a pseudo-cubic coordination of tin, however, still with Pb/Sn order, different of that present in α-and o-$PbSnF_4$, and that order disappears at the $\beta \leftrightarrow \gamma$ transition.

3.3. DISORDERED PHASES IN THE SNF_2/MF_2 SYSTEMS

The $Pb_{1-x}Sn_xF_2$ solid solution ($0 \leq x \leq 0.30$), $PbSn_4F_{10}$, μ-γ-$PbSnF_4$, and the $Ca_{1-x}Sn_xF_2$ solid solution have a cubic fluorite-type diffraction pattern, and show no sign of lattice distortion or superstructure, despite a substantial tin(II) content. The absence of superstructure and the conservation of the F Bravais lattice in the cubic phases show unambiguously that the two metals are fully disordered on the (4a) Wyckoff site of the Fm3m space group. This, combined with the absence of lattice distortion, suggests that the coordination of the two disordered metals is cubic, like in the fluorite-type MF_2 However, if tin(II) takes a cubic coordination, its lone pair is non-stereoactive, since a stereoactive (i.e. axial) lone pair results in site distortion and a lower coordination number. The Mössbauer spectra (fig.5) show that in all the disordered systems, the large quadrupole doublet ($\Delta \approx 1.0$-1.8 mm/s) and the isomer shift ($\delta \approx 3.0$-3.5 mm/s) are typical of covalently bonded tin(II) with a highly axial lone pair. This is not compatible with a cubic coordination.

The presence of a highly stereoactive lone pair forces the tin coordination to be reduced and distorted, and the coordination number to be five at the most. This can be achieved by tin shift towards one of the faces of the F_8 cube, while the lone pair is pointing towards the middle of the opposite face, to give a square pyramidal coordination [22] similar to that observed in black SnO [23]. The strong local distortion at tin, is likely to create fluorine shifts in its vicinity, thereby, creating more minor distortion of the coordination of the other metal. However, in order to keep the overall cubic, Fm3m, crystal lattice, without distortion and without superstructure, M and Sn must be fully disordered, and in addition, the orientation of the Sn lone pair axis must also be disordered about the six possible orientations parallel to the unit-cell axes, to

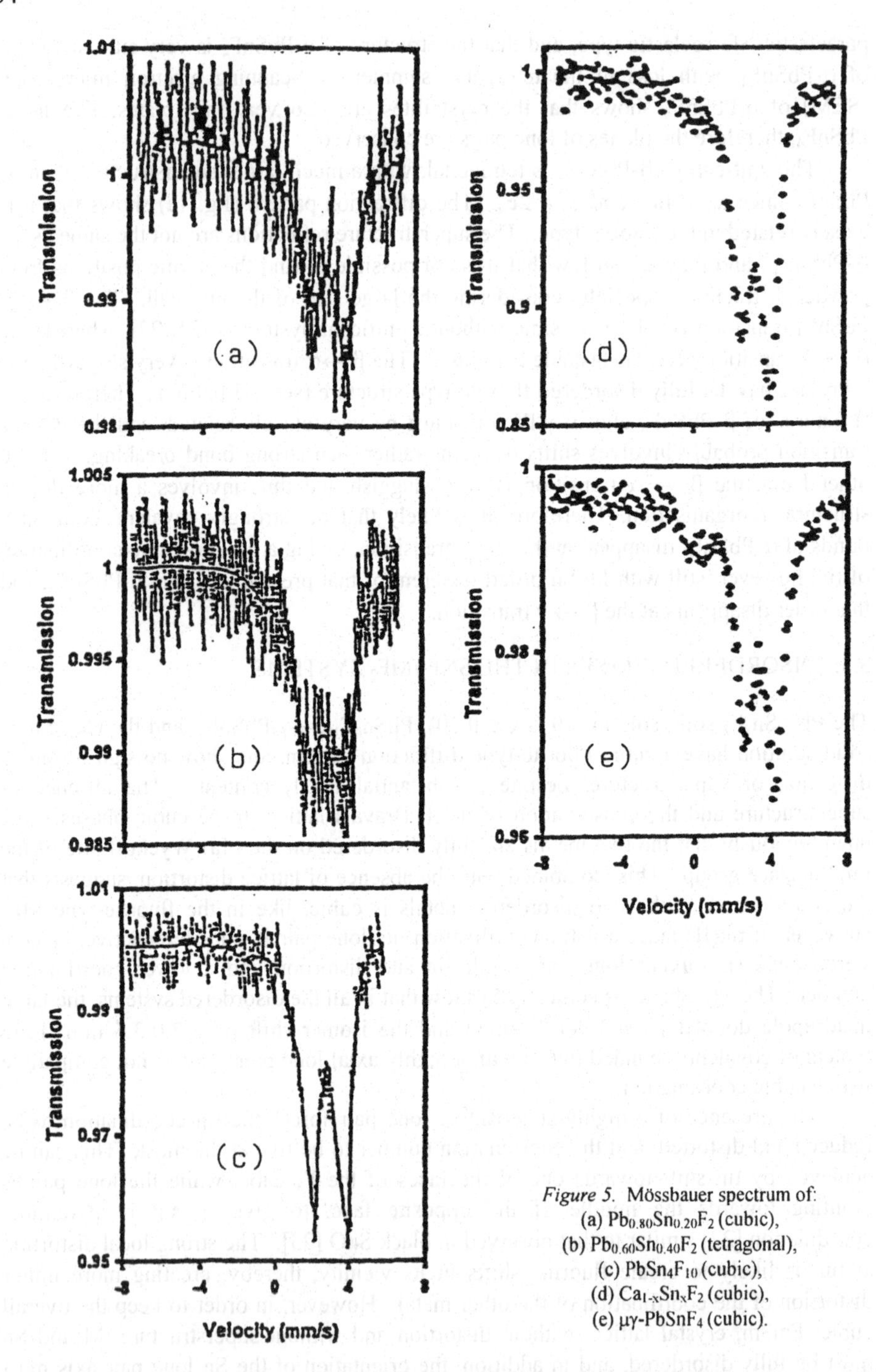

Figure 5. Mössbauer spectrum of:
(a) $Pb_{0.80}Sn_{0.20}F_2$ (cubic),
(b) $Pb_{0.60}Sn_{0.40}F_2$ (tetragonal),
(c) $PbSn_4F_{10}$ (cubic),
(d) $Ca_{1-x}Sn_xF_2$ (cubic),
(e) μγ-$PbSnF_4$ (cubic).

form bonds to the four corners of any of the six faces of the F_8.cubes (fig. 6).

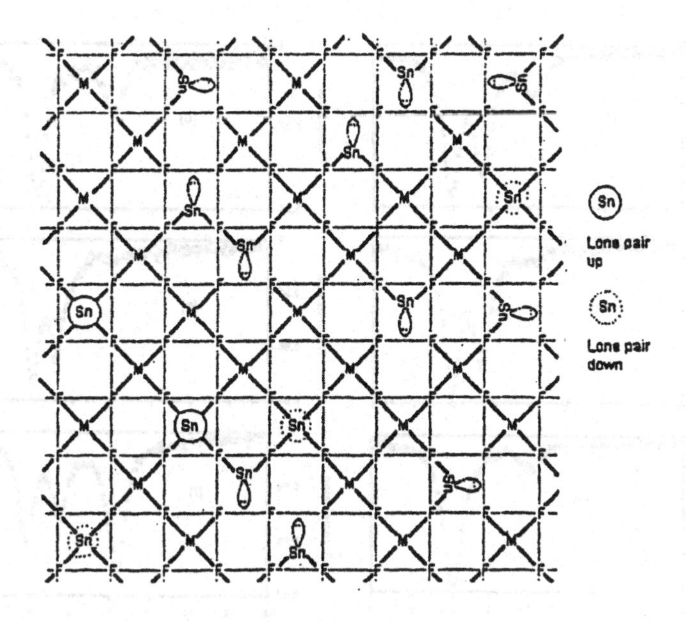

Figure 6. Structure of disordered phases.

The tetragonal $Pb_{1-x}Sn_xF_2$ solid solution ($0.30 < x \leq 0.50$) must have some degree of Pb/Sn disorder since the solid solution is wide. However, the tetragonal distortion of the lattice and the β-$PbSnF_4$ type superstructure show the presence of some ordering of Pb/Sn. The Mössbauer spectrum of the tetragonal solid solution (fig.5b) is characteristic of a strongly stereoactive lone pair, like for the other phases, and therefore, a tetragonal distortion of tin is again most likely to be present.

3.4. CHLORIDE FLUORIDE PHASES

3.4.1. The $Ba_{1-x}Sn_xCl_{1+y}F_{1-y}$ solid solution

The white precipitate obtained when aqueous solutions of SnF_2 and of $BaCl_2{\cdot}2H_2O$ are mixed in the molar ratio X = $BaCl_2/[BaCl_2+SnF_2] > 0.87$, and the product of the high temperature reaction of $BaCl_2$, BaF_2 and SnF_2 in appropriate amounts, give an X-ray diffraction pattern that looks the same as that of BaClF, however, the unit-cell parameter **a** is decreased and **c** increased. A change of unit-cell parameters would not occur if the samples were a mixture of BaClF and amorphous phases. However, since all the samples give a Mössbauer spectrum (fig. 7), they contain tin. Chemical analysis showed that the material has the formula $Ba_{1-x}Sn_xCl_{1+y}F_{1-y}$, with $x \approx 0.15$ and $y \approx 0.09$. Direct reactions at high temperature showed the existence of a very wide solution for $0 \leq x \leq 0.25$ and $-0.25 \leq y \leq 0.25$. The absence of lattice distortion and superstructure compared to unsubstituted BaClF shows that Ba and Sn are fully disordered. In addition, the Cl and F lattices remain ordered, however with full disorder on the site of

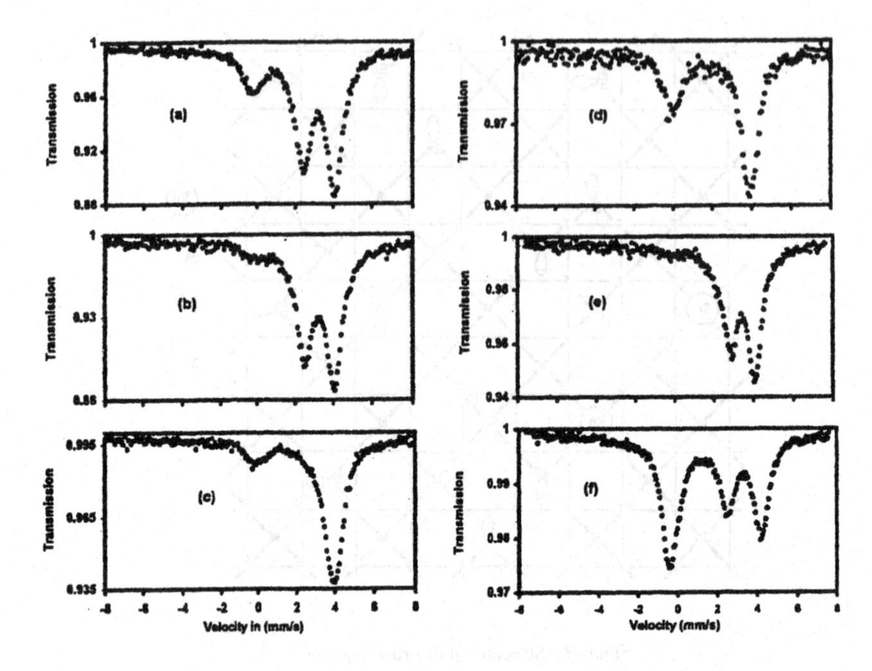

Figure 7. Mössbauer spectrum of: (a) $BaSn_2Cl_2F_4$, (b) $Ba_3Sn_3Cl_4F_8.2.6H_2O$, (c) $Ba_{1-x}Sn_xCl_{1+y}F_{1-y}$ precipitated, (d) $Ba_{1-x}Sn_xCl_{1+y}F_{1-y}$ from direct reaction, (e) $NH_4Sn_2ClF_4$, (f) KSn_2ClF_4.

the deficient anion. Mössbauer spectroscopy shows a single line for tin(II) at 4.0 - 4.1 mm/s, which is characteristic of ionically bonded Sn^{2+}. The small, if any, quadrupole splitting is due mostly to the lattice, since in the Sn^{2+} ion, the tin lone pair is on the spherical $5s^2$ orbital. This situation is surprising since, when tin(II) is coordinated to F or Cl, bonding is about invariably covalent. However, $SnCl_2$ is a well known exception. Surprisingly, precipitation from $SrCl_2/SnF_2$ solutions did not yield related materials.

3.4.2. *The MCl/SnF₂ systems*

Mixing aqueous solutions of MCl (M = Li, Na, K and NH_4) and SnF_2, did not result in a precipitation. Slow evaporation yielded various kinds of crystals. Two new materials, in the form of needles, were obtained, only for M = K and NH_4: MSn_2ClF_4 and $M_3Sn_5Cl_3F_{10}$.The Mössbauer spectrum of MSn_2ClF_4 (fig. 7e, 7d) shows values of isomer shift and quadrupole splitting typical of covalently bonded tin(II) with a highly axial lone pair. The crystal structure of both types of compounds confirm the high stereoactivity of the lone pair.

3.5. LONE PAIR STEREOACTIVITY AND CONDUCTIVITY MECHANISM

Ionic conduction in solids is possible when there are easy pathways in the crystal structure for long range motion of ions. The high fluoride ion conductivity of the MF_2

fluorite-type has been attributed to the presence of empty F_8 cubes (half of the cubes are empty) which can be used as interstitial sites in a Frenkel defect mechanism (fig. 4 & 6). However, Dénès has shown that it is an oversimplification, since it does not explain why β-PbF_2 is a much more efficient conductor than BaF_2, and it also does not explain why the conductivity of $MSnF_4$ is three-orders of magnitude higher, even though in the α-$PbSnF_4$ type, there are no empty F_8 cubes (fig. 4) [24]. Transport number measurements have shown that the conductivity of these materials is purely ionic ($\tau_e < 10^3$) [6]. On the other hand, tin(II) materials containing the largest halides (Br and I) are often semiconductors, and $CsSnBr_3$ becomes a metal conductor at high temperature, due to the transfer of the tin(II) lone pair to the conduction band. A stereoactive lone pair is strongly held in a hybridized orbital. Transfer to the conduction band would require a change or removal of tin hybridization. Such a drastic change would require a large amount of energy and would result in a major structural transformation with changes in bonding, in coordination number, and also a change of the shape of the polyhedra of coordination. Such a drastic change has never been observed. It results that a stereoactive lone pair does not contribute significantly to electrical conduction, and the materials are insulators or ionic conductors. On the other hand, if the lone pair is not stereoactive, i.e. when it is located on the unhybridized *5s* orbital, bonding is usually ionic and the lone pair is available for charge transport. In this case, the materials are usually semiconductors, or they can also be mixed conductors or metal conductors. Since Mössbauer spectroscopy gives an excellent evaluation of the lone pair stereoactivity, it can also be used for a rapid evaluation of the major charge carrier in fluorides. When the lone pair is stereoactive (δ = 3.0-3.5 mm/s, Δ = 1.0-1.8 mm/s), the lone pair is not a charge carrier, and if the material is a good electrical conductor, fluoride ions are likely to be the major charged mobile species.

PbClF is an ionic conductor, however it is a poor one, with both F^- and Cl^- ions being responsible for its conduction, by use of poorly efficient Schottky defects [25]. Since the $Ba_{1-x}Sn_xCl_{1+y}F_{1-y}$ solid solution has the same structure as PbClF, it can be assumed that it has no more opportunities than PbClF for a large number of Frenkel defects or other types of structural features that would boost ionic mobility. Since incorporation of Sn(II) in the fluoride matrix of the fluorite type MF_2, by partial M/Sn substitution, results in a large increase of the fluoride ion conductivity up to three orders of magnitude, one can wonder whether the same can happen for the PbClF structure. The Mössbauer spectra show that the tin lone pair is not stereoactive in these materials. Being located on the unhybridized *5s* orbital, it is available for transfer to the conduction band, and thus give rise to significant conduction by electron motion. However, only conductivity and transport number measurements can give definite answer.

4. Conclusion

The SnF_2/MF_2, $SnF_2/BaCl_2$, and SnF_2/MCl systems have complex systems of structures and phase transitions, involving order/disorder phenomena, and possible changes to the tin(II) valence electronic structure. The crystal structure and the tin valence electronic structure are key factors that determine whether the material is an insulator or an electrical conductor, and what are the mobile species. Crystal structure determination is made very difficult by the lack of suitable single crystals, the very high preferred orientation present in polycrystalline samples, and the difficulty of removing it, since milling often induces phase transitions. Another difficulty is the very low X-ray scattering factor of the mobile ions (F^-) compared to the heavy metals. Mössbauer spectroscopy has proven to be a convenient method for the evaluation of the stereoactivity of the tin lone pair, and thus can provide invaluable help with their structure determination and the evaluation of the possible charge carriers.

Acknowledgements

The Natural Science and Engineering Research Council of Canada and Concordia University are acknowledged for financial support. AM is grateful to the financial support of the General Secretariat of Education of Libya.

References

1. Dénès, G., Pannetier, J., Lucas, J., and Le Marouille, J.Y. (1979), *J. Solid State Chem.* **30**, 335 - 343.
2. Dénès, G., Pannetier, J., and Lucas, J. (1980), *J. Solid State Chem* **33**, 1 - 12.
3. Pannetier, J., Dénès, G., Durand, M., and Buevoz, J.L. (1980), *J. Physique* **41**, 1019 - 1024.
4. Donaldson, J.D. and Silver, J. (1973), *J. Chem. Soc. A*, 666.
5. Wyckoff, R. G. (1982) *Crystal Structures*, **1**, Interscience New York, p.432.
6. Dénès, G., Birchall,T., Sayer, M., and Bell, M.F. (1984), *Solid State Ionics* **13**, 213 – 219.
7. Dénès, G., Milova, G., Madamba, M.C., and Perfiliev, M. (1996), *Solid State Ionics* **86-88**, 77-82.
8. Claudy, P., Letoffe, J.M., Vilminot, S., Granier, W., Al Ozaibi, Z., and Cot, L. (1981), *J. Fluorine Chem.* **18**, 203.
9. Barrett, J., Bird, S.R.A., Donaldson, J.D., and Silver, J. (1971), *J. Chem. Soc.* **(A)**, 3105.
10. Dénès, G. (1988), *J. Solid State Chem.* **77**, 54 - 59.
11. Ruebenbauer, K. and Birchall, T. (1979), *Hyperfine Interact* **7**,125.
12. Monnier, J., Dénès, G., and Anderson, R.B. (1984), *Canad. J. Chem Eng.* **62**, 419 – 424.
13. Dénès, G., Pannetier, J., and Lucas, J. (1975), *C.R. Acad Sci Paris Ser C* **280**, 831 – 834.
14. Birchall, T., Dénès, G., Ruebenbauer, K., and Pannetier, J. (1981), *J. Chem Soc., Dalton Tran.*, 2296 – 2299.
15. Dénès, G. and Laou, E. (1994), *Hyperfine Interact.* **92**, 1013 - 1018.
16. Donaldson, J.D. and Senior, B.D. (1967), *J. Chem. Soc. A*, 1821 - 1825.
17. Birchall, T., Dénès, G., Ruebenbauer, K., and Pannetier, J. (1986), *Hyperfine Interact.* **29**, 1331 - 1334.
18. Dénès, G., Tyliszczak, T, Yu, Y.H., and Hitchcock, A.P.(1991), *J. Solid State Chem* **91**, 1 - 15.
19. Galy, J., Meunier, G., Aström, A., and Andersson, S. (1975), *J Solid State Chem* **13**, 142 - 159.
20. Brown, I.D. (1974), *J Solid State Chem* **11**, 214 - 233.
21. Pannetier, J., Dénès, G., and Lucas, J. (1979) *Mater. Res Bull* **14**, 627 - 631.
22. Dénès, G., Tyliszczak, T, Yu, Y.H., and Hitchcock, A.P.(1993) *J. Solid State Chem* **104**, 239 - 252.
23. Pannetier, J. and Dénès, G. (1980) *Acta Cryst.* B **36**, 2763 - 2765.
24. Dénès, G. (1995) Solid State Ionics IV, *Mat. Res. Symp. Proc.* **369**, 295 - 300.
25. Saiful Islam, M. (1989) Solid State Ionics, *Mat. Res. Symp. Proc.* **135,** 295 - 300.

THE $Ba_{1-x}Sn_xCl_{1+y}F_{1-y}$ SOLID SOLUTION

The First Time incorporation of Divalent Tin in the PbClF Structure, and the Very Rare Occurrence of the Sn^{2+} Ion in a Chloride Fluoride Matrix

ABDUALHAFEED MUNTASAR AND GEORGES DÉNÈS
Concordia University, Department of Chemistry and Biochemistry,
Laboratory of Solid State Chemistry and Mössbauer Spectroscopy,
Laboratory for Inorganic Materials, Montréal, Québec, Canada

Abstract

A tetragonal distortion (P4/nmm) of the β-PbF_2 fluorite type is obtained by 50% substitution of F by Cl, and it gives the PbClF structure. In PbClF, both anions contribute to the conduction, however it is a poor conductor, but it is purely ionic. We have found that one can substitute up to 25% Ba by Sn and, up to 15% Cl by F ($y<0$) or up to 25% F by Cl ($y>0$), to give a $Ba_{1-x}Sn_xCl_{1+y}F_{1-y}$ solid solution ($0 \leq x \leq 0.25$) which has the same structure as BaClF (PbClF type), with full Ba/Sn disorder, whereas the anion sites remain ordered like in unsubstituted BaClF. However, there is disorder between –yF and (1+y)Cl on the Cl site if $y<0$, or between yCl and (1-y)F on the F site if $y>0$. Surprisingly, tin(II) is present in the ionic Sn^{2+} stannous ion form instead of being covalently bonded. This is established from the Mössbauer parameters ($\delta \approx 4.05$mm/s, $\Delta \approx 0$). This also makes it possible to have a significant electron conductivity from the non-hybridized tin (II) lone pair.

1. Introduction

Divalent metal fluorides crystallize in the fluorite type structure for the largest metals (Ca, Sr, Ba, Cd, Hg, Pb and Eu). In this structure, the metal is in a cubic environment, and the coordination of fluoride is tetrahedral. Figure 1 gives a slice half a unit-cell thick of $BaSnF_4$, BaF_2, and BaClF In BaF_2, BaF_8 cubes alternate with empty F_8 cubes along the three axes of the unit-cell. The cubes not containing a metal ion are potential interstitial sites for the formation of fluoride ion Frenkel defects, and this has been used to explain their high fluoride conductivity. However, Dénès has shown that this hypothesis alone does not explain the conductivity trend within the fluorite type fluorides [1]. When half of the metal in SrF_2, BaF_2 and β-PbF_2 is replaced by tin(II), highly layered $MSnF_4$ materials with ...M M Sn Sn M M... order are obtained, which requires doubling the *c* parameter [2]. The tin lone pair axis is parallel to *c* and all the lone pairs cluster in planes parallel to *(a,b)*, creating very effective cleavage planes which result in a highly lamellar structure, and give rise to a high degree of preferred

M. Miglierini and D. Petridis (eds.), Mössbauer Spectroscopy in Materials Science, 39–48.

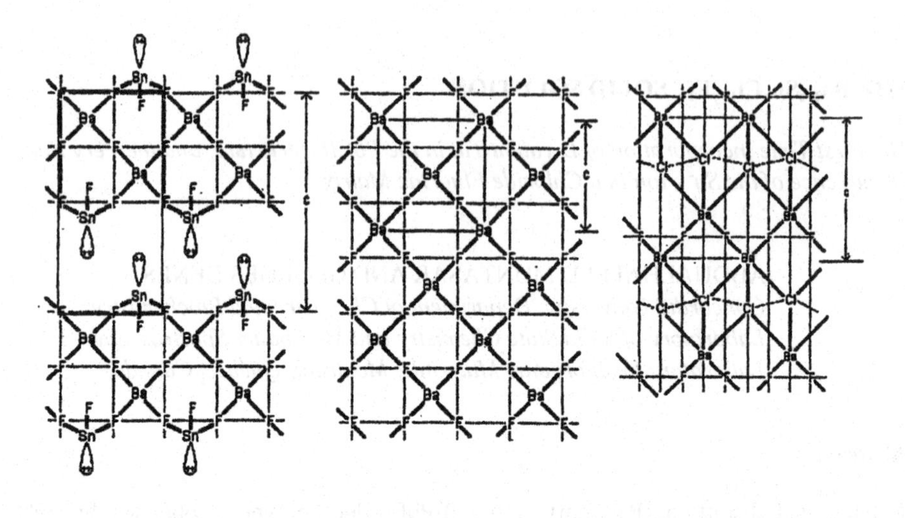

$BaSnF_4$ BaF_2 BaClF

Figure 1. Projection a slice, half a unit-cell thick, of the Structures of $BaSnF_4$, BaF_2, and BaClF onto the (a,b) plane in the BaF_2 axes.

orientation [3]. In BaClF, half of the fluoride ions are replaced by chloride ions, and due to the size difference, the two types of anions order to give the PbClF type structure [4]. The ...F Cl F ... order also results in a tetragonal distortion [5,6]. Each second plane of fluorine parallel to *(a,b)*, has been replaced by a corrugated sheet of chlorine that form one short axial Ba-Cl bond parallel to *c* and four longer equatorial Ba-Cl' bonds on the same side. On the other side, half of each BaF_8 cube remains and forms four Ba-F bonds. The polyhedron of coordination of barium is a $BaF_4ClCl'_4$ monocapped square antiprism,. The tin coordination in $BaSnF_4$ is a distorted $SnFF'_4E$ octahedron where F forms a strong covalent bond, the four F' form ionic bonds to an $[ESnF]^+$ cation, and E is the stereoactive tin lone pair, which occupies a hybrid orbital opposite to the axial fluorine. Tin is sp hybridized.

Before we started this work, there was no report of metal substitution by tin(II) in the PbClF type, or of chloride fluorides of barium and tin(II). SnClF is orthorhombic (Pnma space group). It has a polymeric structure containing $SnClF_3E$ (E= lone pair) groups which are highly distorted pseudotrigonal bipyramid, with three-coordinated fluorine and terminal chlorine [7]. The main goal of this work was to explore the possible existence of such materials, and study their structures and properties.

2. Experimental

All preparation were carried out using SnF_2 (99%) from Ozark Mahoning, $BaCl_2·2H_2O$ (analytical grade) from American Chemicals, BaF_2 (99%) from Allied Chemicals and Dye Corporations, HF (40%) from Mallinckrodt and doubly distilled water. All crystalline materials were checked by X-ray diffraction and thermal analysis (DTA/TGA) and were found to conform to the expected formula. X-ray powder diffraction was carried out on a Philips diffractometer equipped with a PW 1050/25 focussing goniometer which has been automated with the SIE112 system from Sietronics. The SIE112 software was used for data accumulation and processing. Phase identification was carried out using the JCPDS database and the μPDSM software by Fein-Marquat.

Precipitation reactions were carried out by adding a 1.274M solution of $BaCl_2·H_2O$ to a 1.500M solution of SnF_2, or vice versa on stirring. The white precipitate formed was filtered, washed with cold water and dried in air at ambient temperature. The stoichiometry of the reaction mixture was defined by the molar fraction of Ba, $X = Ba/(Ba+Sn)$, which was varied from 0.057 to 0.950. Direct reaction at high temperature were carried out by heating the appropriate mixture of reactants in sealed copper tubes under nitrogen according to a procedure designed by Dénès [9]. Barium chloride dihydrate was dehydrated at 200°C for 2 hours prior to moving it to the glove box.

All the materials prepared were studied by X-ray powder diffraction, and selected samples were also investigated by thermal analysis (DTA/TGA) and by ^{119}Sn Mössbauer spectroscopy. Mössbauer spectroscopy was carried out using an Elscint driving system, a scintillation detector equipped with a (Tl)NaI crystal, a Pd foil to absorb X-ray lines, and a Tracor Northern TN 7200 multichannel analyzer equipped with a built-in amplifier and a single channel analyzer. The γ-ray source was 5 mCi $Ca^{119m}SnO_3$ supplied by Amersham. Commercial $CaSnO_3$ from Alfa and α-SnF_2 from Ozark Mahoning were used for velocity calibration. All measurements were carried out at ambient conditions in the constant acceleration mode [8]. Spectrum fitting was carried out using the GMFP software [10].

3. Results and Discussion

3.1. MATERIALS OBTAINED BY PRECIPITATION

A white precipitate is always obtained on contact between the solution of SnF_2 and of $BaCl_2·2H_2O$. The composition of the material obtained varies with the fractional molar ratio X of the reactants [8]. In addition to the known phases $BaSn_2F_6$ and $BaSnF_4$, several hitherto unknown chloride fluorides of tin(II) and barium were produced. Two of these have been isolated and analyzed; $BaSn_2Cl_2F_4$ (fig. 2b) and $Ba_3Sn_3Cl_4F_8·2.6H_2O$. In addition, for a molar fraction X of Ba in the reaction mixture larger than 0.870, the precipitate has the same diffraction pattern as that of BaClF (fig. 2a & 2c). Therefore, it appears like if BaClF has been obtained. However, the Mössbauer spectrum of the sample shows the presence of a non-negligible amount of tin(II) (fig. 3c) and a minor amount of tin(IV), the latter being due to surface oxidation of the particles [12]. Therefore, the material obtained is not pure BaClF. In addition, the tin spectrum is not similar to that usually observed for tin(II) fluorides or chlorides, e.g. $BaSnF_4$ (fig. 3a), since it is a single line at ca. 4 mm/s for the BaClF like phase, whereas the spectrum of

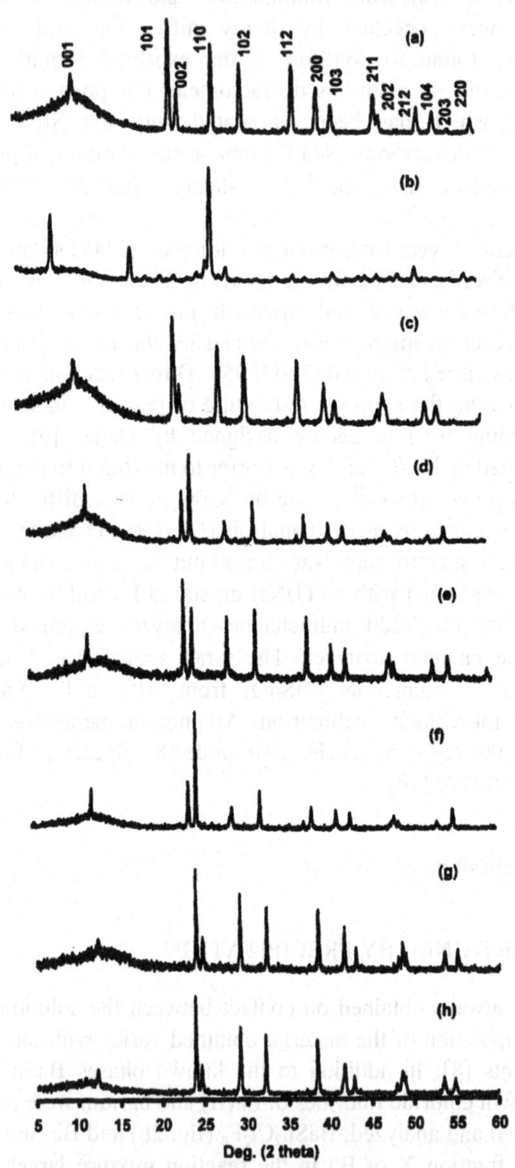

Figure 2. X-ray powder diffraction of (a) BaClF, (b) BaSn2Cl2F4, (c) solid solution from precipitation, (d) solid solution from direct reaction (y > 0), (e) solid solution from direct reaction (y < 0), (f) solid solution from direct reaction that has a very weak Mössbauer signal, (g) solid solution from direct reaction oxidized (aged), (h) solid solution from precipitated not oxidized after ageing.

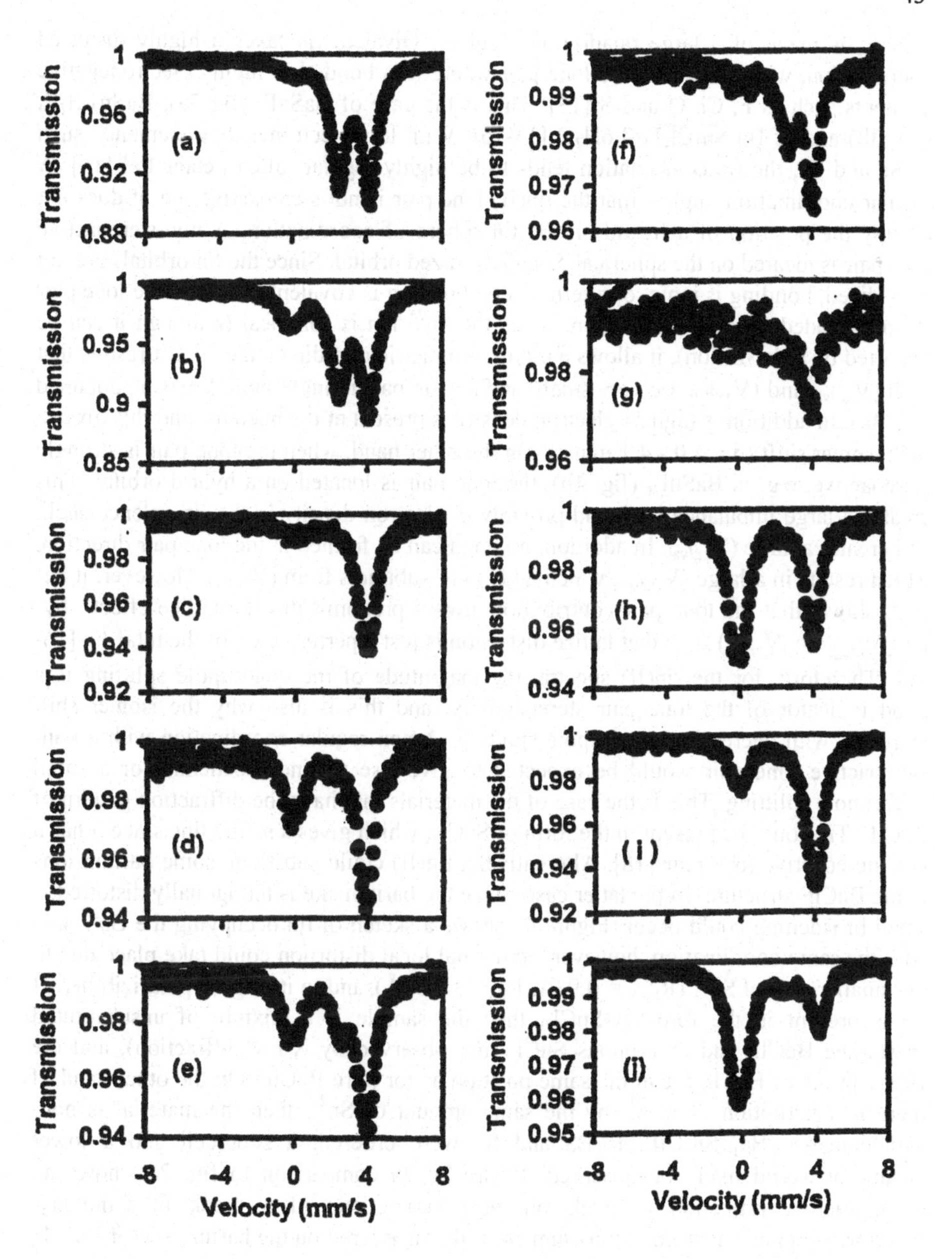

Figure 3. Mössbauer spectrum of: (a) $BaSnF_4$, (b) $BaSn_2Cl_2F_4$, and the $Ba_{1-x}Sn_xCl_{1+y}F_{1-y}$ solid solution: (c) precipitated, (d) x = 0.11 & y = 0, (e) x = 0.20 & y = 0.10, (f) x = 0.20 & y = -0.12, (g) x = 0.225 & y = 0.10, (h) x = 0.11 & y = 0 aged 37.5 months, (i) precipitate aged 45 months, (j) x = 0.225 & y = 0.10 aged 21 months.

$BaSnF_4$ is made of a large quadrupole doublet. Divalent tin takes a highly distorted coordination, with a stereoactive lone pair where it is bonded to highly electronegative elements such as F, Cl, O and S [13]. This is the case of $BaSnF_4$ (fig 3a), $BaSn_2Cl_2F_4$ (fig. 3b) and of $Ba_3Sn_3Cl_4F_8{\cdot}2.6H_2O$ [14-15]. With less electronegative elements, such as Se and Te, the tin coordination tends to be highly regular, often octahedral [16]. A regular coordination implies that the tin(II) lone pair is not stereoactive, i.e. it does not occupy the position of a ligand on the tin sphere of coordination. A non-stereoactive lone pair is located on the spherical *5s* unhybridized orbital. Since the tin orbitals are not hybridized, bonding is ionic, or alternatively, bonding is covalent, however the lone pair is not included in the.hybridization. Since the Sn^{2+} ion is spherical (although it can be distorted by polarization), it allows a highly symmetric coordination of tin. It results that both $(V_{zz})_{val}$ and $(V_{zz})_{latt}$ are very small, and a Mössbauer single line ($\Delta \approx 0$) is obtained (fig. 4a). In addition, a high *5s* electron density is present at the nucleus, and this gives a high isomer shift ($\delta \approx 4.0$ - 4.1 mm/s). On the other hand, when the lone pair is strongly stereoactive, e.g. in $BaSnF_4$ (fig. 4b), the lone pair is located on a hybrid orbital. This creates a large imbalance in *p*, and possibly *d*, electron density in the tin valence shell, and creates a large $(V_{zz})_{val}$. In addition, no bond can be formed in the lone pair direction, which results in a large $(V_{zz})_{latt}$, which adds to or subtracts from $(V_{zz})_{val}$. However, it has been shown that the lone pair contribution always predominates the lattice effect by far ($|V_{zz_{val}}| >> |V_{zz_{latt}}|$) such that lattice distortion is just a perturbation of the total V_{zz} [16-19]. Therefore, for the tin(II) spectra, the magnitude of the quadrupole splitting is a good indicator of the lone pair stereoactivity, and this is also why the isomer shift increases with decreasing quadrupole splitting. A non-regular coordination with a non-stereoactive lone pair would be expected to give rise to line broadening or a small quadrupole splitting. This is the case of the materials that have the diffraction pattern of BaClF. Tin could be present in the form of $SnCl_2$, which gives a single line since it has a non-stereoactive lone pair [18]. Alternatively, tin(II) could substitute some barium ions in the BaClF structure. In the latter case, since the barium site is tetragonally distorted, a small broadening could occur. Figure 4c shows a sketch of tin occupying the Ba^{2+} site, with the same coordination, however, additional local distortion could take place due to the smaller size of Sn^{2+} ($R_{Sn^{2+}} = 0.93$Å, $R_{Ba^{2+}} = 1.35$Å), and to its higher polarizibility. If tin is present in the form of $SnCl_2$, then the sample is a mixture of unsubstituted crystalline BaClF and amorphous $SnCl_2$ (not observed by X-ray diffraction), and the Bragg peaks of BaClF are at the same position as for pure BaClF. On the other hand, if there is substitution of xBa^{2+} by the same amount of Sn^{2+}, then the material is non-stoichiometric $Ba_{1-x}Sn_xClF$. If Ba and Sn were ordered, a supercell and a lower symmetry would likely be observed. Figure 2c, in comparison to fig. 2a, shows no superstructure reflection or peak-splitting characteristic of a break in symmetry, therefore it appears that tin and barium are fully disordered on the barium site of BaClF, and there is no change of lattice symmetry. Furthermore, a close examination of the peak position shows that the peaks of the new materials (fig. 2c) are shifted compared to BaClF (fig. 2a): e.g. (101) and (002) have moved towards one another, and so have (200) and (103). The unit-cell parameters have changed: *a* has decreased by 0.91% and *c* has increased by 0.97%,. The decreases of *a* can be understood by substituting large Ba^{2+} by smaller Sn^{2+}. Chemical analysis shows that 45% of Ba has been replaced by tin.

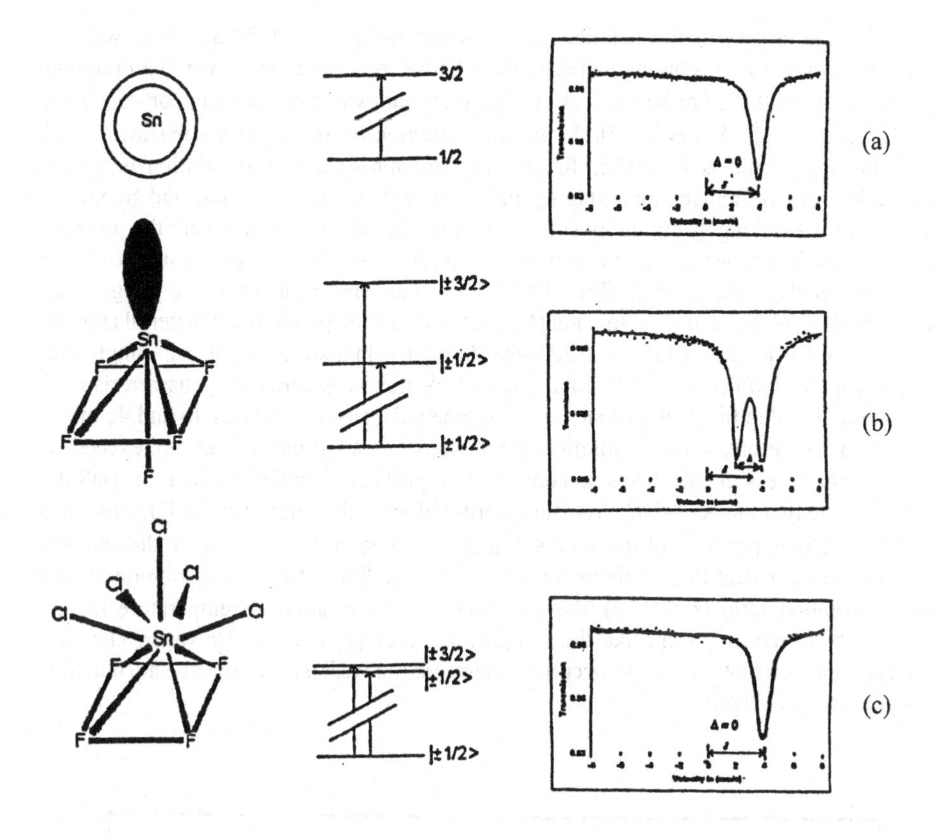

Figure 4. ^{119}Sn Mössbauer spectrum of divalent tin as a function of electronic structure and bonding type: (a) Sn^{2+} ion with a $5s^2$ non-stereoactive lone pair, (b) covalently bonded Sn(II) with a highly stereoactive lone pair in $BaSnF_4$,(c) Sn^{2+} ion substituting Ba^{2+} in BaClF.

In addition, surprisingly it also shows that 9% of fluorine has been replaced by chlorine. This makes the material non-stoichiometric both for cations and anions. Its formula should be expressed as$Ba_{1-x}Sn_xCl_{1+y}F_{1-y}$ with $x = 0.15$ and $y = 0.09$. Other BaClF like precipitates gave a similar composition. We did not observe any substitution of Cl by F for the precipitates (i.e. $y < 0$). Substituting F^- anions by larger Cl^- ions probably explains the increase in the *c* unit-cell parameter.

3.2. MATERIALS OBTAINED BY DIRECT REACTION

Direct reaction of appropriate amounts of BaF_2, $BaCl_2$ and SnF_2 were carried out at 300 °C, 500 °C, 600 °C and 800 °C, according to the following reaction:

$$\frac{1-2x-y}{2}BaF_2 + xSnF_2 + \frac{1+y}{2}BaCl_2 \rightarrow Ba_{1-x}Sn_xCl_{1+y}F_{1-y} \quad (1)$$

A wide range of solid solution was obtained for $0 \leq y \leq 0.25$ at $x = 0$, with the minimum value of y becoming increasingly negative as x increases, down to a minimum $y \approx -0.15$ at $x \approx 0.07$. For larger x values, the minimum value of y remain constant at ca. -0.15 up to $x = 0.18$. For $x > 0.18$, the solid solution becomes rapidly narrower, such that the upper limit is $x = 0.25$ (fig. 5). The domain of the solid solution is weakly dependent on the temperature of preparation, with both the minimum and maximum values of y moving to more positive (or less negative) values when the reaction temperature is increased. The diffraction patterns (fig. 2d - 2f) are very similar to that of the precipitated phase (fig. 2c). The Mössbauer spectrum shows a single line characteristic of the stannous Sn^{2+} ion (fig. 3d), like for the precipitated material (fig. 3c) provided $x < 0.13$. For $0.13 < x < 0.22$, regardless of y, in addition to the Sn^{2+} single line, a quadrupole doublet ($\delta = 3.15$ mm/s, $\Delta = 1.29$ mm/s) is observed, characteristic of covalently bonded tin(II) with a stereoactive lone pair (cases of figures 4b and 4c mixed together). Therefore, the concentration of Sn^{2+} ions in the barium sites cannot exceed ca. 13%. The excess tin(II) takes a coordination probably similar to that of $BaSnF_4$, however the two kinds of tin(II) remain disordered up to the limit of the solid solution at $x \approx 0.25$. The upper limit of the solid solution is located at the point where the amounts of ionic and covalent tin are about the same. Near the limit of the solid solution at high tin substitution ratio ($x > 0.22$), the Mössbauer signal at ambient temperature is very weak, which makes it look like the samples are very poor in tin. However, chemical analysis showed that no tin loss occurred, therefore the weak signal must be attributed to a low recoil-free fraction.

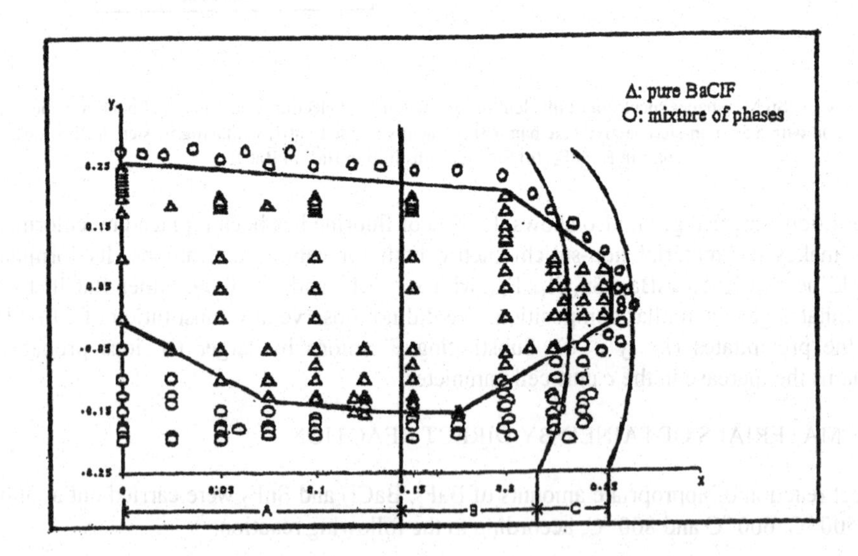

Figure 5. Domain of solid solution at 500°C. A: single line at $\delta \cong 4.07$ mm/s (Sn^{2+} ion only), B: single line at $\delta \cong 4.07$ mm/s & Sn(II) covalent ($\delta \cong 3.15$ mm/s, $\Delta \cong 1.29$ mm/s), C: very low recoil-free fraction.

Most likely, with increasing tin content on the barium site, more and more local distortions occur, and this, combined with the presence of two tin coordinations, creates a soft lattice around tin. At these high x values, the stability of the material is probably decreasing, since the stability limit of the solid solution is very close, and therefore, lattice softening could be expected, somehow similar to that observed near some phase transitions. For samples prepared by direct reaction, the shift of the Bragg peaks relative to BaClF is smaller than for the precipitates, regardless of x and y (fig. 2d - 2f). Therefore, although the samples prepared by the two methods appear identical up to $x \approx 0.13$, the position of Cl substituting F might not be the same in both. For $y < 0$, the Cl corrugated layer is partially substituted by F, again with a minor influence on the unit-cell parameters (fig. 2e).

3.3. PHYSICAL AND CHEMICAL PROPERTIES

3.3.1. Sensitivity to Oxidation

About every tin (II) compound gives a tin(IV) Mössbauer signal of variable strength. This is due to surface oxidation to tin(IV), most likely SnO_2 ($\delta \approx 0$). The very high recoil-free fraction of SnO_2 relative to Sn(II) halides make the Sn(IV) signal much larger than its real abundance in the sample, e.g. ca. 13 times in α-SnF_2 [9]. Large single crystals give no tin(IV) signal, which confirms that oxidation occurs at the surface of the particles [19]. Figures 3h to 3j show the Mössbauer spectrum of the solid solution after ageing in air at ambient temperature (fig. 3d, 3c, and 3g, respectively, before ageing). The materials prepared by direct reaction (fig. 3d, and 3g) show considerable oxidation after ageing for 37.5 months and 21 months respectively (fig. 3h, and 3j). Most particularly, the sample at high x value has a very low recoil-free fraction when freshly prepared (fig. 3g). However, after ageing, it gives a very strong tin (IV) signal (fig. 3j), which is further confirmation that no tin loss occurred. In contrast with the materials prepared by direct reactions, the precipitates undergo only a mild oxidation (fig. 3c $\rightarrow$ fig. 3i) after 45 months ageing. This different behaviour is not understood at this point, however, it is another difference between the materials prepared by the two methods, the other being the larger Bragg peak shift for the precipitates. The X-ray diffraction patterns show no change upon aging, even though a large amount of tin(IV) is observed, therefore no crystalline SnO_2 is formed. Presumably, tin(IV) is randomly distributed with tin(II) on the barium sites.

3.3.2. Potential charge carriers and conducting properties

Electrical conduction in our materials could occur, either by electron motion if they are *semiconductors*, or by ionic motion if they are *ionic conductors*. If both mechanisms take place at the same time, they are called *mixed conductors*. Since the $Ba_{1-x}Sn_xCl_{1+y}F_{1-y}$ solid solution has the PbClF structure, it might be a poor ionic conductor like PbClF. However, since the presence of covalently bonded tin(II) in many MF_2 strongly enhances their fluoride ion conductivity, particularly in stoichiometric $MSnF_4$ and $PbSn_4F_{10}$, and in non-stoichiometric $Pb_{1-x}Sn_xF_2$, an increase of fluoride ion mobility could happen in the $Ba_{1-x}Sn_xCl_{1+y}F_{1-y}$ solid solution for $x > 0.13$, where some of the tin(II) is covalently bonded. In addition, since all or some of the tin(II) has a non-stereoactive lone pair, a fraction of the 5s electron density might be transferred to the conduction band and give rise to semiconduction. This is not possible in covalently bonded tin, since the lone pair is locked in a hybrid orbital. However, only conductivity and transport number measurements can ascertain the conducting properties.

4. Conclusions

In BaClF, substitution on both the cation and the anion sites gives the fully disordered $Ba_{1-x}Sn_xCl_{1+y}F_{1-y}$ solid solution. The materials can be prepared either by precipitation or by high temperature reactions. The properties of the materials, including the unit-cell parameters, vary with the method of preparation. The precipitated materials oxidize much more slowly on aging. By high temperature reaction, up to 13% of Ba^{2+} can be substituted by Sn^{2+} ions, and an additional 13% by covalently bonded tin(II). Above 22% Sn, near the limit of solubility, the tin sublattice becomes very soft. The $5s^2$ unhybridized non-stereoactive lone pair is a potential charge carrier, whereas the covalently bonded tin(II) may increase the fluoride ion conductivity of the material. Ionic conductor $BaSnF_4$ was prepared for the first time by the aqueous route.

Acknowledgements

The Natural Science and Engineering Research Council of Canada, Concordia University, and the General Secretariat of Education of Libya, are acknowledged for financial support.

References

1. Dénès, G. (1995) Solid State Ionics IV, *Mater. Res. Symp. Proc.*, **369**, 295.
2. Birchall, T., Dénès, G., Ruebenbauer, K., and Pannetier, (1986) *Hyperf. Inter.* **29**, 1331.
3. Dénès, G., Yu, Y.H., Tyliszczak, T., and Hitchcock, A.P.(1991), *J. Solid State Chem.* **91**, 1.
4. Wyckoff, R.G. (1982) *Crystal Structures* **1**, Interscience, New York, p. 432.
5. Flahaut, J. (1974), *J. Solid State Chem.* **9**, 124.
6. Dénès, G, Pannetier, J., and Lucas, J. (1975), *C.R. Acad. Sc. Paris, Ser C* **280**, 831.
7. Geneys, C., Vilminot, S. and Cot, L. (1976) *Acta Cryst.* B **32**, 3199.
8. Dénès, G. and Muntasar, A. (1996), *Mater. Structure.* **3**, 246.
9. Dénès, G. (1988), *J. Solid State Chem.* **77**, 54.
10. Ruebenbauer, K. and Birchall, T. (1979), *Hyperf. Interact* **7**, 125.
11. Claudy, P., Letiffe, J.M., Vilminot, S., Granier, W., Al Ozaibi, Z., and Cot, L. (1981), *J. Fluorine Chem.* **18**, 203.
12. Dénès, G. and Laou, E. (1994), *Hyperf. Inter.* **92**, 1013.
13. Greenwood, N.N. and Gibb, T.C. (1971) *Mössbauer Spectroscopy*, Chapman and Hall, London.
14. Muntasar, A. and Dénès, G. (1996), in I. Ortalli (ed.) *ICAME-95 Conf Proceeding*, Societa Italiana di Fisica **50**, 127-130.
15. Dénès, G., Muntasar, A., and Zhu, Z. (1996), *Hyperf. Inter(C)* **1**, 468-471.
16. Donaldson, J. D. and Silver, J. (1973), *J.Chem. Soc.*(**A**) 666.
17. Barrett, J., Bird, S. R. A., Donaldson, J., and Silver, J. (1971), *J. Chem. Soc.* (**A**), 3105.
18. Donaldson, J.D. and Silver, J. (1969), *J. Chem. Soc.* (**B**), 2358.
19. Birchall, T., Dénès, G., Ruebenbauer, K., and Pannetier, J. (1981), *J. Chem. Soc. Dalton Trans.* 2296.

THE MECHANISM OF β-Fe_2O_3 FORMATION BY SOLID-STATE REACTION BETWEEN NaCl AND $Fe_2(SO_4)_3$

R. ZBORIL [1], M. MASHLAN [2] AND D. KRAUSOVA[1]

[1)]*Department of Inorganic and Physical Chemistry, Palacky University, Svobody 8, 771 46 Olomouc, Czech Republic.*

[2)]*Department of Experimental Physics, Palacky University, Svobody 26, 771 46 Olomouc, Czech Republic.*

Abstract

Mössbauer spectroscopy, x-ray powder diffraction, thermal analysis and elemental analysis were used for the study of the mechanism of reaction between NaCl and $Fe_2(SO_4)_3$ in the oxidizing atmosphere (air). It was found that the reaction starts at 400°C in isothermal conditions and is accompanied by the liberation of Cl_2. The initial reaction proceeds in the molar ratio of NaCl:$Fe_2(SO_4)_3$=3:1. β-Fe_2O_3, $NaFe(SO_4)_2$, $Na_3Fe(SO_4)_3$ and Na_2SO_4 were identified as solid products of the initial reaction. At the second stage both the double sulfates were transformed into β-Fe_2O_3 and Na_2SO_4, the gaseous SO_3 was liberated simultaneously. Thermally less stable $NaFe(SO_4)_2$ decomposes before $Na_3Fe(SO_4)_3$. The hexagonal α-Fe_2O_3 is formed in this process only as the product of the structural transformation of β-Fe_2O_3. This beta-alpha transformation accompanies both the initial reaction and the secondary conversions of the intermediates. The Mössbauer parameters of the reaction products are enlisted.

1. Introduction

A few processes of cubic β-Fe_2O_3 formation are mentioned in the literature [1-8]. β-Fe_2O_3 is a body-centered cubic (bixbyite) structure with Ia3 space group and a lattice parameter a_o=9.393Å [1,4-6]. 24 Fe^{3+} ions in the cubic unit cell have C_2 symmetry (*d* site) and 8 ions have C_{3i} symmetry (*b* site) [5-7]. The Neel temperature of magnetic transition was observed between 100K [8] and 119K [4]. β-Fe_2O_3 is thermally metastable and it transforms into α-Fe_2O_3 at 770K [1,6,8].

β-Fe_2O_3 is magnetically disordered at room temperature and its room temperature Mössbauer spectrum shows only the pure quadrupole splitting [1].

The formation mechanism of β-Fe_2O_3 particles prepared by calcination of a mixture of NaCl and $Fe_2(SO_4)_3$ has been studied only by Ikeda *et al.* [4]. They suggested that the process consisted of three stages. At the first stage, $NaFe(SO_4)_2$ was formed by the reaction of NaCl and $Fe_2(SO_4)_3$. The exact course of this initial reaction was not specified and it remained somewhat unresolved. At the second stage, according to Ikeda

M. Miglierini and D. Petridis (eds.), Mössbauer Spectroscopy in Materials Science, 49–56.

et al., $NaFe(SO_4)_2$ was decomposed to β-Fe_2O_3 and $Na_3Fe(SO_4)_3$. At the last stage, α-Fe_2O_3 and Na_2SO_4 were formed by the reaction of β-Fe_2O_3 and $Na_3Fe(SO_4)_3$. Ikeda *et al.* [4] used the washing of the reaction products with distilled water for the isolation of the β-Fe_2O_3 particles. Unluckily, the exact experimental conditions for the preparation of pure β-Fe_2O_3 without the contamination of α-Fe_2O_3 were not reported.

In this work a totally new mechanism for the solid-state reaction between NaCl and $Fe_2(SO_4)_3$ is suggested and the optimal heating conditions for β-Fe_2O_3 preparation are also described.

2. Materials and Methods

As starting materials, NaCl and $Fe_2(SO_4)_3{\cdot}5H_2O$ (Sigma-Aldrich) were used. $Fe_2(SO_4)_3$ was prepared by dehydration of $Fe_2(SO_4)_3.5H_2O$ at 400°C for 6 hours in a nitrogen atmosphere. The compounds were mixed in various molar ratios and heated in air under various temperature and time conditions. Room temperature Mössbauer spectra were collected using a Mössbauer spectrometer in constant acceleration mode with a $^{57}Co(Rh)$ source. The XRD7 (Seifert-FPM, Germany) diffractometer was used for measurements of the x-ray powder diffraction patterns. The measurements were realized at Kα-Cu wavelength in the diffraction angles range 7-68° 2θ (with the step of 0,02° 2θ and integration time 2s). A Ni absorption filter and a secondary carbon cylindrical monochromator were used. The individual phases were identified by means of PDF2 (ACDD) database [9]. For thermal analysis measurements the Q1500D Derivatograph (MOM Budapest, Hungary) was used. It records simultaneously the DTA, TG and T curves (temperature increase of 2.5°C/min., temperature range of 20-900°C). Chloride ions were determined by the mercurimetric titration with sodium nitroprusside as indicator.

3. Results and Discussion

3.1. THE COURSE OF THE PRIMARY REACTION

The determination of the initial molar ratio of NaCl:$Fe_2(SO_4)_3$ as well as the quantitative determination of all reaction products plays an important role in the understanding of the full reaction mechanism.

We have studied the course of the primary reaction as a function of the molar ratio of both initial compounds. According to the thermal analysis results, the reaction process starts at 400°C. The starting mixtures were heated at this temperature for 2 hours in the air and were analyzed by x-ray powder diffraction and Mössbauer spectroscopy.

The $Fe_2(SO_4)_3$ excess was identified by XRD in the products obtained by the calcination of mixtures with a molar ratio of NaCl:$Fe_2(SO_4)_3$ < 3. The Mössbauer spectra for these ratios presented a typical $Fe_2(SO_4)_3$ singlet with an isomer shift value of IS_{Fe}=0.50mm/s.

On the other hand, the XRD measurements showed an excess of NaCl in the products obtained by the calcination of the mixtures with a molar ratio of $NaCl:Fe_2(SO_4)_3 > 3$. The amount of chloride ions in these samples was determined by mercurimetric titration.

The product obtained by calcination of $NaCl:Fe_2(SO_4)_3 = 3:1$ mixture at 400°C for two hours did not contain reactants. This fact was in accordance with the XRD and Mössbauer spectroscopy results (Tab.1 and Tab.2). The absence of chloride was confirmed by the mercurimetric determination (0%). β-Fe_2O_3, $NaFe(SO_4)_2$, $Na_3Fe(SO_4)_3$ and Na_2SO_4 reaction products were unambiguously detected by XRD. Three iron phases were also revealed by Mössbauer spectroscopy results.

TABLE 1. XRD phases identification of the product obtained by calcination of $NaCl:Fe_2(SO_4)_3 = 3:1$ mixture at 400°C for 2 hours.

Compound	Crystal system	Unit cell parameters [Å]	Number of the card from the PDF2 database
$Na_3Fe(SO_4)_3$	hexagonal	a=b=13.415; c=9.025	[39-243]
$NaFe(SO_4)_2$	monoclinic	a=8.02; b=5.14; c=7.18; β=92.12°	[27-718]
β-Fe_2O_3	cubic	a_o=9.404	[39-238]
Na_2SO_4	orthorhombic	a=6.97; b=8.95; c=5.61	[24-1132]

TABLE 2. The results of the RT Mössbauer spectroscopy for the product obtained by calcination of $NaCl:Fe_2(SO_4)_3 = 3:1$ mixture at 400°C for 2 hours.

β-Fe_2O_3 doublet			$Na_3Fe(SO_4)_3$ doublet			$NaFe(SO_4)_2$ doublet		
IS [mm/s]	QS [mm/s]	A [%]	IS [mm/s]	QS [mm/s]	A [%]	IS [mm/s]	QS [mm/s]	A [%]
0.37	0.74	50.5	0.46	0.05	29.4	0.46	0.41	20.1

IS- isomer shift related to metallic iron, QS-quadrupole splitting, A- percentage of subspectrum area

The same isomer shift (0.46mm/s) was observed for both of the double sulfates. The hexagonal $Na_3Fe(SO_4)_3$ corresponds to a doublet with a very low value of quadrupole splitting (0.05mm/s). The monoclinic $NaFe(SO_4)_2$ corresponds to the doublet with a higher value of quadrupole splitting - about 0.4mm/s. The room temperature spectrum of β-Fe_2O_3 can be successfully fitted by a single quadrupole doublet with an isomer shift of 0.37mm/s and a quadrupole splitting of 0.74mm/s, though two nonequivalent sites of ferric ions are present in the cubic unit cell. The low temperature Mössbauer spectra are more useful for the elucidation of this spinel structure, but this was not our aim.

The percentages of the subspectra areas showed (Tab.2) that the product of the primary reaction contained approximately 50% ferric ions in the form of β-Fe_2O_3, 30%

in the form of $Na_3Fe(SO_4)_3$ and 20% in the form of $NaFe(SO_4)_2$. The percentages of the double sulfates remained unchanged also in products obtained by the calcination of the mixture with a molar ratio of $NaCl:Fe_2(SO_4)_3$ = 3:1 at 400°C for a longer time (3-24hours). The β-Fe_2O_3 percentage gradually decreased with the calcination time, but this decrease was compensated by the α-Fe_2O_3 formation. The total content of ferric oxide (alpha + beta) also remained constant.

The ratio of ferric ions of 5:3:2 corresponds to the molar ratio of $Fe_2O_3:Na_3Fe(SO_4)_3:NaFe(SO_4)_2$=5:6:4 ($Fe_2O_3$ contains two ferric ions in contradistinction to both double sulfates). This molar ratio of reaction products is in agreement with the weight ratio obtained from the XRD measurements.

Finally we can summarize that the solid state reaction between NaCl and $Fe_2(SO_4)_3$ in air proceeds via redox mechanism according to the following equation (1):

$$30NaCl + 10Fe_2(SO_4)_3 + 15/2O_2 \rightarrow 6Na_3Fe(SO_4)_3 + 4NaFe(SO_4)_2 + 5\beta\text{-}Fe_2O_3 + 4Na_2SO_4 + 15Cl_2$$

TG measurement in isothermal conditions (400°C) showed that the weight decrease after reaction was 14.2%. This experimental value corresponds closely to the theoretical decrease of 14.3%, calculated from equation (1).

It is worth mentioning the important fact that the structural transformation of β-Fe_2O_3 to α-Fe_2O_3 occurs already at the temperature of 400°C and accompanies the primary reaction.

For comparison, Ikeda *et al.* [4] did not explain at all the course of the primary reaction. They studied the reaction mechanism with the initial molar ratio of $NaCl:Fe_2(SO_4)_3$ = 2:1, in other words with the excess of $Fe_2(SO_4)_3$. Moreover the conversion of this redundant $Fe_2(SO_4)_3$ to α-Fe_2O_3 could result in incorrect conclusions on the secondary conversions of the double sulfates.

3.2. THE SECONDARY CONVERSIONS OF $NaFe(SO_4)_2$ AND $Na_3Fe(SO_4)_3$

The second stage of the multistage process includes the conversions of $NaFe(SO_4)_2$ and $Na_3Fe(SO_4)_3$. Both double sulfates are thermally stable up to 440°C. At this temperature, the conversion of $NaFe(SO_4)_2$ starts.

The transformation mechanism of $NaFe(SO_4)_2$ at 440°C is illustrated in Fig. 1. The quantitative results of Mössbauer spectroscopy show that ferric ions from $NaFe(SO_4)_2$ were transferred to Fe_2O_3, while the percentage of ferric ions corresponding to $Na_3Fe(SO_4)_3$ remained unchanged. The conversion of $NaFe(SO_4)_2$ to β-Fe_2O_3 proceeded gradually at 440°C and its rate was almost the same as the rate of transformation of β-Fe_2O_3 to α-Fe_2O_3. After the increase of heating temperature to 480°C, the rate of $NaFe(SO_4)_2$ conversion significantly increased (Tab.3). No x-ray diffraction peaks corresponding to $NaFe(SO_4)_2$ were detected for the sample heated at 480°C for two hours. On the other hand, the quantitative increase of β-Fe_2O_3 and Na_2SO_4 content was significant in comparison with the XRD pattern of the product obtained by calcination at 400°C for 2 hours. According to these results, the conversion of $NaFe(SO_4)_2$ can be expressed by the following equation:

$$2NaFe(SO_4)_2 \rightarrow \beta\text{-}Fe_2O_3 + Na_2SO_4 + 3SO_3 \qquad (2)$$

Indications of the transformation of $NaFe(SO_4)_2$ to $Na_3Fe(SO_4)_3$ and β-Fe_2O_3, as mentioned by Ikeda *et al.*[4], were not found.

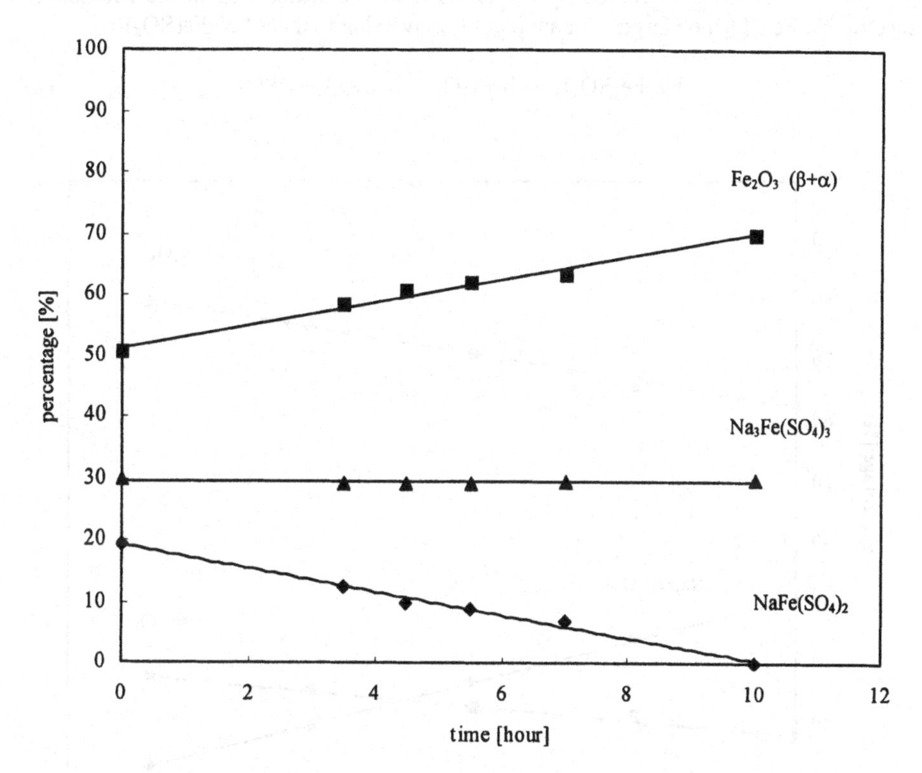

Figure 1. The transformation of $NaFe(SO_4)_2$ at constant temperature 440°C (starting material: the mixture with the molar ratio of NaCl:$Fe_2(SO_4)_3$ = 3:1 heated at 400°C for 2hours). The percentage of ferric ions was determined from the areas of Mössbauer spectra.

TABLE 3. The distribution of ferric ions in the samples obtained:
A) by calcination of the mixture with a molar ratio of NaCl:$Fe_2(SO_4)_3$ = 3:1 at 400°C for 2 hours
B) by calcination of the sample A) at 480°C for 2 hours.

Compound	The percentage in Sample A	The percentage in sample B
β-Fe_2O_3	50.5	63.7
α-Fe_2O_3	0	7.4
$Na_3Fe(SO_4)_3$	29.4	28.9
$NaFe(SO_4)_2$	20.1	0

The thermal decomposition of $Na_3Fe(SO_4)_3$ started at 500°C. Also in this case Fe^{3+} ions from $Na_3Fe(SO_4)_3$ were transferred to β-Fe_2O_3 (Fig.2). The slow transformation of β-Fe_2O_3 to α-Fe_2O_3 is exhibited by the increase of the sextet area in the Mössbauer spectra. $Na_3Fe(SO_4)_3$ undergoes an analogous conversion to that of $NaFe(SO_4)_2$:

$$2Na_3Fe(SO_4)_3 \rightarrow \beta\text{-}Fe_2O_3 + 3Na_2SO_4 + 3SO_3 \tag{3}$$

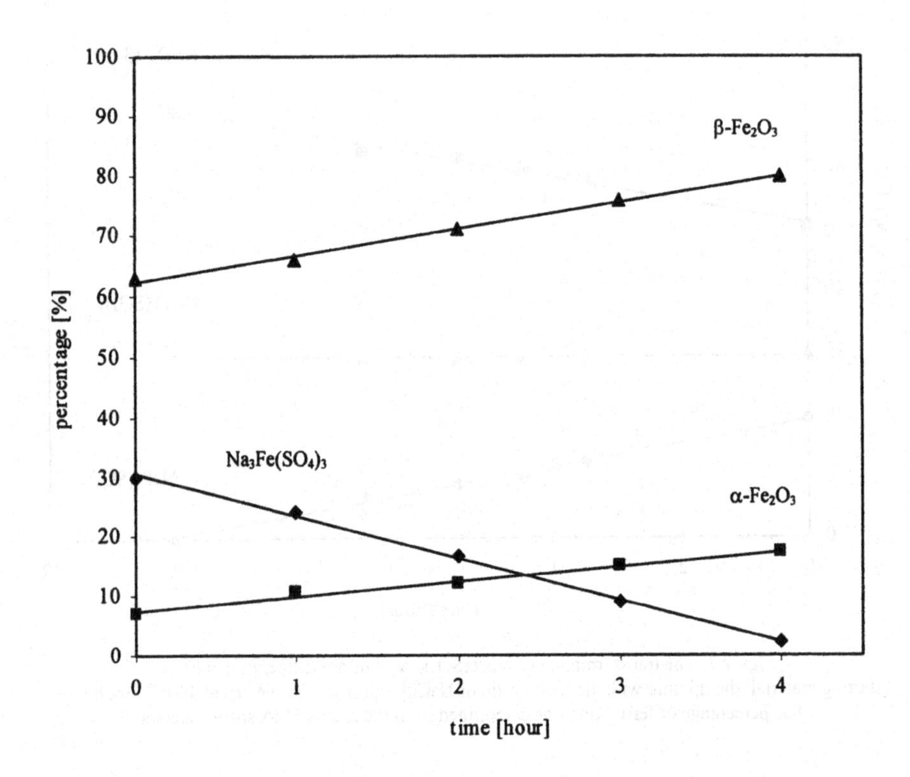

Figure 2. The transformation of $Na_3Fe(SO_4)_3$ at constant temperature 500°C (starting material: mixture with the molar ratio of NaCl:$Fe_2(SO_4)_3$ = 3:1 heated at 480°C for 2hours). The percentage of ferric ions was determined from the areas of Mössbauer spectra.

The possibility of the mutual reaction of $Na_3Fe(SO_4)_3$ with β-Fe_2O_3 to α-Fe_2O_3 and Na_2SO_4, as reported by Ikeda *et al.* [4], was refuted. Our argument is based on the fact that the decrease of the $Na_3Fe(SO_4)_3$ percentage is accompanied by the simultaneous increase of β-Fe_2O_3 percentage, the α-Fe_2O_3 content increases only gradually (Fig. 2). Japanese authors mixed $Na_3Fe(SO_4)_3$ with β-Fe_2O_3 in a molar ratio of 1:1 and heated it at 700°C. This high temperature of calcination was the probable cause of the incorrect interpretation of the results, because the decomposition of $Na_3Fe(SO_4)_3$ to β-Fe_2O_3 and Na_2SO_4 as well as the structural transformation of β-Fe_2O_3 to α-Fe_2O_3

occurred very quickly at 700°C. Thus, the incorrect assumption that α-Fe_2O_3 and Na_2SO_4 are produced by the mutual interaction of $Na_3Fe(SO_4)_3$ with β-Fe_2O_3 could arise.

3.3. β-$Fe_2O_3 \rightarrow \alpha$-$Fe_2O_3$ TRANSFORMATION, FINAL REACTION STAGE

As mentioned above, our experimental results showed that the transformation of β-Fe_2O_3 to α-Fe_2O_3 occurs already at 400°C. Beta-alpha transformation accompanies both the primary reaction and the secondary conversions of the intermediate phases.

The transformation of β-Fe_2O_3 to α-Fe_2O_3 also plays a key role in the process of β-Fe_2O_3 preparation by the calcination of a mixture of NaCl with $Fe_2(SO_4)_3$. Both reactants and the reaction products except for Fe_2O_3 can be dissolved in distilled water, and β-Fe_2O_3 can be isolated by filtration. It is important to inhibit the beta-alpha transformation and select the optimal heating conditions to achieve the maximum purity of β-Fe_2O_3.

We achieved the best quality of β-Fe_2O_3 by calcination at 400°C for 0.5 hour. A NaCl excess had a favorable influence because it accelerated β-Fe_2O_3 formation by the primary reaction. After dissolution and filtration, the product did not contain any lines of α-Fe_2O_3 in XRD pattern as well as in the room temperature Mössbauer spectrum.

It is important to mention one more interesting fact. Not only ferric oxide but also the second final product, sodium sulfate, has a polymorph character. The temperature-induced transformations of Na_2SO_4 and Fe_2O_3 polymorphs conclude the reaction process at higher temperatures in accordance with the XRD results (Tab.4).

TABLE 4. XRD phases identification of the product obtained by calcination of $NaCl:Fe_2(SO_4)_3 = 3:1$ mixture at 600°C for 2 hours.

Compound	Crystal system	Unit cell parameters [Å]	The number of the card from the PDF2 database
Na_2SO_4	orthorhombic	a=9.8211; b=12.3076; c=5.8623	[37-1465]
Na_2SO_4	orthorhombic	a=6.97; b=8.95; c=5.61	[24-1132]
β-Fe_2O_3	cubic	a_o=9.404	[39-238]
α-Fe_2O_3	hexagonal	a=b=5.036; c=13.749	[33-664]

For simplicity we can ignore the structural transformations of the final products and express the global course of the reaction between NaCl and $Fe_2(SO_4)_3$ according to the summarizing equation:

$$6NaCl + 2Fe_2(SO_4)_3 + 3/2O_2 \rightarrow 2Fe_2O_3 + 3Na_2SO_4 + 3Cl_2 + 3SO_3 \quad (4)$$

The dynamic TG measurement (20-900C°) of the mixture of NaCl and $Fe_2(SO_4)_3$ with the molar ratio 3:1 showed that the global weight decrease, after the termination of all reactions, was 35.1%. This experimental value is in a very good agreement with the theoretical value 35.2%, calculated from equation (4).

4. Conclusion

In this study the mechanism of reaction between NaCl and $Fe_2(SO_4)_3$ in air was successfully determined using transmission Mössbauer spectroscopy and x-ray powder diffraction. The reactants react in the molar ratio of NaCl:$Fe_2(SO_4)_3$=3:1. The reaction proceeds via redox mechanism. The atmospheric oxygen functions as the oxidizing agent and the global course of the process can be expressed by the equation (4). This process includes three solid-state reactions (1), (2), (3) proceeding step by step in the narrow temperature range from 400 to 500°C.

All reactions are accompanied by the structural transformation of β-Fe_2O_3 to α-Fe_2O_3. The reaction studied is suitable for the isolation of pure β-Fe_2O_3 by the dissolution of other reaction products with distilled water and by filtration of the solid phase. Taking into account the above-mentioned beta-alpha transformation, the choice of the minimum calcination temperature (400°C) and the shortest calcination time is important for the achievement of the maximum purity of β-Fe_2O_3.

Acknowledgements

Financial support from the Grant Agency of The Czech Republic under Project 203/96/1664 and from the internal grant of Palacky University is gratefully acknowledged.

References

1. Ben-Dor, L., Fischbein, E., Felner, I. and Kalman, Z. (1977) β Fe_2O_3: Preparation of thin films by chemical vapor deposition from organometalic chelates and their characterization, *J. Electrochem. Soc.* **124**, 451-457.
2. Muruyama, T. and Kanagawa, T. (1996) Electrochromic properties of iron oxide thin films prepared by chemical vapor deposition, *J. Electrochem. Soc.* **143**, 1675-1677.
3. Gonzales-Carreno, T., Morales, M.P. and Serna, Ç.J. (1994) Fine beta Fe_2O_3 particles with cubic structure obtained by spray pyrolysis, *J. Mater. Sci. Lett.* **13**, 381-382.
4. Ikeda, Y., Takano, M. and Bando, Y. (1986) Formation mechanism of needle-like α-Fe_2O_3 particles grown along the c axis and characterization of precursorily formed β-Fe_2O_3, *Bull. Inst. Chem. Res., Kyoto Univ.* **64**, 249-258.
5. Bauminger, E.R., Ben-Dor, L., Felner, I., Fischbein, E., Nowik, I. and Ofer, S. (1977) Mössbauer effect studies of β-Fe_2O_3, *Physica* **86-88*B***, 910-912.
6. Ben-Dor, L. and Fischbein, E. (1976) Concerning the β phase of iron (III) oxide, *Acta Cryst. B* **32,** 667.
7. Wiarda, D., Wenzel, T., Uhrmacher, M. and Lieb, K.P. (1992) Hyperfine interaction of ^{111}Cd impurities in Mn_2O_3, Mn_3O_4 and β-Fe_2O_3, *J. Phys. Chem. Solids* **53**, 1199-1209.
8. Wiarda, D. and Weyer, G. (1993) Mössbauer investigations of the antiferromagnetic phase in the metastable β-Fe_2O_3, *Int. J. Mod. Phys. B* **7**, 353-356.
9. Powder Diffraction File 1997, International Center for Diffraction Data, Pennsylvania, U.S.A.

MÖSSBAUER MEASUREMENTS OF SOLID SOLUTIONS $(Fe_xCr_{1-x})_2O_3$, 0<x<1

ERNST-GEORG CASPARY[1] AND TOMÁŠ GRYGAR[2]
[1)]*Institute of Physics; Academy of Sciences, V Holešovičkách 2*
CZ-18200 Prague, Czech Republic
[2)]*Institute of Inorganic Chemistry; Academy of Sciences*
CZ-25068 Řež, Czech Republic

1. Structure of Cr_2O_3(corundum) and α-Fe_2O_3 (hematite)

The crystal structure of both oxides is given by hcp (hexagonal closed packed) O^{2-} ions, where the cations occupy octahedral holes (figure 1) only. The layers of the hcp structure are denoted by A and B. The x-y coordinates of the cations correspond to the C layer of the fcc close-packed structure. The relation of cations to anions is 2:3 . Therefore not all of the possible octahedral sites are filled, which gives rise to the hexagonal substructure (see figure 2). In figure 3a the hexagonal elementary cell of corundum is drawn. From the six C-layers plotted only three are necessary in the case of corundum structure. The cation sites are marked as dots. The unoccupied octahedral sites (compare figure 1) are arranged in such a way, that they are on different x-y positions in neighbouring layers. As indicated in figure 3a, the electrostatic repulsion of neighbouring cations gives rise to a small deviation from the ideal structure. These deviation will be further ignored.

TABLE 1. Ionic radii and electronic configurations.

Ion	radius [A]	electron configuration.
O^{2-}	1.32	$2s^22p^6$
Fe^{3+}	0.67	$3d^54s^0$
Cr^{3+}	0.64	$3d^34s^0$

TABLE 2. Next neighbours in an ideal structure (see figure 3).

Distance	octahedral sites	cations	common O^{2-} neighbours
$(2/3)^{1/2}$ a = 0.82 a	2	1	3
a = diameter of O^{2-}	6	3	2
$(a^2+h^2)^{1/2}$ a = 1.29 a	12	9	1

The ionic radii and the electronic configuration of the cations are given inTable 1. In Table 2 the cation neighbours up to the third nearest neighbours are listed. Farther cation neighbours do not share a common oxygen ion. A common oxygen ion may provide an

M. Miglierini and D. Petridis (eds.), Mössbauer Spectroscopy in Materials Science, 57–62.

easy path for supertransferred magnetic hyperfine fields.

While the crystalline structures of corundum and hematite are equivalent, the magnetic structures are not (see figure 4). In both cases (in the case of hematite only at temperatures higher than 260K) the spins are orientated perpendicular to the $\underline{3}$ axis of the structure. But in corundum the spins are ordered up-down-up-down while they are ordered up-down-down-up in hematite (see figure 4). Therefore, in contrast to corundum, which is a pure antiferromagnet, in hematite a small ferromagnetic component may occur.

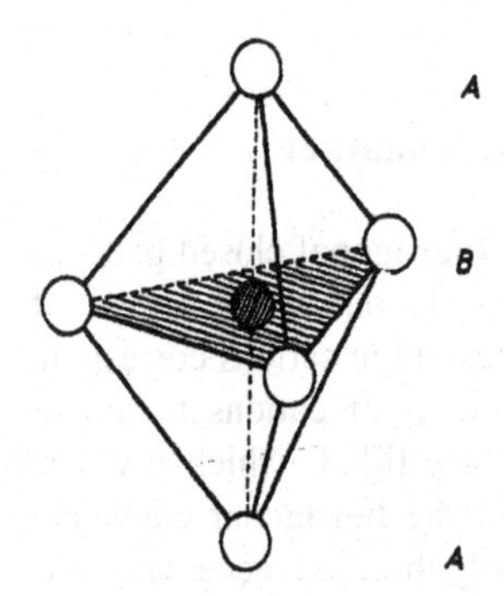

Figure 1. Octahedral sites.

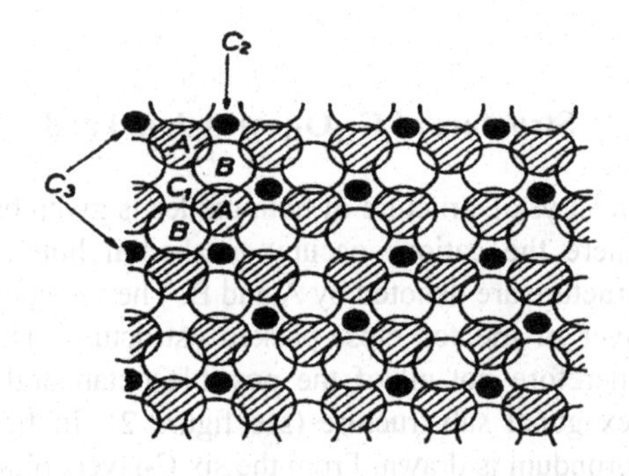

Figure 2. Hexagonal substructure.

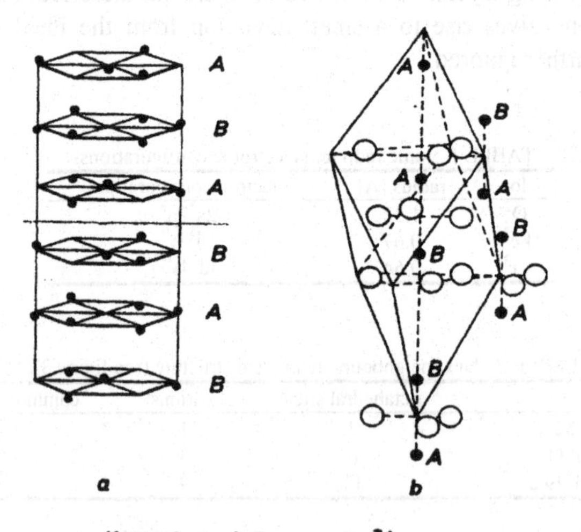

Figure 3. Crystalline structures.

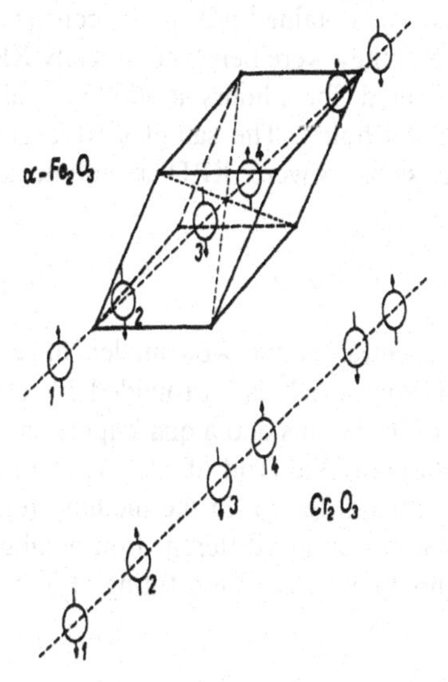

Figure 4. Magnetic structures.

2. Preparation of the Samples

Ferric nitrate and chromic chloride were dissolved in water. The solutions were precipitated by 15% excess of sodium hydroxide. The suspension was shortly boiled and then left to cool. The precipitates were filtered, washed with distilled water and then

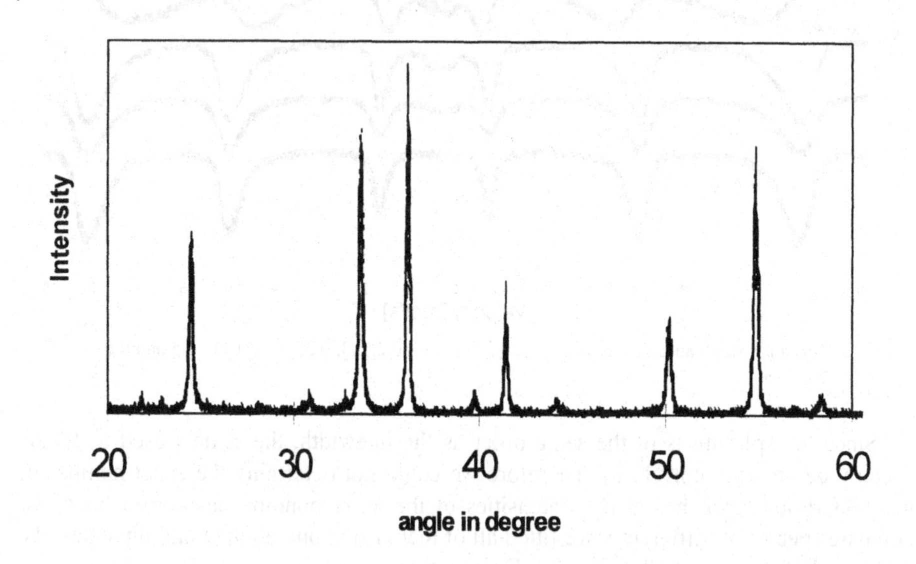

Figure 5. X-ray pattern of heated samples.

dried at laboratory conditions. The obtained precursors consisted of irregular aggregates of nanometer-sized particles and they were hence completely XRD amorphous.

The dry precursors were heated for 2 hours at 1000°C in air and then left to cool to room temperature within several hours. The samples were ground in an agate mortar before the measurements. The samples were XRD crystalline according to XRD.

3. Measurements

For all samples Mössbauer spectra for the iron nuclei, have been measured at room temperature. The spectra of iron which is surrounded by chrome ions only show a doublet with an isomer shift of 0.35mm/s and a quadrupole splitting of 0.38 mm/s . For pure hematite, one sextet with hyperfine field of 51.7 T, isomer shift of 0.37 mm/s and quadrupole splitting of -0.2 mm/s is found. In the medium regime, the sextet seems to split into a number of lines, representing different iron neighbourhoods (see figure 6). For $0.35<x<0.45$ we used up to 9 sextets for fitting. For $x<0.2$ the sextets are not detectable any more.

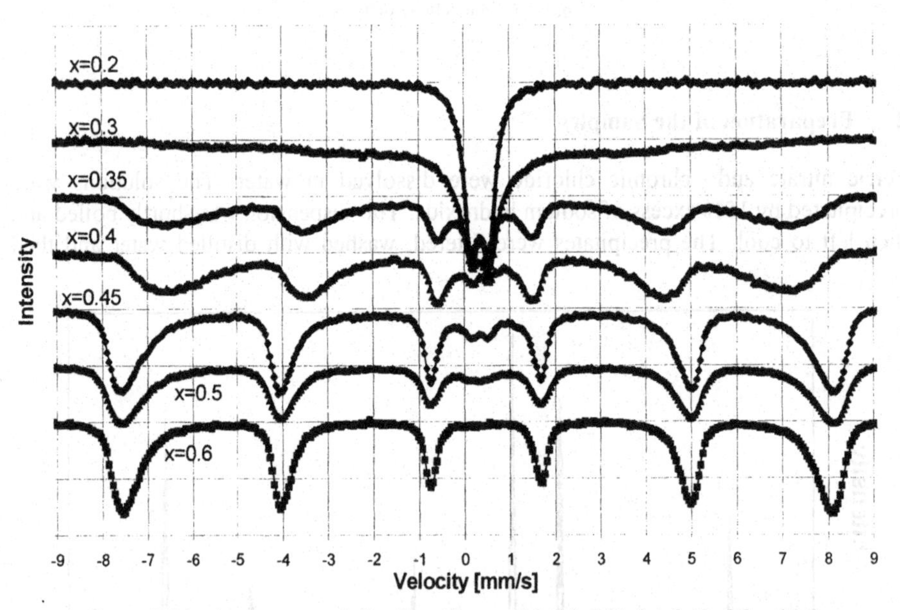

Figure 6. Mössbauer spectra of $(Fe_{2x}Cr_{2-2x})O_3$; x = 0.2, 0.3, 0.35, 0.4, 0.45, 0.5 and 0.6.

Since the splitting is of the same order as the linewidth, the sextets used to fit the spectra are strongly correlated. Therefore we could not determine the exact number of neighbourhoods and the relative intensities of the corresponding absorption lines. To compare spectra of different x we fitted all of them using one doublet and three sextets. Anyway the intensity of all sextets has been zero for $x\leq 0.2$.

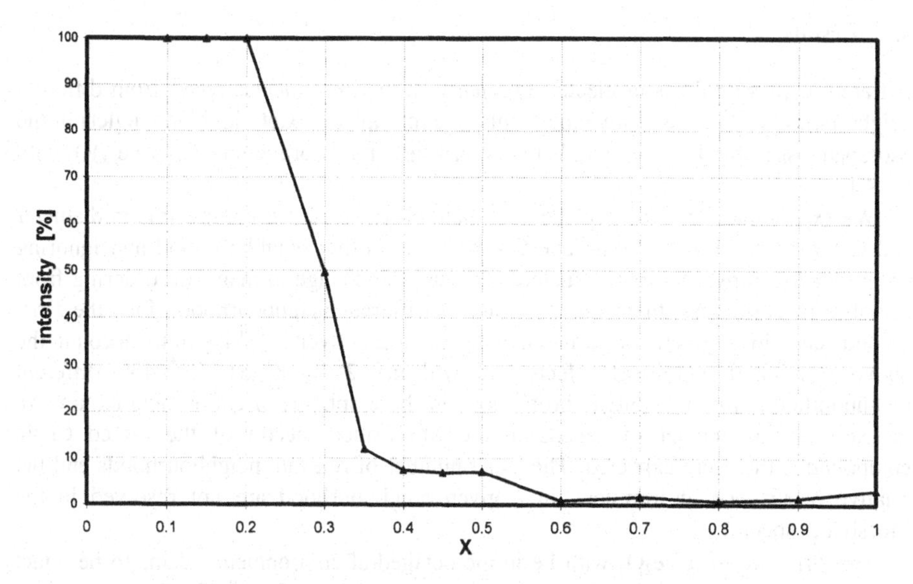

Figure 7. Dependence of the relative intensity of the doublet on x.

In figure 7 the intensity of the doublet relative to the total intensity (intensity of the doublet and the three sextets) is plotted against the concentration of iron x. A doublet or singlet with low intensity seemed to be present even in the Mössbauer spectra of samples with high iron concentrations (x>0.7). The isomer shift seems to be slightly lower (0.2 to 0.3 mm/s) but no reliable fit of the quadrupole interaction was possible. Since the effect is visible even for x = 1 (pure hematite) this part of the spectra must originate from other defects than the substitution of chrome.

Hyperfine fields of the three sextets are plotted in figure 8 as a function of x.

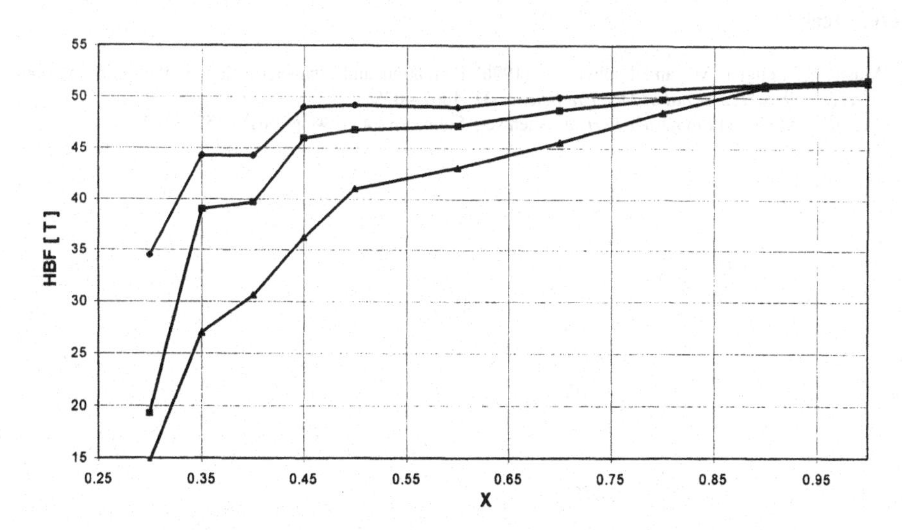

Figure 8. Dependence of the magnetic hyperfine field on x.

4. Results

Different hyperfine fields or electric gradients on the iron nucleus result from different neighbourhoods. We therefore expect only one multiplet (sextet, doublet, singlet) in the Mössbauer spectra of iron in pure hematite or nearly pure corundum ($(Cr_{2-2x}Fe_{2x})O_3$ with $x << 1$).

We expect that Cr^{3+} will replace Fe^{3+} randomly, since it is comparable in diameter and charge. Accepting maximal one Cr-ion in the farther neighbourhood and ignoring possible anisotropy, increasing dislocations and the change in magnetic ordering from hematite to corundum, there are 2, 3, and 4 different neighbourhoods for the first, second, and third sphere of cation neighbours respectively. Taking into account the above mentioned additional effects, we will get $2^2=4$, $8^2=64$, $20^2=400$ different neighbourhoods for substituted chrome ions in these spheres of cation neighbours. An increase of the number of sextets in the Mössbauer spectra of the mixed Cr-Fe compound is therefore expected. The exact number of relevant neighbourhoods and the relative occurrence of iron ions in a given neighbourhood are not resolved in the Mössbauer spectra.

The structure of α-Fe_2O_3 with Fe in the octahedral environment, seems to be intact for most of the iron cations, down to a concentration of $x>0.45$. For lower iron concentrations the sextet broadens and a well resolved sextet disappears at $x \approx 0.3$. In addition, the sextet seems to split into three or more sextets for $x<0.8$. For $0<x<0.8$, a paramagnetic doublet is observed.

We therefore propose that at least three different neighbourhoods coexist in the region $0.3<x<0.8$, one of them being isostructural with Fe_2O_3 and one with Cr_2O_3. The sextet with the strongest hyperfine field correspond to hematite like neighbourhood. Sextets with lower hyperfine fields correspond to neighbourhoods with one or more (but not all) iron ions substituted by chrome.

References

1. Musič, S., Lenglet, M., and Popovič, S. (1996) Formation and characterisation of the solid solution (Cr_xFe_{1-x}), $0\leq x\leq 1$, *Journal of Material Science* **31**, 4067-4078.
2. Krupička, S. (1973) *Physik der Ferrite*, Viehweg+Sohn, Braunschweig ISBN: 352083123.

CONVERSION ELECTRON MÖSSBAUER SPECTROSCOPY

Development and Chemical Applications

KIYOSHI NOMURA
Graduate School of Engineering, The University of Tokyo,
Hongo 7-3-1, Bunkyo-ku, Tokyo, 113 Japan

1. Introduction

Conversion electron Mössbauer spectrometry (CEMS) has been developed and widely used for surface study of solid materials, which contain iron (^{57}Fe) and tin (^{119}Sn). Table 1 shows the development of CEMS. At the initial stage, an electron energy analyzer was experimentally used to certificate the principle of CEMS, but the luminosity was very low. The electron energy analyzer was not useful for daily measurement of CEMS. CEMS, X-ray MS (XMS) and scattering γ-ray MS (GMS) are effectively used in order to analyze the chemical states of surface layers with the thickness from several tens of nm to several tens of mm non-destructively. Integral CEMS (ICEMS) has been most widespread since a simple gas flow counter was developed. Low temperature CEMS, Auger electron MS (AEMS) and glancing angle CEMS (GACEMS) have been developed recently. Depth selective CEMS (DCEMS) has been again used to analyze thin films with more precise depth although enriched ^{57}Fe is often used. Some reviews concerning CEMS have been published in the various fields [43]. The surface of solid improved by ion implantation has been most often analyzed by CEMS [44] because the ion implanted layer can be easily controlled to the observed layer. There are relatively not so many chemical applications of CEMS. The development of CEMS and the analytical applications to (i) corrosion layers, (ii) chemical coatings of steel, (iii) oxide films prepared by spray pyrolysis, (iv) transparent conductive films and (v) diffused tin in the bottom surface of glass are presented in this article. The advantage and feasibility of CEMS are hereby discussed from the viewpoint of the chemical applications.

2. Principles of CEMS

The emission probability of electrons and photons in relaxation process after Mössbauer effect (100%) is as follows. (1) In the case of ^{57}Fe, 7.3 keV K conversion electron (80%), 13.6 keV L conversion electron (9%), 5.6 keV KLL Auger electron (53%), re-emitted 14.4 keV γ-ray (10%) and 6.4 keV characteristic K_α X-ray (27%) are produced. Total probability of electrons emitted is 142 %. (2) In the case of ^{119} Sn, 19.8 keV L conversion electron (84%), 23.0 keV M conversion electron (13%), 2.8 keV Auger

M. Miglierini and D. Petridis (eds.), Mössbauer Spectroscopy in Materials Science, 63–78.

electron (75%), 23.8 keV γ-ray (16%), and 3.6 keV X-ray (9%) are produced. Total emission probability of electrons is 172%.

The emitted electron energy (E_K) corresponds to the photon energy minus binding energy ($E-E_B$). The conversion coefficients, N_e/N_γ, are 8.3 for ^{57}Fe, and 5.1 for ^{119}Sn. It is clear from total emission probability of electron that the detection of electrons is advantageous for ICEMS of thin film and surface layers of solids.

TABLE 1. Development of CEMS.

Year	Authors	Contents	Ref.
1958	R.L. Mössbauer	Discovery of Mössbauer Effect.	[1]
1961	E. Kankeleite	^{182}W CEMS by a magnetic wedge spectrometer.	[2]
1964	K.P.Mitrofanov, V.S. Shpinel	^{119}Sn CEMS by a double lens β-spectrometer.	[3]
1968	J.H.Terrell, J.J.Spijkerman	A back scatter 2π gas counter for ^{57}Fe CEMS and XMS.	[4]
1969	Ts. Bonchev et al.	^{119}Sn DCEMS.	[5]
1972	B.Keisch	A back scattering gas counter for GMS.	[6]
1973	U.Baverstam et al.	^{57}Fe and ^{119}Sn DCEMS by electromagnetic field spectrometer.	[7]
1974	Y.Isozumi et al.	Duel gas counters for ^{57}Fe CEMS and XMS.	[8]
1974	C.M.Yagnik, et al.	A back scatter 2 counter for ^{119}Sn CEMS.	[9]
1975	J.M. Thomas et al.	Experimental estimation of depth information of CEMS.	[10]
1975	J. Bainbridge	A quantitative treatment of CEMS.	[11]
1975	B.D.Sawicka, et al.	Ion implantation study by CEMS.	[12]
1976	M.J. Tricker et al.	Angular effect of CEMS using a He counter.	[14]
1976	G.Weyer	Parallel-plate avalanche counter for CEMS.	[30]
1977	Tsv Bonchev et al.	DCEMS by a magnetic solenoid β-spectrometer.	[13]
1977	D.Salomon et al.	^{181}Ta CEMS.	[15]
1978	M.J. Graham et al.	A relationship between the oxide thickness and relative peak's area.	[16]
1978	D.Liljequist et al.	Theoretical expression of ^{57}Fe DCEMS by a Monte Carlo method.	[17]
1978	A. Proykova et al.	Theoretical expression of ^{119}Sn CEMS by a Monte Carlo method.	[18]
1979	M. Grozdanov et al.	Low energy electrons MS.	[19]
1979	Y.Isozumi, et al.	High temperature CEMS by a gas counter.	[20]
1980	T. Shigematu, et al.	High resolution electrostatic electron spectrometer for DCEMS.	[21]
1981	J.A.Sawicki, et al.	^{57}Fe CEMS by electron multiplier at 78K and 4K (1986).	[22]
1982	P.H.Smit, R.P.Van Stapele,	Analytical method of DCEM spectra from ICEM spectra.	[23]
1983	S. Staniek, et al.	^{57}Fe Auger electron MS (AEMS).	[24]
1983	J.A.Sawicki et al.	^{197}Au CEMS.	[25]
1983	Y. Isozumi et al.	Low temperature CEMS by a gas counter at 77K.	[26]
1984	M.A.Andreeva, R.N.Kuzmin,	Theory of CEMS at total external reflection.	[29]
1985	J.J.Bara, B.F.Bogacz	^{57}Fe DCEMS by X-ray and electron coincidence of gas counter.	[27]
1985	J.Korecki, U. Gradmann	In situ CEMS for analysis of monolayer magnetism.	[28]
1985	J.C. Frost et al.	^{57}Fe GICEMS under total external reflection.	[39]
1986	K.Nomura Y.Ujihira	Simultaneous measurement of CEMS and XMS at 100 K.	[31]
1989	Zs. Kajcsos, et al.	A low noise scintillation detector for UHV-ICEMS (>3keV).	[32]
1990	Klingelhofer, E.Kankeleite	LEEMS by an orange type magnetic spectrometer.	[33]
1990	G. Weyer	PPAC for ^{119}Sn DCEMS.	[34]
1991	K. Fukumura et al.	A gas counter for CEMS at 15 K.	[35]
1991	J.A.Sawicki et al.	^{193}Ir CEMS.	[36]
1991	B.R.Bullard, et al.	^{183}W, ^{191}Ir, ^{159}Tb CEMS.	[37]
1992	A.P.Kuprin and A.A.Novakova,	DCEMS by changing high voltage of gas counter.	[38]
1993	J.A. Sawicki et al.	^{170}Yb CEMS.	[40]
1997	A.L. Kolmetsky et al.	Air scintillation counter for CEMS.	[41]
1998	K. Nomura et al.	^{119}Sn GACEMS experiment.	[42]

3. Scattering Mössbauer Spectrometry and Approximately Observed Depths

Scattering techniques of Mössbauer spectrometry (MS) are divided into the following subgroups as shown in Table 2. The depth information of solid surface layers obtained by CEMS depends on the energy of electrons detected. Using a highly resolved electron energy spectrometer, the thinner layers on the surface can be analyzed layer by layer. This method is called DCEMS. However, the detection efficiency of electrons is extremely small. DCEMS requires careful energy analysis in a high vacuum chamber. On the contrast, ICEMS are easily performed by detecting all resonant electrons without the discrimination of electron energy. ICEMS has been often applied to the practical analysis of samples containing natural abundance of ^{57}Fe (2.17%) and ^{119}Sn (8.58%) as well as to low temperature CEMS and GACEMS.

In order to save measuring time and to observe the different states among top surfaces, intermediate and deep layers, the simultaneous measurement of CEMS, XMS and GMS is very useful from the view point of practical measurement.

TABLE 2. Scattering modes and probing depths.

Scattering Modes	Probing depth for Fe (or Sn)	Appropriate detectors
A). Scattering γ-ray MS (GMS)	20 μm (100 μm)	Kr/Xe gas filled counter (NaI(Tl) scintillator)
B). Scattering X ray MS (XMS or CXMS)	< 10 μm	2π Ar gas flow counter.
C). Conversion electron MS (CEMS)		
a) Integral CEMS (ICEMS) At low and high temperatures	< 300 nm(<3 μm)	2π He gas flow counter gas filled counter/ electron multiplier
b) Depth selective CEMS		
i) Layer by layer analysis by ICEMS (destructive method)	> 100 nm(200 nm)	2π He gas flow counter
ii) DCEMS (using energy loss of K conversion electrons)	< 100 nm (<500 nm/L electron)	electron multiplier with electron energy analyzer
iii) Auger electron MS (AEMS) (<600 eV)	< 5 nm (<10 nm)	electron multiplier in high vacuum chamber
iv) Low Energy Electron MS (LEEMS)	(<15 eV) 5-10 nm (10-20 nm)	electron multiplier
v) Glancing Angle CEMS (GACEMS)	< 50 nm (<300 nm)	2π He gas flow counter
vi) Grazing Incident CEMS (GICEMS) (near total external reflection)	< 2-3 nm (3-5 nm)	2π He gas flow counter

3.1. INTEGRAL CEMS

3.1.1. *Probing Depth and Sensitivity*

65% and 96% of total resonant electrons come from iron surface of 60 nm depth and 300 nm depth, respectively [10]. X-ray photoelectron (XPE) and γ-ray photoelectron (GPE) contribute to whole conversion electrons in addition to Auger, K-conversion, and L-conversion electrons [17]. The practical relationship between iron oxide thickness and relative peak area of Mössbauer spectra is as follows [16].

$$d(nm) = 1.95x10^2 \ln (1 - 0.01P)$$

where, d is the thickness (nm) of iron oxide film and P the percentage of Mössbauer peaks due to iron oxide.

The range of low energy electrons (1-10keV) in solid can be estimated using Feldman's expression [45].

$$R(nm) = 25 (A / \rho Z^{n/2}) E^n$$

where, n is 1.2/(1-0.29logZ), E is the energy of electron (keV), R is range (nm), A is atomic or molecular weight, Z is atomic number or electron number of a molecular, and ρ is density (g/cm^3).

The thickness of samples observed by ICEMS can be roughly estimated by above equations. The detection limit is 10^{14} atoms/cm^2 for ^{57}Fe, and 10^{15} atoms/cm^2 for ^{119}Sn.

3.1.2. *Detectors for CEMS*

(a) Gas counter. A back-scatter 2 π proportional He gas counter has high detection efficiency for low energy electrons of about 10 keV, where He gas is insensitive to incident X-ray and γ-ray [4]. CEM spectra can be easily available by flowing He gas mixed with several % CH_4 gas as a quencher. XM spectra also can be gotten by setting Lucite plate (3 mm thick) in front of an incident window in order to filter X-ray emitted from the source and by flowing Ar/CH_4 gas. Ar gas can count 6.3 keV Fe X-rays more effectively than 14.4 keV incident γ-rays. The range of 7.3 keV electrons in He gas is at highest 5 mm. The mean energy of electron emitted is less than 7.3 keV, and the appropriate distance is about 2mm for Fe CEMS. The detection efficiency and S/N are of most importance to improve the counting efficiency. The sensitive area of sample is close to the anode wire, and the higher applied voltage gives the less S/N ratio because the non-resonance electrons come from counter wall and sample. Further, much attention should be taken to the electronic contact between counter body and sample because the charging-up of sample during long time measurement affects the electric field in a counter and reduces the gas multiplication. The charging-up effect in He counter is avoided by setting mesh electrode on the sample surface or by covering thin carbon coating on the surface.

Parallel plate avalanche counter (PPAC) [30], which consists of flat plate electrodes of graphite and sample, is advantageous for insulating materials. Any organic polyatomic gas can be used as counting gas. Usually acetone vapor is preferred. ^{119}Sn DCEMS using PPAC is possible by discriminating electron energy and controlling gas pressure [34]. However, PPAC was not so spread as a proportional counter because the setting of a sample and the control of pressure are a little complicated.

When a gas counter is operated at low temperature, the gas multiplication decreases, and some quenchers are condensed on the surface. However, a gas filled counter of He gas mixtures with 5% CO at pressure 1 atm was operated at 77K. The purified He, Ne and mixtures of He+Ne, He+5%N_2 or He+ several % CO, were used for operating at the lower temperatures between 15 and 300 K [35]. The discharge of applied high voltage should be taken attention to at low temperature.

(b) Electron multiplier. Channeltrons have relatively large efficiency for low energy electrons. Electron detectors such as channeltron and channelplate have inherently no energy resolution, and need to work in vacuum chamber. Sawicki et al. reported early that both a sample and a channel electron multiplier installed are cooled in a conventional Dewar of a liquid He or liquid N_2 container for ICEMS [22].

At high temperatures, a gas flow counter was installed in a furnace [20]. Then another type of gas counter was developed so as to heat only the sample up to highest 740 K [46]. At higher temperature less recoilless fraction, quenching gas more reactive to the sample surface and thermal electron increases the background counts. However, the gas reactivity can be utilized to monitor *in situ* gas sensing of sensor materials because various flammable gases can be mixed in He gas as a quenching gas [47].

3.2. DCEMS

3.2.1. *An Electron Energy Aanalyzer*

DCEM spectra can be generally obtained in high vacuum by selecting electrons with a specific energy range. Counting rate and spectra's quality depend on geometry and properly shaped collimator. A cylindrical mirror type electrostatic electron spectrometer, a retarding-field electron spectrometer, a magnetic sector β-spectrometer and an orange type magnetic electron spectrometer have been developed for DCEMS [44]. A low noise scintillation detector with thin phosphor layer (SCD) was used to detect the electrons in high vacuum and cooling system [32]. SCD is superior to channeltron by one order of magnitude in higher than 3 keV electrons, and useful for an electrostatic cylindrical mirror analyzer. The orange type spectrometer [33] is improved by equipment of 40 electron multipliers, and is expected to reach a high efficiency (about 15%) without degradation of energy resolution.

Klingelhofer and Meisel reported that thin surface layers (equivalent to about 1.0 to 3.0 nm Fe) are well detectable by DCEMS and KLL Auger electron MS [48]. It was found from LMM Auger electron MS that an enhancement of the surface signal with regard to the bulk was by more than 2 as expected due to the very small mean free path of the Auger electrons of only 1.0 to 2.0 nm. The intensity of the low energy electrons (about 10 eV) was high unexpectedly. LEEMS by detection of secondary electrons does not reflect the information of so thin surface as AEMS, but is fascinating for high efficiency.

3.2.2. *A 2π Proportional Gas Counter*

Smit and Stapele [23] showed the analytical method to obtain DCEM spectra from ICEM spectra, coupled with etching procedure. By repeatedly removing thin layers from a sample and subsequently measuring CEM spectrum, a set of spectra can be obtained, from which it is possible to extract the spectra belonging to the removed layers.

The energy resolution of a proportional counter is too poor to select 7.3-6.4 keV conversion electrons from the photoelectrons emitted from 6.4 keV X-ray. However, the coincidence of X-ray and conversion electron was utilized for ^{57}Fe DCEMS, using an assembly of He and Ar gas flow proportional counters[27]. The counting rates became less than 1/25. The pronounced depth selectivity of CEM spectra was successfully achieved by selecting the angle of the incident γ-ray in addition to selecting the energy of the emitted electrons. On the other hand, depth information changed depending on the applied voltage [38]. Hereby DCEMS of rough structures using a gas counter can be obtained by selecting the energy of detected electrons and by changing the supplied voltage.

3.2.3. *GACEMS and GICEMS*

GACEMS is useful for analysis of layer structures and orientation of components in thin films. The influence of surface roughness on the surface selectivity in ^{57}Fe GACEMS is small under the approximate condition that the electron-emission angle relative to the surface is larger than the average surface inclination angle associated with the surface roughness [49]. M.A. Andreeva and R.N. Kuzmin treated theory of CEMS at total external reflection [29]. ^{57}Fe GICEM spectra were observed by J.C. Frost et al. [39-(1)]. S.M. Irakaev et al. also set up high performance GICEMS [39-(2)] although it took a month to get one spectrum using 100 mCi ^{57}Co and enriched ^{57}Fe sample. At glancing angles of less than 10 mrad, the selective observation of a surface layer with a mean depth < 10 nm is possible. The drastically changed spectra were obtained near the total external reflection (critical angle q_c: 0.21 °(3.8mrad)).

GACEMS are compared with conventional CEMS and GICEMS as shown in Table 3. Because L conversion electron energy of ^{119}Sn is high, the penetration depth (3μm) is relatively deep. ^{119}Sn GACEMS is useful to increase the surface sensitivity[42]. The θ_c of total external reflection is 0.12° for ^{119}Sn. ^{119}Sn GICEMS would be realized if synchrotron orbital radiation could be available as 23.8 keV photons.

TABLE 3. Comparison of CEMS, GACEMS and GICEMS methodology.

Methods	normal CEMS (Fe/Sn)	GACEMS (Fe/Sn)	GICEMS(Fe/Sn)
System	simple	relatively easy	precise
Incident angles of γ rays to surface	90/45 fixed	<10 °	θ_c 0.21(/0.12°)
Distance between detector and source	<10 cm	20 - 40 cm	> 50 cm
Activity of source used	25mCi(/>1mCi)	50mCi(/>10mCi)	>100mCi(/>25mCi)
Area irradiated	several cm^2	4-10 cm^2	10-25 cm^2
Peak Int. / Back ground	small	relatively large	large
Surface layers observed	<300nm(/0.3μm)	50nm(/< 200nm)	<5nm(/<10nm)
Depth selectivity	poor	changeable	high
Feasibility	very easy	relatively easy	difficult

4. Chemical Applications of CEMS

4.1. CORROSION STUDY

4.1.1. *Wet Corrosion of Iron and Steel in a Solution*

Iron and steel are corroded by the moisturizing and the impurities in the atmosphere or by the dry process in high temperatures. It is relatively easy to identify the corrosion products such as Fe_3O_4, α-Fe_2O_3 , α-FeOOH, and $Fe(OH)_2$ by CEMS. However, the careful analysis of paramagnetic peaks of Fe(III) species such as γ-

FeOOH, β-FeOOH and small particles of iron oxides is necessary, and so low temperature measurement is sometimes needed. CEMS and XMS are useful for analysis of poor crystalline and intermediate products, which can hardly be identified by X-ray diffraction method (XRD).

XMS and GMS can be applied to nondestructive analysis of thick rust layers, but, are not so sensitive for thin layers such as the interface between solution and solid surface as well as adhesive layers rusted beneath a coating unless the enriched ^{57}Fe atoms are doped at the interface.

The initial corrosion products formed in a solution change easily during drying for *ex situ* measurement of CEMS or XMS. Great attention should be paid to drying because unstable ferrous species disappeared often during drying. For examples, when steel was dipped in NO_3^- solution, γ-FeOOH layers were first deposited on the surface, and, the magnetite layers were grown together with paramagnetic ferrous species beneath the γ-FeOOH layers. The paramagnetic ferrous species could not be found unless the samples were cooled and dried in vacuum [50].

Low temperature CEMS was most effectively used for the following corrosion study. β-FeOOH, γ-FeOOH and FeOCl were found in the rust of steel dipped in sea for two months [51]. β -FeOOH is known to be formed via the hydrolysis of $FeCl_2$ and $FeCl_3$, but iron chlorides are not formed in neutral solution of NaCl like sea water. -FeOOH had not been recognized in the CEM spectra of mild steel corroded in a 3%NaCl solution for a short period.

To clarify the formation of β-FeOOH, steel sheets were corroded in 3% NaCl solution under different pH, temperatures and concentrations of dissolved oxygen (DO), and the oxide layers produced were characterized by CEMS at dry ice temperature (195K). The Neel temperature of β-FeOOH and γ-FeOOH is 293K and 50K, respectively. γ-FeOOH was usually formed in open NaCl solution, which contains normally 5 to 6 ppm DO. It was found from CEMS that β-FeOOH could be easily formed by retarding DO to 1 ppm [52].

The pH of the solution shifted to higher value with immersion time for several days. The intermediate compounds of ferrous species were detected by drying the frozen samples. The poor crystalline β-FeOOH is found to be produced with a large value of Δ and a little small magnetic field as compared with crystalline β-FeOOH.

$$1/2O_2 + H_2O + 2e \rightarrow 2OH- \qquad (1)$$
$$Fe \rightarrow Fe(OH)^+ \rightarrow Fe(OH)_2^+ \qquad (2)$$
$$Fe(OH)_2^+ + 2OH^- \rightarrow Fe(OH)_4^- \qquad (3)$$
$$2Fe(OH)_4^- \rightarrow 2FeOOH + 2H_2O + 2OH^- \qquad (4)$$

The supply of OH^- is suppressed in the solution of low DO concentration, and chloride ions are easily coordinated to Fe(III) ion during the process of equations (3) and (4).

$$Fe(OH)_2^+ + (2 - x)OH^- + x\,Cl^- \rightarrow Fe[(OH)_{4-x}\,Cl_x^-] \qquad (5)$$
$$2Fe[(OH)_{4-x}\,Cl^-_x] \rightarrow 2\beta\text{-}FeOOH(Cl^-_x) + 2H_2O + 2(1-x)OH^- \qquad (6)$$

Chloride ions contribute to formation of β-FeOOH in the polymerization process according to above (6).

4.1.2. *Environmental Corrosion of Weathering Steel*

The stable corrosion layers of weathering steel protect the bulk corrosion. The surface oxides have been analyzed by CEMS and XMS [53]. α-FeOOH was formed in small particles, which exhibited superparamagnetic relaxation. α -FeOOH presents beneath layers of γ-FeOOH and ferrihydrite. Ferrihydrite was found during a simulated atmospheric corrosion of steel by Leidheiser and Czako-Nagy [54]. It is a poorly ordered hydrous iron oxide with the bulk composition of $5Fe_2O_3\ 9H_2O$, and is a precursor to -FeOOH. Corrosion products of weathering steel exposed for 13 years were analyzed by XMS and CEMS [56]. γ-FeOOH, fine particles of α-FeOOH and ferrihydrite were produced as the main rust. After exposure to wet-dry cycles in an SO_2 polluted atmosphere, poorly crystallized α -FeOOH and quasi-amorphous ferrihydrite were identified as the main corrosion products by J. Davalos et al. [55]. The rust of weathering steel had different particle size from that of pure iron. Superparamagnetic α-FeOOH was formed only on weathering steel.

Fig.1 shows the CEM XM and GM spectra of as-prepared weathering steel. The peaks of α-FeOOH are not recognized but the peaks of α-Fe_2O_3 and Fe_3O_4 in the XM and GM spectra. It is clear from the relative intensity of DCEMS that α-FeOOH and α-$Fe_2\ O_3$ exist in the top and intermediate layers. The thickness of α-FeOOH is less than 100nm. After one year, the paramagnetic iron components were produced as dominant rust. It is still a question why the stable rusts are formed after exposure in wet and dry air atmosphere.

4.1.3. *Oxide Layers of Stainless Steels*

Oxide films on SUS316 (18%Cr+8%Ni+3%Mo+Fe balance) produced at high temperature or by wet oxidation are prominent for solid state pH sensor [59]. The shinny color of thin oxide films changed from golden yellow to blue, depending on heating temperatures. The oxide layers on austenitic stainless steels heated up to 1073 K for 1 hour were analyzed by CEMS [57]. The magnetic ferrite compounds were produced in the top rust layer on stainless steel heated up to 873 K, whereas paramagnetic iron species in the intermediate layer above 873 K. Cr was concentrated, but Fe and Ni were depleted on the surface at high temperatures. The paramagnetic peaks were due to a superparamagnetism of fine α-Fe_2O_3 dispersed in Cr_2O_3 top layers. When austenitic stainless steel was heated for several hours, fine α-Fe_2O_3 grains became large in chromium depleted layers [58]. CEMS of stainless steel oxidized in chromium and sulfate solution gave only paramagnetic peaks of iron oxyhydroxide.

4.1.4. *Black Oxide Layers of Steel*

Black coating on steel is obtained by treating in alkaline solution ($NaOH+NaNO_3$ $+NaNO_2$) at the boiling temperature (130 to 150 °C). In the CEM spectra, a broad doublet (IS = 0.40 mm/s, QS = 0.74 mm/s, FWHM = 1.2 mm/s) was observed. The doublet peaks became magnetically split peaks of magnetite at liquid N_2 temperature. With the increase of dipping time, the peak intensity of the broad doublet increased. The

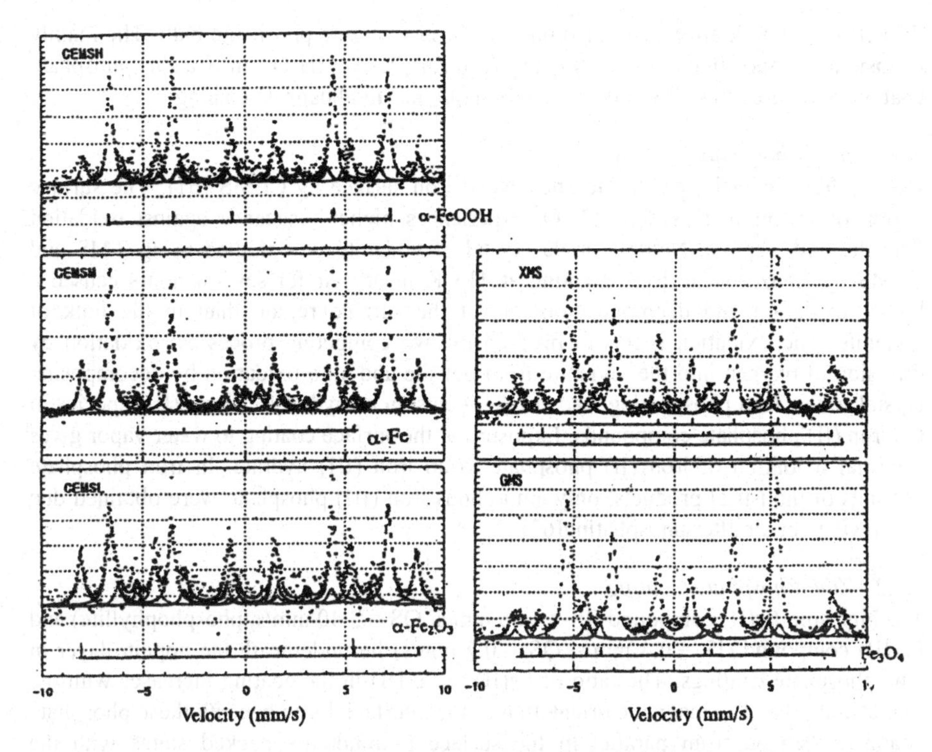

Figure 1. CEM, XM and GM spectra of as prepared weathering steel, obtained with a back scatter gas flow proportional counter. CEMSH, CEMSM, and CEMSL were simultaneously obtained by selecting the electrons with high, middle, and low energy ranges, using a He+5%CH_4 gas counter. XMS and GMS were obtained by discriminating X-ray and γ-ray peaks using a Ar+5%CH_4 gas counter.

particle size (7 to 10 nm) of magnetite formed in nascent stage did not grow, but the number of magnetite particles increased in the surface layer with immersion time. Fine magnetite affects little to the scale of refined steel goods and contributes to the corrosion resistance.

4.2. PHOSPHATING OF STEEL

Phosphating of metal and alloy is widely used to obtain the corrosion resistant surface and to treat the metal surface before painting. Phosphating of steel in different phosphate bathes leads to various coatings that differ in morphology and composition. Orthorhombic hopeite, $Zn_3(PO_4)_2$ $4H_2O$, and/or monoclinic phosphophillite,

$Zn_2Fe(PO_4)_2$ $4H_2O$, are precipitated on steel surface in zinc phosphate bath. The mainly deposited compositions are scholzite, $(Ca_2,Zn)(PO_4)_2$ $2H_2O$, in Ca-Zn phosphate coating, and hureaulite, $(Mn,Fe)_5H_2$ (PO_4) $4H_2O$, in Mn phosphate coating.

4.2.1. *Iron Phosphate Coating*

Berry (1979) investigated on the phosphated iron surface by CEMS [61]. The surface region of vivianite, $Fe_3(PO_4)_2$ $8H_2O$, experiences higher resistance against oxidation than the bulk sample. Natural single crystal of vivianite was studied using TMS and CEMS by Hanzel et al. [62]. Heating at 413 K in dry air for several hours caused a higher oxidation and decomposition rate at the surface region than in the bulk of vivianite. The oxidation seems to proceed by two competing processes: oxidation by diffusion of oxygen into the vivianite from outside and auto-oxidation by decomposing crystal water. Iron phosphating leads to the inclusion of hydrolyzed iron (III) species in the iron (II) phosphate compounds. Exposure of the surface coating to water vapor gives a partial oxidation of iron (II) phosphate to an iron (III) hydrate. In iron phosphate coatings of industrial products, only amorphous iron (III) phosphate were obtained due to blowing hot air after phosphating[63].

4.2.2. *Zinc Phosphate Coating*

The high spin Fe(II) compound (IS = 1.26 mm/s, QS = 3.40 mm/s, phosphophyllite) and Fe(III) compound (IS = 0.4 mm/s, QS = 0.7 mm/s) lie underneath the hopeite layer in zinc phosphate coatings. The ratio of Fe(II) to Fe(III) in the coating increases with the immersion time. The magnetic orientation of the interfacial iron beneath these phosphate coatings changed from parallel to the surface to randomly packed states with the increase of the immersion time. The roughness of interface can be estimated nondestructively from analysis of the peak ratio of iron substrate sextet in CEM spectra of various coatings.

When zinc phosphate coatings on steel were heat-treated above 600°C, zinc and phosphor were sublimated from the coatings[65]. It was found that the substrate iron contributes to the sublimation of zinc and phosphor.

4.2.3. *Manganese Phosphate Coatings*

Mössbauer spectra of Mn phosphate coatings are composed of three doublets of paramagnetic iron (II) species and a doublet of iron (III) species with a small intensity [66]. The Mössbauer parameters correspond to those of iron (II) substituted in three sites of octahedral crystal structure of hureaulite. The summation of Mn and Fe contents in the coatings is almost 34% irrespective of the phosphating time. The steel surface is covered with Mn hureaulite at the initial stage of phosphating, with Fe and Mn hureaulite at the second, and with Mn hureaulite finally.

4.2.4. *Converted Tin Phosphate Coatings*

Tin phosphate is insoluble, and can not be directly deposited on steel in phosphate bath. Converted tin phosphate coatings are obtained by dipping Zn and Mn phosphate coatings on steel into hot $SnCl_2$ solution. Sn(II) phosphates were observed as main components together with Sn(IV) species in CEM spectra of converted tin phosphate coatings [67]. Ionic properties of the converted coatings can be estimated from the isomer shift and quadrulpole splitting of stanneous tin products. On the other hand, only Sn(IV) species were observed in CEM spectra of tin phosphate coatings prepared by Ar sputtering of stannous phosphate target [68]. Dry process of phosphating is superior to chemical process for protecting against corrosion resistance.

4.3. OXIDE FILMS PREPARED BY SPRAY PYROLYSIS

4.3.1. *Iron Oxide Films on Transparent and Electric Conductive Glass*

Iron oxide films were prepared by spray pyrolysis of ferric chloride alcohol solution on electric conductive glass coated with SnO_2, ITO, ZnO_2, and WO_3. A sextet of Hematite, α -Fe_2O_3, appeared in CEM spectrum of SnO_2 coated glass. In CEM spectra of ITO and WO_3 coated glass, a doublet appeared in addition to a sextet of Hematite. In CEM spectrum of ZnO coated glass, only a doublet appeared. These doublets are considered to be due to fine particles of Hematite, dispersed in the ITO, ZnO_2 and WO_3 oxide films because these oxides are soluble in acid solution of spray.

4.3.2. *Tin Oxide Films for Gas Sensor*

SnO_2 has been utilized as gas sensor materials. Spray pyrolysis can easily make films of various tin based oxides. The gas sensitivity of metal mixed tin oxides with and without catalyst was investigated [70] and these tin based films were characterized by CEMS [71]. The chemical states changed little even if SnO_2 films were treated in H_2 atmosphere, but the addition of the other metal oxides allowed the chemical change of tin due to the degree of spillover of H_2 gas. In CEM spectra of Ni mixed tin oxide films with Pd catalyst, Sn(0) and Sn(II) peaks were observed in addition to Sn(IV)O_2 peaks after treating H_2 atmosphere, whereas only SnO_2 peaks were observed in CEM spectra of Mn mixed tin oxides with Pd.

Sn-Bi mixed oxide film with CO selectivity was characterized in He gas with different quenching gases at different temperatures by CEMS [72]. In CEM spectra of Sn-Bi oxide film heated up to 500°C, reduced Sn (II) species could not be observed, but it was found from *in situ* CEMS that the Mössbauer parameters of Sn(IV) peaks changed drastically at the temperature, where gas sensitization started to increase.

4.4. TRANSPARENT AND CONDUCTIVE ITO FILM

ITO films are useful for solar cell and electric device. By doping several % Sn atoms in In_2O_3, ITO films show high electric conductivity. The doped Sn atoms, however, are not always substituted in the In_2O_3 lattice. ITO films with various contents of Sn were analyzed by CEMS, assuming that a doublet is due to soluble Sn(IV), and that a singlet due to the insoluble Sn(IV). The approximate relation between the relative peak ratio and the solubility was obtained by CEMS [73]. Recently the doping effect was more precisely analyzed by decomposition of Sn(IV) peaks to three components. It was found that electrically inactive Sn(IV) components correspond to 7-fold and 8-fold coordinated Sn dopants and that the intensity increase from 4 at % of Sn doping concentration [74].

4.5. TIN STATES IN BOTTOM SURFACE OF FLOAT GLASS

Plate glass is produced by floating molten glass in a tin fused bath. Tin diffused in the glass results sometimes in blooming after heating, and the different composition between bottom and upper surfaces may disturb the uniform printing of an electrode on glass. Principi et al. tried to analysis the tin states of float glass by DCEMS using a He gas counter [75]. They showed that the population ratio of Sn(II) to Sn(IV) in top surface was higher than deep layers. William's groups also have recently characterized the tin states in float glass by transmission MS at 78 K [76,77]. The tin states analyzed so far were not always consistent.

By combining with the depth profile of tin concentration (Fig. 2), we tried to analysis tin states again by CEMS as shown in Fig.3. Tin concentration decreased with deep layers, but in around 7 μm depth, the mount of Sn concentration was observed. On the bottom surface, both Sn(II) and Sn(IV) species were detected, but in the 3μm etched surface, the major component was Sn(II) species. At the deeper layers than 6m, where the tin concentration changed from the exponential decrease, almost all Sn(IV) species were observed. Sn atoms were fixed as IV valence states in the bulk. Sn(IV) species observed on the top surface is considered to be produced by the air oxidation of Sn(II) species after putting out from reducing atmosphere. Inherently Sn(II) species behavior as glass network modifier and can easily diffuse into the bulk. A small amount of Sn(IV) species are considered to fix as glass network former. These are supported also from the fact that the Debye temperature of Sn(II) in glass is lower than that of Sn(IV) [77].

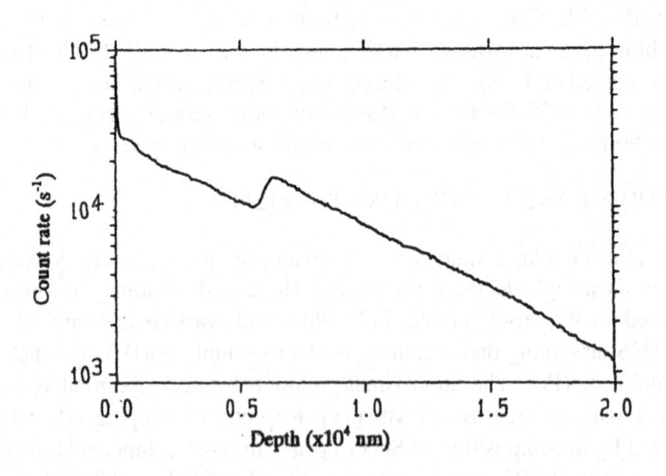

Figure 2. Depth profile of Sn in bottom surface of float glass by SIMS.

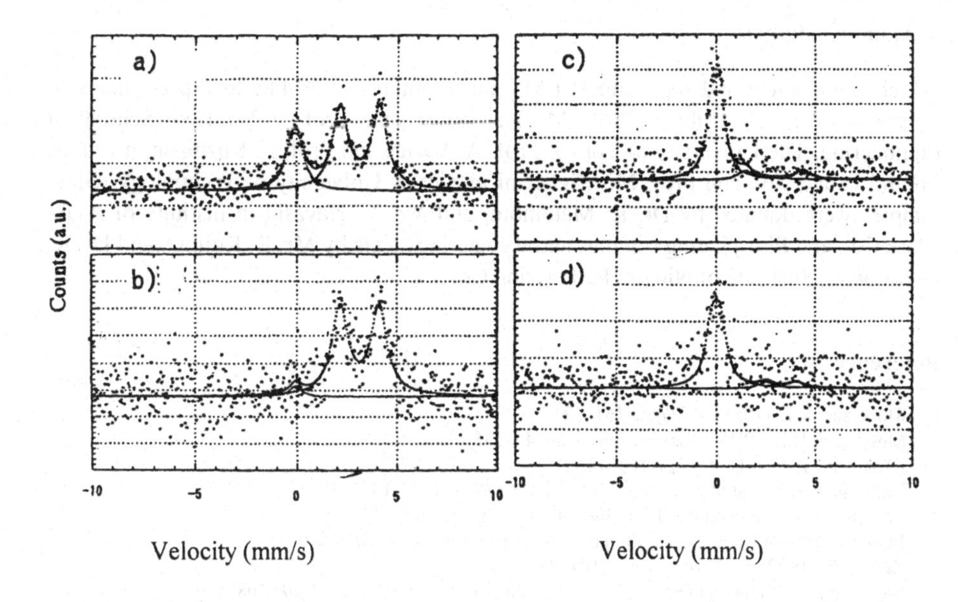

Figure 3. ^{119}Sn CEM spectra of etched bottom surface of float glass: a) bottom, b) 3μm etched, c) 6μm etched, and d) 9μm etched surfaces.

5. Conclusions

CEMS has been widely used in various fields, based on development of detection techniques of electrons. Especially ICEMS and XMS using a He gas counter have been used for characterization of surface products of chemically treated steel and industrial materials. It is further expected that GACEMS, DCEMS and AEMS would be more often used for layer by layer analysis of the thinner solid surface. ICEMS, DCEMS and XMS would become more powerful by simultaneous measurement at lower temperatures, and further by combining with another analytical methods such as EPMA, XRD and X-ray fluorescent analyzer. A simple counter can be easily improved according to the purpose of a study. *In situ* CEMS and DCEMS could afford us more detailed and useful information on the chemical structure and orientation of intermediate products in the surface layers of materials exposed in various environments.

6. Acknowledgment

For encouragement and supporting CEMS study, author would like to express thank to Emeritus Prof. Y. Ujihira, Prof. M. Nakazawa and Prof. T.Terai in School of Engineering, The University of Tokyo, Prof. A. Vertes and Prof. E. Kuzmann in Etovos Lorand University, and Prof. M. Miglierini in Slovak University of Technology. Glass samples were donated by Dr. K. Matumoto, and Mr. Y. Hayashi in Institute of Asahi Glass Co., and Phosphating samples and weathering steels by Mr. R. Kojima, and Dr. K. Kurosawa in Institute of Nippon Parkerizing Co.

Reference

1. Mössbauer R.L. (1958), *Z. Physic* **151**, 124;
   Mössbauer, R.L. (1958), *Naturwissenschften* **45**, 538.
2. Kankeleite, E. (1961), *Z. Phys.* **164**, 442.
3. Mitrofanov, K.P. and Shpinel, V.S. (1964), *Soviet Phys. JETP* **13**, 233.
4. Terrell, J.H. and Spijkerman, J.J. (1968), *Apply. Phys. Let.* **13**, 11.
5. Bonchev, Ts., Jordanov, A. and Minkova, A. (1969), *Nucl. Instr. Meth.* **70**, 35.
6. Keisch, B. (1972), *Nucl. Instr. Meth.* **104**, 237.
7. Baverstam, U., Bohm, Ringstrom, B. and Ekdahl, T.(1973) *Nucl. Instr. Meth.* **108**, 439.
8. Isozumi, Y., Lee, D.I. and Kadar, I. (1974) *Nucl. Instr. Meth.* **120**, 23.
9. Yagnik, C.M., Mazak, R.A. and Collins, R.L. (1974) *Nucl. Instr. Meth.* **114**, 1.
10. Thomas, J.M., Tricker, M.J. and Winterbottom, A.P. (1975) *J. Chem. Soc. Farad. Trans.* 1708.
11. Bainbridge, J. (1975) *Nucl. Instr. Meth.* **128**, 531.
12. Sawicka, B.D., Sawicki, J. and Stanek, J. (1975) *Nucl. Instr. Meth.* **130**, 615.
13. Bonchev, Ts., Minkova, A., Kushev, G. and Grozdanov, M. (1977), *Nucl. Instr. Meth.* **147**, 481.
14. Tricker, M.J., Freeman, A.G. and Winterbottom, A.P. (1976), *Nucl. Instr. Meth.* **135**, 117.
15. Salmon, D., West, P.J. and Weyer, G. (1977), *Hyperfine Interac.* **5**, 61.
16. Graham, M.J., Mitchell, D.F. and Channing, D.A. (1978), *Oxidation of Metals* **12**, 247.
17. Liljequist, D., Ekdahl, T. and Baverstam, U. (1978), *Nucl. Instr. Meth.* **155**, 529.
18. Proykova, A., Minkova, A., Slavov, B. and Bonchev, Ts. (1978), *Bulg. J. Phys.* **3**, 248.
19. Grozdanov, M., Bonchev, Ts. and Lilkov, V. (1979), *Nucl. Instr. Meth.* **165**, 231.
20. Isozumi, Y., Kurakado, M. and Katano, R. (1979), *Nucl. Instr. Meth.* **166**, 407.
21. Shigematu, T., Pfannes, H.D. and Keune, W. (1989), *American Phys. Soc.* **45**, 1206.
22. Sawicki, J.A., Tyliszczak, T. and Gzowski, O. (1981), *Nucl. Instr. Meth.* **190**, 433.
23. Smit, P.H. and van Stapele, R.P. (1982), *Appl. Phys.* **A28**, 113.
24. Staniek, S., Shigematsu, T., Keune, W. and Pfannes, H.D. (1983), *J. Magn. Magn. Mater.* **35**, 347.
25. Sawicki, J.A., Tyliszcak, T., Stanek, J., Sawicka, B.D. and Kowalski., J. (1983), *Nucl. Instr. Meth.* **215**, 567.
26. Isozumi, Y., Kurakado, M. and Katano, R. (1983), *Nucl. Instr. Meth.* **204**, 571.
27. Bara, J.J. and Bogacz, B.F. (1985), *Nucl. Instr. Meth. in Phys. Res.* **A238**, 469.
28. Korecki, J. and Gradmann, U. (1985), *Phys. Rev. Lett.* **55**, 2491.
29. Andreeva, M.A. and Kuzmin, R.N. (1984), *Phys. Stat. Sol.* (b) **125**, 461;
   Topalov, P. and Proykova, A. (1985), *Nucl. Instr. Meth. in Phys. Res.* **A236**, 142.
30. Weyer, G. (1976), *Mössbauer effect Methodlogy* **10**, 301.
31. Nomura, K. and Ujihira, Y. (1986), *Bunseki Kagaku* **35**, 748.
32. Kajcsos, Zs., Sauer, Ch., Holzwarth, A., Kurz, R., Zinn, W., Ligtenberg, M.A.C. and Van Aller, G. (1988), *Nucl. Instr. Meth. in Phys. Res.* **B34**, 384.
33. Klingelhofer, G. and Kankeleite, E. (1990), *Hyperfine Interactions* **53**, 1905.
34. Weyer, G. (1990), *Hyperfine Interactions* **58**, 2561.
35. Fukumura, K., Nakanishi, A., Kobayashi, T., Katano, R. and Isozumi, Y. (1991) *Nucl. Instr. Meth. in Phys. Res.* **A301**, 482.

36. Sawicki, J.A., Sawicka, B.D. and Wagner, F.E. (1991), *Nucl. Instr. Meth. in Phys. Res.* **B62**, 253.
37. Bullard, B.R., Mullen, J.G. and Schupp,G. (1991), *Phys. Rev.* **43**, 7405 and 7416.
38. Kuprin, A.P. and Novakova, A.A. (1992), *Nucl. Instr. Meth. in Phys. Res.* **B62**, 493.
39. Frost, J.C., Cowie, B.C.C., Chapman, S.N. and Marshall, J.F. (1985), *Appl. Phys. Lett.* **47**, 581, Irakaev, S.M., Andreeva, M.A., Semrenov, V.G., Belozerskii, G.N. and Grishin, O.V. (1993), *Nucl. Instr. Meth. in Phys. Res.* **B74**, 545.
40. Sawicki, J.A., Niesen, L. and De Waard, H. (1993), *Nucl. Instr. Meth. in Phys. Res.* **B83**, 454.
41. Kholmetskii, A.L., Misevich, O.V., Mashlan, M., Chudakov, V.A., Anashkevich, A.F. and Gurachevskii, V.L. (1997), *Nucl. Instr. Meth. in Phys. Res.* **B129**, 110.
42. Nomura, K., Ujihira, Y. and Nakazawa, M. (1998), *Hyperfine Interactions (C)* **3**, 261.
43. Spijkerman, J.J. (1971) CEMS, in I.J. Gruverman (ed.), *Mössbauer Effect Methodology Vol. 7*, Plenum Press, New York, p.85;
Huffman, G.P. (1976) Theoretical expressions for analysis of multilayer surface films by electron re-emission Mössbauer spectroscopy, in I.J. Gruverman (ed.) *Mössbauer Effect Methology Vol.10*, Plenum Press, New York, p.209;
Wagner, F.E. (1976), *J. Phys. (Paris) Colloq.* C6 **37**, 673;
Simmons, G.W. and Leidheiser, Jr., H. (1976) *Applications of Mössbauer Spectroscopy Vol. 1*, Academic press, New York, 85;
Tricker, M.J. (1981) CEMS and its recent development, in J.G. Stevens and G.K. Shenoy (eds.), *Mössbauer Spectroscopy and its Chemical Applications*, Adv. in Chem. Ser. No.194, American Chemical Society, Washington DC;
Ujihira, Y. (1985) Analytical applications of CEMS, *Rev. Anal. Chem.* **8**, 125;
Belozerski, G.N. (1993) *Mössbauer Studies of Surface Layers*, Elsevier Science Pub. Amsterdam;
Ujihira, Y. and Nomura, K. (1996) *Analyses of corrosion products of steel by CEMS*, Signpost, India.
44. Nomura, K., Ujihira, Y. and Vertes, A. (1996), *J. Radioanaly. Nucl. Chem.* **202**, 103.
45. Feldman, C. (1960), *Phys. Rev.* **117**, 455.
46. Inaba, M., Nakagawa, H. and Ujihira, Y. (1981), *Nucl. Instr. Meth.* **180**, 131.
47. Nomura, K., Sharma, S.S. and Ujihira, Y. (1993), *Nucl. Instr. Meth. in Phys. Res.* **B76**, 357.
48. Klingelhofer, G. and Meisel, W. (1990), *Hyperfine Interactions* **57**, 1911.
49. Ismail, M. and Liljequist, D. (1986), *Hyperfine Interactions* **42**, 1509.
50. Ujihira, Y. and Nomura, K. (1980), *Journal de Physique* **41**, C-391.
51. Peev, T., Georgieva, M.K., Nagy, S. and Vertes, A. (1978), *Radiochemi. Radioanal. Lett.* **33**, 265.
52. Nomura, K., Tasaka, M. and Ujihira, Y. (1988) *Corrosion* **44**, 131.
53. Cook, D.C. (1986) *Hyperfine Interact.* **28**, 891.
54. Leidheiser Jr, H. and Czako-Nagy, I. (1984), *Corros. Sci.* **24**, 569.
55. Davalos, J., Gracia, M., Marco, J.F. and Gancedo, J.R. (1991), *Hyperfine Interactions* **69**, 871.
56. Nomura, K. and Ujihira, Y. (1986), *Hyperfine Interactions* **29**, 1467.
57. Nomura, K. and Ujihira, Y. (1990), *J. Mater. Sci.* **25**, 1745.
58. Stewart, I. and Tricker, M.J. (1986), *Corrosion Sci.* **26**, 1041.
59. Nomura, K. and Ujihira, Y. (1988), *Analy. Chem.* **60**, 2564.
60. Nomura, K. and Ujihira, Y. (1985) *Applications of the Mössbauer Effect*, Gordon and Breach, Sci. Publ., New York, p1185.
61. Berry, F.J. (1979), *J. Chem. Soc. Dalton Trans.* 1736.
62. Hanzel, D., Meisel, W. and Gutlich, P. (1990), *Solid State Commun.* **76**, 307.
63. Nomura, K., Ujihira, Y. and Kojima, R. (1982), in Proceedings of ICAME91, Phys. Sci. Spec. Vol. , Indian National Science Academy, New Delhi, 311.
64. Nomura, K., Ujihira, Y., Matushima, Y., Kojima, R. and Sugawara, Y. (1980), *Nippon Kagaku Kaishi* 1372.
65. Nomura, K. and Ujihira, Y. (1983), *J. Analytical and Applied Pyrolysis* **5**, 221.
66. Nomura, K. and Ujihira, Y. (1982), *J. Mater. Sci.* **17**, 3437.
67. Nomura, K., Ujihira, Y. and Kojima, R. (1992), *J. Mater. Sci.* **27**, 2449.
68. Nomura, K., Ujihira, Y., Takai, O. and Kojima, R. (1991), *Hyperfine Interactions* **69**, 549.
69. Nomura, K. and Ujihira, Y. (1986), *Hyperfine Interactions* **29**, 1474.
70. Nomura, K., Sharma, S.S., Ujihira, Y., Fueda, A. and Murakami, T. (1989). *J. Mater. Sci.* **24**, 937.
71. Sharma, S.S., Nomura, K. and Ujihira, Y. (1991), *J. Mater. Sci.* **25**, 4104.
72. Sharma, S.S., Nomura, K. and Ujihira, Y. (1992), *J. Appl. Phys.* **71**, 2000.

73. Nomura, K., Ujihira, Y., Tanaka, S. and Matsumoto, K. (1988), *Hyperfine Interactions* **42**, 1207.
74. Yamada, N., Shigesato, Y., Yasui, I., Li, H., Ujihira, Y. and Nomura, K. (1998), *Hyperfine Interactions* **112**, 213.
75. Principi, G., Maddalena, A., Gupta, A., Geotti-Bianchini, F., Hreglich, S. and Verita, M. (1993), *Nucl. Instr. Meth. Phys. Res.* **B76**, 215.
76. William, K.F.E., Johnson, C.E., Greengrass, J., Tilley, B.P., Gelder, D. and Johnson, J.A. (1977), *J. Non-Crystalline Solids* **211**, 164.
77. Johnson, J.A., Johnson, C.E., William, K.F.E., Holland, D. and Karim, M.M. (1995), *Hyperfine Interactions* **95**, 41.

^{57}Fe IMPLANTED IN DIAMOND

Studied by Conversion Electron Mössbauer Spectroscopy

K. BHARUTH-RAM[1], D. NAIDOO[1] AND G. KLINGELHÖFER[2]
[1)]Physics Department, University of Durban-Westville, Durban 4000, South Africa
[2)]Institut für Kernphysik, Technische-Hochschule Darmstadt, D-62489 Darmstadt, Germany

Abstract

Recent conversion electron Mössbauer Spectroscopy (CEMS) studies of ^{57}Fe implanted in diamond are reviewed. Two techniques have been used in these studies. Firstly, in-beam Mössbauer (IBMS) measurements have been used to study the lattice sites of ^{57}Fe in diamond under low dopant concentrations ($\leq 10^{11}$ cm^{-2}). Secondly, source-based CEMS measurements have been performed on synthetic single crystal diamonds produced by both high-temperature high-pressure (HTHP) processing and chemical vapour deposition (CVD). In the latter investigations 70 keV ^{57}Fe ions were implanted in the samples to a dose of 5 x 10^{14} cm^{-2}, and the spectra were studied up to annealing temperatures of 1470 K. The IBMS spectra were resolved into two symmetric doublets and a weak singlet, while the CEMS spectra for the HTHP and CVD samples were resolved into two doublets and two singlets (S_1 and S_2). All three measurements show that only about 10% of the probe atoms are at sites of high symmetry, the remainder are at highly perturbed configurations. The isomer shift of the singlet in the IBMS spectra is in good agreement with theoretical calculations for tetrahedral interstitial Fe. The singlets in the source based measurements at higher doses ($\geq 10^{14}$ cm^{-2}) have rather large isomer shifts attesting to the considerable lattice strain in the vicinity of the probes. The measurements also gave consistent isomer shifts for one of the doublets (δ = 0.00(4) mm/s), which was in agreement with the calculations for substitutional Fe. The large quadrupole splitting ($\approx$2 mm/s), however, caution against any firm conclusions.

1. Introduction

Diamond has several properties such as its wide band gap (5.4 eV), high thermal conductivity, high mobility for both *p*- and *n*-type carriers, and hardness, that makes it an ideal base for semiconducting devices provided effective *p*- and *n*-doped layers can be produced. While boron doped *p*-type semiconducting diamond exist in nature and has also been produced in the laboratory attempts at *n*-type doping of diamond have met

M. Miglierini and D. Petridis (eds.), Mössbauer Spectroscopy in Materials Science, 79–86.

with limited success. The low solubility of potential dopants in diamond precludes thermal diffusion as a means of incorporating dopant ions in the diamond lattice. An alternative method is offered by ion implantation which is free of any limitations imposed by solubility considerations and offers precise control of dopant species, depth and concentration. However, the implantation process is invariably accompanied by lattice damage, which is compounded in semiconductors by the large variety of defects which can be formed. Hence, several techniques, and in particular Mössbauer Spectroscopy [1-5], have been used to investigate the implantation sites and diffusion of dopant atoms in diamond, and to study the immediate neighbourhood of the implanted atoms, and their annealing characteristics.

There exist several early Mössbauer studies on ^{57}Co implanted in diamond. Sawicka *et al.* [1,2] obtained spectra dominated by a broad doublet, which suggested that the majority of the implanted Fe atoms were in highly damaged regions in the host matrix. A systematic study of the amorphization of the diamond lattice produced by the implantation of different doses of ^{57}Co was conducted by de Potter and Langouche [4]. Their data showed that implantation doses $\geq 10^{15}$ Co cm^{-2} results in the formation of a completely amorphous layer, which does not anneal up to temperatures of 1173 K. Mössbauer spectra obtained for a dose of 10^{14} ^{57}Co ions/cm^2 were resolved into two doublets and two singlets. The singlets had quite large isomer shifts indicative of the considerable lattice strain around the Fe atoms.

In recent studies the implantation sites of ^{57}Fe implanted directly in diamond have been investigated using conversion electron Mössbauer Spectroscopy (CEMS). Bharuth-Ram *et al.* (6,7) used the in-beam Mössbauer spectroscopy (IBMS) technique to study the lattice sites of ^{57}Fe in synthetic high temperature high pressure produced (HTHP) diamond under low dopant concentrations ($\leq 10^{11}$ cm^{-2}), and carried out measurements at sample temperatures of 300 - 800 K. A drawback of IBMS is that the measurements are made in the "as implanted" state and at the sample temperature; an annealing sequence on the implanted sample cannot be performed. Consequently, source-based measurements were then made on HTHP diamonds implanted with 70 keV ^{57}Fe ions to a dose of 5×10^{14} cm^{-2}, and the CEMS spectra were studied after annealing the samples up to 1470 K [8]. Comparative CEMS studies were also done on diamond samples produced by chemical vapour deposition (CVD) [9]. These measurements are reviewed and a comparative study is made of the results obtained.

2. IBMS Investigations

In the IBMS method a heavy ion beam (Ar or Xe) on an enriched ^{57}Fe foil is used to Coulomb excite the Mössbauer level at 14.4 keV and recoil implant the excited ^{57}Fe* nuclei into the host matrix. The kinetic energy imparted to the recoiling nuclei ranges up to several MeV, and results in a uniform implantation profile with a depth extending to 10μm. The sample, implanted with the excited ^{57}Fe* nuclei, acts as a source, and, hence, low implantation doses ($\leq 10^{11}$ cm^{-2}) are sufficient to obtain spectra with good statistics. This coupled with the large implantation range results in extremely low dopant concentrations ($\leq 10^{15}$ cm^{-3}). Single atom implantation is thus achieved, there is no interaction between neighbouring probe atoms, and overlap of implantation induced damage cascades is avoided. Recoilless emission of the 14.4 keV γ rays from ^{57}Fe occurs within the lifetime ($t_{1/2}$ = 100 ns) of the Mössbauer state after the implanted probes have come to rest. Hence, the probes cannot form clusters, precipitate, or escape from the sample during the time window of the measurement. However, as stated above, thermal treatment of the host sample after implantation, such as an isochronal annealing sequence, is not possible.

2.1. EXPERIMENTAL DETAILS

The IBMS measurements were made at the VICKSI Accelerator at the Hahn-Meitner-Institute, Berlin, where the ^{57}Fe nuclei were produced in their 14.4 keV Mössbauer state by bombarding an enriched ^{57}Fe foil with a pulsed 110 MeV ^{40}Ar beam. Diamond targets of large surface area (20mm x 20mm) were assembled from 5.0mm x 5.0mm x 2mm thick "tiles" of synthetic high-temperature-high-pressure (HTHP) produced type Ib single crystal diamonds. Two diamond targets, mounted on either side of the primary beam, allowed the recoiling $^{57}Fe^*$ nuclei to be implanted into the target samples while the ^{40}Ar beam passed through without striking the samples. Conversion electron spectra were collected with small, light-weight gas-filled avalanche counters mounted directly onto conventional Mössbauer drives. Spectra were collected for sample temperatures of 300 K, 600 K, 700 K and 800 K.

2.2. RESULTS

In-beam Mössbauer measurements permit the study of single isolated implants at various temperatures on the chracteristic time scale of the Mössbauer state. The measurements of Bharuth-Ram et al. [6,7] were specifically aimed at studying Fe implants in diamond where Fe solubility is vanishingly small. The CEMS spectra, displayed in Fig.1 for sample temperatures of 300 K, 600 K and 800 K, consist of three components, a single line which at 300 K has an isomer shift $\delta = +0.16(3)$ mm/s, and two doublets D_1 and D_2, with isomer shifts $\delta_1 = +0.04(3)$ mm/s and $\delta_2 = -0.51(2)$ mm/s and quadrupole splittings $\Delta E_Q = 2.01$ mm/s and 2.19 mm/s, respectively. The areal fractions of the components are listed in Table I.

The isomer shift of the singlet is in good agreement with theoretical calculations for interstitial Fe, and shows that implantation into diamond results in 10% Fe in fairly undisturbed interstitial sites and about 90% in highly damaged regions. IBMS studies in Si and Ge [10] also show small fractions of interstitial Fe and the rest in damaged regions.

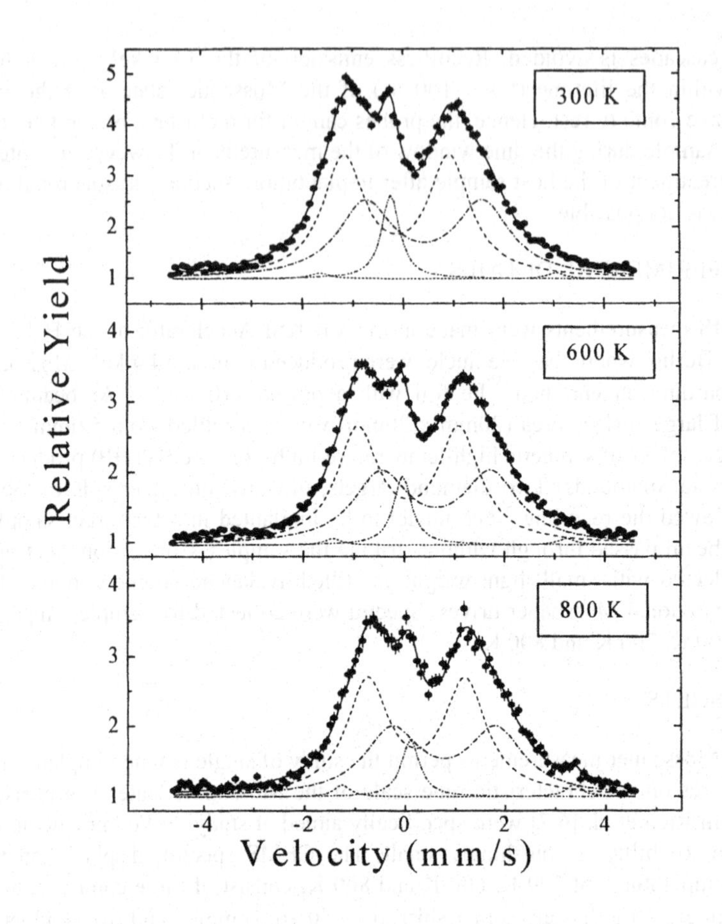

Figure 1. In-beam Mössbauer spectra of ^{57}Fe in diamond.

TABLE 1. Areal fractions (f) and isomer shifts (δ) of the components observed in the CEMS spectra on ^{57}Fe implanted in diamond: a) IBMS measurements, b) HTHP diamond and c) CVD diamond.

	a) IBMS		b) HTHP diamond		c) CVD diamond	
	f(%)	δ(mm/s)	f(%)	δ(mm/s)	f(%)	δ(mm/s)
D_1	50(2)	+0.04(3)	34(3)	-0.02(4)	30(3)	-0.04(4)
D_2	42(2)	-0.51(3)	59(3)	+0.41(4)	57(3)	+0.44(4)
S_1	8(2)	+0.16(3)	5(3)	-1.61(5)	8(2)	-1.50(5)
S_2			2(1)	+0.86(5)	5(2)	+1.16(5)

3. Source-based CEMS Studies

The source-based CEMS studies [8,9] were undertaken to study the annealing characetristics of lattice damage induced by Fe implantation. For direct comparison with the IBMS measurements the diamond sample used here was also a single crystal diamond synthesized in the HTHP process. Comparative studies were also done on a CVD diamond sample. Isotopically separated ^{57}Fe ions were implanted into the samples at an energy of 70 keV and to a dose of 5 x 10^{14} cm^{-2}, i.e. just below the amorphization threshold. The HTHP diamond was mono-crystalline and the CVD sample poly-crystalline. However in the latter case the average crystallite size was $\approx$ 100 μm, which is several orders of magnitude larger than the estimated mean implantation range (310 Å) and straggle (90Å) of the Fe probe atoms in diamond. The probes were therefore expected to sample essentially single crystal environments, except for those implanted at the grain boundaries. CEMS measurements were made on the as-implanted samples and after annealing the samples for 10 min. in vacuum at annealing temperatures, T_A, of 600 K, 950 K and 1470 K.

3.1. RESULTS

CEMS spectra, as-implanted and after annealing at T_A = 950 K and 1470 K, are shown on Fig. 2, for the HTHP and CVD diamond samples. The spectra were resolved into two doublets and two singlets, as has been observed previously for natural diamond samples[3,4]. For the CVD sample the spectrum collected after the annealing at 1470 K shows a considerable reduction in the resonance absorption effect, indicating that a large fraction of the implanted Fe have diffused out of the sample.

The areal fractions and isomer isomer shifts of the components are listed in Table 1, together with the results of the IBMS measurements. The isomer shifts are expressed as source shifts relative to α-Fe. The doublets for both samples had quadrupole splittings of 1.90(10) mm/s.

The intensities of the components observed in the CEMS spectra for the HTHP and CVD diamond samples are plotted in Fig. 3, as a function of the annealing temperature T_A. As in the case of the IBMS results, over 90% of the resonance strength for both samples was in the two doublets. In the single crystal HTHP diamond the doublets show no observable temperature dependence. The two singlet components contribute about 8% of the intensity. Above 950 K singlet S_2 disappears, and S_1 increases in intensity from 5% to 7%, and its isomer shift decreases from -1.61 mm/s to -1.16 mm/s.

In the CVD diamond the Mössbauer components show a behaviour similar to that of the HTHP sample at annealing temperatures $T_A \leq 950$ K. However, at 1470 K the population of the doublets show a dramatic decrease, falling to below 10%. The strong depopulation of these components suggest that their major contributions were due to Fe trapped at grain boundaries in the polycrystalline sample, or at damage centres close to crystallite surfaces. Annealing makes an appreciable effect on the intensities of the singlet components. At T_A= 1470 K, S_1 increases from 5% to 9%, S_2 disappears and a small (3.5%) singlet component appears with an isomer shift $\delta = 0$ mm/s.

4. Discussion

Theoretical calculations of the isomer shifts of Fe at interstitial sites in diamond give $\delta = +0.22$ mm/s. The isomer shift of the singlet observed in the in-beam measurements is in good agreements with this value, and has been assigned to interstitial Fe. In the source based measurements the high symmetry singlet components have quite large isomer shifts, presumably reflecting the considerable lattice strain around the Fe atoms introduced by the much higher implantation dose.

For substitutioanl Fe the calculations give an isomer shift of δ = -0.19 to +0.09 mm/s. The isomer shift of one of the doublet components (D_1 in Table 1) observed in all three measurements is in agreement with this. It is therefore tempting to associate this component with substitutional Fe, with the large electric field gradient ($\approx 1.0 \times 10^{18}$ V/cm^2) being due to lattice strain caused by the large mismatch of the Fe/C mass and atomic radius. Lending credence to such an assignment are the results of Emission

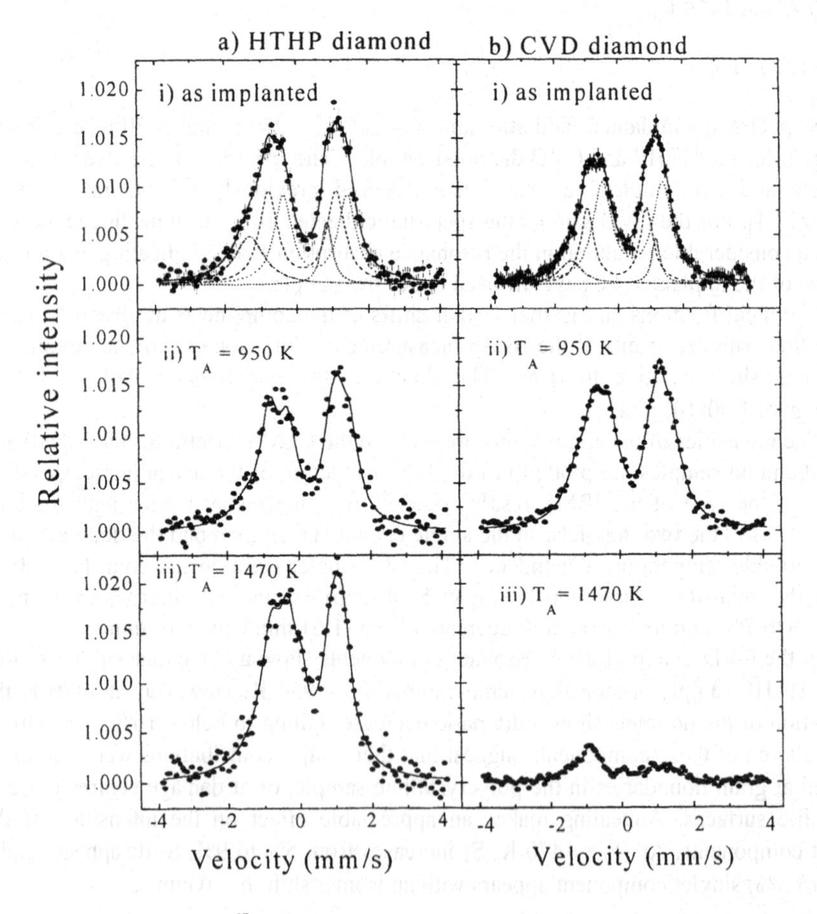

Figure 2. CEMS spectra of ^{57}Fe in a) HTHP and b) CVD diamond, as-implanted and after annealing at temperatures of 950 K and 1470 K

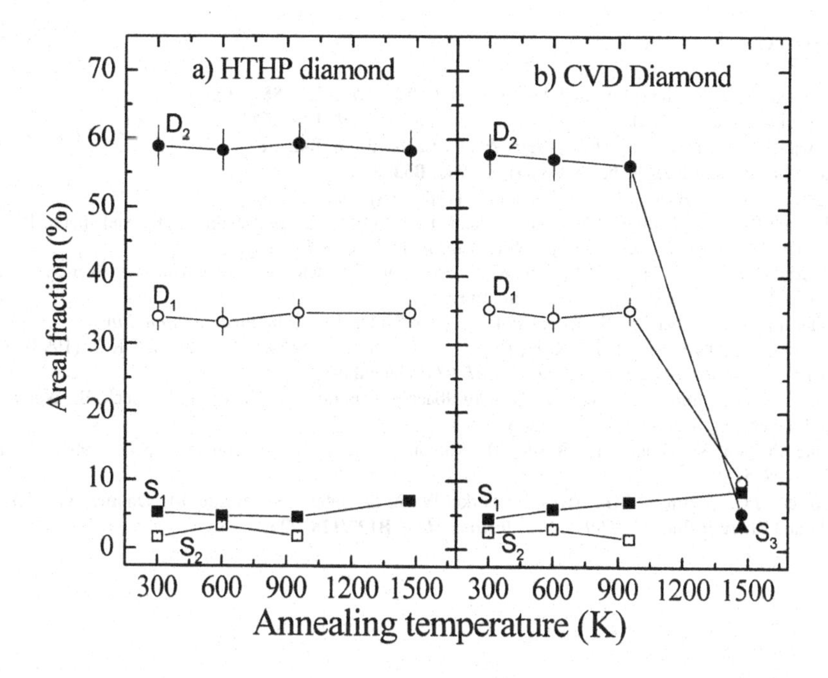

Figure 3. Areal fraction of components as a function of annealing temperature, observed in the CEMS spectra for the HTHP and CVD diamond samples

Channeling and β-γ Perturbed Angular Correlation measurements on ^{73}As implanted in diamond [11, 12], which unequivocally show substitutional fractions of 55%, but with considerable lattice damage around the implanted As atoms that does not heal even after annealing at 1670 K.

5. Conclusions

CEMS spectra of Fe implanted in diamond show three components at low dose implantation, a single line and two doublets. At higher dose an additional singlet component is present. The in-beam Mössbauer measurements show that on implantation about 10% of the Fe atoms take up tetrahedral interstitial sites in the diamond lattice while the remainder are at highly damaged environments. In the source based CEMS measurements also, about 8-10% of the Fe atoms are at high symmetry sites, but their large isomer shifts suggest considerable lattice strain around these Fe atoms. Annealing studies in these measurements show that even after annealing at 1470 K the fractions at the damaged sites show no evidence of lattice recovery.

The isomer shift of one of the doublet components is consistent with that for substitutional Fe. However, the large quadrupole splitting cautions against any firm conclusions.

References

1. Sawicka, B.D., Sawicki, J.A., and de Waard, H. (1981), *Phys. Lett.* **85A**, 303.
2. Sawicki, J.A. and Sawicka, B.D. (1982), *Nucl. Instrum. Meth.* **194**, 465.
3. Sawicki, J.A. and Sawicka, B.D. (1990), *Nucl. Instrum. Meth.* **B46**, 38.
4. de Potter, M. and Langouche, G. (1983), *Z. Phys.* **B53**, 89.
5. Latshaw, G.L., Russel, P.B., and Hanna, S.S. (1980), *Hyp. Int.* **8**, 105.
6. Bharuth-Ram, K., Hartick, M., Dorn,C., Held, P., Kankeleit, E., Sellschop, J.P.F., Sielemann, R., and Wende, L. (1996) in I. Ortalli (ed.), *Proceedings ICAME95*, (SIF, Bologna,).
7. Bharuth-Ram, K., Hartick, M., Kankeleit, E., Dorn, C., Held, P., Sielemann, R., Wende, L. and Sellschop, J.P.F. (1998), *Phys. Rev B* (in print).
8. Bharuth-Ram, K., Naidoo, D., Klingelhöfer, G., and Butler, J.E. (submitted), *J. Appl. Phys.*
9. Naidoo, D., Bharuth-Ram, K., Klingelhöfer, G., Butler, J.E., Schaaf, P., and Lieb K.P. (1998) in E. Baggio-Saitovitch (ed.), *Proceedings ICAME 97* (to be published)
10. Kubler, J.E., Kumm, A.E., Overhof, H., Schwalbach, P., Hartick, M., Kankeleit, E., Keck, B., Wende, L., and Sielemann, R. (1993), *Z. Phys.* **B93**, 155
11. Bharuth-Ram, K., Quintel, H., Restle, M., Ronning, C., Hofsäss, H., and Jahn, S.G. (1995), *J. Appl. Phys.* **78**, 5180.
12. Correia, J.G., Marques, J.G., Alves, E., Forkel-Wirth, D., Jahn, S.G., Restle, M., Dalmer, M., Hofsäss, H., and Bharuth-Ram, K. (1997), *Nucl.Instrum Meth.* **B127/128**, 723.

IRON STATES FOLLOWING COBALT DECAY IN RUTILE SINGLE CYSTAL

Iron States in Rutile

U.D. WDOWIK AND K. RUEBENBAUER
Mössbauer Spectroscopy Laboratory
Institute of Physics and Computer Science, Pedagogical University
PL-30-084 Cracow, ul. Podchorazych 2, Poland

1. Introduction

Rutile (TiO_2) belongs to $P4_2/mnm$ space group. The crystal structure of rutile is schematically shown in Fig. 1. The cation sublattice constituted of titanium

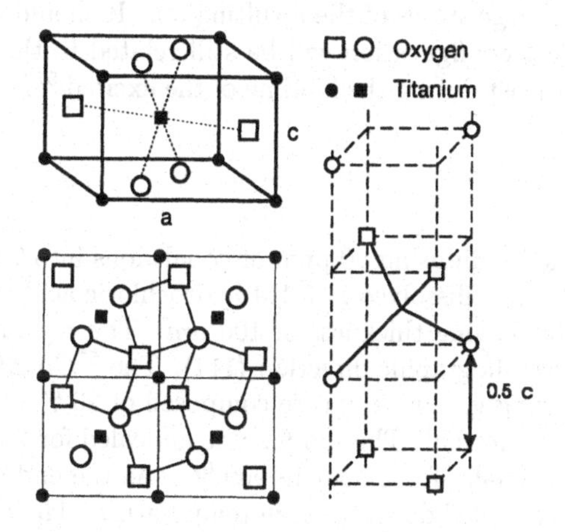

Figure 1. Crystal structure of rutile.

ions is body-centered tetragonal. Each titanium ion is surrounded by a slightly distorted octahedron of six oxygen ions. The octahedra are linked to each other forming chains parallel to the c-axis. Between the chains are regions of low electron density, "open channels" (easy diffusion channels). At regular intervals of $\frac{c}{2}$ along the channel are potential minima which are possible interstitial sites for cations. The open channels cause anisotropy in the diffusion process and may allow fast diffusion of smaller ions parallel to the tetragonal axis [1,2].

M. Miglierini and D. Petridis (eds.), Mössbauer Spectroscopy in Materials Science, 87–96.

The present paper is concerned with the study of the charge states and environments of ^{57}Fe ions in TiO_2 using the Mőssbauer effect as a microscopic probe to measure charge states at the ^{57}Fe nuclei.

Features of the Mőssbauer spectroscopy allow to use it as a tool to study the behavior of dopants in the host lattice. A considerable fraction of investigations have focused on ^{57}Fe, with both stable iron and radioactive ^{57}Co as dopants. Although the behavior of iron as a dilute impurity is quite well understood now, however, there still exists much uncertainty about the relationship between the parent cobalt ion and the iron daughter ion which results from the ^{57}Co electron-capture decay. Because of intrinsic crystal defects the resulting spectra of ^{57}Co-doped crystal may not relate to properties of the ideal crystal. It is well known that very dilute dopants may lead to results quite different from those for percent dopants [3,4].

Unfortunately, the Mőssbauer spectroscopy cannot be done with absorbers containing iron in the ppm range. Therefore, to obtain ppm spectra, rutile has to be doped with ^{57}Co, the latter decaying to ^{57}Fe. Such Mőssbauer sources can give some information about the behavior of iron in the ppm range. There are, however, some complications because the iron enters the lattice not directly, but rather as a cobalt ion, and the subsequent Auger processes frequently produce a number of different charge states of the resulting ion. It should be also noted that the surrounding defect configuration will be still related to the Co impurity and not to the Fe ion; at least during the lifetime of the excited ^{57}Fe Mőssbauer state.

2. Experimental

The source was made by diffusing 50 ppm of cobalt ions into TiO_2 single crystal. 5 mCi of carrier-free ^{57}Co dissolved in 0.1M hydrochloric acid was deposited onto the sample, the latter having thickness of 100 μm. The crystal surface was perpendicular to the crystallographic direction $\langle 111 \rangle$. The ^{57}Co diffusion anneal was performed under flowing gaseous mixture composed of 20 % O_2 and 80 % He at pressure being close to normal. The sample was annealed for 2 hours in 100 ^0C, 4 hours in 400 ^0C, and finally for 1 hour in 900 ^0C. At the end of diffusion anneal the crystal was slowly cooled down to room temperature. The cooling time was of the order of 10 hours.

Iron contamination was governed predominantly by a decay product of ^{57}Co. Concentration of other impurities was determined to be less than 1 ppm. In order to check the influence of impurity doping on crystal properties, X-ray diffraction patterns were measured on pure (undoped) and doped (with 50 ppm of non-radioactive Co) rutile. Non-radioactive control sample was prepared in the same manner as the active one. Results of X-ray measurements indicate that doping the crystal with 50 ppm of Co does not disturbe either lattice parameters or thermal vibration ellipsoids of rutile.

Emission experiments were carried out in a Mőssbauer furnace equipped with the internal single-axis goniometer vs temperature and sample orientation. The

TABLE 1. Valence and spin states of ^{57}Fe in rutile-structure TiO_2.

Fe charge configuration	Spin configuration	Location
1. High Spin $Fe^{3+}(S=\frac{5}{2})$	$[Ar]\,3d^5 4s^0$ ↑ ↑ ↑ ↑ ↑	lattice
2. High Spin $Fe^{2+}(S=2)$	$[Ar]\,3d^6 4s^0$ ↑↓ ↑ ↑ ↑ ↑	lattice
3. Low Spin $Fe^{2+}(S=0)$	$[Ar]\,3d^6 4s^0$ ↑↓ ↑↓ ↑↓	channel(RT)/lattice (HT)
4. High Spin $Fe^{1+}(S=\frac{3}{2})$	$[Ar]\,3d^7 4s^0$ ↑↓ ↑↓ ↑ ↑ ↑	channel

heating process covered the temperature range from 297 K up to 1388 K. After each HT run the sample was quenched to RT and the Mőssbauer effect was again examined as a function of crystal orientation. The sample was always kept under the flowing mixture of 20 % O_2 and 80 % He at ambient pressure. The sample (source) temperature was controlled by Pt-Pt10%Rh thermocouple. Temperature remained stable within 1 K during each run (3 to 4 days per each spectrum). Enriched $K_4{}^{57}Fe(CN)_6 \times 3H_2O$ single line absorber was used. The absorber was kept at RT and was driven by 5 % round-corner triangular reference function in front of the Kr-filled proportional detector. All spectra were collected in 4096 channels of MsAa-1 spectrometer [5]. The sample was lying on the Pt foil.

All isomer shifts (including second-order Doppler effect) are given with respect to iron metal. Convention of positive velocity for source approaching absorber is adopted. Thus, the increase in the electron density at the source nucleus results in a positive change in the shift.

3. Room Temperature Mössbauer Spectra

3.1. RESULTS

The Mőssbauer emission spectra obtained at RT for sample being quenched from various initial (HT) temperatures are shown in Fig. 2. The spectra were fitted to multiple sites described by the hyperfine Hamiltonians assigned to each site. In all cases, i.e., either for RT or HT spectra, the convergence was easily achieved giving the value of quality-fit parameter (MISFIT [6]) which ranged from 0.13% to 4.3% for RT and HT (1286 K) data, respectively. The angular dependence of the emitted γ-rays was observed neither for RT nor HT. A detailed explanation of such behavior will be given and discussed further below. Hence, spectra collected with the beam parallel to $\langle 111 \rangle$ direction are shown solely.

Room temperature spectra of quenched sample, Fig. 2, indicate a variety of charge and spin states of ^{57}Fe, the latter resulting from EC decay of ^{57}Co. The large body of data taken with $(^{57}Co)TiO_2$ source at RT allowed to identify oxidation/spin states of iron in the host rutile matrix as well as to distinguish between substitutional and interstitial sites in which the impurities could be located. Particular configurations of ^{57}Fe in rutile are summarized in Table 1, while their hyperfine parameters are listed in Table 2.

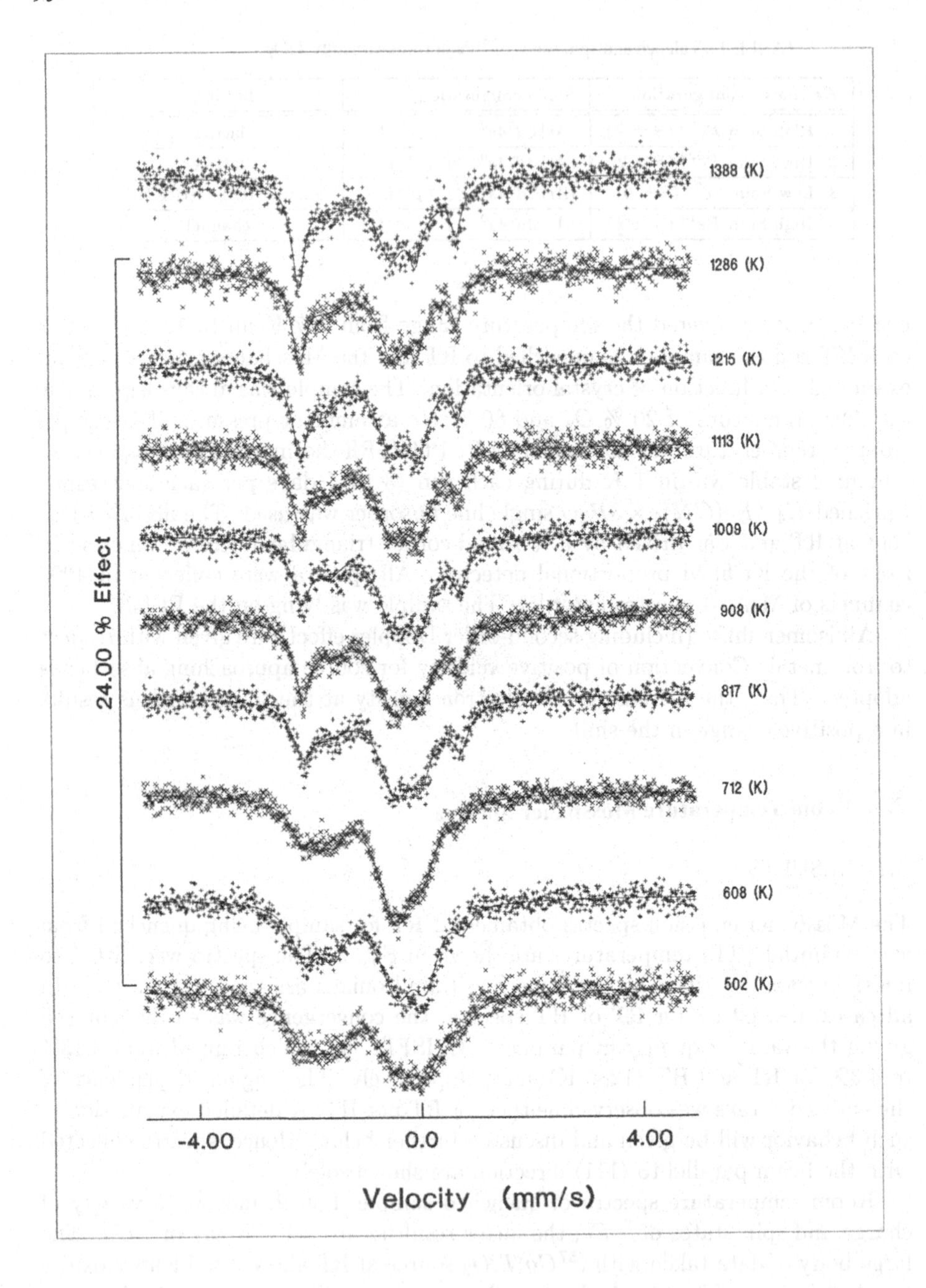

Figure 2. Typical RT Mössbauer spectra of 50ppm ^{57}Co-in-rutile source. The line drawn through the data points represents a fit in Lorentzian approximation valid due to the fact that for (^{57}Co)TiO_2 sample the source resonant thickness can be neglected.

TABLE 2. Hyperfine parameters of various ^{57}Fe configurations in rutile at RT.

Fe charge configuration	Isomer Shift [mm/s] with respect to iron metal	Quadrupole Splitting [mm/s]
1. Fe^{3+} ($S = \frac{5}{2}$)	-0.46(2)	-
2. Fe^{2+} ($S = 2$)	-0.77(1)	1.72(4)
3. Fe^{2+} ($S = 0$)	0.34(1)	0.73(2)
4. Fe^{1+} ($S = \frac{3}{2}$)	-2.12(1)	-

3.2. DISCUSSION

An impurity is able to enter the host TiO_2 lattice substitutionally and/or interstitially. Therefore, multiple valence states of ^{57}Fe represent different and distinguishable local environments in which the impurities could reside in the host lattice. Electron capture decay converts the parent $^{57}Co^{2+}$ located at the regular lattice positions either to Fe^{2+} or Fe^{3+} daughter ions. The oxidation state of the daughter ion depends on the impurity surrounding, i.e., it is mainly related to a variety of possible defects. Almost stoichiometric rutile is to some extent oxygen-deficient material and the tetravalent titanium interstitials $Ti^{\cdot\cdot\cdot\cdot}$ are present as well.

Trivalent iron is a result of Co EC decay, provided the parent atom is isolated and located in undisturbed (at least locally ideal) substitutional cation site. In other words, the anion sublattice in the vicinity of impurity is complete. Two Fe^{3+} ions substitute for Ti^{4+} ion resulting in a single oxygen vacancy (VO^{2-}) maintaining charge neutrality. However, defects induced in the crystal are of nonlocal (extrinsic) character. Oxygen vacancies which accompany trivalent iron impurities are at the next-nearest-neighbor oxygen positions at least. Since Fe^{3+} do not feel the effect of far distant oxygen vacancies the spectral component due to trivalent iron was observed to remain practically unsplit. A quadrupole interaction is either very small or partly canceled, however, it influences the linewidth of this component. Moreover, Fe^{3+} does not exist as a channel residing impurity. It is generally accepted and also confirmed by diffusion studies [1,2] that ions with radii less than 0.73 $\AA$ (for example Fe^{3+}) cannot occupy channel sites of rutile.

Room temperature spectra show also a presence of regular substitutional divalent iron for which the excited state, $I = \frac{3}{2}$, is split by a quadrupole interaction. The isomer shift and the quadrupole splitting are in the range characteristic for Fe^{2+} with spin configuration $S = 2$ both. Co^{2+} residing at distorted oxygen octahedral surrounding, and being disturbed by the near-neighbor oxygen vacancy decays to a high spin Fe^{2+}. The broadened lines observed for this valence state suggest the existence of non-equivalent iron-vacancy complexes and possible slightly different cobalt-vacancy distances.

The most interesting result is a presence of unusual $Fe^{1+}(S = \frac{3}{2})$ charge state. The resonance line corresponding to "exotic" monovalent iron was observed to have

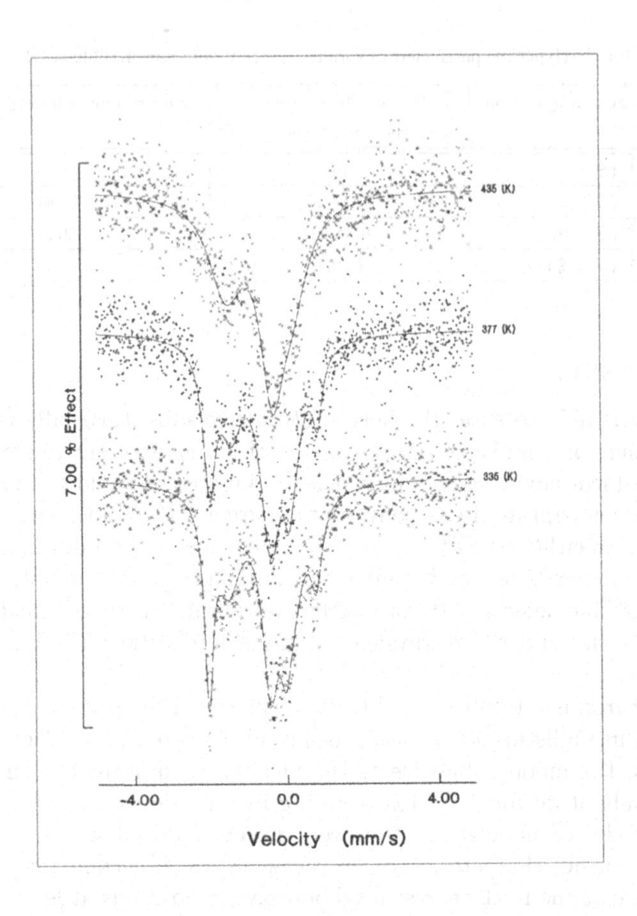

Figure 3. Spectra indicating channel disappearance of the channel contribution upon increasing temperature.

an isomer shift of -2.12 mm/s at RT. This very large shift indicates a charge state of +1 for the iron in a $3d^7$ configuration. The "exotic" iron is present solely after quench and at elevated temperatures not exceeding 400 K (see Fig. 3 for details). Monovalent iron is suggested to represent an isolated state. It is present only in a perfect, i.e., undisturbed tetrahedral channel surrounding. Therefore, it produces a single resonance line of a large isomer shift, the latter indicating a very low electronic charge density at the iron nucleus. Fe^{1+} is a result of radioactive decay of parent cobalt ion diffusing rapidly along the open channels [1] and trapped into the "ideal" (free of intrinsic defects) interstitial potential minima by the abrupt quench. It should be noted, that Fe^{1+} represents a metastable state and it has been previously observed solely by emission Mőssbauer spectroscopy [7-11], the latter having the suitable time window of observation. It has been detected in single crystals being previously quenched and having the Co dopants in the ppm range diffused into.

For a Co impurity "frozen" in the channel disturbed by the oxygen vacancy at the neighborhood the daughter is Fe^{2+} with $S = 0$. Hence, this component represents some associated state, i.e., it is due to the existence of a local defect. One should note that this ion cannot be regarded as an ordinary substitutional one since its spectral component disappears at elevated temperatures (Fig. 3).

It was observed, that contributions corresponding to interstitials are practically temperature independent while the quench is performed from initial temperatures exceeding ca. 700 K. The appropriate thermal treatment is able to erase the previous thermal history of the sample. In most cases a linewidth reduction is observed, and the narrow lines suggest well-defined sites upon such an erasure.

The absence of a change in a doublet symmetry/asymmetry with sample orientation relative to the direction of the γ-ray beam is due to multiple vacancy-impurity configurations. A lattice distortion in the vicinity of impurity leads to nearly global cubic symmetry, and hence accounts for the observed effect.

4. High Temperature Mössbauer Spectra

4.1. RESULTS

Typical high temperature spectra are shown in Fig. 4. One can see, especially from Fig. 3, that signal coming from impurities residing in the interstital sites becomes unobservable above 400 K. This is due to enormous diffusivity of cobalt dopants traveling along the channels by the interstitialcy/channel mechanism [1]. Iron was found to be in two valence states, namely, Fe^{3+} and Fe^{2+} and it is located at regular lattice positions. Divalent iron exists in two spin configurations, i.e., $S = 2$ and $S = 0$. Upon raising temperature high enough high spin Fe^{2+} tends to convert to low spin Fe^{2+} and the former one disappears above ca. 800 K. This effect is related to the increased covalency with temperature. The quadrupole splitting corresponding to high spin Fe^{2+} is temperature dependent. Population of $3d$ orbitals equilibrates with increasing temperature accounting for the observed decrease in quadrupole splitting.

Fe^{3+} seems to be a quite stable iron valence state in a rutile lattice, since it was encountered in the whole experimental temperature region studied. An apparent reduction to divalent iron may be seen above 1000 K (see. Fig. 5).

No significant broadening of linewidths corresponding to substitutional impurities was detected, i.e., no lattice diffusion was observed.

4.2. DISCUSSION

The most remarkable observation is disappearance of the signal coming from the channel at about 400 K with a simultaneous drop in spectral area (see Fig. 6). Hence, one can conclude that the iron being a decay product of the channel located cobalt is a fast diffuser as well. A gradual "conversion" of $Fe^{2+}(S = 2)$ into

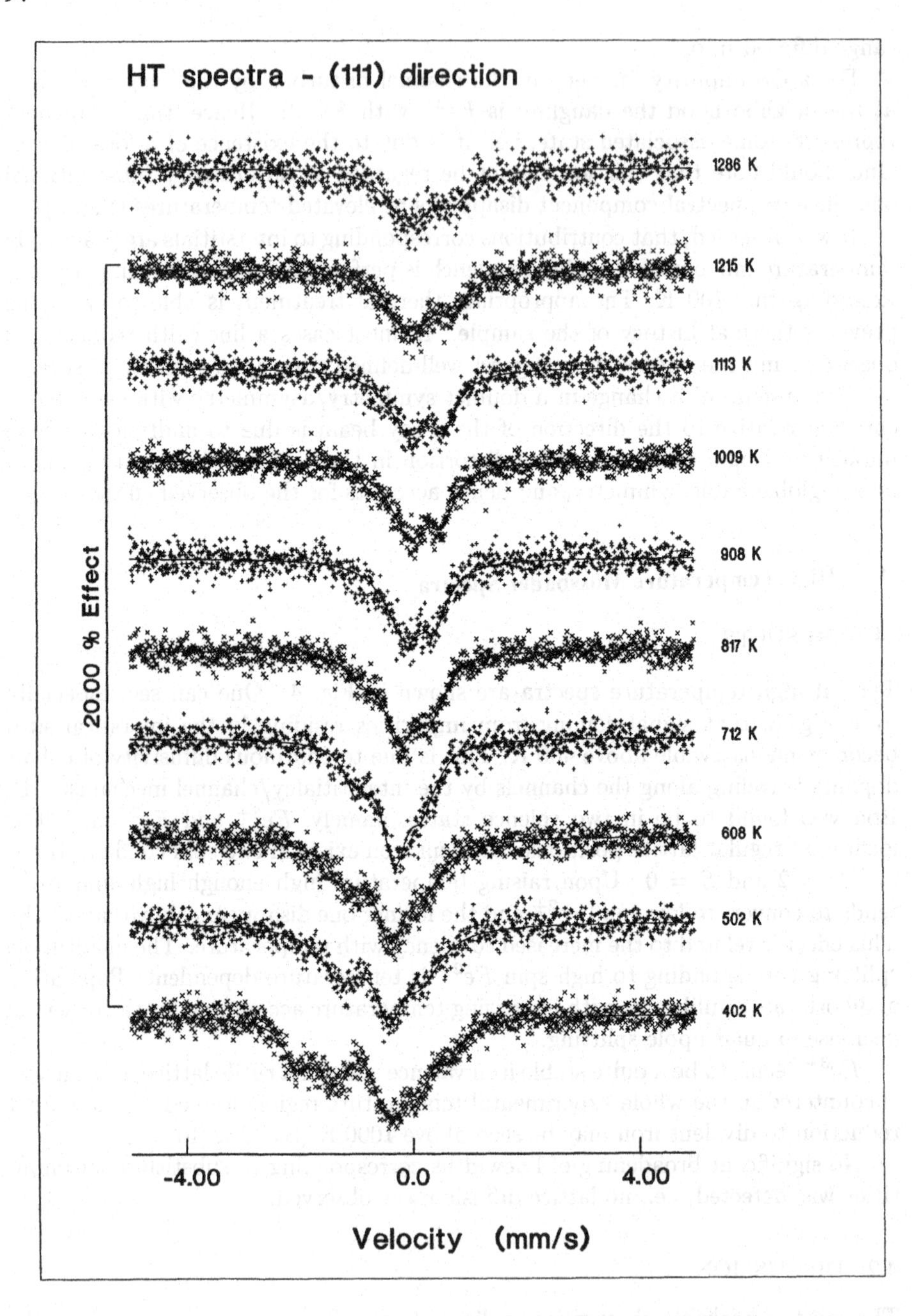

Figure 4. Typical HT Mössbauer spectra of ^{57}Co in rutile *vs.* temperature.

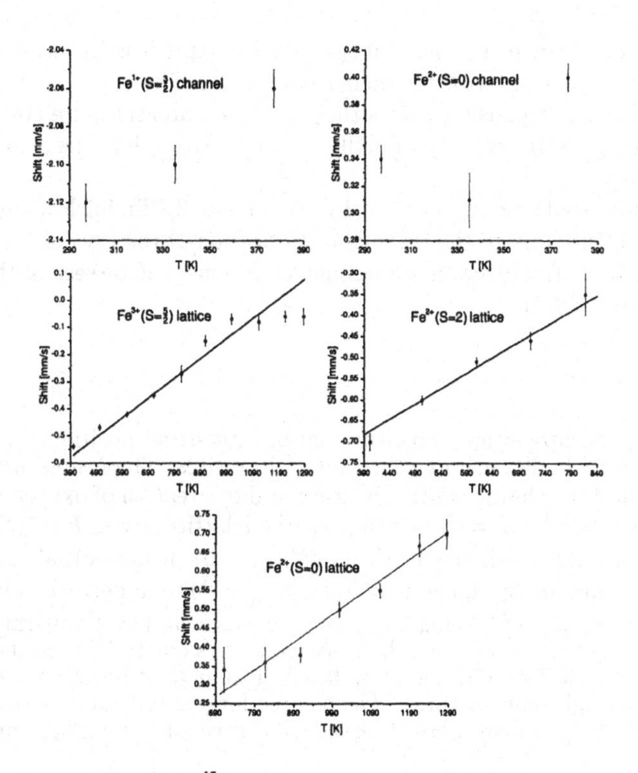

Figure 5. Shifts of particular ^{57}Fe configurations *vs.* temperature. The solid lines represent the slope 7.31 x 10^{-4} mm/sK due to the Dulong-Petit rule.

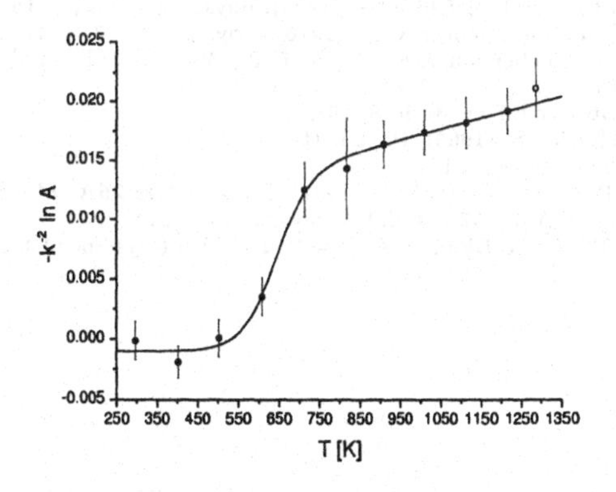

Figure 6. Temperature dependence of the total area of the 14.4-keV γ radiation from ^{57}Fe in the TiO_2 single crystal. Symbol A stands for the total area normalized by the mean value of resonant fraction obtained from RT measurements and k denotes the wave number of the emitted Mössbauer radiation.

$Fe^{2+}(S = 0)$ at perturbed lattice sites could be explained by invoking defect (oxygen vacancy) mobility with the increasing temperature.

A lattice thermal expansion makes the crystal apparently more covalent, and hence Fe^{3+} isomer shift moves gradually towards Fe^{2+} shift, i.e., an apparent reduction occurs.

A quadrupole splitting experienced by $Fe^{2+}(S = 2)$ diminishes with the increasing temperature due to the thermal equilibration of the crystal field levels.

The onset of anharmonicity in the recoilless fraction could be seen at the highest temperatures (see Fig. 6).

5. Conclusions

Cobalt ions can occupy either substitutional or interstitial positions in the TiO_2 lattice with slightly higher energy at the interstitial sites. These sites are occupied by Co^{2+} ions, and the charge neutrality is assured by creation of oxygen vacancies. Cobalt decays to $Fe^{3+}(S = \frac{5}{2})$ in unperturbed lattice sites, $Fe^{2+}(S = 2)$ in lattice sites associated with the VO^{2-}, $Fe^{2+}(S = 0)$ in interstitial sites having adjacent VO^{2-}, and finally to metastable $Fe^{1+}(S = \frac{3}{2})$ in unperturbed interstitial sites. Substitutional Fe^{2+} remains in $S = 2$ state at low temperatures, and converts gradually to $S = 0$ state at high temperatures due to the defects mobility. Interstitial Co is the fast diffuser as well. A host matrix becomes increasingly covalent at very high temperatures. The sample forgets previous thermal history at about 700 K. All processes were observed to be reversible upon heating/cooling.

References

1. Sasaki, J., Peterson, N.L., and Hoshino, H. (1985), J.Phys.Chem.Solids **46**, 1267.
2. Hoshino, H., Peterson, N.L., and Wiley, C.L. (1985), J.Phys.Chem.Solids **46**, 1397.
3. Stampfl, P.P., Travis, J.C., and Bielefeld, M.J. (1973), phys.stat.solidi (a) **15**, 181.
4. Sandin, T.R., Schroeer, D., and Spencer, C.D. (1976), Phys.Rev.B **13**, 4784.
5. Kwater, M., Pochroń, M., Ruebenbauer, K., Terlecki, T., Wdowik, U.D., and Górnicki, R. (1995), Acta Physica Slovaca **45**, 45.
6. Ruby, S.L. (1973), Mőssbauer Effect Meth. **8**, 263.
7. de Coster, M. and Amelinckx, S. (1962), Phys.Letters **1**, 254.
8. Mullen, J.G. (1963), Phys.Rev. **131**, 1415.
9. Chappert, J., Frankel, R.B., and Blum, N.A. (1967), Phys.Letters **25A**, 149.
10. Frankel, R.B. and Blum, N.A. (1967), Bull.Am.Phys.Soc. **12**, 24.
11. Chappert, J., Frankel, R.B., and Blum, N.A. (1967), Bull.Am.Phys.Soc. **12**, 352.

MÖSSBAUER SPECTROSCOPY IN MINERALOGY AND GEOLOGY

Activities of the Mössbauer laboratory in Bratislava

J. LIPKA
Department of Nuclear Physics and Technology, Slovak University of Technology, 812 19 Bratislava, Ilkovičova 3, Slovakia

Abstract

A great advantage of Mössbauer spectroscopy (MS) compared to macroscopic classical techniques lies in the possibility of separating, in a very simple way, magnetic and nonmagnetic sublattices. This may be the case of ions at non-equivalent crystallographic sites, ions in different charge states (for example Fe^{2+}/Fe^{3+}, fluctuating valences, etc.) or magnetically different ions. Typical materials in that respect are magnetic oxides and magnetic fraction of minerals, intermetalic compounds and other materials interesting from a fundamental as well as a technological point of view. In the physics of minerals, Mössbauer spectroscopy gives some insight into electronic structure of inhomogenous mixed-valent compounds. Through the determination of hyperfine parameters, different iron-bearing components and also different types of iron present in a mineral can be distinguished in the sample. This leads to variety of applications in geological and mineralogical branches such as paleomagnetism, meteorites, earth science, extraterrestrial samples, etc.

1. Introduction

Mössbauer spectroscopy is a useful analytical tool in earth sciences. It is applicable to elements having Mössbaeur-active isotopes like iron, tin, gold etc. It gives information about the oxidation state and magnetic state of these elements and about the short-range crystal chemistry of crystalline and amorphous materials containing Mössbauer nuclides.

Iron is the most abundant element in geological and planetological environments among the elements that can be studied by Mössbauer spectroscopy. The precise characterisation of the valence state and the coordination polyhedra of iron allows geologists and mineralogists to solve several current problems of natural inorganic systems. These systems may vary very much in size, from the local crystal chemistry of individual mineral species to the evolution trends of continents.

M. Miglierini and D. Petridis (eds.), Mössbauer Spectroscopy in Materials Science, 97–106.

The aim of this contribution is to present an account of the current applications of Mössbauer spectroscopy in the study of minerals. Besides the Mössbauer Effect Reference and Data Journal (Stevens et al., 1983, 1998 [1, 2]) which lists all papers published on Mössbauer spectroscopy, a number of reviews of applications to the study of clay minerals has been published in recent years (e. g., Bancroft 1973; 1979 [3, 4]; Coey 1975 [5]; Goodman 1980 [6]; Fripiat 1982 [7]; Vandenberghe 1991 [8]; Vertes et.al. 1998 [9]; de Souza Jr. 1998 [10]). The number of bibliographical references in [2] is more than 2 600, and the number of different minerals investigated by MS is more than 400. Therefore the references given in this contribution are restricted to illustrate the activities of our Mössbauer laboratory in this field.

2. Application of Mössbauer Spectroscopy in Qualitative Analysis

2.1. IDENTIFICATION OF OXIDATION STATES

The most common application of Mössbauer spectroscopy is in the identification of the oxidation states of iron. High spin ions Fe^{2+} and Fe^{3+} can be readily distinguished by the magnitudes of both their isomer shifts and their quadrupole splittings.

In Fig. 1 the regions of isomer shift values at room temperature for high spin Fe^{3+} and Fe^{2+}, and low spin FeII and FeIII in "ionic" compounds and minerals are displayed for various co-ordination numbers. The values vary somewhat with co-ordination number, site symmetry and the type of ligand, but normally there is little or no ambiguity in assigning the oxidation state of iron minerals from the isomer shift.

The relation between quadrupole splitting and various electronic states and coordinations of Fe is not a simple matter. In particular, the valence contribution is

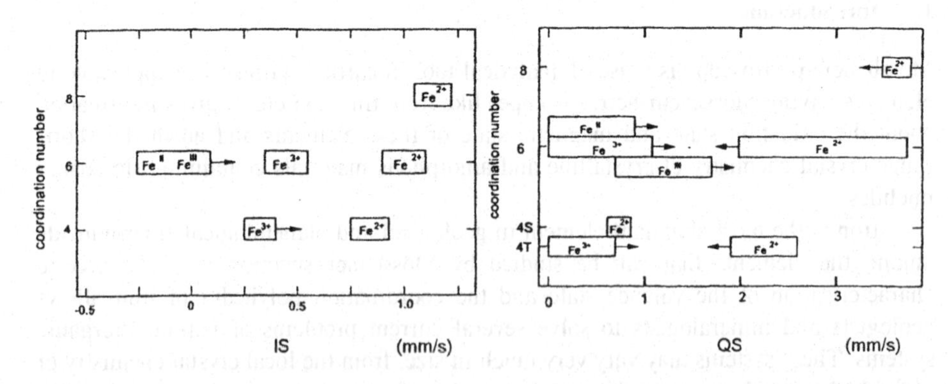

Figure 1. Dependence of the coordination number for high-spin and low-spin Fe compounds on the isomer shift, IS, relative to metallic iron (left side) and quadrupole splitting, QS (right side). Arrows indicate that values outside the boxes have been observed.

governed by many other effects such as ligand field symmetry, covalence, and spin orbit coupling. In Fig. 1 the regions of co-ordination number versus quadrupole splitting values are displayed for high and low spin iron compounds and minerals. It is observed here that many regions are broad and strongly overlap, hampering to some extent an unambiguous determination of oxidation state and co-ordination.

2.2. IDENTIFICATION OF SPECIFIC MINERAL SPECIES

Typical parameters for Fe in some minerals (Table 1) show that there is a large overlap of parameters from one species to another. Thus, Mössbauer spectroscopy has small potential for identifying a specific mineral without some help from other techniques. On the other hand, parameters for certain other groups of minerals may be distinct enough for Mössbauer spectroscopy to be used as a basis for their identification [11 - 15].

Typical and thoroughly studied minerals are those ordered magnetically, e.g. magnetic oxides and magnetic fractions of minerals. For each sublattice, the various Mössbauer parameters are determined individually from which individual moments, ordering temperature, magnetic structure, and substance can be obtained [16, 17].

Four common magnetic structures are shown schematically in Table 2 together with a summary of the features which can be used to identify the type of structure from Mössbauer data.

Another important consequence of the possibility to separate individual subspectra is the determination of the degree of occupation of sites by cations. In general, X-ray diffraction method presents a competing technique to Mössbauer spectroscopy.

However, many samples appear to be of low crystallinity. This is indicated by poor diffraction intensity. On the other hand, Mössbauer spectroscopy can be successfully applied to amorphous or poorly crystallised materials [18].

There are no problems in distinguishing among different magnetically ordered

TABLE 1. Typical Mössbauer parameters for some layer silicates at ambient temperatures (adapted from [7]. Isomer shifts IS are relative to Fe metal.

Silicate	Fe^{3+}				Fe^{2+}			
	IS (mm/s)	QS (mm/s)	IS (mm/s)	QS (mm/s)	IS (mm/s)	QS (mm/s)	IS (mm/s)	QS (mm/s)
Kaolinite	0.36	0.52						
Chamosite	0.38	0.78			1.14	2.57	1.12	2.20
Muscovite	0.40	0.72			1.13	3.00	1.12	2.20
Illite	0.33	0.65	0.38	1.21	1.14	2.75		
Glauconite	0.36	0.40	0.35	1.05	1.15	1.7-2.7		
Nontronite*	0.36	0.30	0.36	0.62				
Montmoril-lonite	0.32	0.52	0.37	1.15	1.07	2.90		
Biotite	0.40	0.52	0.38	0.92	1.11	2.62	1.09	2.18

*) Sometimes also a 3rd component is observed with IS ≈ 0.2 mm/s and QS ≈ 0.5 ÷ 0.6 mm/s.

TABLE 2. Four common magnetic structures and their identification from magnetically split Mössbauer spectra (adapted from [31]).

Ferromagnetic	Antiferromagnetic	Ferrimagnetic	Canted antiferromagnetic (weak ferro-magnetic)
↑ ↑ ↑ ↑	↑ ↑ ↓ ↓	↑ ↑ ↓ ↓	↗ ↗ ↘ ↘
No splitting of the spectral lines in an applied field	Spitting of the spectral lines in an applied field directed along the antiferro-magnetic axis. Sharp spin reorientation	Distinct magnetically split spectra with different hyperfine fields	Splitting of the spectral lines in an applied field , directed along the antiferromagnetic axis. Continuous spin reorientation

phases when they are present in a well crystallised form, with low degree or without any substitution. Both substitution and the presence of small, the so-called superparamagnetic particles, make the situation more complicated. In these cases, it is necessary to perform other supplementary measurements at different temperatures down to liquid helium without and with an external magnetic field applied [19].

The study of magnetic order in silicate minerals is also not completely solved. Magnetic hyperfine spectra contain much more information, particularly concerning the electronic structure of the ferrous ion, than it is available from paramagnetic quadrupole doublets. Spectra which are unresolved in the paramagnetic state may be clearly separated below the magnetic ordering temperature. The minerals provide an area for application of the well-established ideas of magnetism in insulating 3d compounds, particularly in quasi-one or two-dimensional compounds. There is scope here for work on synthetic minerals which would be better model compounds. When the magnetism of pure minerals is properly understood, low temperature studies throw light on the problem of cation disorder in well-chosen materials.

The parameters and magnetic properties for some highly-crystalline minerals are given in Table 3.

In general, Mössbauer spectroscopy can be performed either in transmission mode (TMS) or in backscattering mode by detecting the decay products of deexcitation of the nuclei in the specimen which have been resonantly exited by the source radiation.

In the latter case, one can use conversion electron Mössbauer spectroscopy (CEMS), conversion X-ray Mössbauer spectroscopy (CXMS) or re-emitted gamma-ray Mössbauer spectroscopy (RGMS).

The mean escape depth of the radiation varies with its energy in the range from 20 up to 20.000 nm. The major advantage of backscattering geometry is the possibility

TABLE 3. Mössbauer parameters for some crystalline oxides and hydroxides of iron (adapted from [7]). Abbreviations stand for: A-antiferromagnetic, Fe - ferromagnetic, Order.T. - Curie resp. Neel temperatures, Exp.T. - temperature of measurement, H - hyperfine field, IS - isomer shift, QS - quadrupole splitting.

Sample	Magn. Properties	Order. T. (K)	Exp. T. (K)	H (T)	IS (mm/s)	QS (mm/s)
α-Fe_2O_3	Fe	956	300	51.6	0.36	
			77	52.7	0.48	
β-Fe_2O_3		107	295		0.324	0.72
					0.331	0.97
			4.2	49.6		
				51.9		
γ-Fe_2O_3	A		300	48.8	0.27	
				49.9	0.41	
Fe_3O_4			300	49.3	0.25	
				46.0	0.65	
α-FeOOH	A	393	300	38.4	0.37	
			77	50.4	0.48	
β-FeOOH	A	295	300	-	0.37	0.55
					0.38	
			77	47.3	0.52	
				46.3	0.48	
				43.7	0.48	
γ-FeOOH	A	73	77	-	0.48	0.55
			4.2	46.0	0.51	
α-FeOOH	Fe	(455)	296	38.0	0.35	
		(420)	80	52.5	0.7	
				50.5	0.7	

to study iron-rich samples *in situ* without any special preparation of the sample, e.g. the surface of Mars.

We have studied impactite from the Monturaqui crater in Chile using both transmission and backscattering geometry. The available impactites are of dimensions approximately 5 x 3 x 4 cm and consist of black-grey core and reddish-brown rind. Corresponding Mössbauer spectra are given in Fig. 2.

We have found that iron in Monturaqui impactite exists in magnetically ordered phases: slightly substituted maghemite, nonstoichiometric magnetite and goethite which is at room temperature partly in superparamagnetic state. The rest of iron is in paramagnetic phases: silicate glass and some grains of olivine [20, 21, 22].

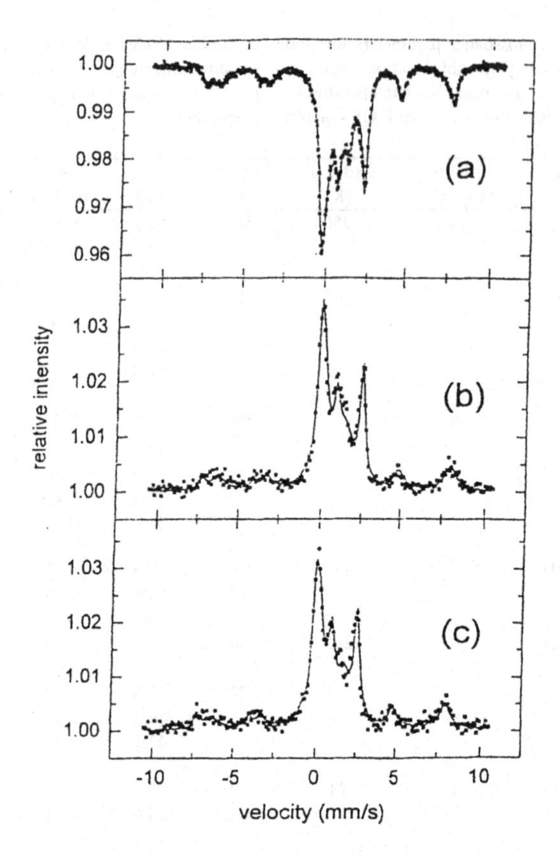

Figure 2. Room temperature Mössbauer spectra of Monturaqui impactite obtained in: transmission (a) and backscattering geometries from polished (b) and unpolished (c) sides.

3. Quantitative Analytical Applications

Because of the nature of the Mössbauer technique, any quantitative information obtained from Mössbauer spectra measurements must relate solely to the distribution of iron.

Thus, in a mixed phase system it is only possible to obtain quantitative distribution of iron components between the phases. Without a knowledge of the iron contents of each phase it is impossible to determine the amount of each phase present. The quantitative analytical applications are therefore extremely limited. But there are areas, particularly in the determination of ferrous/ferric ratios in unstable phases, where the technique is able to make invaluable contribution.

One major difficulty in this type of work concerns the possibility that the recoil-free fraction is not identical for all components in the system being investigated. Most

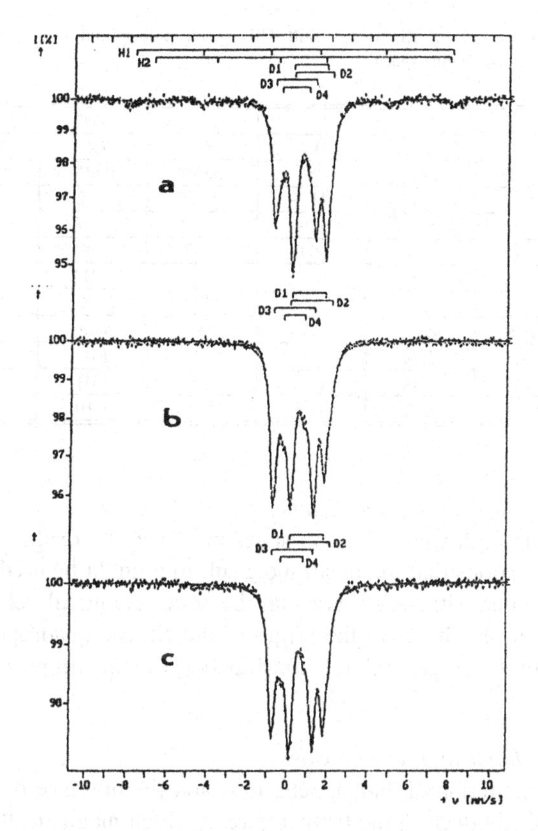

Figure 3. Room temperature Mössbauer spectra of selected allanite samples - (a) sample A1, (b) -sample A2, and (c) sample A3.

of the workers have ignored this effect which can be significant especially at room temperature measurements.

Examples included here are our results on allanite where Mössbauer spectroscopy has been successfully used [23, 24]. Allanite is a characteristic accessory mineral in many granites, granodiorites, monzonites and syenites and occurs in larger amounts in some limenites scarns and in pegmanites. The structure of allanite can be expressed as: $(Ca, Ce)_2(Fe^{2+},Fe^{3+})Al_2O.OH[Si_2O_7][SiO_4]$ where Ce stands for all rare earth elements. The Mössbauer spectra in Fig. 3 show four doublets assigned to Fe^{2+}(doublet D_1) and Fe^{3+} (D_3) in M3 positions, and Fe^{2+} (D_2) and Fe^{3+} (D_4) in M1 sites. In the sample A1 small amount of magnetic phase (probably magnetite) is present. Mössbauer parameters and corresponding areas of the four doublets are given in Table 4.

3.1. APPLICATION IN STRUCTURAL ANALYSIS

This is a very wide field that has developed over the past 20 years. Some aspects of this type of application are briefly discussed below [7].

TABLE 4. Mössbauer parameters (IS - isomer shift, QS - quadrupole splitting) and relative areas (A_{rel}) of four components in allanite spectra.

Sample	D1			D2		
	IS	QS	A_{rel}	IS	QS	A_{rel}
	(mm/s)	mm/s	%	mm/s	mm/s	%
A1	0.97	1.65	41.5	1.10	1.92	7.5
A2	0.97	1.65	33.4	1.10	2.07	8.7
A3	0.97	1.64	36.0	1.07	2.00	10.8
Sample	D3			D4		
	IS	QS	A_{rel}	IS	QS	A_{rel}
	mm/s	mm/s	%	mm/s	mm/s	%
A1	0.26	2.00	6.5	0.24	1.07	9.0
A2	0.26	1.98	45.1	0.26	1.01	12.8
A3	0.26	1.99	42.1	0.25	1.10	11.1

3.1.1. *Determination of Co-ordination Number*

Isomer shift allows the high spin oxidation states of iron to be readily distinguished. There have been suggestions that the quadrupole splitiing might be used as a guide to coordination number, but, although there may be some empirical relationship over some group of samples, the theory of the origin of the electric quadrupole interaction eliminates the possibility of general relationship between quadrupole splitting and coordination number.

3.1.2. *Distribution of Impurity Ions in Oxides*

Both isomorphous substitution of diamagnetic ions and the presence of small particle sizes can have a marked effect on the temperature at which magnetically split spectra are obtained at which 6-peak spectra are obtained from magnetically ordered minerals. As it was mentioned above (part 2.2.) Mössbauer spectroscopy allows to study also nonstoichiometric or superparamagnetic compounds.

3.1.3. *Identification of Sites Containing Octahedral Iron in Silicates*

If the co-ordinating groups or degree of distortion from cubic symmetry vary for different crystallographic sites, then the electric field gradients at those sites will also vary.

Generally accepted assignments of components in the Mössbauer spectra to the crystallographic sites have been derived through the study of many samples of each mineral species.

3.1.4. *The Study of Mineral Alteration Reactions*

Mössbauer spectroscopy has the ability to identify the crystallographic sites that contain iron in minerals as well as the oxidation states of that iron. It has considerable potential in the study of mineral reactions such as weathering effects of temperature and pressure, redox reactions, manufacturing technology etc. [25 - 30].

4. Conclusions

Mössbauer spectroscopy can be used to determine the oxidation states of iron in minerals and to identify the presence of some mineral species in samples of unknown composition, particularly when these species exhibit magnetic ordering. Small particle sizes and isomorphous substitutions can drastically alter the Mössbauer spectra at a particular temperature. In quantitative analysis, only a distribution of iron can be determined accurately when spectra are obtained over a range of temperatures.

In the silicate minerals, it is possible to distinguish iron-containing sites according to their co-ordination numbers and their electric field gradients.

Acknowledgement

This work was partly supported by National grant VEGA 1/4286/97

References

1. Stevens, J.G., Stevens, V.E., and Gettys, W.L. (1983) *Mössbauer Effect Reference and Data Journal*, Mössbauer Effect Data Center, The University of North Carolina, at Asheville USA.
2. Stevens, J.G., Khasanov, A.M., Miller, J.W., Pollak, H., and Li, Z. (1998) *Mössbauer Mineral Handbook*, Mössbauer Effect Data Center, The University of North Carolina at Asheville, USA.
3. Bancroft, G.M. (1973) *Mössbauer Spectroscopy: an Introduction for Inorganic Chemists and Geochemists*, McGraw-Hills, London.
4. Bancroft, G.M. (1979) Mössbauer Spectroscopic Studies of the Chemical State of Iron in Silicate Minerals, *J. Physics (Paris) Colloq.* **C2, 40**, 464-471.
5. Coey, J.M.D. (1975) Clay Minerals and their Transformations Studies with Nuclear Techniques: the Contribution of Mössbauer Spectroscopy, in *1st Conference on Mössbauer Spectroscopy*, Crakow, Poland.
6. Goodman, B.A., (1980) Mössbauer Spectroscopy, in: J. W. Stucki and W. L. Banwart (eds.), *Advanced Chemical Methods for Soil and Clay Minerals Research*, D. Reidel, Dordrecht, 1-92.
7. Goodman, B. A. (1982) Mössbauer Spectroscopy, in J. J. Fripiat (ed.) *Advanced Techniques for Clay Minerals Analysis*, Elsevier Scientific Publishing Company, Amsterdam-Oxford-New York, 113-137.
8. Vandenberghe, R.E. (1991) *Mössbauer Spectroscopy and Applications in Geology*, International Training Centre, Ghent, Belgium.
9. Kuzman, E., Nagy, S., Vertes, A., Weiszburg, T.G., and Garg, V.K. (1998) Geological and Mineralogical Applications of Mössbauer Spectroscopy, in: A.Vertes, S. Nagy , and K. Suvegh (eds.) *Nuclear Methods in Mineralogy and Geology: Techniques and Applications*, Plenum Press, New York.
10. Souza Jr.,P.A. (in press) Automation in Mössbauer Spectroscopy Data Analysis, *Laboratory Robotics and Automation 1*.
11. Broska, I., Uher, P., and Lipka, J. (1998) Brown and Blue Schorl from the Spis-Gemer Granite, Slovakia: Composition and Genetic Relations, *Journal of the Geological Society* **43**, 9-16.
12. Helgason, O., Sternhorson, S., Morup, S., Lipka, J., and Knudsen, J. E. (1976) Mössbauer Studies of Icelandic Lavas, *Journal de Physique Colloq.* **37**, C6-829 - 832.
13. Orlicky, O., Fytikas, M., Benka, J., Lipka, J., Mihalikova, A., and Toman, B. (1988) Magnetic and Petrological Investigation of Selected Andesitodacides and Olivine Basanites from Greece, *Geologica Carpatica* **39**, 489-504.
14. Balaz, P. and Lipka, J. (in press) Application of Mössbauer Spectroscopy to Study of Sulphidic minerals, *Journal of Mineralogy*.
15. Lipka, J., Balaz, P., Boroska, F., Turcaniova, L., Toth, I., and Grone, R. (1997) Mössbauer Study of the Slovak Brown Coal, *Czechoslovak Journal of Physics* **47**, 537-540.
16. Lipka, J., Madsen, M.B., Orlicky, O., Koch, C.J.W., and Morup, S. (1988) A Study of Titanomagnetites in Basaltic Rocks from Nigeria, *Physica Scripta* **38**, 508-512.

17. Orlicky, O., Cano, F., Lipka, J., Mihalikova, A., and Toman, B. (1993) Magnetic Minerals of Basaltic Rocks: A Study of their Nature and Composition, *Geologica Carpatica* **43**, 287-293.
18. Sitek, J., Prejsa, M., Cirak, J., Hucl, M., and Lipka, J. (1978) Mössbauer Study of the Amorphous $Pd_{70}Fe_{10}Si_{20}$ Alloy upon Transition to the Crystalline State, *Journal Noncrystal Solids* **28**, 255-259.
19. Morup, S., Topsoe, H., and Lipka, J. (1976) Modified Theory for Mössbauer Spectra of Superparamagnetic Particles: Application to Fe_3O_4, *Journal de Physique Colloq.* **37**, C6-287-290.
20. Lipka, J., Jensen, H.G., Knudsen, J.M., Madsen, M.B., Bentzon, M.B., Koch, C.B., and Morup, S. (1992) A Mössbauer Study of an Impactite from the Moturaqui Crate, *Hyperfine Interaction* **70**, 965-968.
21. Lipka, J., Madsen, M.B., Koch, C.B.W., Miglierini, M., Knudsen, J.M., Morup, S., and Bontzon, M.B. (1994) The Monturaqui Impactite and the Iron in it, *Meteoritics* **29**, 492.
22. Lipka, J., Rojkovic, I., Toth, I., and Seberini, M. (1997) First Mössbauer Study of the Slovak Meteorite Rumanova, *Czechoslovak Journal of Physics* **47**, 529-532.
23. Lipka, J., Petrik, I., Toth, I., and Gajdosova, M. (1995) Mössbauer Study of Allanite, *Acta Physica Slovaca* **45**, 61-66.
24. Petrik, I., Broska, I., Lipka, J., and Siman, P. (1995) Granitoid Allanite - (Ce): Substitution Relations, Redox Conditions and REE Distribution (on the example of I-type Granitoids Western Carpathious, Slovakia), *Geologica Carphatica* **46**, 79-94.
25. Balaz, P., Bastl, Z., Briancin, J., Ebert, I., and Lipka, J. (1992) Surface and Bulk Properties of Mechanically Acticated Zink Sulphide, *Journal of Materials Science* **27**, 653-656.
26. Balaz, P., Ebert, I., Lipka, J., and Sepelak, V. (1992) Structural Changes in Mechanically activated innabar, *Journal of Materials Science Letters* **11**, 754-456.
27. Balaz, P., Achimovicova, M., Sepelak, V., Bastl. Z., and Lipka, J. (1994) Solid State Properties and Leaching of Mechanically Activated Tetrahidrite, *Acta Metallurgica Sinica (English Edition)* **7**, 79-83.
28. Balaz, P., Hauert, R., Krack, M., and Lipka, J. (1994) Spectroscopical Study of Mechanically Activated Pyrite *International Journal of Mechanochemistry and Mechanical Alloying (IJMMA)* **1**, 107-111.
29. Tkacova, K., Stefulova, N., Lipka, J., and Sepelak, V. (1995) Contamination of Quartz by Iron in Energy-intensive Griding in Air and Liquids of Various Polarity, *Powder Technology* **83**, 163-171.
30. Balaz, P., Bastl, Z., Havlik, T., Lipka, J., and Toth, I. (1997) Characterisation of Mechanosynthetised Sulphides, *Material Science Forum* **235-238**, 217-222.
31. Dickson, D.P.E. and Berry, F.J. (1986) *Mössbauer Spectroscopy*, Cambridge University Press, Cambridge, London, New York, New Rochelle, Melbourne, Sydney.

HIGH-CHROMIUM FERRITIC STEELS

Microscopic Reasons for Brittleness and Hardness Studied by Mössbauer Effect

S. M. DUBIEL[1], J. CIEŚLAK[1] AND B. SEPIOL[2]
[1)] *Faculty of Physics and Nuclear Techniques, University of Mining and Metallurgy (AGH), Al. Mickiewicza 30, PL-30-059 Kraków, Poland*
[2)] *Institut für Materialphysik, Universität Wien, Strudlhofgasse 5, A-1090 Wien, Austria*

Abstract

Relevance of the Mössbauer Spectroscopy (MS) in the investigation of microscopic reasons for brittleness and hardness of high-chromium ferritic steels is demonstrated and discussed. It is shown that MS is a suitable tool to study (i) the formation of the sigma phase and (ii) the phase decomposition into Fe- and Cr-rich phases, the two phenomena responsible for degradation of the steel properties. The uniqueness of MS in the investigation of the latter phenomenon is underlined and its ability of making distinction between nucleation and growth, on one hand, and spinodal process, on the other, is emphasized.

1. Introduction

Thanks to their outstanding heat-, creep-, and acid-resistant features high-chromium ferritic steels have found a wide application in various branches of industry, especially those involving high temperature wear and corrosion [1, 2]. Since their introduction on the market, there has been a continuous increase of their production reaching the level of over 12 mln ton in 1991, and estimated 18 mln ton in the year 2000. The properties of these materials are a sensitive function of composition, microstructure and alloying conditions [1-6], and they may be affected by two microscopic alternative phenomena: (1) the precipitation of the σ phase, which is hard and brittle, and (2) the α/α' phase decomposition, resulting in an unwanted embrittlement (the so-called „475° C embrittlement"). Although both phenomena occur on the microscale, they result in a macroscopic degradation of materials.

As the high-chromium steels are composed to over 95% of Fe-Cr alloy, investigation of this system is believed to supply the key information related to the brittleness of the steels. Consequently, Fe-Cr alloys have been regarded as model alloys, and they were intensively investigated with respect to the phenomena (1) and (2) both experimentally [7-28] and theoretically [29-34]. Despite that, neither the phase boundaries of the α/α' and α/σ equilibrium nor the mechanisms responsible for their

M. Miglierini and D. Petridis (eds.), Mössbauer Spectroscopy in Materials Science, 107–118.

formation are precisely known. Concerning the former, there are considerable differences regarding the eutectic temperature, the α/σ phase boundaries, the width of the stable miscibility gap and its metastable extension in various versions of the phase diagram [35-38]. With regard to the latter, one does not know yet whether the σ phase nucleates directly form the α phase, or rather it grows via intermediate phases as evidenced by the recent study [24]. Also the process of the phase decomposition into Fe-rich (α) and Cr-rich (α') phases is far from being well understood and it needs further investigation. In particular, it remains unclear whether the decomposition proceeds only by the spinodal mechanism or nucleation and growth has to be included to fully describe the formation of the resulting miscibility gap.

Various experimental methods such as small angle X-ray [20, 22, 28] and neutron [18, 23, 27] scattering, atom probe field ion [25, 26] and electron [13, 25] microscopy and Mössbauer Spectroscopy (MS) [7-12, 14-15, 17, 21, 24] were applied in the investigation of the issue. It seems, however, that MS is the most appropriate. Its particular advantage stems from the fact that it is par excellence a microscopic method and it enables detection of a new phase at the level of a unit cell provided its concentration > 1%. This feature is especially valuable when studying early stages of phase formation because the data obtained at these stages is crucial for the understanding mechanisms of phase formation. It is, however, especially well-suited to study the process of the phase decomposition. The resulting phases, α and α', are hardly „seen" by other methods because of (i) similar factors for electron and X-ray scattering as well as (ii) very close values of the lattice parameter (difference < 5%). On contrary, spectral parameters of MS (hyperfine field, *H* and isomer shift, *IS*) sensitively depend on the number of Cr atoms in the vicinity of the probe Fe nuclei. This makes a quantitative analysis of the phase decomposition possible and enabled a precise determination of the miscibility gap boundaries [21]. Recent model studies gave evidence that MS can be also used to make a distinction between the two mechanisms responsible for the phase decomposition [34].

The paper describes the application of MS in the study of the two phenomena responsible for the brittleness and hardness.

2. The Sigma Phase

The σ phase, whose crystallographic structure was for the first time determined for the Fe-Cr alloy [39], is very hard (microhardness ~ 1200 HV) and brittle. It has a tertragonal unit cell with 30 atoms distributed among five sites from which two are occupied exclusively by Fe atoms, one by Cr atoms and the remaining two by both kinds of atoms [40-42]. The actual site occupation depends both on the alloy composition and thermal history of the sample. An Fe-Cr alloy can be 100% transformed into the σ phase by an isothermal annealing provided Cr content lies within the range of ~ 45 - 50 at% and the temperature of annealing within ~ 800 - 1090 K. The exact limits for the existence of the phase in the system have, however, to be determined. Over fifty examples of binary σ phases and many examples of ternary σ phases have been reported and reviewed [43].

Although the first identification of the σ phase was done by the X-ray diffraction method, it can also be easily identified and studied by MS. The distinction between σ and α, from which it precipitates, can be in MS based on the room temperature Mössbauer spectra. As the latter phase is at 295 K magnetic [35] ($T_c \approx 585$ K for an α-$Fe_{53}Cr_{47}$ alloy) , its corresponding ^{57}Fe Mössbauer spectrum exhibits a characteristic six-line pattern - see Fig. 1a. On the other hand, $T_c \approx 55$ K for a σ-$Fe_{53}Cr_{47}$ alloy, hence the sample at RT is non-magnetic, and, consequently its ^{57}Fe Mössbauer spectrum has a shape shown in Fig. 1b. There is also a pronounced difference in the isomer shift, ΔIS, between the two spectra: $\Delta IS = IS_{\sigma} - IS_{\alpha} \approx -0.11$ mm/s [12] which means the Fe-site charge-density is in the σ phase by ~ 0.06 s-like electron larger than in the α phase. The clear-cut difference in the Mössbauer spectra for the two phases enables a quantitative phase analysis of the σ phase formation.

A fraction of the σ phase precipitated, A_{σ} can be easily determined from the Mössbauer spectrum based on the following formula:

$$A_{\sigma} = S_{\sigma} / S \qquad (1)$$

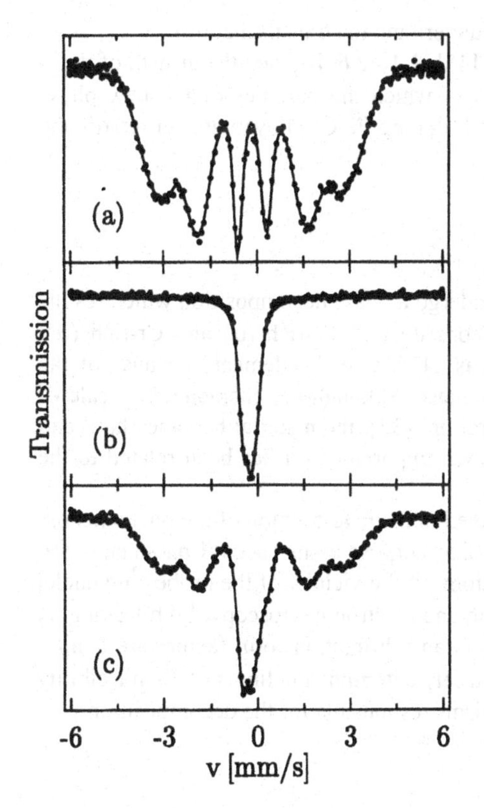

Figure 1 ^{57}Fe Mössbauer spectra recorded at 295 K on the $Fe_{53.8}Cr_{46.2}$ alloy : (a) α phase, (b) σ phase, and (c) α + σ phases.

where S_{σ} is the spectral area associated with the subspectrum originating from the σ phase, and S is the total spectral area. The formula (1) is valid assuming the Lamb-Mössbauer factor has the same value for the two phases. Based on such an approach the quantitative phase analysis can be made. An example of the ^{57}Fe Mössbauer spectrum recorded at 295 K on a two-phase sample is shown in Fig. 1c.

Influence of alloying elements on the existence and the kinetics of the σ phase formation is of particular interest. It is generally believed that addition of a third element into an Fe-Cr alloy accelerates the α-σ transformation. Examples of this kind are Mo [44] and Ti [45]. The A_{σ} -values found with formula (1) for an udoped (open circles) and Ti-doped (full circles) $Fe_{53.8}$ $Cr_{46.2}$ alloy annealed at $T = 973$ K can be seen in Fig. 2. The solid lines represent there the best-fits to the data in terms of the Johnson-Mehl-Avrami equation:

$$A_{\sigma} = 1 - \exp\left[-(kt)^{n}\right] \qquad (2)$$

where k is a time constant (related to the activation energy), and n is a form factor

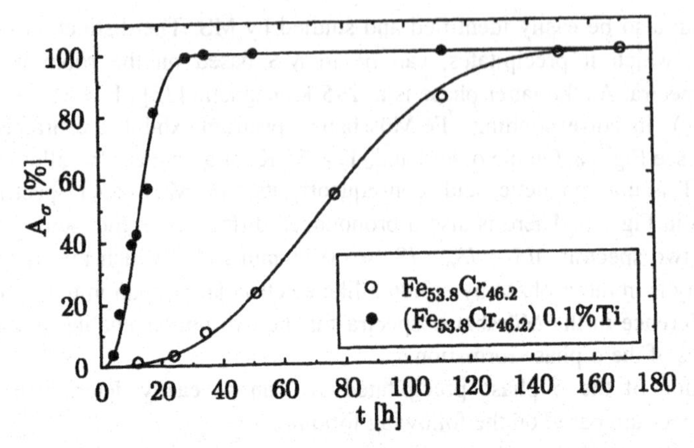

Figure 2 The fraction of the σ phase, A_σ, precipitated in the $Fe_{53.8}Cr_{46.2}$ (open circles) and in the $Fe_{53.8}Cr_{46.2}$ 0.1%Ti (full circles) samples versus ageing time, *t*, at 973 K [45].

(which is used as a tracer of the mechanisms underlying the transformation).

The values of the *n*-parameter found in [45], 2.4 - 2.6, indicate the growth of the σ phase is controlled by the diffusion process in which the particles of the new phase nucleate at a constant rate. On the other hand, doping Fe-Cr alloys with Sn retards the nucleation of the σ phase significantly [46].

3. Phase Decomposition

Fe-Cr alloys aged at T lower than ~ 820 K undergo a phase decomposition which results in a formation of interconnected structure consisting of Fe-rich (α) and Cr-rich (α') phases. This phenomenon has been known as „475° C embrittlement" because of the temperature at which the embrittlement is highest. Although the characteristic scale of the regions involved is of the order of several nm [32], the material becomes hard and brittle. Knowledge of the microstructure is then important as it has been related to the macroscopic mechanical properties.

Among various experimental methods used in the investigation of the phenomenon, MS seems to be particularly well-suited for that purpose as its spectral parameters (*H, IS*) are very sensitive to the presence of Cr atoms in the vicinity of the probe ^{57}Fe nuclei [12]. This contrasts with the X-ray diffraction and electron microscopy, both lacking an adequate resolution because lattice parameters and diffraction form factors are similar for α and α'. Using MS one can not only precisely determine the limits of the miscibility gap [17, 21], but also investigate the mechanisms responsible for the decomposition.

3.1. LIMITS OF THE MISCIBILITY GAP

Distinction between α and α' can be based on the room temperature ^{57}Fe Mössbauer spectra. Following the Fe-Cr phase diagram [35], all samples containing less than ~

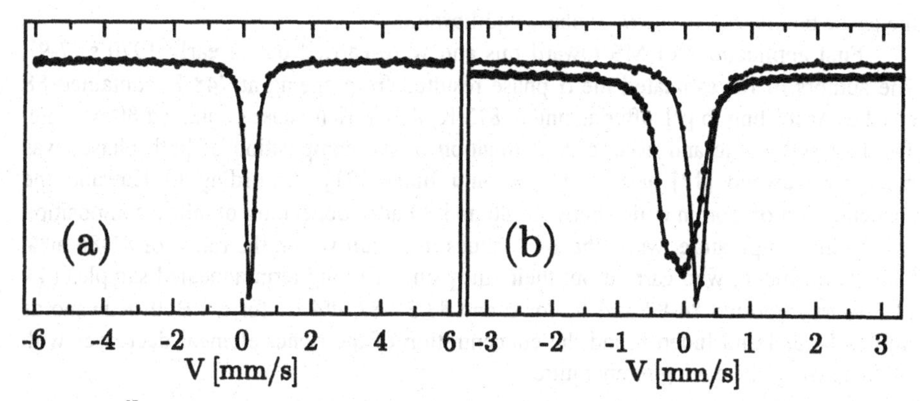

Figure 3. ^{57}Fe Mössbauer spectra recorded at 295 K on (a) $Fe_{10}Cr_{90}$ alloy (α' phase) and (b) $Fe_{53.8}Cr_{46.2}$ (σ phase) + $Fe_{10}Cr_{90}$ (α' phase).

67at% Cr are at room temperature magnetic, while those with more than ~ 67 at% Cr are paramagnetic. Consequently, their corresponding ^{57}Fe Mössbauer spectra look like those presented in Fig.1a and 3a, respectively. It is clear that in that way one can not only identify the two phases and determine their relative abundance using formula (1), but, in addition, it is possible to precisely determine their composition, hence the limits of the miscibility gap. For this purpose one takes advantage of the monotone relationship between: (a) the average hf field, $<H>$, and the chromium content, x_{Cr} - see Fig. 4a, and (b) the isomer shift, *IS*, and the iron content, x_{Fe} - see Fig.4b. From (a) the composition of α and from (b) that of α' can be determined with an accuracy of ~ 1% [21].

It is worth mentioning here that there is a clear - cut difference between the ^{57}Fe Mössbauer spectra for α' and σ phases recorded at 295 K - see Fig.3b. Consequently, one can easily identify the phases with MS and use the method for the determination of

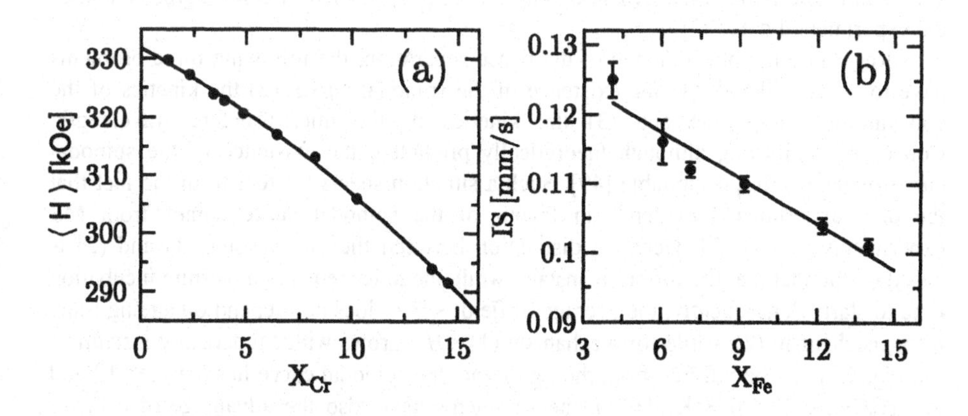

Figure 4. The average hf field, $<H>$, for Fe-Cr alloys versus Cr content, x_{Cr} (a), and the isomer shift, *IS*, versus Fe content, x_{Fe} (b) [21].

the phase boundaries of the α‘/σ phase equilibrium.

First applications of MS toward this end were carried out in early 1970’s [7-9]. The authors of [8] estimated the α phase resulted from ageing at 748 K contained 88 at% Fe. According to [9], after ageing at 813 K, the Fe-rich phase contained 80 wt% Fe. The first systematic and precise determination of the composition of both phases was made by Kuwano [17] and by Dubiel and Inden [21]. According to Kuwano the concentration of iron in α lies between 86 and 90 at%, depending on alloy composition and ageing temperature, while the x_{Cr} - values in α‘ fall within the range of 81-86 at%. Dubiel and Inden, who carried out their study on very long term annealed samples (4 - 11.5 years) reported 84-87 at% Fe for α and 83.5 - 88 at% Cr for α‘. Both Kuwano as well as Dubiel and Inden found the concentration of the richer element decreases with the increase of the ageing temperature.

3.2. MECHANISMS OF DECOMPOSITION

Fe-Cr system has a miscibility gap in the centre of its phase diagram. Alloys with compositions within the gap are not stable and undergo a phase separation at elevated temperatures [47]. According to theoretical considerations, two mechanisms may be responsible for the process: (1) spinodal and (2) nucleation and growth. The sign of d^2G/dx^2, where G is the Gibbs free energy, and x is the composition, defines the regions where the two mechanisms are active. If $d^2G/dx^2 > 0$, the decomposition goes via spinodal, if $d^2G/dx^2 < 0$, via nucleation and growth. The locus of inflection points in the (T,x) plane, $d^2G/dx^2 = 0$, defines the spinodal curve.

Classical theory of the spinodal process was formulated by Cahn [47]. In his model the composition fluctuations are treated as a series of Fourier components which grow exponentially as the process of decomposition proceeds. The wavelength of the resulting microstructure should in this case remain constant. This, however, disagrees with observations, according to which both the size of the microstructure as well as the amplitude of composition grow in accord with the power low [33]. An alternative approach to the issue is the dynamic Ising model [48]. It gives a better agreement with the experimental data [33].

Regarding the phenomenon of the phase separation, the following three problems have to be considered: (1) the existence of the spinodal curve, (2) the kinetics of the concentration amplitude and (3) the kinetics of the microstructure wavelength. Concerning point (1), although theoretically predicted, the existence of the spinodal curve remains still questionable [49]. Such a situation seems to stem from the fact that the only experimental evidence in favour of the spinodal curve comes from MS experiments [8,9,11,17]. Here the distinction between the mechanisms (1) and (2) is based on the fact that the former is instant, while the latter requires a definite incubation time to start. Consequently, the average hf field, $<H>$, does not depend on ageing time, t, for mechanism (2), while for mechanism (1) $<H>$ grows with t practically instantly - see Fig. 5. It can be inferred from the figure that the spinodal curve lies between 12 and 20 at%Cr for T = 688 K. *In situ* measurements have also the advantage of a better resolution than those at room temperature. This is clear from Fig.6a showing $<H>$ versus chromium concentration as measured at 668 K (dashed line) and at 295 K. The

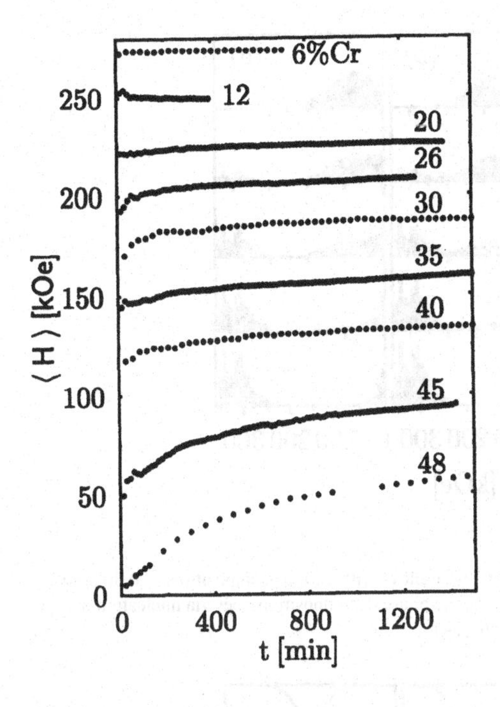

Figure 5. The average hf field, $<H>$, versus ageing time, t, for Fe-Cr alloys with various content of Cr shown obtained from ^{57}Fe spectra measured *in situ* at 668 K.

former has a larger curvature, especially in the range of the equiatomic composition, which makes the *in situ* measurements more sensitive and useful. Examples of the corresponding ^{57}Fe Mössbauer spectra can be seen in Fig. 6b and 6c. They give a clear evidence in favour of the *in situ* measurements.

To verify whether these observations and arguments are sound, Cieślak and Dubiel have recently carried out model calculations [34]. Various quantities pertinent to the ^{57}Fe MS such as: hf field distribution histograms, Mössbauer spectra and the average hf fields were calculated assuming the phase separation occurs either via mechanisms (1) or (2). The results obtained give a clear evidence that MS is potentially able to make a unique distinction between the two modes of phase decomposition, hence to determine the spinodal curve. A set

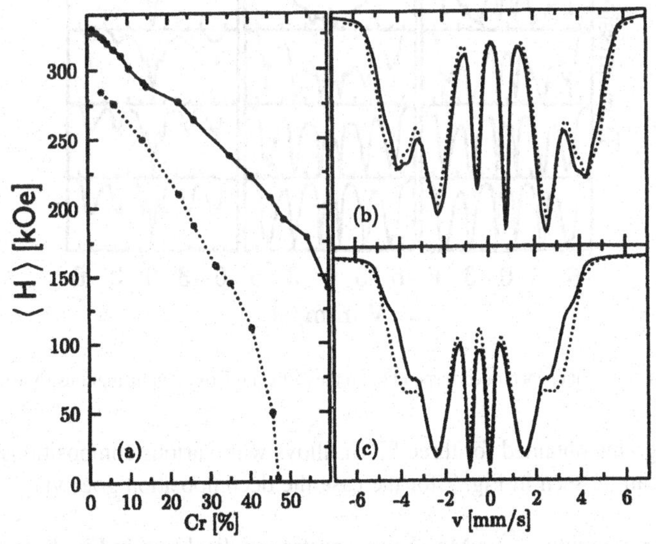

Figure 6. (a) Average hf field, $<H>$, versus chromium content for Fe-Cr alloys as derived from ^{57}Fe spectra measured at 295 K (solid line) and *in situ* at 688 K (dashed line), and ^{57}Fe spectra recorded (b) at 295 K on untreated sample (solid line) and on the one aged for 30 h at 668 K (dotted line), and (c) *in situ* at 688 K after ageing for 15 min (solid line), and for 30 h (dotted line).

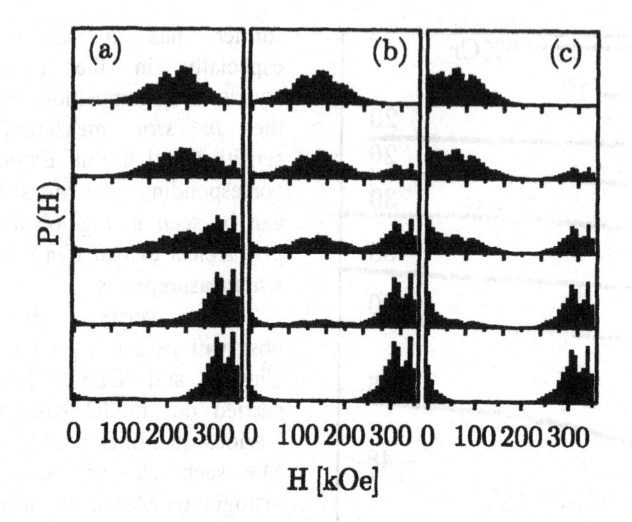

Figure 7. Hyperfine field distribution histograms for Fe-Cr alloys with various compositions: (a) 30 at%Cr, (b) 50at%Cr and (c) 70 at%Cr calculated for the case the phase decomposition goes via nucleation and growth [34].

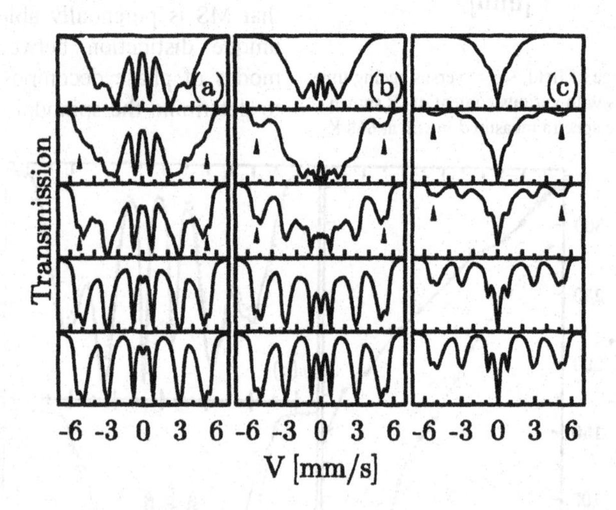

Figure 8. ^{57}Fe Mössbauer spectra corresponding to the hf field distribution histograms shown in Fig. 7 [34].

of the histograms obtained for three Fe-Cr alloys with various compositions (30, 50 and 70 at%Cr) can be seen in Fig. 7 for the case the decomposition goes via nucleation and growth.

The corresponding ^{57}Fe Mössbauer spectra are displayed in Fig. 8. It can be easily seen in the two figures that the characteristic feature for this mode of phase decomposition is that certain peaks and lines (marked by arrows) do not change their position, but they do change their intensity as the process of decomposition proceeds. On the contrary, for the spinodal process, the peaks and the lines (indicated by arrows) change continuously both their position and intensity - see Figs. 9 and 10. This behaviour makes the distinction between the two mechanisms possible.

Another spectral quantity that can be taken as a good measure for the distinction

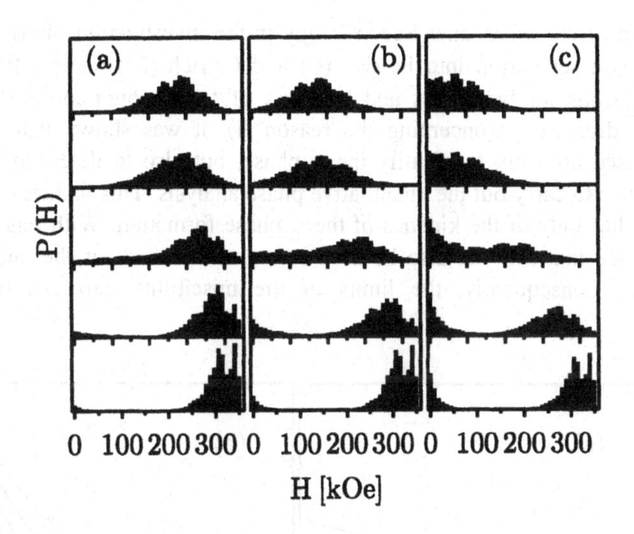

Figure 9. Hyperfine field distribution histograms for Fe-Cr alloys with various compositions: (a) 30 at%Cr, (b) 50 at%Cr and (c) 70 at%Cr expected in the case the phase separation goes via spinodal [34].

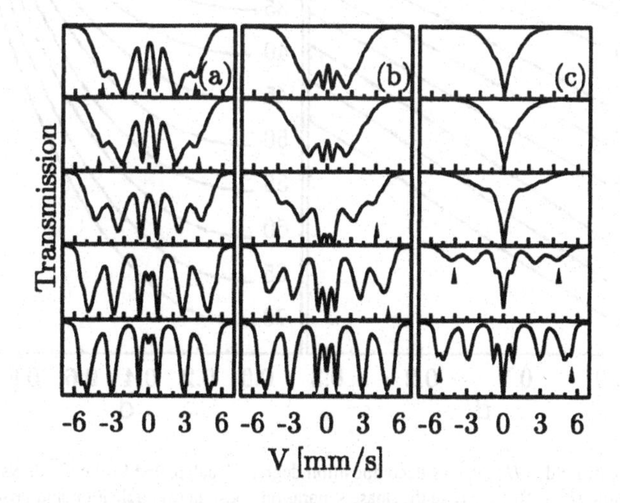

Figure 10. ^{57}Fe Mössbauer spectra corresponding to the hf field distribution histograms shown in Fig.9 [34].

between the mechanisms (1) and (2) is the average hf field, <*H*>. As shown in Fig. 11a, <*H*> is a linear function of the decomposition degree, *d*, for nucleation and growth, and nonlinear for the spinodal - see Fig.11b. The largest difference in the <*H*> - values occurs at the middle stages of the phase separation which means that at these stages the distinction between the mechanism (1) and (2) is easiest.

4. Summary

The application of the Mössbauer Spectroscopy in the investigation of (i) the σ phase and (ii) the phase separation into Fe-rich (α) and Cr-rich (α') phases, the main two microscopic reasons for brittleness and hardness of ferritic high-chromium steels, is presented and discussed. Concerning the reason (i), it was shown that MS can be conveniently used not only to identify the σ phase, but also to determine its relative abundance, hence to carry out the quantitative phase analysis. This feature of MS makes it suitable for the study of the kinetics of the σ phase formation. With regard to (ii), it was demonstrated that MS can also be used as a good method for the quantitative α' phase analysis. Consequently, the limits of the miscibility gap can be precisely

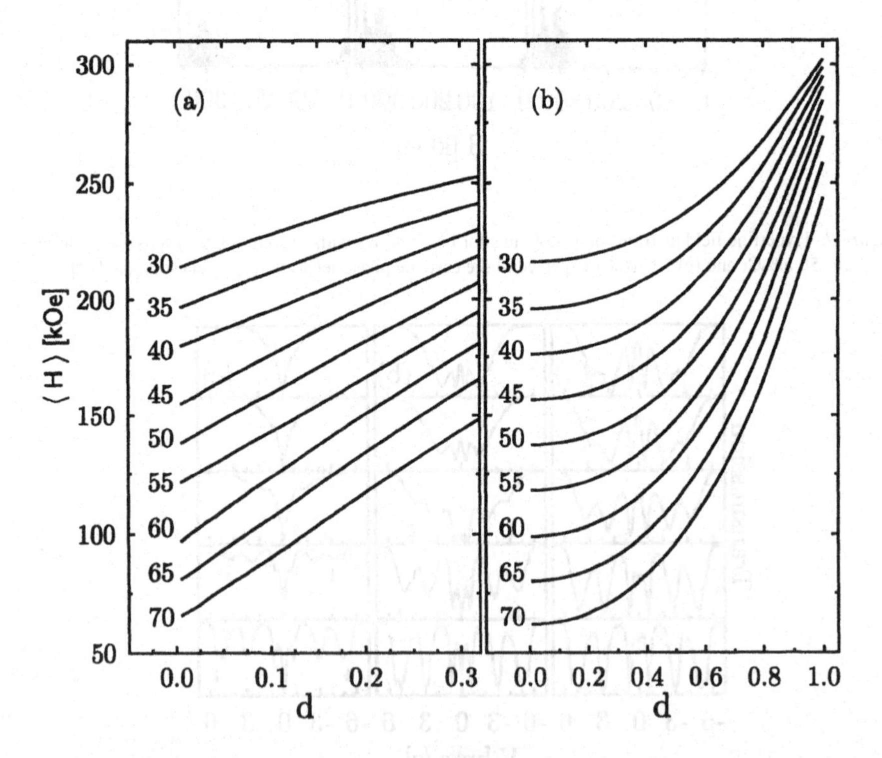

Figure 11. Average hf field, <*H*>, versus decomposition degree, *d*, calculated for Fe-Cr alloys with various content of chromium shown for the case the phase separation goes via (a) nucleation and growth and (b) spinodal. The figures labelling the lines stand for the content of chromium [34].

determined with MS. Furthermore, the relevance of MS in the study of the mechanisms involved in the phase separation is demonstrated and discussed. In particular, it was shown that MS is a suitable method for making the distinction between nucleation and growth, on one hand, and spinodal process, on the other.

References

1. Oppenheim, R. (1982) Güteeigenschaften des Superferrit X1CrNiMoNb 2842 (Remanit 4575) als Sonderstahl Chemie-Apparatbau, *Thyssen Edelst. Techn. Ber.* **8,** 97-110.
2. Heubner, U. (1987) Nickel alloys and high-alloy special stainless steels - Materials summary and metallurgical principles, in U. Heubner (ed.), *Nickel alloys and high-alloy special stainless steels,* Expert-Verlag, Sindelfingen, pp.15-47.
3. Wolff, I.M. and Ball, A. (1991) Ductility of high-chromium super-ferritic alloys - I. Microstructural influences on the ductile-brittle transition, *Acta Metall. Mater.* **39,** 2759-2770.
4. Wolff, I.M. and Ball, A. (1991) Ductility of high-chromium super-ferritic alloys - II. Plastic deformation and crack-zone shielding, *Acta Metall. Mater.* **39,** 2771-2781.
5. Wollf, I.M. and Ball, A. (1991) Substitutional alloying and deformation modes in high-chromium ferritic alloys, *Metall. Trans.* **23A,** 627-638.
6. Sauthoff, G. and Speller, W. (1982) Deformation behaviour of a Fe-Cr-Si alloy with various distributions of embrittling sigma phase precipitate, *Z. Metallkde.* **73,** 35-42.
7. Ettwig, H.-H. and Pepperhoff, W. (1970) Untersuchungen über das metastabile System Eisen-Chrom mit Hilfe des Mössbauer -Effektes, *Arch. Eisenhüttenwes.* **41,** 471-474.
8. Chandra, D. and Schwartz, L.H. (1971) Mössbauer effect studies of spinodal decomposition in Fe-Cr, in I. J. Gruvermann (ed.) *Mössbauer Effect Methodology*, vol. 6, Pergamon Press, New York, pp. 79-92.
9. De Nys, T. and Gielen, P. M. (1971) Spinodal decomposition in Fe-Cr system, *Metall. Trans.* **2,** 1423-1428.
10. Sumimoto, Y., Moriya, T., Ino, H., and Fujita, F. (1973) The Mössbauer effect of Fe-V and Fe-Cr phase, *J. Phys. Soc. Jpn.* **35,** 461-468.
11. Solomon, H.D. and Levinson, L.M. (1978) Mössbauer effect study of '475° C' embrittlement of duplex and ferritic stainless steels, *Acta Metall.* **26,** 429-442.
12. Dubiel, S.M. and Żukrowski, J. (1981) Mössbauer effect study of charge and spin transfer in Fe-Cr, *J. Magn. Magn. Mater.* **23,** 214-228.
13. Ishimasa, T. (1981) Electron microscopic studies on the Fe-Cr σ-phase, *J. Sci. Hiroshima Univ. A* **45,** 29-57.
14. Frattini, R., Longworth, G., Matteazi, P., Principi, G., and Tiziani, A. (1981) Mössbauer studies of surface sigma phase formation in an Fe-45Cr alloy heated in low partial oxygen pressure, *Scripta Metall.* **15,** 873-877.
15. Japa, E.A., Starzynski, J., and Dubiel, S.M. (1982) Mössbauer effect study of a sigma phase fromatiom in Fe-45.5at%Cr alloy, *J. Phys. F: Metal Phys.* **12,** L159-161.
16. Brenner, S.S., Miller, M.K., and Soffa, W.A. (1982) Spinodal decomposition in iron-32 at% chromium at 470° C, *Scripta Met.* **16,** 831-836.
17. Kuwano, H. (1985) Mössbauer effect study of the mechanism of the phase decomposition in Fe-Cr alloys, *Trans. Jpn. Inst. Met.* **26,** 482-491.
18. Furusaka, M., Ishikawa, Y, Yamaguchi, S., and Fujino, Y. (1986) Phase separation process in FeCr alloys sudied by neutron small angle scattering, *J. Phys. Soc. Jpn.* **55,** 2253-2269.
19. LaSalle, J.C. and Schwartz, L.H. (1986) Further studies of spinodal decomposition in Fe-Cr, *Acta Metall.* **34,** 989-1000.
20. Villar, R. and Cizeron, G. (1987) Evolusions structurales developees au sein de la phase sigma Fe-Cr, *Acta Metall.* **35,** 1229-1236.
21. Dubiel, S.M. and Inden, G. (1987) On the miscibility gap in the Fe-Cr system: a Mössbauer study on long term annealed alloys, *Z. Metallkde* **78,** 544-549.
22. Simon, J.P. and Lyon, O. (1989) Phase separation in a Fe-Cr-Co alloy studied by anomalous small angle x-ray scattering, *Acta Metall.* **37,** 1727-1733.

23. Bley, F. (1992) Neutron small-angle scattering study of unmixing in Fe-Cr alloys, *Acta Metall. Mater.* **40,** 1505-1517.
24. Dubiel, S.M. and Costa, B.F.O. (1993) Intermediate phases of the α-σ phase transition in the Fe-Cr system, *Phys. Rev. B* **47,** 12227-12231.
25. Miller, M.K., Anderson, I.M., Bentley, J., and Russell, K.F. (1996) Phase separation in the Fe-Cr-Ni system, *Appl. Surf. Sci.* **94/95,** 391-397.
26. Miller, M.K. and Russell, K.F. (1996) Comparison of the rate of decomposition in Fe-45%Cr, Fe-45%Cr-5%Ni and duplex stainless steels, *Appl. Surf. Sci.* **94/95,** 398-402.
27. Miller, M.K., Stoller, R.E. and Russell, K.F. (1996) Effect of neutron-irradiation on the spinodal decomposition of Fe-32% Cr model alloy, *J. Nucl. Mater.* **230,** 219-225.
28. Shek, C.H., Shao, Y.Z., Wong, K.W., and Lai, J.K.L. (1997) Spatial fractal characteristic of spinodal decomposition in Fe-Cr-Ni duplex stainless steels, *Scripta Met.* **37,** 529-533.
29. Brauwers, M. (1977) Occurance of the sigma phase compound from cluster model, *J. Phys. F: Metal. Phys.* **7,** 921-927.
30. Williams, R.O. (1981) Interface formation during spinodal decomposition, *Acta Metall.* **29,** 95-100.
31. Cerezo, A, Hyde, J.M., Miller, M.K., Petts, S.C., Setna, R.P., and Smith, G.D.W. (1992) Atomistic modelling of diffusional phase transformations, *Phil. Trans. R. Soc. Lond. A* **341,** 313-326.
32. Miller, M.K, Hyde, J.M., Hetherington, M.G., Cerezo, A., Smith, G.D.W., and Elliott, C.M. (1995) Spinodal decomposition in Fe-Cr alloys: Experimental study at the atomic level and comparison with computer models, *Acta Metall. Mater.* **43,** 3385-3401; ibid **43,** 3403-3414; ibid **43,** 3415-3426.
33. Hyde, J.M., Sutton, A.P., Harris, J.R.G., Cerezo, A., and Gardiner, A. (1996) Modelling spinodal decomposition at the atomic scale: beyond the Cahn-Hilliard model, *Modelling Simul. Mater. Sci. Eng.* **4,** 33 -54.
34. Cieślak, J. and Dubiel, S.M. (1998) Nucleation and growth versus spinodal decomposition in Fe-Cr alloys: Mössbauer effect modelling, *J. Alloys Comp.,* **269,** 208-218.
35. Hansen, M. (1958) *Constitution of binary alloys,* McGraw Hill, New York, p.527.
36. Kubaschewski, O. (1982) *Iron binary phase diagrams,* Springer Verlag, Berlin, p. 31-34.
37. Anderson, J.O. and Sundman, B. (1987) Theromdynamic properties of the Fe-Cr system, *Calphad* **11,** 83-92.
38. Massalski, T.B., Okamoto, H, Subramanian, P.R., and Kacprzak, L. (1990) *Binary alloy phase diagrams,* ASM International, Materials Park, OH.
39. Bergman, G. and Shoemaker, D. (1954) The determination of the crystal structure of the σ phase in the iron-chromium and iron-molybdenum systems, *Acta Crystal.* **7,** 857-865.
40. Yakel, H.L. (1983) Atom distributions in sigma phases, *Acta Crystal.*B **39,** 20-28.
41. Gupta, A., Principi, G., and Paolucci, G. M. (1990) Iron sites in the FeCr ordered sigma phase, *Hyp. Inter.* **54,** 805-810.
42. Sluiter, M.H.F., Estarjani, K., and Kawazoe, Y. (1995) Site occupation reversal in the Fe-Cr σ phase, *Phys. Rev. Lett.* **75,** 3142-3145.
43. Hall, E.O. and Algie, S.H. (1966) The sigma phase *Metall. Rev.* **11,** 61-88.
44. Waanders, F.B., Vorster, S.W. and Pollak, H. (in press) The influence of temperature on the formation of sigma phase in Fe-Cr alloys containing Mo, *Hyp. Inter.*
45. Błachowski, A., Cieślak, J., Dubiel, S.M., and Sepiol, B. (in press) Kinetics of the σ phase formation in an $Fe_{53}Cr_{47}$ 0.1at%Ti alloy, *Phil. Mag. Lett.*
46. Costa, B.F.O. and Dubiel, S.M. (1993) On the influence of tin on the formation of the σ phase in the Fe-Cr system, *Phys. Stat. Sol. (a)* **139,** 83-94.
47. Cahn, J. W. (1968) Spinodal decomposition *Trans. Met. Soc. AIME* **242,** 166-180.
48. Penrose, O. (1991) A mean-field theory equation of motion for the dynamic Ising model, *J. Stat. Phys.* **63,** 975- 986.
49. Ustinovshikov, Y., Shirobokova, M., and Pushkarev, B. (1996) A structural study of the Fe-Cr system, *Acta Mater.* **44,** 5021-5032.

MICROSTRUCTURAL ANALYSIS OF NUCLEAR REACTOR PRESSURE VESSEL STEELS

Positron Annihilation and Mössbauer Spectroscopy Study

V. SLUGEŇ
Department of Nuclear Physics and Technology FEI STU Bratislava, Ilkovičova 3, 81219 Bratislava, Slovak Republic

1. Introduction

One of the most fundamental tasks of nuclear-reactor safety research is assessing the integrity of the reactor's pressure vessel (RPV). Properties of RPV steels and influence of thermal and neutron treatment on these properties are routinely investigated by macroscopic methods such as Charpy V-notch and tensile tests. It turns out that the embrittlement of steel is a very complicated process depending on many factors (thermal and radiation treatment, chemical compositions, preparing conditions, ageing, etc.). A number of semi-empirical laws, based on macroscopic data, have been established but unfortunately, these laws are never completely consistent with all data and do not provide the desired accuracy. Therefore, many additional test methods [1-10] have been developed to unreveal the complex microscopic mechanisms responsible for the RPV-steel embrittlement.

The present study is based mainly on experimental data of positron-annihilation spectroscopy (PAS) and Mössbauer spectroscopy (MS) obtained from measurements on eight different types of RPV steels used in Eastern as well as in Western nuclear-power plants (NPP), and from transmission electron microscopy (TEM) on the Russian RPV-steels 15Kh2MFA and 15Kh2NMFA.

MS is a powerful analytical technique because of its specificity for one single element and because of its extremely high sensitivity to changes in the atomic configuration in the near vicinity of the probe isotopes (in this case ^{57}Fe). MS measures hyperfine interactions and these provide valuable and often unique information about the magnetic and electronic state of the iron species, their chemical bonding to co-ordinating ligands, the local crystal symmetry at the iron sites, structural defects, lattice-dynamical properties, elastic stresses, etc. [11,12]. In general, a Mössbauer spectrum shows different components if the probe atoms are located at lattice positions, which are chemically unequivalent. From the parameters that characterise a particular Mössbauer sub-spectrum it can, for instance, be established whether the corresponding probe atoms reside in sites which are not affected by structural lattice defects, or whether they are located at defect-correlated positions. In this respect, however, it is almost imperative to combine Mössbauer measurements with other research methods, which preferably are

M. Miglierini and D. Petridis (eds.), Mössbauer Spectroscopy in Materials Science, 119–130.

sensitive to the nature of the defect properties. Combining the results of MS with those of TEM and PAS (and possibly of various other techniques [13]) on the same samples seems to be a promising approach, and hence efforts in that respect are recommended since they may be of significant technological importance for future developments of nuclear-reactor elements which are likely to be exposed to severe radiation damage.

2. Lifetime PAS study of RPV-steels

In total 8 non-irradiated specimens from different commonly used RPV-steels were studied using PAS lifetime technique since 1996. JRQ and EGF are Japanese and German ferric steels, often used as the reference base metal in the construction of Western nuclear reactors. The sample 73W is an American Linde0124 weld metal. Beside these, two surveillance specimen base materials from Belgian nuclear power plants (NPP) Doel (D4) and Chooz (CH) were studied. With the aim to compare Western types of RPV steels to eastern types, 15Kh2MFA (specimens YA and YTA) and 15Kh2NMFA (XTA) RPV-steels were measured as well. Specimens XTA and YTA were prepared in laboratory conditions as a proper simulation coarse grain area of the heat-affected zone (HAZ) of welded joint (T_{max} = 1300 °C, $\Delta t_{8/5}$= 30 s) [14].

The chemical composition of the RPV-steel specimens is given in Table 1.

TABLE 1. The chemical composition of the studied RPV-steel specimens [15].

CODE	TYPE OF STEEL	CONTENS OF ALLOYING ELEMENTS IN RPV SPECIMENS (WT.%)											
		C	Si	Mn	Mo	Ni	Cr	Cu	P	S	V	Co	Total
EGF	22NiMoCr37	0.22	0.23	0.88	0.51	0.84	0.39	0.080	0.006	0.004			3.160
D4	A508B Cl.3	0.20	0.28	1.43	0.53	0.75	-	0.055	0.008	0.008			3.261
CH	18MND5	0.18	0.26	1.55	0.50	0.65	0.18	0.140	0.007	0.002			3.469
JRQ	A533B Cl.	0.18	0.24	1.42	0.51	0.84	0.14	0.140	0.017	0.004			3.491
73W (weld)	Linde 0124	0.10	0.45	1.56	0.58	0.60	0.25	0.310	0.005	0.005			3.860
YA	15Kh2MFA	0.14	0.31	0.37	0.58	0.2	2.64	0.090	0.014	0.017	0.27	0.019	4.650
YTA (HAZ)	15Kh2MFA	0.14	0.24	0.40	0.72	0.14	2.93	0.110	0.013	0.017	0.31	0.011	5.031
XTA (HAZ)	15Kh2NMFA	0.18	0.24	0.52	0.62	1.26	2.22	0.080	0.010	0.013	0.08	0.008	5.231

Several types of analyses and fitting approaches have been performed on the measured data. The optimal decomposition of lifetimes and the proper physical model for interpretation was searched using a reasonable combination of the program POSITRONFIT and MELT. From the lifetime distribution (MELT) analyses, it was clear that the spectra are so complicated, that beside the source (kapton) correction only 1 or 2 components as a maximum, could be seriously considered.

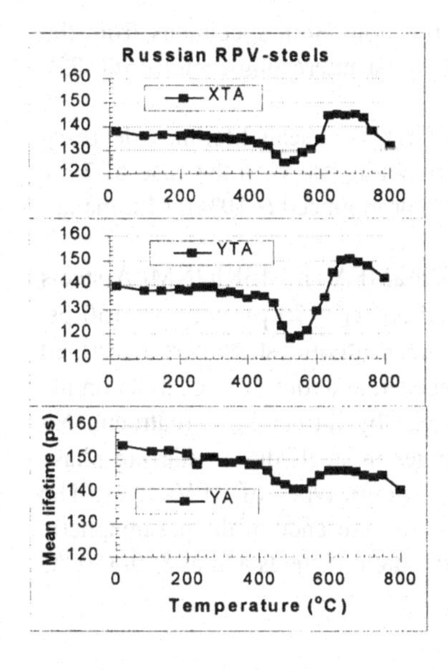

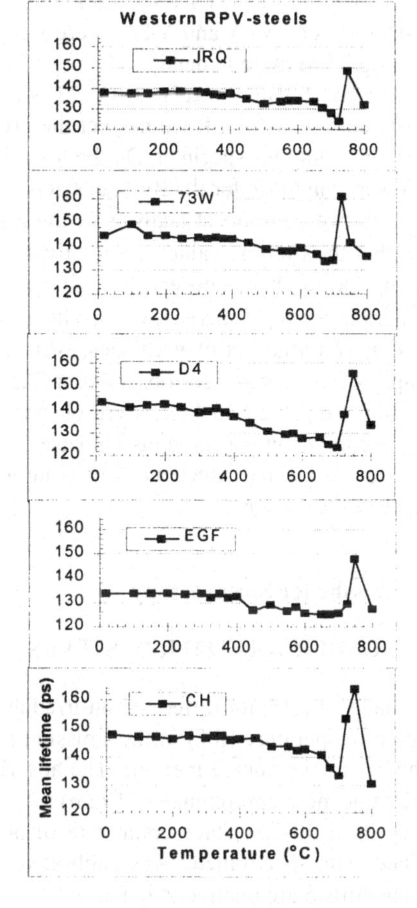

Figure 1. Evolution of the mean positron lifetime for Russian- and Western-types of RPV-steels after suitable source correction.

Therefore, we performed a source correction by fixing not only the lifetime of the kapton (τ_K = 382 ps) but also its intensity (I_K = 13.5%). This approach was taken according to our extensive MELT analysis, supported also by other works [16,17] describing the proper kapton correction by low-alloyed steels measurements. We neglected the long-lifetime component at the average value of about τ_{LT}=1500 ps and I_{LT}=0.3 %. Its presence can be explained as a contribution from a thin surface oxide layer [8] and was almost constant.

Due to very complicated defect structure in the studied steels, first we analysed all the spectra with only one mean lifetime. Comparisons of the results for all studied RPV-steel specimens are given in Figure 1.

At the first glance there are significant differences between the Western and the Russian types of steels, but also between individual specimens within these two groups. The mean lifetime in the Russian steels is more sensitive to temperature changes (up to 15 %, see Figure 1). After the minima the mean lifetime increases, in case of XTA and YTA specimens to significantly higher values than are the initial ones. It is caused due to different heat treatment parameters (temperature level and the dwell time) because samples YA and YTA have the same chemical composition. Both specimens from the simulated HAZ (XTA and YTA) were studied in the most interesting region (500-720 °C) using TEM more in detail.

For the Western steels (JRQ, 73W, EGF and CH), the mean lifetimes decrease slower (up to ~ 5-6 %, if we neglect the probable artefact at 100°C in the case of 73W specimen). Only for specimen D4 the lowering is more pronounced (~10%). Minima are shallower than those for the Russian steels.

The two component analysis of the Russian 15Kh2MFA and 15Kh2NMFA steels show that the lifetime values are relatively stable at about τ_1 = 70 ps and τ_2 = 150 ps. The significant change observed at YTA steel in the mean lifetime study was confirmed also here in the rapid increase of τ_2. The maximum reached at about 525 °C, hints on the new cluster formation of vacancies, which are removed by further heat treatment. No "jumps" of τ_2 values were observed at 725 °C. Changes in steels due to starting phase transition from b.c.c to f.c.c. were observed on all Western-types of RPV-steels. The same phenomenon appeared also in case of MS, when the presence of the paramagnetic austenite was clearly detected at 750°C in the Western batch of steels and at 800-850 °C in the Russians steels.

3. Mössbauer Spectroscopy

3.1. NON-IRRADIATED RPV-STEELS

Mössbauer spectra for different non-irradiated RPV-steel specimens have been recorded at room temperature (RT) in transmission geometry using a constant-acceleration drive with a triangular source motion. The absorbers were prepared by polishing small disks (initial thickness approximately 1 mm) to a thickness of 50 μm or less. The spectra were run until an off-resonance count rate of at least 1.10^6 for the unfolded spectrum was reached. The spectrometer was calibrated with respect to natural iron, *viz.*, all quoted isomer shifts δ are relative to α-Fe.

The obtained spectra are typical for steels with a low alloy-element concentration. With the purpose of presenting only the most significant differences between the various steel samples, the results for the following selection of RPV-steels will be discussed in more detail: a typical Eastern RPV steel (AX1) from the VVER-440 type NPP Bohunice (Slovakia), a typical Western-type RPV reference steel (JRQ) and a Linde124 weld metal (73W). The respective compositions are included in Table 1.

3.1.1. *Results*

The MS spectra of the three RPV-steel samples are shown in Figure 2a-c. They were fitted by the help of model-independent hyperfine-field (H_{hf}) distributions. The MS spectra of the JRQ and 73W weld material can be described by a hyperfine-field distribution (HFD), with H-values ranging from 25 to 35 T, and a quadrupole-splitting distribution (QSD) in the range 0.0 to 1.0 mm/s. For the magnetic component one single adjustable value for the isomer shift δ and one for the quadrupole shift $2\varepsilon_Q$ were used. They were both found to be zero within the experimental error limits. The line width of the elementary subspectra for the sextet distribution was iterated to be 0.28 mm/s for

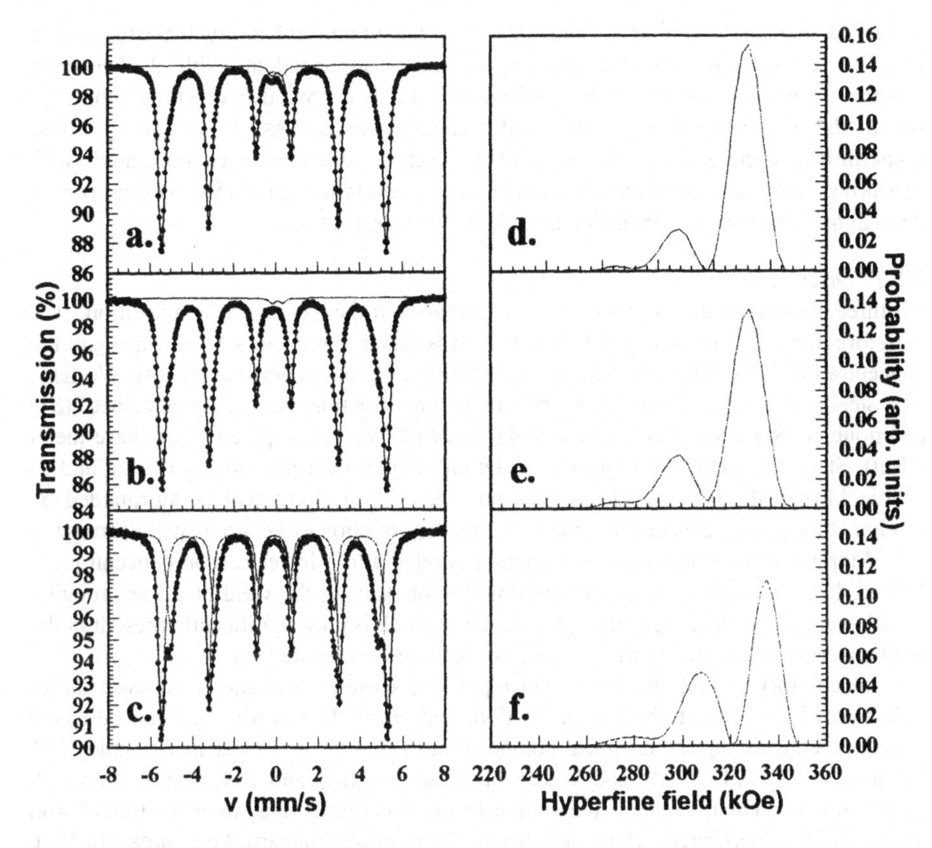

Figure 2. Mössbauer spectra and the derived H_{hf}-distribution profiles for the JRQ(a,d), 73W (b,e) and AX1 (c,f) specimens.

JRQ and 0.29 mm/s for 73W. The small area fraction of the doublet component, present in both JRQ and 73W, did not permit a precise determination of its line width and was therefore kept fixed at 0.3 mm/s.

In contrast to the MS of the Western-type steels, the spectrum of AX1 was fitted more adequately by two independent H_{hf}-distributions, the first one ranging between 32 and 35 T, and the second one between 23 and 32 T. Moreover, a slight asymmetry in the absorption lines of the second component could be noticed and this was accounted for by including a linear H_{hf}-δ correlation in the fitting procedure. The quadrupole shifts obtained for both components were again close to 0.0 mm/s. The line widths for the elementary sextet spectra were 0.27 mm/s and 0.30 mm/s, respectively. The doublet mentioned above is completely absent in the Eastern base material.

The resulting H_{hf}-distribution profiles (DP) (see Figure 2d-f) show one dominant maximum, and a number of secondary peaks. To determine the maximum-probability hyperfine fields corresponding to these peaks, the DP for the three steels have been described by a superposition of gaussian-shaped profiles. For both JRQ and 73W three gaussians were found to be adequate, while the DP for the Eastern steel had to be described with at least four components. The maximum-probability hyperfine fields $H_{hf,i}$, i = 1 to 4, and the corresponding δ_i values are listed in Table 2. The most prominent gaussian, characterised by a field of ~33.2 T, is obviously due to Fe atoms for which the nearest-neighbour shell only contains Fe atoms. It is the so-called 'unperturbed component'. The additional weaker gaussians are associated with 'perturbed' components which are associated to iron atoms surrounded by one, two or three alloying elements in their first neighbour shell, respectively.

3.1.2. *Discussion*

All three selected materials show small differences in their chemical composition. It is therefore tempting to relate the different Mössbauer parameters to changes in the concentration of the alloying elements. Noticeable differences between the two Western-type steel samples (see Table 2) are the smaller area fractions for the 'unperturbed' (S_1) and doublet (S_D) components of the weld metal (73W) as compared to the base metal (JRQ). Since the weld metal has a considerably higher concentration of Cr, Cu and Si (Table 1) than the base metal, a larger number of iron atoms will be surrounded by foreign atoms, and, consequently, the 'unperturbed' fraction of Fe-atoms will decrease.

Moreover, the smaller doublet fraction is related to a lower carbon concentration. Indeed, both the carbon level and the doublet intensity in the weld metal are roughly 50% lower than in the base metal. This doublet, which is only significantly present in the JRQ material, is ascribed to Mn and/or Cr-substituted cementite [19].

In contrast to both Western steel types, the eastern steel shows an even larger reduction of the 'unperturbed' area fraction, and no doublet at all. This can be easily explained considering the chemical composition (Table 1). It is clear that sample AX1 has higher levels of Cr, Mo and V than both samples JRQ and 73W, while almost all other alloying elements remain at a similar level. This results in a larger fraction of iron atoms in the 'perturbed' state and hence in a lower 'unperturbed' area fraction. Notwithstanding the fact that the carbon level is of the same order of magnitude as for the Western-type steel, no doublet could be detected for AX1. In the Eastern-type steel mainly $Cr_{23}C_6$, Cr_7C_3 and VC carbides are formed in which almost no Fe is incorporated, thus rendering these carbides invisible for the Mössbauer effect. The relatively large alloy-element concentration (especially Cr) found for this steel sample explains also why the H_{hf} and δ values for the 'perturbed' component deviate substantially from the 'ideal' values found for α-Fe (H_{hf} = 33.0 T, δ = 0.0 mm/s).

TABLE 2. Resulting maximum-probability hyperfine fields ($H_{hf\ 1,...,4}$) derived from the analysis of the distribution profiles, together with the corresponding isomer shifts ($\delta_{1,...,4}$) and relative area of the 'unperturbed' component (S_1). The average quadrupole splitting (QS), isomer shift (δ_D) and area fraction (S_D) of the doublet, which is present only in the Western-type steels, are also given.

Sample	H_{hf1} (T)	H_{hf2} (T)	H_{hf3} (T)	H_{hf4} (T)	δ_1 (mm/s)	δ_2 (mm/s)	δ_3 (mm/s)	δ_4 (mm/s)	S_1 (%)	QS (mm/s)	δ_D (mm/s)	S_D (%)
JRQ	33.2	29.0	27.9	-		-0.001		-	84.0	0.49	0.17	2.3
73W	33.2	30.7	29.0	-		0.001		-	76.7	0.50	0.18	0.9
AX1	33.5	30.7	28.1	25.2	0.001	-0.011	-0.027	-0.044	59.4	-	-	-

3.1.3. *Conclusion*
Only few studies, indicating that Mössbauer spectroscopy is a potentially interesting tool to investigate the microstructural aspects of irradiation embrittlement of RPV steels, have been performed so far [9,20]. For this reason it was decided to investigate three RPV-steel samples with slightly different chemical compositions to determine the applicability of the Mössbauer effect to the problem of RPV steel embrittlement. The results show that the distribution analysis of the Mössbauer spectra enables to distinguish between the different steel types. Small differences in carbon concentration between Western base (JRQ) and weld (73W) metal is reflected in the small area fraction of the cementite doublet for the weld metal. Differences between Eastern and Western type RPV steels are reflected in the overall shape of the derived H_{hf}-DP. The larger fraction of the 'perturbed' area for the Eastern steel, the differences in H_{hf} and δ values and the absence of a carbide doublet subspectrum are all due to the fact that the overall alloy-element concentration (especially for Cr and V) for the Eastern steel is larger than for Western-type steels.

Notwithstanding these encouraging results, a lot of work remains to be done. Additional experiments on a number of samples of different composition, irradiated as well as non-irradiated, have to be performed to gain more insight in the exact nature of the relationship that different alloy elements have with the relevant Mössbauer parameters, such as area fractions, isomer shifts and maximum-probability hyperfine-field values.

3.2. ISOCHRONAL ANNEALING OF RPV-STEELS

A total of seven specimens from four different, commonly used RPV-steels were selected for this study: JRQ and HSST03 are A533B Cl. 1 ferric steels that often serve as base metal in the construction of Western nuclear reactors. From both materials irradiated as well as non-irradiated specimens were annealed (JRQ616, X51: irradiated, Q228, X132: non-irradiated). The fifth and sixth steel samples, irradiated (73W) and non-irradiated (732W) are Linde0124 weld metal. With the aim of comparing Western types of RPV steels to Eastern ones, Russian 15Kh2MFA RPV steel was studied as well (YA). This second generation Cr-Mo-V type of RPV steel has been commercially used in VVER-440 nuclear reactors since the seventies.

Samples have been irradiated for about eight weeks up to a nominal fluency of $\phi=5.10^{19}$n/cm^2 (E_n > 1 MeV) in the CALLISTO PWR irradiation rig of the BR2 research reactor at the Belgian Nuclear Research Centre (SCK-CEN Mol, Belgium). Isochronal annealings of chosen specimens (range of temperatures: 20 - 1000 °C, step 50 °C) were performed in a vacuum of better than 3.10^{-3} Pa. Samples were annealed for 30 minutes at each temperature and afterwards cooled down in vacuum to room temperature using a "low- stress cooling speed" (100 °C/h).

In the first stage special attention went to the temperature region up to 700°C. At higher temperatures a new additional singlet component (isomer shift of about -0.298 mm/s) is clearly observable. Its presence can be explained as a paramagnetic austenit formation, which is due to structural changes from b.c.c. to f.c.c. starting at about 720°C. In case of Russian 15Kh2MFA steel, this singlet was observed beyond 800°C.

It can be concluded from this part of the study that in general the thermal treatment, the irradiation and the post-irradiation heat treatment do not drastically affect the Mössbauer parameters. Currently, some small differences observed for the hyperfine fields and fractional areas of the sextets are further examined in more detail, however, the changes are almost within the range of statistical errors. Results relevant to the comparison between irradiated and non-irradiated RPV-steel specimens are listed in Table 3.

TABLE 3. Some relevant RT MS parameters of irradiated and non-irradiated Western-type RPV-steels. $H_{hf,i}$ are the maximum-probability hyperfine fields derived from the H_{hf}-distribution profiles and S_i are the corresponding area fractions. The area S_3 belong to the doublet fraction.

Type of RPV-steel	Treatment	H_{hf1} (T)	H_{hf2} (T)	S_1 (%)	S_2 (%)	S_3 (%)
JRQ steel plate 16.616 (A533B Cl.1)	as received	33.1	30.5	81.6	15.6	2.8
	irradiated	33.1	30.5	80.6	16.4	3.0
HSST03 (A533B Cl.1)	as received	33.2	30.5	79.1	18.3	2.6
	irradiated	33.1	30.4	77.5	19.6	2.9
Linde 0124 weld	as received	33.1	30.4	80.4	18.2	1.4
	irradiated	33.1	30.4	80.0	18.2	1.8

4. Integral Low-Energy Electron Mössbauer Spectroscopy Applied to Thermally Treated Russian RPV-Steels

The integral low-energy electron Mössbauer spectroscopy (ILEEMS) is known to be sensitive only to a thin surface layer of a few nanometers. Specimens in the form of small disks (5 mm diameter, thickness of about 0.8 mm) were polished to the condition "like mirror" and measured in a high-vacuum (~10^{-4} Pa) chamber containing a channeltron electron detector. By applying a positive bias of 200 V to the channeltron's cathode, it was found possible to detect the low-energy electrons with a reasonable efficiency. The same model as for the transmission MS was used for fitting the ILEEMS

TABLE 4 Some relevant ILEEMS and metallurgy parameters for thermally treated Russian RPV-steels (at RT). The 3-sextets fit, line width of the most pronounced component fixed at 0.23 mm/s, another widths and A1/A3, A2/A3 ratios coupled together. HJ is thecalculated Hollomon-Jaffe parameter.

Types of RPV-steel		HJ	$H_{hf,1}$ (T)	$H_{hf,2}$ (T)	$H_{hf,3}$ (T)	δ_1 (mm/s)	$\delta_{2,3}$ (mm/s)	Γ_1 (mm/s)	$\Gamma_{2,3}$ (mm/s)	S_1 (%)	S_2 (%)	S_3 (%)	Hard-ness
15Kh2MFA	yt35	14640	33.5	31.4	28.3	0.003	-0.009	0.23	0.30	52	39	9	354.9
15Kh2MFA	yt38	16240	33.4	31.3	28.6	0.004	-0.001	0.23	0.35	50	41	9	386.6
15Kh2MFA	yt44	16805	33.5	30.8	28.5	0.001	-0.008	0.23	0.39	47	42	11	392.7
15Kh2MFA	yt51	17490	33.5	31.1	27.0	0.004	-0.002	0.23	0.30	55	37	8	395.6
15Kh2MFA	yt61	19860	33.4	30.7	28.5	0.003	-0.004	0.23	0.32	57	37	6	265.7
15Kh2MFA	yt62	20850	33.4	30.76	28.5	0.000	0.008	0.23	0.33	57	35	8	203.8
15Kh2NMFA	xt2	14070	33.4	31.0	28.4	0.005	-0.001	0.23	0.33	56	37	7	394.1
15Kh2NMFA	xt5	15460	33.4	31.1	28.0	0.005	0.002	0.23	0.35	56	37	7	414.6
15Kh2NMFA	xt14	17920	33.4	31.2	28.0	-0.003	-0.008	0.23	0.30	52	40	8	339.4
15Kh2NMFA	xt30	20850	33.3	30.6	28.6	0.003	0.001	0.23	0.28	69	28	3	201.9

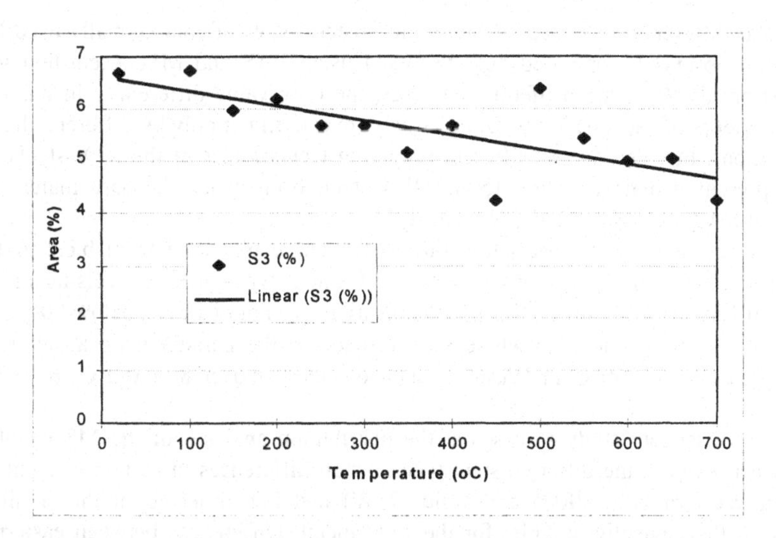

Figure 3. Areas under the third sextet of MS spectra obtained from isochronal annealing experiment (15Kh2MFA specimen) as obtained from transmission MS experiments.

spectra. The linewidth of *the unperturbed* component was fixed at a value of 0.23 mm/s.

ILEEMS was applied on 15Kh2MFA (YT) and 15Kh2NMFA (XT) steels. The chemical composition and preparing technology of used specimens is identical with composition of YTA and XTA specimens shown in Table 1. These HAZ specimens were prepared especially for the next TEM measurements.

The ILEEMS results (see Table 4) indicate slight but significant changes in the areas S_1, S_2, S_3 beneath the sextets and basically confirm the transmission MS measurements performed on 15Kh2MFA specimens. A small gradual decrease in S_3 (which refers to two or more alloying atoms in the nearest neighbourhood) with increasing annealing temperature is probably due to the release of some of the alloying elements as a result of the higher temperature and the precipitation of these elements as carbides or other non-magnetic compounds (see Figure 3).

Nevertheless the area S_2 of the second sextet (*perturbed* component) in dependence to Hollomon-Jaffe parameter could be correlated with the hardness curve. Basically, it is in agreement with our hypothesis about physical explanation of this MS component.

The results further confirm the ability of ILEEMS to detect changes in the surface microstructure due to different thermal treatment.

5. Conclusion

Methods as PAS LT, MS and ILEEMS were applied to the RPV-steel investigation. In total eight different PRV-steels were studied using some of these methods with the aim to extract some novel information about the microstructural changes caused by thermal, irradiation and post-irradiation heat treatment.

Clear differences between Western and Eastern types of commercially used RPV-steel were observed, compared and discussed using PAS and MS. According to the results from the PAS mean-lifetime analyses, the observable differences in the values and behaviours of successive annealing curves are caused not only by different chemical compositions, but also due to different preparing technologies of the RPV-steels. This fact is presented mainly on the 15Kh2MFA steel, from which the base material and simulated HAZ were studied simultaneously.

Changes in steel microstructure due to the phase transition from b.c.c to f.c.c., starting at ~ 725-750°C, were observed for all Western-types of RPV steels using PAS . In case of Russian RPV steels, this phase transition was not observed below 800°C. The same is reflected in the MS, where the presence of the paramagnetic austenite was clearly detected at 750°C in Western batch of steels and at 800-850°C in Russians steels.

The Mössbauer study shows that the distribution analyses of the MS enables to distinguish between the different steel types. Small differences in carbon concentration between Western base (JRQ) and weld (73W) metal is reflected in the small area fraction of the cementite doublet for the weld metal. Differences between eastern and western type RPV steels are apparent in the overall shape of the derived H_{hf}-DP. The larger fraction of the 'perturbed' area for the eastern steel, the differences in H_{hf} and δ values, and the absence of a carbide doublet subspectrum are all due to the fact that the overall alloy-element concentration (especially Cr and V) for the eastern steel is larger than for western-type steels.

From Mössbauer measurements performed on the western-type of RPV-steels A533B Cl.1 (JRQ, HSST03) and Linde 0124 weld (73W), we can conclude that thermal treatment (up to 700 °C), irradiation (up to ϕ=5.10^{19}n/cm^2 , E_n > 1 MeV) and post-irradiation heat treatment did not affect MS substantially.

On the other hand, the first MS measurements performed on the Russian RPV-steel 15Kh2MFA, irradiated with comparable neutron fluency (ϕ=6,7.10^{19}n/cm^2 (E_n > 0.5 MeV) in the framework of "Surveillance Specimen Program", recently revealed observable changes in the relevant parameters of the MS [26].

RPV embrittlement (limiting factor in the lifetime of vessels of today's NPP) is a more pronounced problem in Eastern (Russian) types of nuclear reactors (VVER-440). It is due to the narrower gap between the outside surface of the core barrel and the inside surface of the RPV than in western RPV's. The neutron flux and consequently neutron fluency on the RPV wall is generally higher on VVER-440 type reactors than in other equivalent types. This influence of neutron flux (even neutrons of energy over 0.1 MeV) on RPV embrittlement is much more impressive than contributions from a coolant temperature or an operational pressure in the primary circuit. Therefore, PAS, MS and TEM measurements on several RPV specimens irradiated in the NPP Jaslovské Bohunice (Slovakia) during their one-, two-, three- and five-year stays into the operating nuclear reactor, started in the framework of the "Extended Surveillance Specimen Program" (1994) is in progress. New additional information about microstructural changes in thermally and irradiation-treated RPV materials are expected in the next years (neutron diffraction should be included for the study in the near future).

It was confirmed that HAZ is the most sensitive place for thermal and neutron embrittlement in the reactor. PAS LT measurements on the successively annealed HAZ specimens (XTA, YTA) have shown the rapid increase in the vacancy-type defect formation in the temperature region 525-600 °C. Thereforc these specimens were studied using TEM in more detail.

According to the obtained results, it is possible to conclude that the applied methods can produce progress in the microstructural study of RPV-steels and in the optimisation of temperature-time regime for the regenerative post-irradiation thermal treatment of RPVs.

Acknowledgement

Author thanks E. De Grave, P.M.A. de Bakker and D. Segers from University of Gent (Belgium) for scientific discussions, recommendations and possibility to perform a part of this work in their lab. Author would like to thank also NATO (Research Fellowship - 1996), IAEA (Research contract 9001/RBF) and SGA (grant No 1/4286/97) for support.

References

1. Phythian, W.J. and English, C.A. (1993), *J. Nucl. Mater.* **205**, 162.
2. Brauer, G, Liszkay, L., Molnar, B., and Krause, R. (1991), *Nuclear Engineering and Design* **127,** 47.
3. Pareja, R., De Diego, N., De La Cruz, R.M., and Del Rio, J. (1993), *Nucl. Technol.* **104,** 52.
4. Lopes Gil, C., De Lima, A.P., Ayres De Campos, N., Fernandez, J.V., Kögel, G., Sperr, H., Triftshäuser, W., and Pachur, D. (1989), *J. Nucl. Mater* **161**, 1.
5. Abdurasulev, Z.P., Arifov, P.U., and Artjumov, N.I. (1985) *Methods for positron annihilation diagnostics and evaluation of spectra,* FAN, Taskent (in Russian).
6. Hautojärvi, P. (1979) *Positron in solids*, Springer-Verlag, Berlin.
7. Valo, M., Krause, R., Saarinen, K., Hautojarvi, P., and Hawthorne, R. (1992) *ASTM STP* 1125, Stoller, Philadelphia.
8. Bečvár, F., Jirásková, Y., Keilová, E., Kočík, J., Lešták, L., Procházka, I., Sedlák, B., and Šob, M. (1992), *Mat. Sci. Forum* Vols.**105-110**, 901.
9. Brauer,G., Matz, W., Liszkay, L., Molnar, B., and Krause R., (1992), *Mat. Sci. Forum*, **97-99**, 379.
10. Brauer, G. (1995), *Mat. Sci. Forum* Vols. **175-178**, 303.
11. Cohen, L. (1980) *Application of Mössbauer spectroscopy*, Volume II. Academic Press, New York.
12. Brauer, G., Matz, W., and Fetzer, Cs. (1990), *Hyperfine Interaction* **56**, 1563.
13. Amaev, A.D.,Dragunov, Y.D., Kryukov, A.M., Lebedev, L.M., and Sokolov, M.A. (1986) Investigation of irradiation emtrittlement of reactor VVER-440 vessel materials, *IAEA specialists meeting proceedings*, Plzen.
14. Šmída, T. and Magula, V. (1988) *Zváranie* **37,** 34 (in Slovak).
15. Haščík, J., Lipka, J., Kupča, L., Slugeň, V., Miglierini, M., Gröne, R., Tóth, I., and Vitázek, K. (1995), *acta physica slovaca*, **45,** 37.
16. Kirkegaard, P., Pedersen, N., and Eldrup, M. (1989) PATFIT 88 - a data processing system for positron annihilation Spectra on Mainframe and Personal Computers. *Risõ, M-2740.*
17. Shukla, A., Peter, M., and Hoffman, L. (1993), *Nucl. Instr. & Meth in Phys. Res.* **A335,** 310.
18. De Bakker, P.M.A. and Segers, D. (1996) *Proceedings from Seminar on Analytical Techniques.* SCK-CEN, Mol .
19. Honeycombe, R.W.K. (1981) *Steels - Microstructure and Properties,* Edward Arnold Publishers Ltd., London.
20. Šron, R.Z. (1980), *Avtomaticeskaja svarka* **7**, 1 (in Russian).

21. Janovec, J. (1988), *Kovové materiály* **26**, 63 (in Slovak).
22. Magula, V. and Janovec, J. (1994), *Ironmaking and Steelmaking* **21**, 223.
23. Haščík, J., Slugeň, V., Lipka, J. Kupča, Ľ., Hinca, R., Tóth, I., Gröne, R., and Uváčik, P. (1998) Progress in investigation of WWER-440 reactor pressure vessel steels by gamma and Mössbauer spectroscopy. *Proceedings from the international conference "Nuclear Option in Countries with Small and Medium Electricity Grids.* 15.-18. June 1998, Dubrovník, Croatia, p. 157.
24. Sawicki, J.A. and Brett, M.E. (1993), *Nucl. Instrum. Meth. Phys. Res.* **B76**, 254.
25. Törrönen, J. (1979) Microstructural Parameters and yielding in a quenched and tempered Cr- Mo-V pressure vessel steel. *Report VTT-22*, Technical Research Centre of Finland.
26. Amarasiriwardena, D.D, De Grave, E., Bowen, L.H., and Weed, S.B. (1986) Quantitative determination of aluminum-substituted goethite-hematite mixtures by Mössbauer spectroscopy. *Clays Clay Minerals* **34**, 250.
27. Hautojärvi, P., Judin, T., Vehanen, A., Yli-Kauppila, J., Girard, P., and Minier; C. (1981) *J. Phys.* ***F 11***, 1337.
28. Van Hoorebeke, L., Fabry, A., van Walle, E., Van de Velde, J., Segers, D., and Dorikens-Vanpraet, L. (1996), *Nucl. Inst. & Meth. in Phys. Res.* **A 371**, 566.
29. Gröne, R., Haščík, J., Lipka, R., Pietrzyk, I., Slugeň, V., and Vitázek, K. (1996) *The Safety of Nuclear Energy* **4(42)**, 293 (in Slovak).
30. Koutský, J. and Kočík, J. (1994) *Radiation Damage of Structural Materials,* Academia, Praque.

MECHANOMIXING IN VARIOUS IRON-CONTAINING BINARY ALLOYS

G. LE CAËR[1], T. ZILLER[1], P. DELCROIX[1] AND J.P. MORNIROLI[2]
[1)] *Laboratoire de Science et Génie des Matériaux Métalliques, UMR CNRS 7584, Ecole des Mines, F-54042 Nancy Cedex, France*
[2)] *Laboratoire de Métallurgie Physique, URA CNRS 234, Université des Sciences et Techniques de Lille-Flandres-Artois, F-59655 Villeneuve d'Ascq Cedex, France*

1. Introduction

Mechanical alloying (MA), which is a dry and high-energy milling process, has attracted considerable interest in recent years owing to the wide range of materials, often with non-equilibrium structures, wich can be synthesized: amorphous, quasicrystalline and nanocrystalline phases, extended solid solutions, alloys of immiscible elements, all sorts of compounds and composites. Although the first steps of mixing of ductile elemental powders are well described, i.e. the formation of lamellar structures which are progressively refined and convoluted by repeated flattening, fracturing and welding actions [1], the mechanisms operating at a nanometer scale, when microstructures appear as homogeneous typically at the scale of scanning electron microscopy (SEM), remain unclear for many systems. Lots of difficulties have indeed to be overcome to characterize milled powders at a nanometer scale by direct observation techniques, for instance to prepare samples for investigations by high resolution electron microscopy (HREM) while avoiding sample evolution. Although Mössbauer spectrometry is less direct than microscopy techniques, it provides nevertheless complementary information, often original, with reliable statistical features which are missing or difficult to obtain with the latter techniques.

For the purpose of scrutinizing the evolution of hyperfine magnetic features during solid state mixing by mechanical methods, the T contents of initial $Fe_{1-x}T_x$ elemental powder mixtures are chosen to yield final ground powders which consist of non-magnetic phases at room temperature (RT). A relative milling time is defined as $\tau = t_m/t_f$ where t_m is the milling time and t_f is the minimum milling time needed to reach a final stationary state. Phenomena which are not observed by diffraction techniques are evidenced by Mössbauer spectrometry studies of the milling time evolution of ball-milled powder mixtures in many, but not in all, binary Fe-T systems (T= Al, Si, V, Cr for instance), notably in Cr-rich FeCr alloys [2-3]. The decreasing volumic fraction of a ground powder which remains magnetic at room temperature gives rise to hyperfine magnetic field distributions (HMFD) which stay stationary in shape after relative milling times $\tau \approx 0.4$-0.6 [2]. The latter distributions consists typically of a narrow peak located at a field close to the field H = 331 kG of b.c.c. Fe at RT, called for simplicity 'α-Fe' peak, with a weight of about 1/3 of the total magnetic contribution and of a broad,

M. Miglierini and D. Petridis (eds.), Mössbauer Spectroscopy in Materials Science, 131–142.

almost featureless band from ≈ 90-100 kG to ≈ 300-320 kG (Fig. 1). Le Caër et al. [2] have proposed that nanometer-sized Fe-rich zones with diffuse interfaces with T-rich zones, characterized at the atomic scale by distributions of lengths of flat regions and of step heights, are responsible for such HFMD stationary shapes. More binary alloys must be investigated using various techniques to ascertain the latter interpretation and to describe the associated morphologies.

The aim of the present paper is thus to characterize the mixing process in $Fe_{0.28}Mn_{0.72}$ and $Fe_{0.5}V_{0.5}$ alloys and to follow the evolution of partially ground $Fe_{0.5}V_{0.5}$ and $Fe_{0.3}Cr_{0.7}$ powders during isothermal annealing, by means of X-ray and neutron diffraction, Mössbauer spectrometry and microprobe analysis. The elements Mn and V were further chosen because the neutron scattering lengths are b_{Mn}= -0.373.10^{-12} cm and b_V = 0.0382.10^{-12} cm respectively while b_{Fe} = 0.954.10^{-12} cm. The average scattering length of the alloy $Fe_{0.28}Mn_{0.72}$ is zero: only phases which are chemically or (/and) crystallographically heterogeneous yield diffraction patterns. As only Fe atoms of FeV alloys contribute to the intensities of the neutron diffraction lines, the formation of ordered phases is easier to observe by the latter technique than it is by X-ray diffraction.

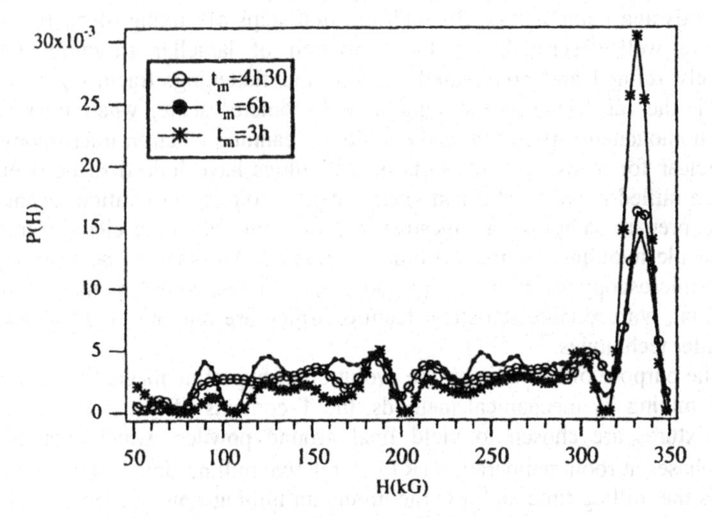

Figure 1. ^{57}Fe Hyperfine magnetic field distributions at room temperature for $Fe_{0.3}Cr_{0.7}$ powder mixtures milled for various times.

2. Experimental

Elemental powders of Fe (<10 µm, purity 99.9+), Mn (44 µm, purity 99+), Cr (44 µm, purity 99+) and V (44 µm, purity 99.5) were used as starting materials. The compositions investigated were $Fe_{0.28}Mn_{0.72}$, $Fe_{0.3}Cr_{0.7}$ and $Fe_{0.5}V_{0.5}$. MA was carried out under argon atmosphere in a Fritsch Pulverisette 7 planetary ball mill using steel vial and balls (7 balls, Ø=16 mm, m=15 g) with a powder-to-ball ratio R = 1/20 and a rotation speed $\omega = 640$ rpm. Milling was interrupted each 45 min for 15 min to let the vials cool down.

Isothermal annealings were performed at 773 K under vacuum in sealed containers.

Neutron diffraction patterns ($\lambda = 0.128$ nm) were obtained using the D1B multicounter instrument at ILL (Grenoble). X-ray diffraction patterns were recorded on a powder diffractometer with INEL CPS 120 type curved detector, using Fe $K_{\alpha 1}$ radiation ($\lambda = 0.19360$ nm) for $Fe_{0.28}Mn_{0.72}$ samples and on a $\theta/2\theta$ mode goniometer using Co $K_{\alpha 1}$ radiation ($\lambda = 0.17889$ nm) for the $Fe_{0.5}V_{0.5}$ powders.

Compositions of powder particles were measured by microprobe analysis. Backscattered electrons and X-ray images were further used to observe microstructural changes during milling.

^{57}Fe Mössbauer spectra were recorded at RT in transmission geometry with a spectrometer operated in the conventional constant acceleration mode. A ^{57}Co source in Rh with a strength of ~ 10 mCi was used. The magnetic components of spectra were analyzed, employing Lorentzian lineshapes, as distributions of hyperfine magnetic fields (HMFD) by a conventional constrained Hesse-Rübatsch method. As usual, the ^{57}Fe isomer shifts (IS) are given with respect to α-Fe at RT.

3. Results

3.1. MICROPROBE ANALYSIS

3.1.1. *As-milled $Fe_{0.28}Mn_{0.72}$ Powders*

Microprobe analyses were performed on about 20 powder particles for each sample (t_m =1, 2, 3 and 5 h). Concentrations of both elements were measured in each particle along a line crossing it and a distribution of Mn contents in the selected particles was obtained for the corresponding milling time. The latter distribution is characterized among others by its mean $<x>$ and its standard deviation σ. A " chemical mixing time " t_{mc} can thus be defined as the minimum time needed to reach a stationary distribution of Mn content of powder particles. The latter characteristic time has not been considered till now in the literature. Neglecting eventual small fluctuations, the stationary distribution is expected to correspond to $<x> = 0.72$ and $\sigma = 0$. Heterogeneities (chemical, structural, morphological) among particles exist therefore and must be considered when analyzing experimental results obtained for milling times shorter than t_{mc}.

During the first MA step, fracturing and welding of Mn and of Fe particles, trapped between colliding balls or between vial wall and balls, explain the occurence of composition fluctuations both between particles and inside particles. Microstructures are progressively refined. For $t_m = 5$ h, the concentration profiles inside particles are almost flat with values close to 28at.% for Fe and to 72at.% for Mn as expected from the initial mixture composition. The milling time dependence of σ normalized by the average Mn content (Fig. 2) yields $t_{mc} \approx 5$ h which is also the minimum milling time t_f

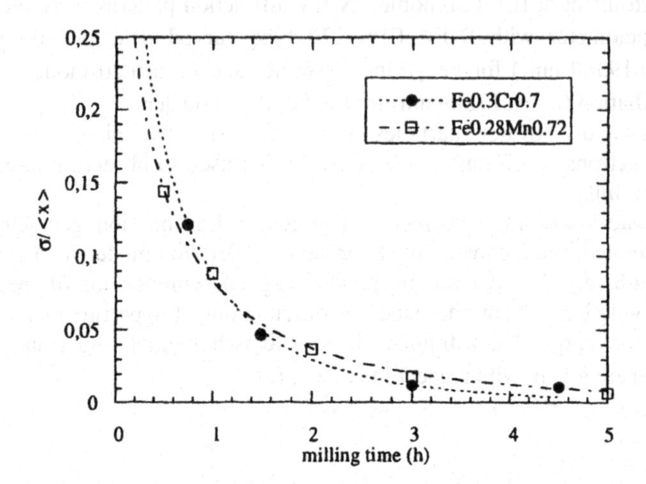

Figure 2. Milling time dependence of the ratio σ/<x> for ground $Fe_{0.30}Cr_{0.70}$ (<x>=0.70) and $Fe_{0.28}Mn_{0.72}$ (<x>=0.72) powders.

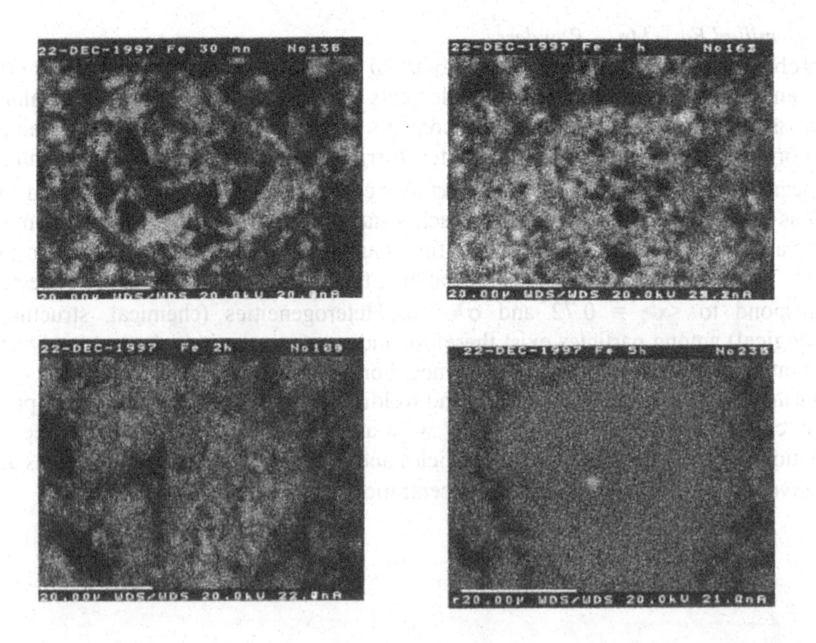

Figure 3. SEM micrographs of $Fe_{0.28}Mn_{0.72}$ powder particles milled for 30 min, 1h, 2h and 5h. Fe appears in white.

needed for complete alloying (stationary structural state) : $t_{mc} \approx t_f$.

For short milling times, SEM micrographs (Fig.3) show spheroidal Mn particles embedded in a Fe matrix instead of lamellar microstructures which are typically observed when ductile metallic powders are ground together. The former morphologies, which are similar to those observed for instance in ground Si-Ge alloys [4], are explained by the brittleness of α-Mn particles wich are fractured but not plastically flattened during milling.

3.1.2. *As-milled* $Fe_{0.30}Cr_{0.70}$ *Powders*

An analysis similar to the previous one was performed for $Fe_{0.30}Cr_{0.70}$ samples. In contrast to the Mn case, $t_{mc} \approx 3$ h is much shorter than $t_f \approx 12$ h. Furthermore, SEM micrographs exhibit the aforementioned lamellar structures wich are progressively refined during MA (Fig.4). A lamellar type microstructure is similarly observed in Fe-V ground powders.

3.2. NEUTRON AND X-RAY DIFFRACTION

3.2.1. *As-milled* $Fe_{0.28}Mn_{0.72}$ *Powders*

Neutron diffraction patterns of $Fe_{0.28}Mn_{0.72}$ powders ground for t_m= 0, 1, 2, 3, 5 and 7 h are shown in figure 5. The Bragg peaks of α-Fe and of α-Mn are clearly distinguishable

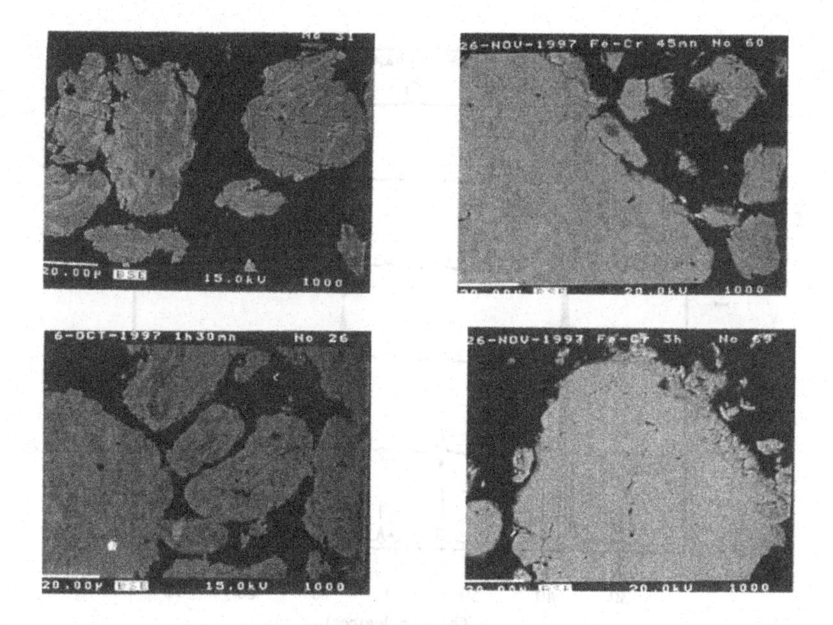

Figure 4. SEM micrographs (backscattered electrons with atomic number contrast) of $Fe_{0.3}Cr_{0.7}$ powder particles milled for 25 min, 45 min, 1h30, and 3h respectively.

in the diffraction pattern of the initial mixture. Their intensities decrease with milling time as the pure metals are progressively consumed by alloying. A γ-type phase (f.c.c.) is formed transiently between $t_m = 1$ and 2 h. Its Bragg peaks, which are very broad and poorly defined, prevent from determining its composition from lattice parameters. Between 2 and 3 hours of grinding, the total intensity diffracted by the latter phase stays almost constant while it decreases for t_m longer than 5 h. For $t_m = 7$ h, diffraction peaks are almost no more visible suggesting that a disordered and homogeneous alloy is formed. We notice than MnO is formed but its proportion does not exceed 3-4 mol.%.

X-ray diffraction patterns allow us to conclude that the final powder consists as expected of a solid solution of Fe in α-Mn. The γ phase is not clearly detected because its Bragg peaks coincide with some α phase peaks. An increase of the intensity of the (422) α-phase line with milling time, until $t_m = 3$ h, agrees nevertheless with the formation of the γ phase, followed by its decomposition, as deduced from neutron diffraction experiments.

3.2.2. *As-milled and Annealed FeV Powders*

Neutron and X-ray diffraction patterns of as-milled FeV powders confirm the results of Fultz et al. [5]: milled powders consist of a mixture of an amorphous phase and of a disordered b.c.c. FeV solid solution. Powders milled for 4 hours were annealed at 773 K for 30 minutes and 4 hours respectively. Neutron superlattice reflections due to the transformation from a disordered A2 (b.c.c.) solid solution to an ordered B2 alloy (CsCl

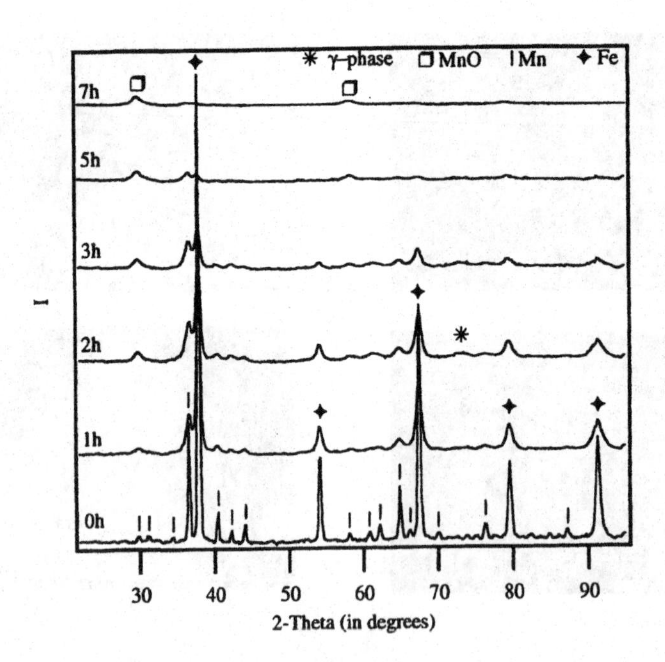

Figure 5. Neutron diffraction patterns of $Fe_{0.28}Mn_{0.72}$ powder mixtures milled for different times.

type structure) are clearly observed while this ordering phenomenon passes unnoticed from XRD patterns (Fig. 6). For the equiatomic alloy, the long-range-order parameter η is defined as $\eta = 1 - 2c$, where c is the Fe concentration on the V sublattice. It varies from $\eta = 1$ for perfect order ($c = 0$) to $\eta = 0$ for complete disorder ($c = 0.5$). From diffraction line intensities, we obtain $\eta = 0.3$ for an annealing time $t_a = 30$ min and $\eta = 0.5$ for $t_a = 4$ h. Using the Scherrer formula, an average size of ordered domains has been deduced from the broadening of Bragg peaks to be of the order of 10 nm.

The equilibrium phase for FeV at 773K is the classical sigma phase. The bcc to σ transformation is however very sluggish [6] as it is in the Fe-Cr system. Metastable b.c.c. solid solutions retained at low temperatures tend to order with a B2 type structure. As shown by theoretical models, the A2 $\rightarrow$ B2 transition is second order around the equiatomic composition [6]. Specimen which are water-quenched from 1403 K show weak B2 superlattice reflections which agree with the strong ordering tendency in such alloys. For $Fe_{0.65}V_{0.35}$ and $Fe_{0.60}V_{0.40}$ alloys, Sanchez et al. [6] observed that the B2 phase is formed in quenched samples down to 700K. MA $Fe_{1-x}V_x$ are disordered b.c.c. alloys which offer also the possibility of investigating the latter disorder-order transition. The fast increase of η reported above suggests that the B2 phase might be formed at much lower temperatures in MA samples than in quenched samples. A detailed investigation of the kinetics of the ordering process at various temperatures will soon be performed.

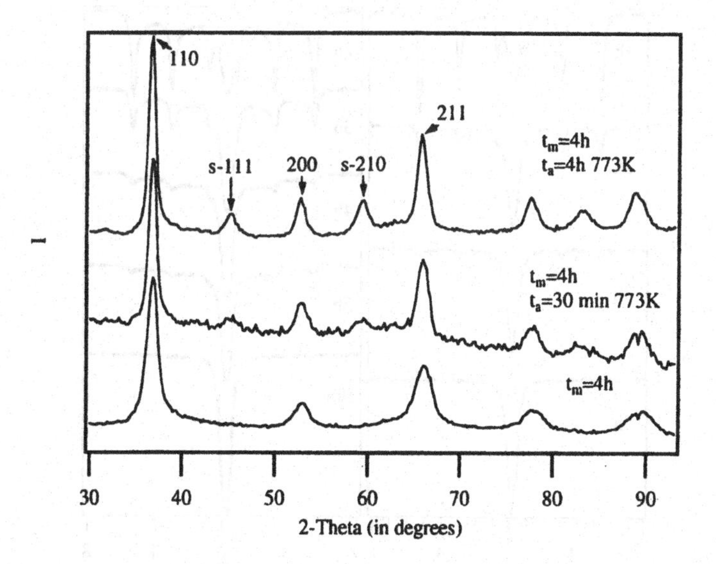

Figure 6. X-ray diffraction patterns of $Fe_{0.5}V_{0.5}$ powder mixtures after annealing at 773 K for 30 min and 4h, respectively.

3.3. MÖSSBAUER SPECTROMETRY

3.3.1. *As-milled $Fe_{0.28}Mn_{0.72}$ Powders*

RT Mössbauer spectra of $Fe_{0.28}Mn_{0.72}$ powders milled for t_m= 1, 2, 3 and 5 hours are shown in figure 7. For t_m= 1 h to 3 h, the spectra show the superposition of a sextet of uncombined α-Fe and of a paramagnetic asymmetric doublet wich is associated with a α-Mn(Fe) phase. The central component is best fitted with four distinct doublets whose hyperfine parameters do not depend on milling time. For classically synthesized α-Mn(Fe) alloys [7], only two doublets are observed. The quadrupole splittings (QS) and isomer shifts (IS_1 = -0.24 mm.s^{-1}, QS_1 = 0.28 mm.s^{-1}, IS_2 = -0.09 mm.s^{-1} and QS_2 = 0.22 mm.s^{-1}) of the four previous doublets agree with the published values. The two supplementary doublets indicate that Fe atoms occupy further the two other crystallographic sites as confirmed by the absence of neutron diffraction lines for ground α-Mn(Fe) (Fig.5).

Figure 8 shows HMFD's associated with the magnetic components of RT Mössbauer spectra. They consists of two main parts :

- a peak at 331 kG corresponding to α-Fe with no Mn atom first nearest-neighbour.
- some small peaks between 200 and 320 kG.

These HMFD's do not show the characteristic features reported by Le Caër and al. for Fe-Cr, Fe-W and Fe-Si systems [2], namely a narrow 'α-Fe' peak as observed here but a broad band from ≈ 90-100 kG to ≈ 300-320 kG with a weight of about 2/3 of the

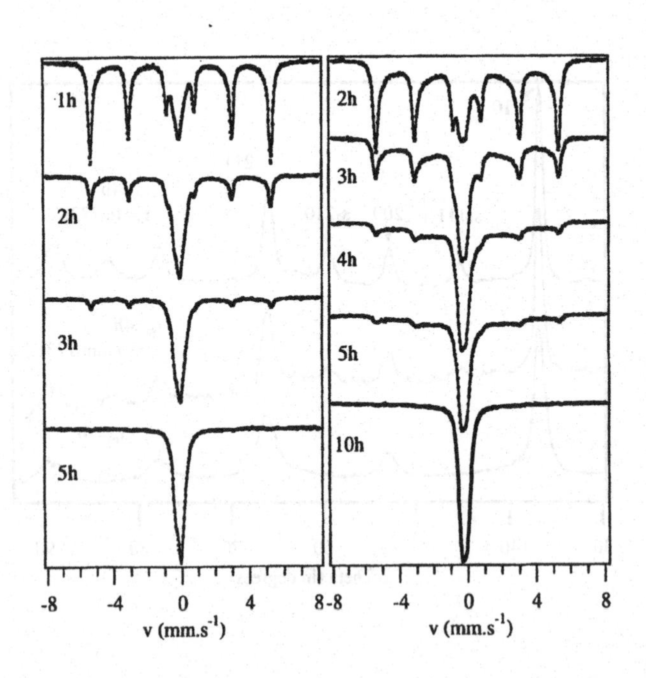

Figure 7. Mössbauer spectra (RT) of $Fe_{0.28}Mn_{0.72}$ (left) and $Fe_{0.5}V_{0.5}$(right) powder mixtures after different milling times.

total magnetic contribution (Fig. 1). Moreover, for the previous systems, the HMFD's remain stationary in shape after relative milling times $\tau \approx 0.4\text{-}0.6$ whereas a different behaviour is observed for $Fe_{0.28}Mn_{0.72}$.

3.3.2. *As-milled FeV Powders*

The whole spectra were analysed as resulting from HMFD's with IS = -0,22 mm.s^{-1} for 0 kG = H = 50 kG and IS = 0 mm.s^{-1} for H > 50 kG. HMFD's of ground $Fe_{0.5}V_{0.5}$ powder mixtures are similar in shape to those of $Fe_{0.7}Cr_{0.3}$ with their characteristic 'α-Fe' peaks and their broad bands [7]. The distributions remain further stationary in shape for relative milling times longer than $\tau \approx 0.4$ (t_m = 4h), a conclusion which was not reported in [7]. The weight (≈ 4/5) of the broad band with respect to the total magnetic contribution is sligthly larger in the stationary state than it is for $Fe_{0.7}Cr_{0.3}$ [2]. Such distribution shapes were attributed to nanometric Fe zones, in which Fe is uncombined or weakly combined with T, with diffuse interfaces with T-rich zones. HFMD shapes similar to those discussed here are measured with high precision by NMR [8] in Co/Cr multilayers among others and attributed to diffuse interfaces with a concentration gradient.

3.3.3. *Annealing of As-milled $Fe_{0.3}Cr_{0.7}$ Powders*

Figure 9 shows RT Mössbauer spectra and HMFD's of $Fe_{0.3}Cr_{0.7}$ powders milled for 4h30 and annealed 3h and 290h at 773K respectively. The relative intensity of the 'α-Fe' peak decreases during the first three hours of annealing while its hyperfine field remains constant at a value of ≈ 332 kG. The HMFD reaches then a plateau shape. For longer annealing times, the 'α-Fe' field increases by about 5 kG while a HMFD

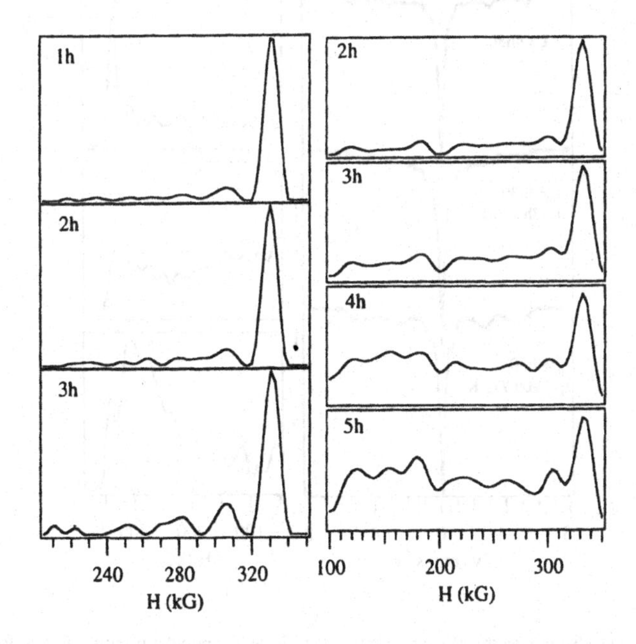

Figure 8. HMFD's (RT) of $Fe_{0.3}Cr_{0.7}$ (left) and $Fe_{0.5}V_{0.5}$ (right) powder mixtures after different milling times.

characteristic of an alloy with about 14 at.% Cr emerges progressively. Both features indicate that a true magnetic Fe-Cr alloy is not formed during the first hours of annealing. For long annealing times (for instance, 290h), the sample consists of a mixture of two phases, one Fe-rich and the other Cr-rich, a trend consistent with the equilibrium phase diagram [9]. Longer annealing times are needed to reach actually the final equilibrium state. A sample ground for 12h and annealed during 290h at 773 K yields a spectrum which is almost identical with the spectrum of fig. 9 (t_a = 290h). This result confirms that a milling time of 4h30 is larger than the chemical mixing time which was indeed determined to be $t_{mc} \approx 3h$ in section 3.1.2.

4. Discussion and Conclusion

The characteristic features of HMFD's extracted from Mössbauer spectra show that the elements mix differently in the solid state when milling powder mixtures of Fe and Mn with an overall concentration $Fe_{0.28}Mn_{0.72}$ than they do when grinding $Fe_{0.3}Cr_{0.7}$ and $Fe_{0.5}V_{0.5}$ powder mixtures. The features of the HFMD's extracted from Fe-Cr spectra , which are similar to those of HFMD's measured directly by NMR in many multilayers ([8] and references therein), have been attributed to nanometric Fe-rich zones with diffuse interfaces with Fe-poor zones [2]. To understand why various binary systems behave differently, we have been led to define a " chemical mixing time " t_{mc} as the minimum time needed to reach a stationary distribution of chemical concentration in all

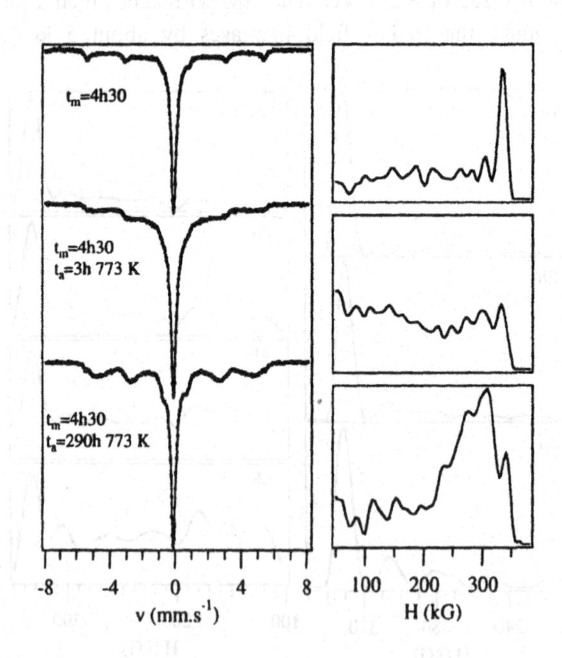

Figure 9. Mössbauer spectra (RT) and HMFD's of $Fe_{0.3}Cr_{0.7}$ powder mixtures milled for 4 h30 and annealed at 773 K.

powder particles in given milling conditions. The chemical mixing time can be determined, as done here, from microprobe analyses. It is quite obvious that a minimum time is in fact needed to mix homogeneously the elements of the initial powder mixture. It is however important to emphasize that t_{mc}, which depends among others on the investigated system, does not necessarily coincides with the minimum synthesis time t_f. The concentration of elements can indeed be basically the same in all powder particles for $t_m = t_{mc}$ while the structures and compositions of the various phases can still evolve with t_m. In our experimental conditions, the chemical mixing time t_{mc} is of the order of 3-4 hours whatever the investigated system. For Fe-Mn alloys, $t_{mc} \approx t_f$, while $t_f \approx 3\ t_{mc}$ for Fe-Cr and for Fe-V alloys (Fig. 10). Differences exist indeed between the Fe-Mn system on the one hand and Fe-V and Fe-Cr systems on the other.

First, Mn particles are brittle and form spheroids in the Fe matrix in contrast to ductile Cr and V particles which form lamellae. Both morphologies are visible by classical microscopy techniques for short milling times. They may however persist down to the nanometric scale and likely explain the possibility to form diffuse interfaces with a large weight in the Fe-V and Fe-Cr systems in contrast to the Fe-Mn system.

Second, Fe is soluble in α-Mn up to 32 at.% while a broad miscibility gap exists in the Fe-Cr phase diagram for Cr contents ranging from ≈ 15 at.% to ≈ 88 at.% at 773 K [9]. The history of particles during milling may account for the experimental results in the Fe-Cr system. The time between two trapping events of a given particle is not simply proportional to the reciprocal of the shock frequency although it is related to it. The latter time has been estimated to be ≈ 10~100 s in classical mills similar to those used here while the impact duration is about few milliseconds [10]. A particle is thus submitted to a very short pulse of stress, accompanied by a temperature increase, which drives the formation of an alloy. Demixion processes can however take place during the long relaxation time which follows the latter shock thanks to the temperature increase, to the large defect concentration and to the nanometric scale at which such phenomena occur. They tend to bring back the system to equilibrium and thus they slow down the alloying process. Longer milling times are required to reach the final stationary state even if the chemical concentration of the considered particle is the final one. Such

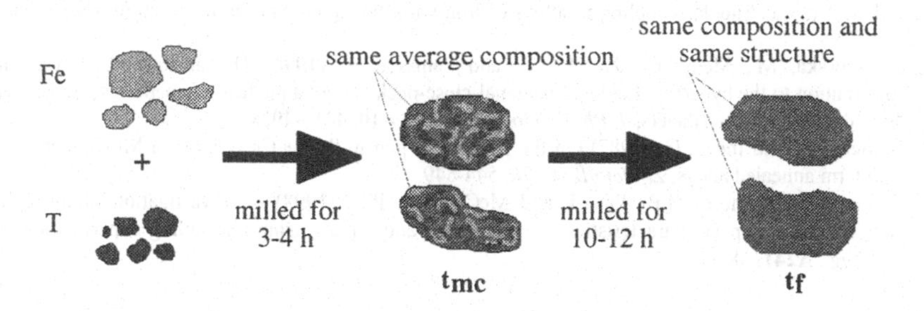

Figure 10. A schematic representation of the mixing process in ground Fe-T elemental powder mixtures : t_{mc} and t_f are respectively the chemical mixing time and the minimum time needed to reach the final stationary state.

phenomena may depend sensitively on the dynamical conditions of milling and on the milling temperature.

The evolution of $Fe_{0.3}Cr_{0.7}$ powders, partly alloyed by milling, during isothermal annealings show that rearrangements of the aforementioned nanometric zones take place without alloy formation for short annealing times. Mössbauer spectra suggest that the density of interfaces of Fe zones increases during that first annealing step, a mechanism which may also play a role in the relaxation step described above. Cr-rich and Cr-poor alloys are formed as expected for longer annealing times. However, even 290h anneals result in HMFD's which show broad and flat bands, although with relatively low intensities, between 50 and 200 kG which are missing in classical alloys heat-treated in similar conditions. Experiments to characterize the former annealed alloys are in progress.

Disordered MA $Fe_{1-x}V_x$ solid solutions order when annealed even at low temperatures. They offer thus the possibility to investigate the metastable Fe-V phase diagram which has been calculated theoretically [6].

To conclude, mixing of elements in the solid state by mechanical means thanks to chaotic transformations of the 'baker' type may be a way to produce original morphologies in some physically notable systems. Such mixtures, ground for $t_m \approx t_{mc}$, may evolve in an unusual way during annealing treatments.

References

1. Gilman, P.S. and Benjamin, J.S. (1976) Mechanical alloying, *Annual review of materials science* **13**, 279-300.
2. Le Caër, G., Delcroix, P., Shen, T.D. and Malaman, B. (1996) Mössbauer investigation of intermixing of Fe and T during ball-milling of $Fe_{1-x}T_x$ (T=Cr, x=0.7, T=W, x=0.5) powder mixtures, *Phys. Rev. B* **54**, 12775-12786.
3. Le Caër, G. and Delcroix, P. (1996) Characterisation of nanostructured materials by Mössbauer spectrometry, *Nanostruct. Mater.* **7**, 127-135.
4. Davis, R.M., McDermott, B. and Koch, C.C. (1988) Mechanical alloying of brittle materials, *Metall. Trans. A* **19A**, 2867-2874.
5. Fultz, B., Le Caër, G. and Matteazzi, P. (1989) Mechanical alloying of Fe and V powders : intermixing and amorphous phase formation, *J. Mater. Res.* **4**, 1450-1455.
6. Sanchez, J.M., Cadeville, M.C., Pierron-Bohnes, V. and Inden, G. (1996) Experimental and theoretical determination of the metastable Fe-V phase diagram, *Phys. Rev. B* **54**, 8958-8961.
7. Kimball, C.W., Phillips, W.C., Nevi, H.M.V. and Preston, R.S. (1966) Magnetic hyperfine interaction and electric quadrupolar coupling in alloys of iron with the alpha-manganese structure, *Phys. Rev.* **146**, 375-378.
8. Malinowska, M., Mény, C., Jedryka, E. and Panissod, P. (1998) The anisotropic first-neighbour contribution to the hyperfine field in hexagonal-close-packed Co : a nuclear magnetic resonance study of diluted alloys and multilayers, *J. Phys. : Condens. Matter* **10**, 4919-4928.
9. Dubiel, S.M and Inden, G. (1987) On the miscibility gap in the Fe-Cr system : a Mössbauer study on long term annealed alloys, *Z. Metallkde.* **78**, 544-549.
10. Huang, H., Dallimore, M.P., Pan, J. and McCormick, P.G. (1998) An investigation of the effect of powder on the impact characteristics between a ball and a plate using free falling experiments, *Mater. Sci. Eng.* **A241**, 38-47.

MÖSSBAUER STUDIES OF BALL MILLED $Fe_{1-x}S_{1+x}$

L. TAKACS[1] AND VIJAYENDRA K. GARG[2]

[1)] *Department of Physics, University of Maryland, Baltimore County, Baltimore, Maryland 21250, USA*

[2)] *Instituto de Física, Universidade de Brasília, 70910-900 Brasília, DF, Brazil*

Abstract

The influence of high-energy ball milling on iron sulphur powder mixtures $Fe_{1-x}S_{1+x}$ (where x = -3, 0, 3, 5, and 10%) with milling times between two hours and 70 hours have been studied by XRD and Mössbauer spectroscopy. Intensive mechanical milling caused changes in the phase composition and properties of the material, which evoked corresponding changes in the Mössbauer spectra. Most of the reaction takes place in a self-sustained manner over a short interval, in our case some time between 2 and 5 hours, as confirmed by XRD. It seems that the reaction is a three step process, consisting of an initial activation period, than an almost instantaneous, self-sustained chemical change, followed by slow homogenization and reaching a uniform final state. The Mössbauer spectra after 0 and 2 hours of milling consists of the sextet of α-Fe and a nonmagnetic doublet. In all other samples the internal magnetic field of the sextet is close to 270 kOe and the Mössbauer parameters of the doublet are close to those of FeS_2. In case of samples with 10% excess S, milling results in the presence of FeS_2. Though we do not know the exact origin of these components yet, the sextet probably reflects the presence of FeS, that is antiferromagnetic at room temperature, and the doublet may originate from FeS_2 in non-crystalline form.

1. Introduction

Mechanical alloying was developed in the late 1970's by Benjamin [1] as a way to circumvent problems encountered in the course of developing oxide dispersion-strengthened superalloys. Since Koch et al. [2] demonstrated that amorphous $Nb_{40}Ni_{60}$ could be formed by ball milling of elemental powder blends, mechanical alloying has emerged as a versatile technique to produce metastable materials in a variety of systems. For example, mechanical alloying [3-7] was used to prepare nanostructured alloys, amorphous materials, metastable solid solutions, dispersion-strengthened alloys, and intermetallic compounds. Ball milling has also been applied to a variety of other chemical processes, e.g. mineral processing [8], inorganic preparation [9], and hydrogels [10]. Self-sustaining combustion is induced by ball milling in a variety of sufficiently

M. Miglierini and D. Petridis (eds.), Mössbauer Spectroscopy in Materials Science, 143–150.

exothermic powder mixtures. For example, combustion occurs when metal chalcogenides are prepared from their elemental components [11,12] or a metal oxide is reduced with a more reactive metal [13, 14]. Although the details of the mechanochemical processes are very much system specific [15], most reactions take place in three distinguishable steps: activation, reaction, and post-reaction milling. During the activation step, the mechanical action of the mill mixes the reactants on an increasingly fine scale. It also generates lattice defects, which serve as fast diffusion paths and active reaction sites, but very little chemical change occurs.

Most of the reaction takes place during the second step. Two different kinetics are possible. In highly exothermic systems, the temperature increase between the colliding balls, probably higher than 100 K [16], can initiate a combustion wave, that propagates through the volume of the vial. If that happens, most of the transformation is complete within a second. The sudden heat release can be detected by measuring the temperature of the milling vial. As ignition represents a critical state of the powder blend, the time it takes to reach this state is a useful parameter [17]. Even if the reaction is not exothermic enough to support the propagation of combustion, the activation period is followed by relatively fast, albeit gradual transformation [12]. Continued milling after the main transformation step results in particle refinement and a more uniform powder.

Sulfides have been recognised as inorganic materials with many technological applications, e.g. in high-energy density batteries, photoelectrolysis, solar energy materials, precursors of superconductor synthesis, diagnostic, luminescence materials, and chalcogenide glasses. Ball milling is a promising method to prepare sulfides. For example, it overcomes the difficulties associated with the large melting point difference between most metals and sulfur. Several authors reported the preparation of sulfides by ball milling [11, 12, 18-25].

The reaction kinetics and even the product phases may depend on the type of the milling equipment, the mass of the powder charge, and the number and the size of the milling balls. For example, ZrB_2 formed gradually in a vibratory mill [26] while combustion was ignited in less than an hour when using a SPEX 8000 shaker mill [27]. Amorphization and crystallization, alloying and de-mixing can occur in the same system, depending on the milling time and the milling conditions [28, 29]. Two groups [23, 25] applied mechanical alloying to prepare FeS. They used different types of milling equipment and reported somewhat different results. Metal chalcogenides [30] and iron sulfides have been of great interest in Mössbauer spectroscopy [31].

We decided to perform further investigations on the Fe-S system using the SPEX 8000 shaker mill. The primary questions were whether the time dependence and the distribution of the product phases are influenced by the higher milling intensity. Some slightly off-stoichiometric samples were also studied to determine if a high defect concentration is tolerated by the FeS phase under the nonequilibrium conditions of ball milling.

2. Experimental

The samples were prepared in a SPEX 8000 shaker mill with round bottom steel vial. The milling charge consisted of 3 g of Fe-S powder blend and eight steel balls, four 6.4 mm and four 12.7 mm in diameter. Milling times varied from 1 to 70 hours. The starting materials were obtained from Alfa Aesar. Precipitated sulfur powder, 99.5% pure with -60 mesh particle size, was used in all experiments. Most preparations used 99.9% pure iron powder with particle size between 6 and 9 microns. Some samples were prepared starting from a 98% pure, -325 mesh powder. The temperature of the milling vial was measured using J-type (chromel-alumel) thermocouples pressed to the top and bottom faces of the milling vial.

X-ray diffraction was performed on a Philips powder diffractometer, utilizing Cu $K\alpha$ radiation, a vertical θ-2θ goniometer, and a solid state detector.

The Mössbauer spectra were recorded in standard transmission geometry using a ^{57}Co source in Rh matrix with an initial activity of 50mCi. The spectra were recorded on samples with $50mg/cm^2$ of powder. A thin alpha iron foil was used to verify the calibration and linearity of the spectrometer. The line width of the standard sample was of the order of 0.26 mm/s. All the spectra were least square fitted.

3. Results and discussion

A series of stoichiometric Fe:S = 1:1 samples were prepared by milling elemental powder blends for different times, t = 2, 5, 10, 12, and 19 hours. Based on the earlier results of Jiang [22], gradual transformation was anticipated and the temperature of the milling vial was not measured during the preparation. However, the X-ray diffractograms - presented in Figure 1 - indicated that most of the reaction took place between 2 and 5 hours of milling time. The X-ray diffraction patterns reported in Fig. 1

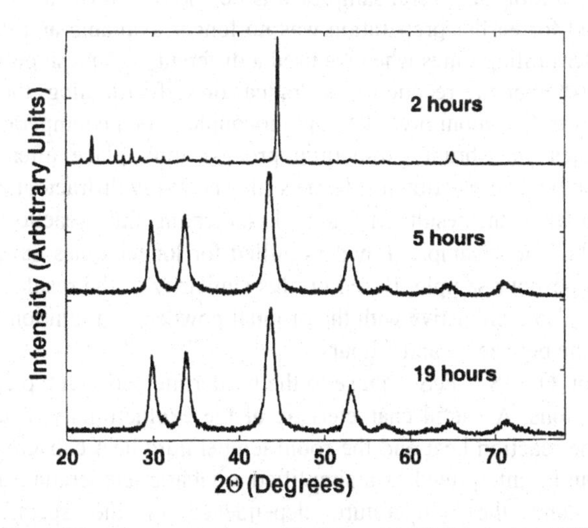

Figure 1. X-ray diffractograms of samples 570 (2 hours) , 572 (5 hours) and 571(19 hours).

are very similar to those measured by Jiang [22]. The lines of Fe and S dominate before reaction (at 2 hours milling time,) while the pattern consists of the lines of the FeS compound almost exclusively after 5 hours and longer milling times. This compound has a NiAs-type hexagonal structure with a = 0.692 nm and c = 0.576 nm. The grain size was estimated using the Sherrer equation. About 9 nm was found after both 5 hours and 19 hours of milling. As this determination neglects lattice strain as a source of line broadening, the 9 nm has to be considered a lower bound of the grain size. The characteristic Mössbauer parameters of the samples studied are tabulated in Table 1.

TABLE 1. Characteristic parameters of $Fe_{1-X}S_{1+X}$ samples.

Sample	Powder Mass (g)	X (%)	Milling Time (h)	Formula	Isomer Shift (1) δ(Fe) mm/s ± 0.03	Isomer Shift (2) δ(Fe) mm/s ± 0.03	Quadrupole Splitting Q.S. mm/s ± 0.03	Internal magnetic Field H_{int} kOe ± 0.02
570	4	0	2	FeS	0.37	-0.00	0.71	330.50
571	3	0	19	FeS	0.34	0.66	0.72	275.16
572	3	0	5	FeS	0.36	0.70	0.72	276.82
573	3	0	10	FeS	0.34	0.57	0.60	267.72
574	3	0	12	FeS	0.38	0.58	0.60	268.20
575	3	10	10	$Fe_{0.9}S_{1.1}$	0.35	0.62	0.70	263.58
576	3	5	10	$Fe_{0.95}S_{1.05}$	0.36	0.61	0.70	264.07
577	3	3	10	$Fe_{0.97}S_{1.03}$	0.36	0.65	0.70	268.58
578	3	(-3)	10	$Fe_{1.03}S_{0.971}$	0.32	0.64	0.60	274.71
579	3	3	70	$Fe_{0.97}S_{1.03}$	0.36	0.61	0.60	269.36

In order to obtain more detailed information on the reaction kinetics between 2 and 5 hours, the preparation of several samples was decided. Unfortunately, the 6-9 micron iron powder used for earlier preparation was no longer available and the reaction took place after shorter milling times when we used a different, 325 mesh powder. In order to clearly decide, whether the reaction was gradual or self-sustaining, the temperature of the milling vial was now monitored with thermocouples. As the temperature recording in Figure 2 indicates, an abrupt exothermic process took place after about 80 min, reflecting the combustive reaction inside the vial. The X-ray diffractograms of these new samples are similar to the results in Figure 1, indicating little reaction before 80 min, similar to the old 2-hour sample. Powders milled for longer times gave diffractograms similar to the old 5-hour sample. Based on this similarity of the results, we assume that the reaction was also combustive with the original powder, and ignition took place at an undetermined time between 2 and 5 hours.

The reaction Fe + S = FeS is close to the borderline between typically gradual and combustive reactions. A useful characteristic of the exothermicity of a reaction is the ratio between the reaction heat and the room temperature heat capacity of the product. This quantity can be interpreted as a simplified adiabatic temperature neglecting phase transformations and the temperature dependence of the specific heat. Using thermochemical parameters from [32] this simplified adiabatic temperature is about 1980 K for the formation of FeS. This value is about the same as the value for the reduction of CuO with Fe, another borderline reaction that was found gradual or combustive depending on the amount of powder and the number and size of the milling balls, even when the same SPEX 8000 mill was employed [33].

The reaction kinetics observed by Jiang [22] was slower, due to the lower intensity milling equipment (Fritsch Pulverisette 5.) According to their Fig. 8 [22] most of the reaction took place between 19 and 43 hours, compared to between 2 and 5 hours in our first series and the 80 minutes ignition time with the 325 mesh powder. Although Jiang et al. [22] do not consider this possibility, a significant fraction of the reaction could take place combustively also in their study. This is suggested by the fact that most of the transformation took place between 20 and 40 hours of milling in their 50 mol % Fe-S mixture, while more gradual changes, extending to longer milling times were observed at other stoichiometries.

The room temperature Mössbauer spectra of samples 573 (10h), 574 (12h), 571 (19h) and 572 (5h), Figure 3 (a), are very similar, the primary difference between them is the intensity ratio of the components. This is similar to the result of Lipka et al. on mechanosynthesized FeS [24]. The same authors obtained very similar results on ball milled chalcopyrite, $CuFeS_2$, with increasing doublet component as a function of processing time. Initial Mössbauer spectra measured at liquid nitrogen temperature show a small increase of the magnetically split fraction, indicating that part of the sample is superparamagnetic. It relaxes fast at room temperature but slowly enough to give a hyperfine split sextet at low temperature.

Off-stoichiometry has little effect on the Mössbauer spectra, except for a small increase of the average hyperfine field with decreasing sulfur content. The X-ray diffraction patterns are also insensitive to the slight changes of the composition. The

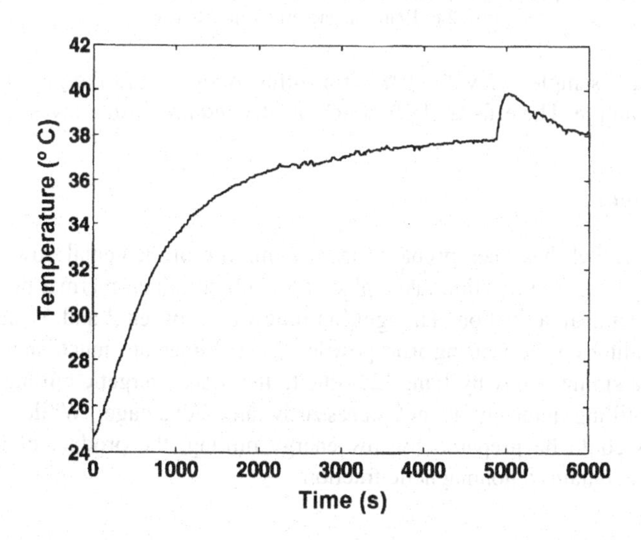

Figure 2. The temperature of the milling vial as a function of milling time.

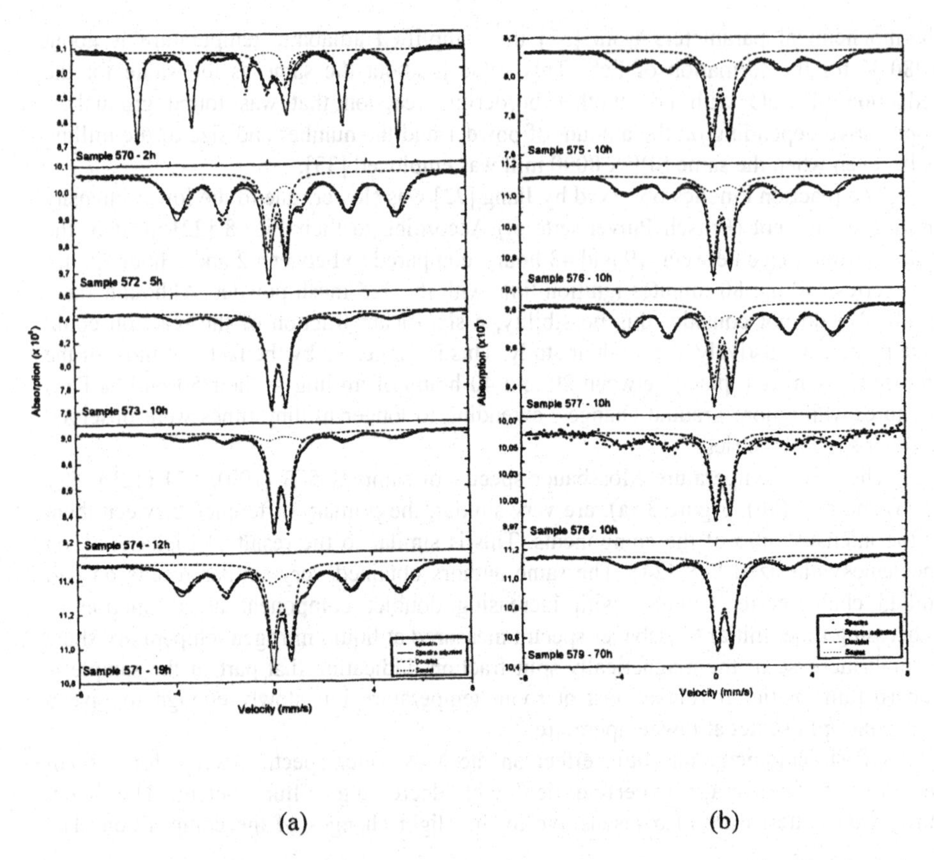

Figure 3. Room temperature Mössbauer spectra of $Fe_{1-x}S_{1+x}$ samples (x = 0), with milling times: (a) from 2 to 19 hours, and (b) 10 and 70 hours.

only exception is sample 575 with 10% extra sulfur. A significant FeS_2 pyrite fraction is found in that sample. The effects of off-stoichiometry require further investigation.

4. Conclusions

The iron sulfide FeS has been prepared from a mixture of elemental powders by high energy ball milling. The reaction takes place as a self-sustained thermal process after a period of mechanical activation. The ignition time was between 2 and 5 hours or about 80 min depending on the starting iron powder. These times are much shorter than the time scale of a similar study by Jiang [22] due to the more energetic milling equipment. The higher milling intensity is not necessarily an advantage. While single-phase magnetic FeS could be prepared by low energy milling, the product of high energy milling always contains a nonmagnetic fraction.

Acknowledgements

Financial support from CNPq (project number VKG-400134/97-7), FAPDF (project number (VKG-193050/96) and from NSF (LT, contract number DMR-9712141) is thankfully acknowledged. The authors also wish to thank the help of Ali Bakhshai with the preparation and diffraction measurements on the samples.

References

1. Benjamin, J.S. (1970) Dispersion strengthened superalloys by mechanical alloying, *Metall. Trans.* **1**, 2943-2951.
2. Koch, C.C., Cavin, O.B., Mckamey, C.G and Scarbrough, J.O. (1983) Preparation of amorphous $Ni_{60}Nb_{40}$ by mechanical alloying, *Appl. Phys. Lett.* **43**, 1017-19.
3. Fecht, H.J. (1996) Formation of nanostructures by mechanical attrition, in A.S. Edelstein and R.C. Cammarata (eds.) *Nanomaterials: Synthesis, Properties and Applications*, Institute of Physics Publishing, Bristol, pp. 89-110.
4. Koch, C.C. (1989) Materials synthesis by mechanical alloying, *Annu. Rev. Mater. Sci.* **12**, 121-43.
5. Weeber, A.W. and Bakker, H. (1988) Amorphization by ball milling. A review, *Physica B* **153**, 93-135.
6. Jiang, J.Z., Lin, R. and Mørup, S. (1998) Characterization of nanostructured α-Fe_2O_3-SnO_2, *Mater. Sci. Forum* **269-72**, 449-54.
7. Suzuki, K. and Sumiyama, K. (1995) Control of structure and formation of amorphous and nonequilibrium crystlline metals by mechanical alloying, *Materials Transactions, JIM*, **36**, 18897.
8. Juhasz, A.Z. and Opoczky, L. (1990) Mechanical Activation of Minerals by Grinding, Pulverizing and Morphology of Particles, Akadémia Kiadó, Budapest, Hungary.
9. Avvakumov, E.G. (1986*) Mechanical Methods of the Activation of Chemical Processes*, Publishing House Nauka, Novosibirsk, in Russian.
10. Kamei, T., Isobe, T., Senna, M., Shinohara, T., Wagatsuma, F., Sumiyama, K. and Suzuki, K. (1998) Reduction and change of magnetic properties of $Co(OH)_2$ on milling with Al, *Mater. Sci. Forum* **269-272**, 247-52.
11. Tschakarov, Chr.G., Gospodinov, G.G. and Bontschev, Z. (1982) Über den Mechanismus der Mechanochemischen Synthese anorganischer Verbindungen, *J. Sol. State Chem.* **41**, 244-52.
12. Takacs, L. and Susol M.A. (1996) Gradual and combustive mechanochemical reactions in the Sn-Zn-S systems, *J. Solid State Chemistry* **121**, 394-9.
13. Schaffer G.B. and McCormick, P.G. (1990) Combustion and resultant powder temperatures during mechanical alloying, *J. Mater. Sci. Letters* **9**, 1014-6.
14. Takacs, L. (1992) Reduction of magnetite by aluminum: a displacement reaction induced by mechanical alloying, *Mater. Lett.* **13**, 119-24.
15. Boldyrev, V.V. (1993) Mechanochemistry and mechanical activation of solids, *Solid State Ionics* **63-65**, 537-43.
16. Maurice, D.R. and Courtney, T.H. (1990) The physics of mechanical alloying: A first report, *Metall. Trans. A* **21A**, 289-303.
17. Schaffer G.B. and McCormick, P.G. (1992) On the kinetics of mechanical alloying, *Metall. Trans. A* **23A**, 1285-90.
18. Kosmac, T., Mauric, D. and Courtney, T.H. (1993) Synthesis of nickel sulfides by mechaanical alloying, *J. Am. Ceram. Soc.* **76**, 2345-52.
19. Balaz, P., Havlik, T., Briancin, J. and Kammel, R. (1995) Structure and properties of mechanically synthesized nickel sulphides, *Scr. Metall. Mater.* **32**, 1357-62.
20. Balaz, P., Havlik, T., Bastl, Z. and Briancin, J. (1995) Mechanosynthesis of iron sulphides, *J. Mater. Sci. Lett.* **14**, 344-6.
21. Balaz, P., Bastl, Z., Havlik, T., Lipka, J. and Toth, I. (1997) Characterization of mechanosynthetized sulphides, *Mater. Sci. Forum* **235-238**, 217-22.
22. Jiang, J.Z., Larsen, R.K., Lin, R., Mørup, S., Chorkendorff, I., Nielsen, K. and West, K. (1998) Mechanochemical synthesis of Fe-S materials, *J. Solid State Chem.* **138**, 114-125.

23. Lin, R. Jiang, J.Z., Larsen, R.K., Mørup, S. and Berry, F.J. (1998) Preparation of iron sulphides by high energy ball milling, *Hyperfine Int. (C)* **3**, 45-48.
24. Lipka, J., Miglierini, M., Sitek, J., Baláz, P. and Tkácova, K. (1993) Influence of mechanical activation on the Mössbauer spectra of the sulphides, *Nucl. Instrum. and Methods in Phys. Res.* **B76**, 183-4.
25. Lipka, J., Baláz, P., Tóth, I., Sitek, J. and Vitázek, K. (1996) Mechanosyntheized iron sulphides. Abstracts of the *Int. Symp. on the Industrial Applications of Mössbauer Effect*, Nov. 4-8, Johannesburg, South Africa, p. 199.
26. Park, Y.H., Hashimoto, H., Abe, T. and Watanabe, R. (1994) Mechanical alloying process of metal-B (M ≡ Ti, Zr) powder mixture, *Mater. Sci. Engin.* **A181/A182**, 1291-5.
27. Takacs, L. (1996) Ball milling-induced combustion in powder mixtures containing titanium, zirconium, or hafnium, *J. Solid State Chemistry* **125**, 75-84.; Combustive mechanochemical reactions with titanium, zirconium and hafnium, *Materials Science Forum* **225-227**, 553-8.
28. Sherif El-Eskandarany, M., Akoi, K., Sumiyama, K. and Suzuki, K. (1997) Cyclic crystaline - amorphous transformations of mechanically alloyed $Co_{75}Ti_{25}$, *Appl. Phys. Lett.* **70**, 1679-81.
29. Magini, M., Burgio, N., Martelli, S., Padella, F., Paradiso, E. and Ennas, G. (1991) Early and late mechanical alloying stages of the Pd-Si system, *J. Mater. Sci.* **26**, 3969-76.
30. Vaughan, D.J. and Craig, J.R. (1978) *Mineal Chemistry of Metal Sulfides*, Cambridge University Press, Cambridge.
31. Garg, V.K., Liu Y.S. and Puri, S.P. (1974) Mössbauer electric field gradient study in FeS_2 (pyrite), *J. Appl. Phys.* **45**, 70-2; Garg, V.K. (1980) Mössbauer studies of iron sulphide minerals, *Revista Brasileira de Fisica* **10**, 535-59; Garg, V.K. and Nakamura, Y. (1991) Magnetic properties of iron marcasite FeS_2, *Hyperfine Interactions* **67**, 447-52, and the references therein.
32. Kubaschewski, O., Alcock, C.B. and Spencer, P.J. (1993) *Materials Thermochemistry*, Pergamon Press, Oxford, (6th ed.)
33. Takacs, L. (1998) Combustion phenomena induced by ball milling, *Mater. Sci. Forum* **269-272**, 513-22.

STRUCTURAL EVOLUTION IN MECHANICALLY ALLOYED Fe-Sn

^{57}Fe and ^{119}Sn Mössbauer Effect Investigations

G.A. DOROFEEV[1], G.N. KONYGIN[1], E.P. YELSUKOV[1],
I.V. POVSTUGAR[2], A.N. STRELETSKII[2], P.YU. BUTYAGIN[2]
A.L. ULYANOV[1], AND E.V. VORONINA[1]
[1)] *Physical-Technical Institute of UrB RAS, 132,Kirov Str., Izhevsk, 426001, Russia*
[2)] *Institute of Chemical Physics RAS, 4, Kosygin Str., Moscow, 117977, Russia*

1. Introduction

Mechanical alloying means the process in which the mixture of different powders is exposed to repeated kneading and re-fracturing by the action of ball-powder collisions until the powder is produced in which each particle has the composition of the initial powder mixture. The process proceeds at low temperature ($T < 0.3\ T_m$) when diffusion is hindered.

Considering mechanical alloying as the process of energy transfer to the substance, storage of this energy in various forms and dissipation of the energy by solid state reactions gives the key to understanding the general principles. However, the variety of stored energy forms in a solid and diversity of the energy dissipation pathways cause the system to pass through a lot of stages in the process of formation of the end product. To understand the mechanical alloying process on the whole all the steps succession is to be studied. Nevertheless, in the literature this problem is not paid enough attention to especially concerning the most initial stages of mechanical alloying.

Since the mechanical alloying processes are connected with reorganization of all the structure of the material including the atomic level, to study them, the techniques having sufficient locality are required. One of such methods is Mössbauer spectroscopy. The most interesting model subject from the point of view of the Mössbauer spectroscopy applications for mechanical alloying research is the *Fe-Sn* system. The presence of both ^{57}Fe and ^{119}Sn Mössbauer isotopes allows to solve the problem comprehensively, examining the atomic environments of each component.

The *Fe-Sn* system is also interesting characterizing the positive mixing enthalpy (ΔH_{mix} = 4-5 kJ/mol [1]), that means the absence of chemical driving force for solid solutions formation. Nevertheless, early studies [2,3] of mechanical alloying in the *Fe-Sn* system with the ratio of components close to 70/30 at.% have shown, that the final product, despite positive mixing enthalpy, is a supersaturated solid solution, and the process goes through the formation of an intermediate product - $FeSn_2$ compound. The rate of

M. Miglierini and D. Petridis (eds.), Mössbauer Spectroscopy in Materials Science, 151–160.

the processes only but not the type of solid state reactions depends on the power intensity of the milling equipment.

Since the mechanical alloying process is athermic it is more correctl to characterize it not by the transformation rate but by the energy yield $G = dN/dD$ [4], where dN is the change in the number of moles of the reaction product, dD is the quantity of the transferred energy which has induced transformation. Hence, the basic characteristic of the milling device is specific power intensity J_m that is the energy dose transferred to a unit powder mass in a unit of time. The energy approach is more useful when possible pathways of transferred energy transformation are analyzed and the numerical estimations of probability of one transformation or another are made. Such an approach is certainly preferable when the results of the experiments carried out with the milling devices different in power intensity are compared. Unfortunately, overwhelming majority of published works in the mechanical alloying/grinding field are carried out without any indication of mechanoreactor power intensity which makes impossible any comparison of the results. In the given work special attention was paid to the attestation mills used, that has enabled to present results in a uniform energy scale.

Thus, the purpose of the present research was both ^{57}Fe and ^{119}Sn Mössbauer study of solid state reactions of mechanical alloying of pure metals *Fe* and *Sn* taken in 68:32 atomic ratio, with detailed consideration of all stage features. The energy approach is used in the analysis of the results.

2. Experimental

The mixture of pure powders of *Fe* and *Sn* (size of particles ≤ 300 μm) in the atomic ratio 68*Fe*/32*Sn* corresponding to the monotectic concentration in the *Fe-Sn* equilibrium diagram and being far from stoichiometric compositions of Fe_3Sn, Fe_5Sn_3, Fe_3Sn_2, *FeSn* and $FeSn_2$ intermetallic compounds within the given system was taken for investigation .

The mechanical alloying was carried out in two types of milling devices:

- Vibratory mill with steel balls (29×∅ 6 mm) and a steel vial of volume 30 cm^3; the mass of the loaded powder - 2 g;
- Pulverizette-7 planetary ball mill with balls (20×∅ 10 mm) and vials made of stainless steel; the mass of the loaded powder - 10 g.

Both mills were tested by the calorimetric method and the power intensity was determined. In the experiments the power intensity J_m of the vibratory mill was varied within of 1.3-7.5 W/g, and Pulverizette-7 mill was 2.0 W/g. The transferred ball-powder energy dose was defined as $D_m = J_m \tau$, where τ is the duration of processing.

γ- Mössbauer transmission spectra were received using $^{57}Co(Cr)$ and $^{119}Sn(CaSnO_3)$ sources at room and liquid nitrogen temperature. The generalizing regular algorithm for the solution of the inverse problem [5] was used to find functions of distribution of hyperfine parameters.

X-ray diffraction patterns of samples were taken on automatic DRON-4 (*Cu* K_α monochromatized radiation) and DRON-3m (*Fe* K_α filtered radiation).

3. Results and discussion

The quantitative phase X-ray analysis results depending on the specific dose of the transferred energy in the mechanical alloying of the $Fe_{68}Sn_{32}$ mixture are presented in Fig. 1. At the early stage of mechanical alloying within 5 kJ/g dose no new phases are formed. Nevertheless, the *β-Sn* quantity decreases appreciably and reaches approximately 20 at.% at 5 kJ/g dose. With further grinding the new phase component - $FeSn_2$ intermetallic compound appears, whose quantity grows very quickly and already with the 7 kJ/g dose runs into the maximal value (≈ 40 at.%). At the same time the *β-Sn* quantity also sharply decreases practically down to zero. With the 10-20 kJ/g dose the final product of mechanical alloying - *α-Fe(Sn)* supersaturated solid solution begins to form. Further, the solid solution becomes predominant right until the full disappearance of pure *α-Fe* and $FeSn_2$ phases (60 kJ/g) and the *α-Fe (Sn)* monophase formation.

Thus, the mechanical alloying in $Fe_{68}Sn_{32}$ mixture proceeds in several stages among which one can resolve the initial stage without new phases formation, the stage of $FeSn_2$ intermetallic compound formation and the final stage of the mixture transformation into the only phase of the *bcc* supersaturated solid solution based on *α-Fe*.

Let us consider the features of mechanical alloying at separate stages in more detail.

3.1. EARLY STAGE OF MECHANICAL ALLOYING

The X-ray diffraction patterns of the $Fe_{68}Sn_{32}$ mixture at the earliest stage of grinding are given in Fig.2,a. On increasing the dose the dispersion of mixture components occurs, which is testified by the broadening of the X-ray reflexes and, as a consequence by

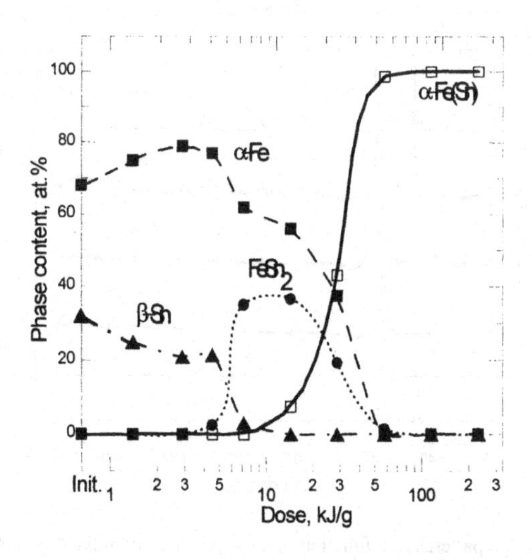

Figure 1. Phase content as a function of dose.

increasing the contact surface area of initial components. However, the contact surface area formed is not enough for alloying yet. Nevertheless, the phase composition does not remain constant and the β-Sn quantity decreases (see Fig. 1).

Fig. 3 illustrates the ^{57}Fe Mössbauer spectra and $P(H)$ functions with different doses of grinding. The top spectrum corresponding to 3 kJ/g, practically does not differ from the spectrum of pure α-Fe, and the effective field is equal to the field in iron. The measurements of the lattice parameters of α-Fe and β-Sn have shown that they remain constant over the whole initial stage. Hence, mutual dissolution of both Fe and Sn does not occur.

In Fig. 4 at the top the ^{119}Sn Mössbauer spectra of the 3 kJ/g dose sample and pure β-Sn for comparison are shown. The asymmetry of the spectrum line with 3 kJ/g dose testifies the presence of the Sn atoms with the unequivalent atomic environment in the sample. In Fig. 4 on the right the function of quadruple splitting distribution density $P(\Delta E_Q)$ calculated from the spectrum is given. The component with $\Delta E_Q \approx$ 0.6-0.9 mm/s besides the peak close to $\Delta E_Q \approx 0$ mm/s obviously corresponding to tin atoms in β-Sn is detected. 0.6-0.9 mm/s value is rather close to quadruple splitting in the $FeSn_2$ com-

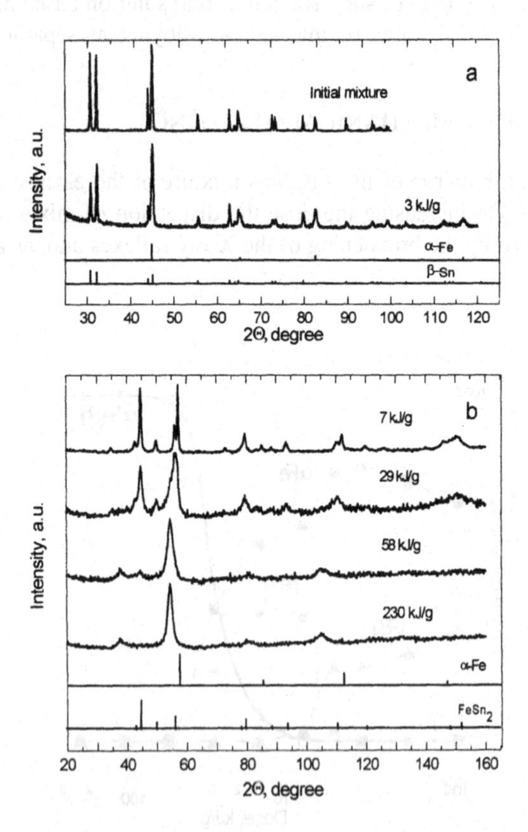

Figure 2. X-ray diffraction patterns as a function of dose. At the bottom the dash patterns of some phases are shown. a - *Cu* K_α; b - *Fe* K_α radiation.

pound. According to the data [6] $\Delta E_Q(FeSn_2)$ = 0.85±0.02 mm/s. Moreover, the isomer shift received (δ = -0.35 mm/s) appeared close to $\delta(FeSn_2)$ = -0.33 mm/s [6, relatively of β-*Sn*]. Thus, in the sample with 3 kJ/g dose the *Sn* atoms are present not only in β-*Sn*, which is obvious, but also in some formations, in which its atomic environment is similar to that in the $FeSn_2$ compound. Let us note that the absence of magnetic splitting in the spectrum of these formations can have the reasons, which will be explained below.

If the grains of the tin-containing formations giving the contribution to the ^{119}Sn Mössbauer spectrum are very small and their atomic structure is distorted, they can not give correct X-ray reflections in the diffraction pattern, the total scattered intensity making the contribution to the diffuse background. Apparently, this is the case in our example (Fig. 2a).

One can assume that at the initial stage of mechanical alloying tin as a softer metal envelops *Fe* particles and then penetrates into them along grain boundaries, forming a thin grain boundary layer with the surrounding *Fe* atoms. The atomic structure of the layer as well as its composition approximately correspond to those of the $FeSn_2$ compound. Hence, the asymmetry of the ^{119}Sn Mössbauer spectrum line can arise (Fig. 4) as well as apparent tin leaving the system registered by the X-ray diffraction.

Let us note that the *Fe-Sn* system is susceptible to tin segregation at the surface as was experimentally shown in [7]. Really, the surface energy of *Sn* (0.675 J/m^2) is less than that of *Fe* (2.475 J/m^2) [1], therefore the placement of tin along Fe grain boundaries results in some drop energy stored in the boundaries.

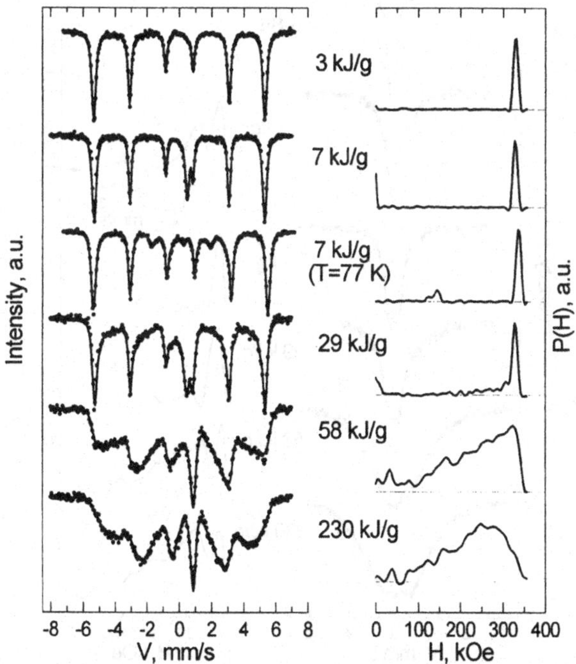

Figure 3. ^{57}Fe Mössbauer spectra with the corresponding hyperfine field distributions *P(H)* as a function of dose. The isomer shifts relatively α-*Fe* at room temperature.

3.2. EXPLOSIVE FORMATION OF $FeSn_2$ INTERMETALLIC COMPOUND

On further crushing, beginning with the dose of 5 kJ/g, a new phase component, the $FeSn_2$ intermetallic compound is formed in a step-wise manner. It is unequivocally fixed by the X-ray diffraction. In the Mössbauer spectra the singlet with the isomer shift $\delta \approx$ 0,5 mm/s in the case with ^{57}Fe (Fig. 3) and the doublet with $\Delta E_Q \approx$ 0,85 mm/s in the case with ^{119}Sn (Fig. 4) correspond to it. $FeSn_2$ has an I4/mcm tetragonal lattice of the $CuAl_2$ type [8]. The Fe sites in this compound have a high symmetry of the nearest environment which gives a zero electrical field gradient. The environment of Sn sites is less symmetric, and electrical field gradient differs from zero. However, $FeSn_2$ is an antiferromagnet at room temperature (T_N = + 107°C [8]). The absence of magnetic splitting in Mössbauer spectra at room temperature is accounted for by magnetic relaxation connected either with a fine grain [9] or with partial atomic disordering of compound [10]. In Fig. 3 the ^{57}Fe spectrum of the sample for 7 kJ/g dose obtained at liquid nitrogen temperature is given. The singlet was split in Zeeman sextet with $H_{eff} \approx$ 150 kOe, which is close to 152±2 kOe for the $FeSn_2$ at given temperature [8].

With 7 kJ/g dose the β-Sn content in the mixture sharply falls practically down to zero, while in step-wise manner the $FeSn_2$ intermetallic compound forms. There are causes to assume that the $FeSn_2$ formation under mechanical alloying has an explosive character. An explosive mechanochemical synthesis usually requires the presence of in-

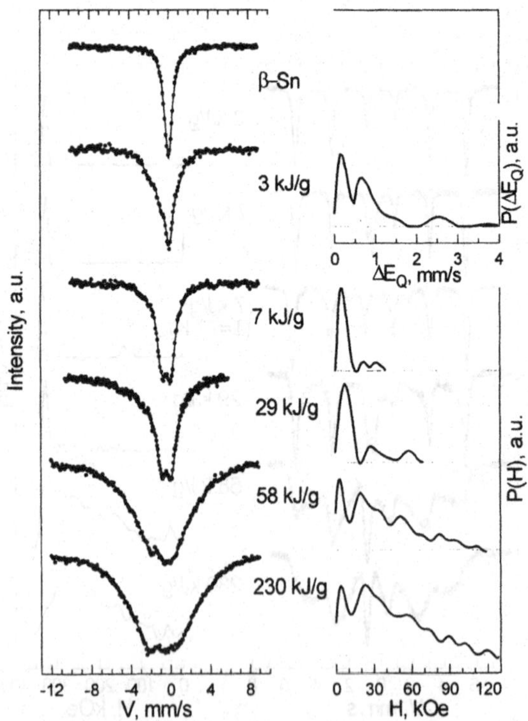

Figure 4. ^{119}Sn Mössbauer spectra with the corresponding hyperfine parameter distributions as a function of dose. The isomer shifts relatively β-Sn at room temperature.

ternal sources of energy, at the expense of which local adiabatic heating stimulating course of reaction can arise [4]. Being aware of the power intensity of the milling device and the heat of the reaction, it is possible to estimate the local jump of the temperature in a microvolume of the powder at the collision moment. The heat released with the formation of the compound portion, will be written down as follows:

$$Q = -\Delta H_{for}^{FeSn2} N, \tag{1}$$

where ΔH_{for}^{FeSn2} (-1 kJ/mole [1]) is the formation enthalpy of the compound, N is the quantity of the product formed. In its turn, in some approximation [4] it is possible to define N as

$$N = \frac{1}{2} G_0 \frac{D^2}{A_s m} \tag{2}$$

where G_0 is the permeability of the boundaries between pure components, D is the dose of the transferred energy, A_s is the work of the surface formation, m is the mass of the powder. Considering, that the heat absorbed by the powder portion at the impact moment, is made up of a transferred energy dose and the heat of the reaction, we receive a final expression for the local heating:

$$\Delta T = \frac{J_m}{c\nu}\left(\frac{-G_0 \Delta H_{for}^{FeSn2} D_m}{2A_s} + 1\right), \tag{3}$$

where J_m and D_m are the specific power intensity and specific dose (the power intensity and the dose referred to a unit of the powder mass) accordingly, c is the specific heat capacity of the powder material, ν is the collision frequency. The calculation by (3) gives for $\Delta T \approx$ 100-150^{0}C. Taking into account, that tin is a fusible metal (T_m = 232^{0}C) one can assert that the given heating is enough to bring it up to melting, or up to pre-melting temperatures, when the amplitude of the atom vibrations grows sharply and the diffusion goes fast [11]. In this interval of time interdiffusion of iron into tin can proceed via interstices to a significant depth, instantly forming $FeSn_2$ by a thermal reaction. The average grain size of the $FeSn_2$ formed, obtained from the X-ray physical broadening line analysis, is 10-20 nm. Using the known relationship $L \approx \sqrt{D_{eff}\tau}$, where L is taken as the thickness of the $FeSn_2$ layer at the joint of initial pure Fe and Sn components, τ is the time of the process, it is possible to estimate the effective diffusivity of iron in tin D_{eff} in mechanical alloying. It is ~ 10^{-18} m^2/s. The received value is by 5 orders higher than the low temperature Fe in Sn interdiffusion coefficient [12].

3.3. SUPERSATURATED SOLID SOLUTION FORMATION

The formation of α-$Fe(Sn)$ supersaturated solid solution is observed beginning with the 10-15 kJ/g dose (Fig. 1), that is when the $FeSn_2$ quantity is maximum. At the same time a great amount of pure α-Fe (50-60 at.%) is still present. This fact is well traced in the ^{57}Fe Mössbauer spectra (Fig. 3 (7-58 kJ/g)). It is seen that in $P(H)$ function in this interval of doses α-Fe component on the background of appearing low-field components of solid solution is singled out everywhere. On the contrary, in the ^{119}Sn spectra the high-field $P(H)$ components correspond to the solid solution (Fig. 4 (29 kJ/g)). Apparently, the first portions of the solid solution occur as a result of dissolution of the grain boundary phase, which appeared at the initial mechanical alloying stage. The intensive formation of the α-$Fe(Sn)$ phase coincides with the beginning of $FeSn_2$ disappearance (see Fig. 1).

The characteristic of the α-$Fe(Sn)$ formation stage is the correlation with the change of the grain size. In Fig. 5 the dose dependence of the average grain size of α-Fe, $FeSn_2$ and α-$Fe(Sn)$ phases is shown. As seen from Figs. 1 and 5 the supersaturated solid solution begins to form quickly, when the grains of all existing phases (α-Fe and $FeSn_2$) are ground down to the limiting (critical) size of 4-5 nm. To this moment all those mechanisms of the stored energy dissipation which accompanied the intermetallic compound formation stage have been exhausted. Namely, the crystallites are ground down to the limiting sizes, the stoichiometric $FeSn_2$ can no longer be formed, as all tin has reacted.

As it was shown in [13] for pure metals the achievement of a crushing grain limit corresponds to the stage of slipping along the grain boundaries (superplasticity mode). This stage seems to be present in a multiphase system as well. In this case friction of the neighbor grains along the boundaries should level the chemical composition probably with new structures formation. The solid state reaction at a final stage is characterized as essentially nonequilibrium and can be caused by the following thermodynamic factors:

- The significant contribution of interfaces to free energy with nanostructure formation, the tendency of the system to lower the surface energy of interfaces by phase transition can be determining;
- The increase of the compound free energy obtained at the first stage at the expense of «mechanical» disordering. Intermetallic compound loses its stability faster under the influence of defects, than solid solution, which is the evidence of a general regularity according to which the structures with higher symmetry are more stable in defect conditions, than the structures with low symmetry.

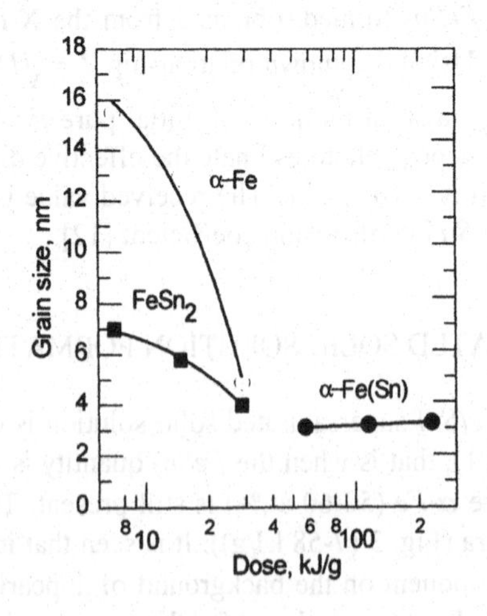

Figure 5. Grain size of phases as a function of dose.

The atomic structure of a final product of mechanical alloying in the *Fe-Sn* system nowadays does not yield to an unequivocal interpretation because of the presence of the line close to angles $2\theta \approx 35\text{-}45^0$ (*Fe* K_α radiation, Fig. 2, b) in the X-ray diffraction patterns besides correct bcc reflexes. Let us note that because of strong broadening of the lines the correct decoding is complicated. The mentioned reflex in work [2] was attributed to $D0_3$ type ordering. Earlier we considered an alternative variant of decoding [3], namely, the hexagonal Fe_5Sn_3 compound ($B8_2$-type). An appropriate stroke diffraction pattern is shown in Fig. 2, b. The ^{57}Fe Mössbauer spectrum of Fe_5Sn_3 represents the resolved superposition of three sextets with H_{eff} at room temperature 171(36), 185(54) and 321(10) kOe [14] (relative intensities of sextets obtained at $T = 77$ K [8] are shown in the brackets). Hence we find $\overline{H}_{eff} = 185$ kOe. H_{eff} calculated from the spectrum of Fig. 3 (230 kJ/g) is equal to 226 kOe. $H_{eff}(^{119}Sn)$ in Fe_5Sn_3 measured at $T = 77$ K is equal to 74 kOe [8]. The calculation from the spectrum of Fig. 4 (230 kJ/g) gives 40 kOe. The Mössbauer spectra form of our samples is rather typical for disordered solid solutions or even for an amorphous phase. Thus, neither spectra shapes nor average hyperfine parameters coincide.

On the basis of work [17], where crystallographical affinity of hexagonal $B8_2$-type structures with a *bcc* structure as the alternating deformed cells of *A2* and *B2* types is shown we conclude that the atomic structure in the end of mechanical alloying is a defective Fe_5Sn_3, unidentified from a defective *bcc* by X-ray.

4. Conclusion

The study of the mechanical alloying of the elemental powder mixture with the $Fe_{68}Sn_{32}$ atom composition shows the following: At the very beginning of mechanical alloying the nucleation of the $FeSn_2$ type formations is found by means of ^{119}Sn Mossbauer spectroscopy, while X-ray diffraction dos not detect any new phase appearance. The formations arise as a result of the *Sn* penetration along *Fe* grain boundaries. Further grinding gives rise to a new phase - the $FeSn_2$ intermetallic compound. The transformation is characterized as an explosive one. The high rate of the reaction is sustained at the cost of the heat released as a result of the transformation as well as the transferred energy in the ball-powder collision process. The given heating is enough to bring a fusible *Sn* up to melting or up to pre-melting temperatures. Under such conditions the fast migration of *Fe* into *Sn* can proceed via interstices. The final product - the supersaturated *bcc* solid solution begins to form intensively at such refining when the grain sizes of $FeSn_2$ and α-*Fe* become of several nanometers.

Acknowledgement

The financial support of the Russian Fund of Fundamental Research (grants 97-03-33483, 96-15-97491 and 98-0332725a) is gratefully acknowledged.

References

1. Boer F.R., Boom R.E.A. (1988) in de Boer and Pettifor (eds.), *Cohesion in Metals Transition Metal Alloys. Vol.1.*, North-Holland, Amsterdam.
2. Cabrera, A.F., Sanchez, F.H., and Mendoza-Zelis, L. (1995) Mechanical alloying of iron and tin powders: a Mössbauer study, *Mater. Sci. Forum* **179-181**, 231-236.
3. Yelsukov, E.P., Dorofeev, G.A., Barinov, V.A., Grigor'eva, T.F., and Boldyrev, V.V. (1998) Solid state reactions in the *Fe-Sn* system under mechanical alloying and grinding, *Mater. Sci. Forum* **269-272**, 151-156.
4. Butyagin, P.Yu. (1998) Mechanical disordering and reactivity of solids, *Chemistry Rev (in press)*.
5. Voronina, E.V., Ershov, N.V., Ageev, A.L., and Babanov, Yu.A.. (1990) Regular algorithm for the solution of the inverse problem in Mössbauer spectroscopy, *Phys. Stat. Sol. (b)* **160**, 625-634.
6. Le Caër, G., Malaman, B., Venturini, G., Fruchart D., and Roques, B. (1985) A Mössbauer study of $FeSn_2$, *J. Phys. F: Met. Phys.* **15**, 1813-1827.
7. Kanunnikova, O.M., Gil'mutdinov, F.Z., and Yelsukov, E.P. (1996) Photoelectron study of $Fe_{1-x}Sn_x$ powders, *Perspektivnyie materialy (in russian)* **6**, 71-76.
8. Trumpy, G., Both, E., Djiega-Mariadassou, C., and Lecocq, P. (1970) Mössbauer-effect studies of iron-tin alloys, *Phys. Rev. B* **2**, 3477-3490.
9. Sanchez, F.H., Socolovsky, L., Cabrera, A.F., and Mendoza-Zélis, L. (1996) Magnetic relaxations in mechanically ground $FeSn_2$, *Mater. Sci. Forum* **225-227**, 713-718.
10. Yelsukov, E.P., Voronina, E.V., Konygin, G.N., Barinov, V.A., Godovikov, S.K., Dorofeev G.A., and Zagainov, A.V. (1997) Structure and magnetic properties of $Fe_{100-x}Sn_x$ ($3.2<x<62$) alloys obtained by mechanical milling, *J. Magn. Magn. Mater.* **166**, 334-348.
11. Boyle, A.J.F., Bunbury, D.St.P., Edwards, C., and Hall, H.E. (1961) Mössbauer effect in tin from 120° K up to melting point, *Proc. Phys. Soc.* **77**, 129-134.
12. Le Caër, G., Matteazzi, P., and Fultz, B. (1992) A microstructural study of mechanical alloying of *Fe* and *Sn* powders, *J. Mater. Res.* **7**, 1387-1395.
13. Fecht, H.-J. (1995) Nanostructure formation by mechanical attrition, *NanoStruct. Mater.* **6**, 33-42.
14. Yamamoto, H. (1966) Mössbauer effect measurement of intermetallic compounds in iron-tin system: Fe_5Sn_3 and $FeSn$, *J. Phys. Soc. Japan* **21**, 1058-1062.

MÖSSBAUER INVESTIGATION OF NITRIDING PROCESSES

Gas Nitriding and Laser Nitriding

PETER SCHAAF AND FELIX LANDRY
Universität Göttingen, II. Physikalisches Institut
Bunsenstrasse 7/9, 37073 Göttingen, Germany

Abstract

Nitriding of metals and steel is well established in surface engineering, and gas nitriding is used most frequently. The addition of oxygen may be advantageous for the nitriding of high chromium steel. Recently, laser nitriding, i. e. the nitrogen take-up from nitrogen gas upon irradiation of a steel surface with short laser pulses, has been discovered and first applications have been found. Mössbauer spectroscopy is one of the first choices in steel research. Here the processes gas nitriding, gas oxinitriding and laser nitriding are compared and it will be demonstrated how Mössbauer spectroscopy in combination with complementary methods (RNRA, GDOS, Nanoindentation) can help to reveal basic mechanisms in both cases.

1. Introduction

Nitriding and nitrocarburizing are among the most versatile surface treatments of steel [1,2,3]: the fatigue, wear and corrosion properties can be improved significantly. The high-cycle fatigue resistance is improved by the diffusion zone consisting of a dispersion of iron and alloying element nitrides in a ferrite matrix. Wear and corrosion properties are improved by the compound layer (white layer) on top of this diffusion zone. The compound layer is normally formed by nucleation of γ', followed by nucleation of ε on top of this γ'. Then, after isolating the substrate from the nitriding atmosphere, growth of this compound layer ε/γ' takes place [4]. The developing phases during nitriding can be predicted on the basis of the Fe-N phase diagram [5,6,7] and its thermodynamics, which are also well known [6,8,9]. Thermodynamics, however, only describes the equilibrium state for the given nitriding conditions. The actual constitution and composition as well as the growth velocity of the nitrided case is determined by kinetics.

Phase analyses of nitrided surfaces can easily be performed by means of Mössbauer spectroscopy. Especially the backscattering geometries of Conversion Electron Mössbauer Spectroscopy (CEMS) and Conversion X-ray Mössbauer Spectroscopy (CXMS) are well suited to analyze treated steel surfaces from the very surface (100-300 nm) up to layers of technical interest (10-30 μm) [10,11,12,13]. Since the Fe-N

M. Miglierini and D. Petridis (eds.), Mössbauer Spectroscopy in Materials Science, 161–172.

system is well established [5,6,7] and technologically important, the knowledge of the Fe-N system and its phases is excellent from a Mössbauer point of view [14,15,16] (and references therein).

The activation of steel surfaces by a pre-oxidation treatment or the addition of oxygen-containing media during the gas nitriding process is well established [17,18,19,20,21]. Nevertheless, the nitriding potential has to be carefully controlled [22,23,24] and the atomistic mechanisms of these benefits are hardly known. An open question is also if the oxygen addition might lead to changes in the structure of the nitrided layer, as seen by Somers and Mittemeijer [25,26,27,28]. A Mössbauer analysis of the influence of oxygen additions during gas nitriding of low alloy and high chromium steels will be presented here.

Meanwhile, it is well established that the irradiation of iron, steel, and other metals with pulses of an excimer laser in a nitrogen atmosphere or air leads to a huge take-up of nitrogen into the irradiated surface [29,30,31,32], although the underlying basic mechanisms are hardly known. In comparison to conventional nitriding methods the use of short laser pulses has several advantages. Due to the small heat affected zone both in depth and lateral dimension, pieces sensitive to heat and of complex shape can be modified. The technique is very effective, extremely fast, and allows an accurate spatial control of the surface treatment without any undesired heating of the substrate. Mössbauer analyses of the influence of the carbon content on the laser nitriding of steel will be presented.

2. Experimental

Pure Iron (Armco, Fe > 99.8%) and carbon steels (C15, C45, C60, C80, C105) were used for the laser nitriding experiments, whereas the gas nitriding experiments were performed with the steel grades C15, 31CrMoV9, X6Cr17 and X5CrNi18.10. From the commercially available rods of 25 mm diameter, slices of 2 mm thickness were cut. Subsequently, these samples were mechanically polished to 1 µm by using first SiC grinding paper (1200, 2400, 4000 mesh) and then diamond paste with a grain size of 1 µm, resulting in a mean roughness of about 10(5) nm for all samples.

A computer-controlled furnace (150 l volume) was available for the gas nitriding experiments [19,21]. The nitriding potential K_N and the oxidation potential (oxygen activity) K_O were measured by solid electrolytic sensors and controlled via gas flow controllers [24,19]. The nitriding conditions were kept constant, only the nitriding duration t_N and the oxygen potential $K_O = p(H_2O)/p(H_2)$ were varied. The nitriding temperature was always $T_N = 823$ K at a nitriding potential of $K_N = p(NH_3)/p(H_2)^{3/2} = 6.0$ [33,34]. The variation of the oxygen activity was performed by adding air and injecting water, reaching values up to and above the oxidation threshold of pure iron ($K_O \geq 0.3$ at 823 K, [20]).

The laser nitriding was carried out employing a XeCl excimer laser with a wavelength of 308 nm, a pulse duration of 55 ns and a pulse energy of 1 J. The raw laser spot was focused to the size of 4x5 mm^2 through an f=200 mm lens, giving a mean energy density of 4 J/cm^2. The whole surface was irradiated by meandering the laser spot over the sample by small shifts after each pulse so that each area of the surface was irradiated with a total of 32 pulses (8 × 4) in pure nitrogen gas at a pressure of 1013 hPa.

Phase analysis was carried out by means of Conversion Electron Mössbauer Spectroscopy (CEMS) [35,36] and Conversion X-ray Mössbauer Spectroscopy (CXMS) [37,38]. The CEMS and CXMS spectra were taken at room temperature by a Simultaneous Triple-Radiation Mössbauer Spectrometer (STRMS) [39,40,41,42] with a $^{57}Co/Rh$ source (≈ 400 MBq) and a constant acceleration drive. The conversion and Auger electrons were detected in a He/CH_4 gas-flow proportional counter; X-rays following conversion were detected in a toroidal Ar/CH_4 gas-flow proportional counter. The spectra were stored in a multichannel scaler with a resolution of 1024 channels [43] and were fitted according to a least squares routine by superimposing Lorentzian lines [44,45]. Typical errors in the relative areas of the subspectra are in the order of 2-3%. The drive velocity was calibrated by measuring a 25 μm foil of α-Fe. Isomer shifts are given relative to α-Fe. It should be noted that the sampling range of the CEMS measurements is about 100-300 nm deep while for CXMS it is 10-20 μm [46,47]. The phase analysis is simplified by a detailed Mössbauer knowledge of the Fe-N system [12,14,15,16,48,49]. Therefore, not only phase fractions but in some cases also their composition can be determined from the Mössbauer spectra, e.g. the nitrogen contents in the austenite and in the ε-nitride [12,14].

Metallography, Glow Discharge Optical Spectroscopy (GDOS), Scanning Electron Microscopy (SEM), Energy Dispersive X-ray Analysis (EDX), Auger Electron Spectroscopy (AES) and surface profilometry have been employed for the characterization of the nitriding results. For selected samples also X-ray Diffraction (XRD) (Cu-K_α, Bragg-Brentano geometry, i.e. an information depth of 2.5-3.5 μm) was applied. Nitrogen depth profiling was performed for the laser nitrided samples by Resonant Nuclear Reaction Analysis (RNRA) [50]. The hardness was measured by the nanoindentation method employing a Fischerscope HV100 with a Vickers diamond [51,52].

3. Results and Discussion

3.1. GAS NITRIDING

The nitriding results for the steels C15 and 31CrMoV9 are summarized in Table 1. All samples of steel C15 show a closed compound layer, independent of the nitriding conditions. The benefit of the oxygen addition (K_O = 0.28-0.30) becomes obvious when looking at the increased nitriding depth and the higher ε content. Nitriding above the oxidation threshold leads to strong oxidation hindering the growth of the nitrided layers. The samples with oxygen addition have a higher surface roughness and a changed surface morphology. Pre-oxidation has no significant effect here.

The corresponding CEM spectra are displayed in Figure 1 and the influence of the oxygen activity can be recognized. All spectra are fitted as a superposition of αα'-Fe and ε-nitride. Samples 1C and 1D with the higher oxygen activities also contain magnetite (Fe_3O_4), a large amount of 35% in sample 1D and only very little in sample 1C. The corresponding CXM spectra do not show any signs of oxides. The Mössbauer phase analyses (CEMS and CXMS) are summarized in Table 2, together with XRD results.

TABLE 1. Influence of the oxygen activity K_O on the nitriding depth d_N, the roughness R_a and the nitrogen content c_N of the nitrided layers for the steels C15 and 31CrMoV9 (t_N = 0.5 h, K_N = 6.0, T_N = 823 K).

Steel	Sample	K_O	d_N [μm]	R_a [μm]	c_N [1)] [gew.%]
C15	1A	<0.1	1.8-2.3	2.4	10.2
	1B	<0.1 [2)]	1.9-2.7	2.6	10.3
	1C	0.28-0.30	2.9-3.6	4.3	11.2
	1D	>0.30	1.3-1.6	5.2	n.b.
31CrMoV9	3A	<0.1	0.0-2.5	2.0	3.2-10.7
	3C	0.28	3.9-5.5	4.5	11.3

[1)] EDX analysis, [2)] preoxidized for 1 h at 623 K

The GDOS analysis shows that the maximum of the oxygen content is located at about 1 μm below the surface. This explains the higher oxide fractions observed with XRD as given in Table 2. From the comparison CEMS-CXMS one can conclude that the oxides are limited to the outer part. In the CXMS spectra no oxides are seen and from the reduced ε-content the nitriding depth can be estimated. The possible presence of cementite or carbides [53,54] cannot be excluded by the Mössbauer results, but they are not observed in XRD either. Although the XRD analysis shows the presence of γ', this phase cannot be resolved in the CEM and CXM spectra. Maybe the larger information depth of CXMS is responsible for that because much more original substrate material is measured and the contrast of the γ' to the combination of the other subspectra is poor [14].

Contrary to C15, the steel 31CrMoV9 already shows considerable passivation. Samples not pre-oxidized and without oxygen addition only show the beginning of a nitriding effect after 0.5 h. Besides local nitride areas, large un-nitrided soft areas can be found. This behavior is dramatically changed by oxygen addition during the nitriding process. A closed 4 μm thick compound layer can be found for K_O = 0.28 after 0.5 h. Thus, the growth rate is even higher than for the unalloyed steel C15. This behavior of

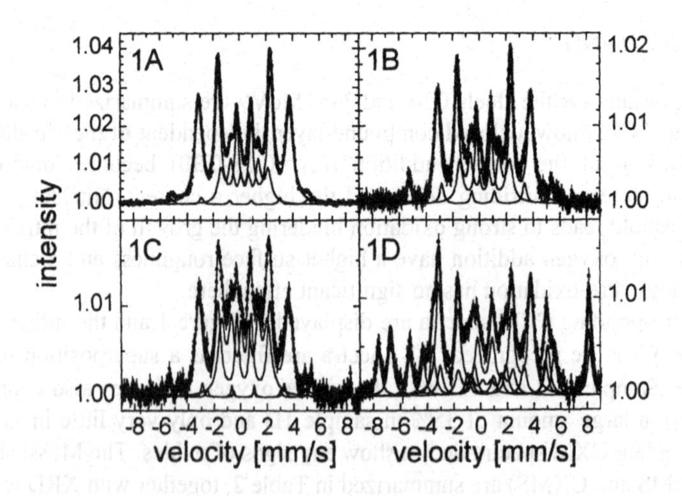

Figure 1. CEM spectra of gas nitrided steel C15 (samples 1A-1D, see also Table 1).

TABLE 2. Results of the phase analyses for nitriding of C15 and 31CrMoV9. The composition x of the ε-Fe_xN is calculated according to [12].

Steel	Sample	Method	α, α'	ε	x	γ'	Fe_3O_4
C15	1A	CEMS	3.0	97.0	2.17	0.0	0.0
		XRD	17.0	45.0		38.0	0.0
		CXMS					
	1B	CEMS	9.5	90.5	2.21	0.0	0.0
		XRD	15.0	48.0		37.0	0.0
		CXMS	83.1	16.9	2.10	0.0	0.0
	1C	CEMS	4.9	93.1	2.33	0.0	2.0
		XRD	9.0	62.0		20.0	9.0
		CXMS	72.2	27.8	2.12	0.0	0.0
	1D	CEMS	13.4	51.5	2.13	0.0	35.1
		XRD	26.0	26.0		9.0	39.0
		CXMS	86.5	13.5	3.00	0.0	0.0
31CrMoV9	3A	CEMS	12.6	87.4	2.12	0.0	0.0
		XRD	67.0	18.0		15.0	0.0
		CXMS					
	3C	CEMS	2.1	92.1	2.18	0.0	5.8
		XRD	0.0	85.0		2.0	13.0
		CXMS	57.4	39.2	2.17	0.0	3.4

low alloy steel has been explained by internal carburization [18], seen in the GDOS profile, which shows a carbon enrichment in the lower part of the compound layer. The oxide layer formed at the beginning of the treatment is dissolved by the formation of nitrides, concluded from the GDOS profile after a 2-h treatment. The maximum concentration of oxygen has been reduced from 28 at.% after 0.5 h down to only 8.4 at.% after 2 h nitriding. This behavior is also represented in the Mössbauer spectra. The sample nitrided without oxygen addition shows large fractions of α-Fe, originating from the incomplete compound layer. The sample nitrided at the oxidation threshold is composed mainly of ε-nitride. This is, in combination with CXMS, a hint to a large thickness of the compound layer.

Stainless steel with a chromium content of 17 wt.% or more can only be homogeneously nitrided at an oxygen activity at or higher than the oxidation threshold [20]. The CEM spectra of nitrided steel X6Cr17 are shown in Figure 2. For nitriding at $K_O = 0.28$-0.30 and 0.5 h the Mössbauer phase analysis exhibits 28% magnetite besides the ε-nitride and original α-Fe. After 4 h nitriding this magnetite fraction has almost disappeared and the surface layer of 100-200 nm thickness, as observed with CEMS, is becoming oxide-free again after the initial oxidation. Even if the oxygen activity is increased above the oxidation threshold (Figure 2c), not much magnetite can be found at the surface after 4 h. The Mössbauer phase analyses for the gas oxinitrided steels X6Cr17 and X5CrNi18.10 are summarized in Table 3.

GDOS and AES analyses show that after 0.5 h nitriding a thin oxide layer covers the surface. Its thickness depends on the oxygen potential and the chromium content of the steel, which determines the oxidation behavior. After 2 h and 4 h, respectively, an iron-nitride is present at the surface. The oxygen maximum is again located near the interface of the compound layer and the diffusion layer. The presented results show that

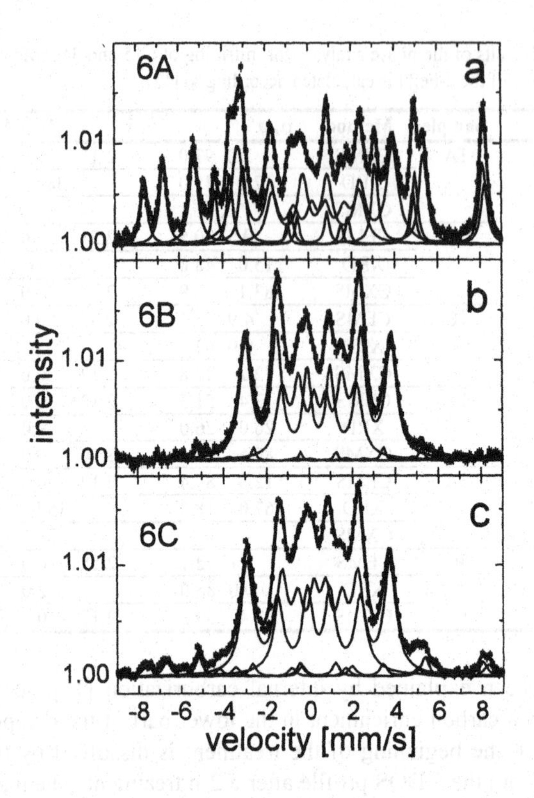

Figure 2. CEM spectra of gas nitrided steel X6Cr17: a) 0.5 h, b) 4 h, both at $K_O = 0.28\text{-}0.30$, c) 4 h at $K_O > 0.30$. (samples 6A-6C, see also Table 3)

TABLE 3. CEM phase analyses of gas nitrided steels X5CrNi18.10 and X6C17 for increasing nitriding time (phase fractions are given in %).

Steel	Sample	t_N [h]	K_O	α/α'	ε	x	Fe_3O_4	FeO
X5CrNi18.10	5A	0.5	0.28-0.30		24.8	2.13	69.7	5.5
	5B	4.0	0.28-0.30	5.0	88.5	2.51	6.5	0.0
X6Cr17	6A	0.5	0.28-0.30	20.2	52.2	2.28	27.6	0.0
	6B	4.0	0.28-0.30	2.8	94.2	2.27	3.0	0.0
	6C	4.0	>0.30	5.1	86.4	2.36	8.4	0.0

the magnetite does not hinder the nitriding. With increasing nitriding time, the oxide layer is transformed into iron-nitride, verified by the Mössbauer results. The steel X5CrNi18.10 (samples 5A and 5B) behaves analogously. This austenitic steel also shows the disappearance of an initial oxide layer. After 4 h the 75% magnetite and wüstite present after 0.5 h decrease to below 7%.

This influence of oxygen addition to the nitriding atmosphere on the mechanisms of the nitriding reactions can be explained by the thermodynamical stabilities in the quaternary Fe-Cr-O-N system. During nitriding with oxygen activities close to the oxidation threshold of pure iron, the reaction starts at the interface gas/steel with an external oxidation of iron during the heating process. This still continues after reaching the nitriding temperature for higher oxygen activities. The formation of chromium oxides, thermodynamically favored, is inhibited by kinetic reasons [20]. By this oxidation, the passivation layer, which is present on the original surface and hinders the nitrogen take-up, is destroyed, requiring increasing oxygen potentials for increasing chromium content. The iron oxide layer formed is permeable for nitrogen. Thus, parallel to the external oxidation the formation of chromium nitrides in the Fe matrix is also possible, i.e. an internal nitriding of the surface region. The oxidation threshold is higher for iron nitrides than for pure iron, as can be deduced from the thermodynamic stability diagram. After finishing the formation of chromium nitrides just above the iron oxide/metal interface, nitriding starts and a diffusion-controlled transformation of the iron oxides into iron nitrides takes place. The time until this reaction starts goes along with the chromium content, e.g. for C15 this time was below 0.5 h. The formation of chromium nitrides reduces the Cr content of the steel matrix, i.e. a transformation into the α phase can take place by this depletion in alloying elements. Indeed, for sample 5B a fraction of 5% α-Fe was observed.

3.2. LASER NITRIDING

Previous Mössbauer investigations of the laser nitriding process showed [14,16,55], that this process leads to the formation of the phases ε and γ (sometimes also some α''). Here, the results for laser nitrided Fe and carbon steels are presented. Figure 3 shows the CEM spectra as obtained from laser nitrided Fe, Ck15, C80 and C105. It immediately becomes obvious that the carbon content of the samples strongly influences the appearance of the Mössbauer spectra.

The transition from the binary Fe-N system to the ternary Fe-N-C system could be expected to cause some difficulties in the Mössbauer phases analysis. Nevertheless, all the spectra are well fitted with the proper subspectra of the phases α, γ, ε and α'' of the Fe-N system. The carbonitride ε-$Fe_x(N_{1-y}C_y)$, which probably evolves here, is hardly investigated systematically, but first results show that the appearance of this phase in the Mössbauer spectra is more influenced by x than by y [56]. Also for the austenite the additional carbon does not change the parameters significantly. A possible presence of the cementite (θ) cannot be excluded due to the strong overlap with the ε subspectra [53], but as thermodynamics favors the formation of the ε phases [57] it can be assumed that no cementite is present in the CEM spectra. Furthermore, additional carbon stabilizes the α and ε phase, whereas it reduces the existence regions of the γ and γ' phases [57].

The resulting Mössbauer phase analyses are summarized in Figure 4. An increase of the ε phase with the carbon content can be observed. The γ decreases the same way as ε increases. The α remains constant, as well as the α''. A small amount of oxide (wüstite) can be found in some cases. The CXMS spectra (not shown) with their larger information depth (10-20 µm) only show the phases α and γ, the latter rising linearly from about 5% for Fe to 20% for C105. This is an indication that the heat-affected zone, i.e. the austenized layer, also increases with the carbon content, due to a lowered austenizing

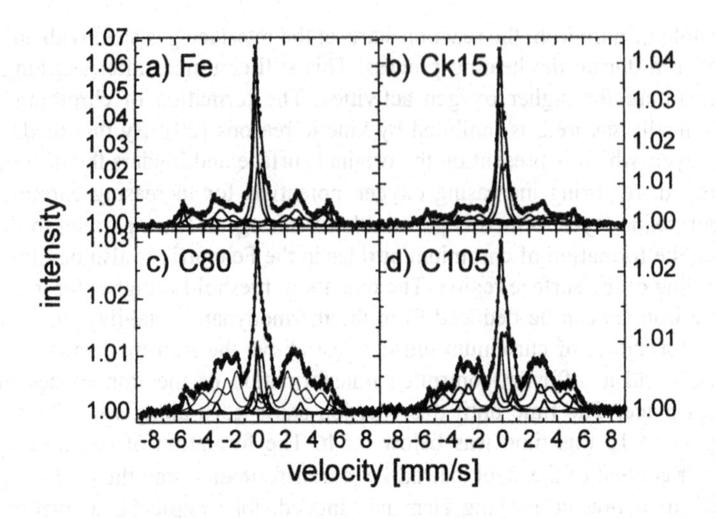

Figure 3. CEM spectra of laser nitrided samples: a) Fe, b) Ck15, c) C80, d) C105.

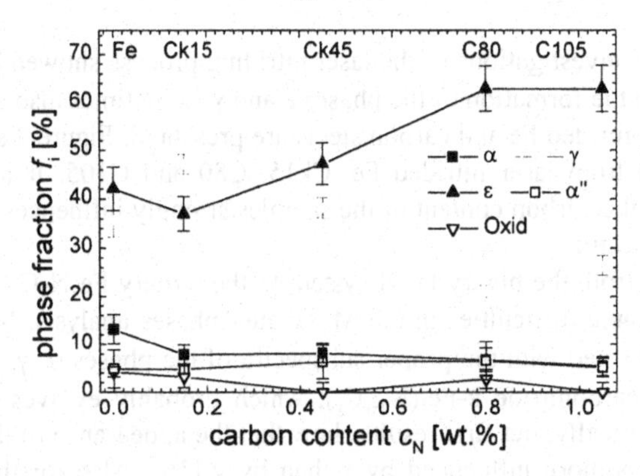

Figure 4. Phase fractions in the surface of laser nitrided steel in dependence of the carbon content as obtained by CEMS.

temperature A_3, which has a minimum at the eutectoid point in the Fe-C system at 0.80 wt.% C [58].

The surface hardness is about the same for all the samples with values around HU 1=5000 N/mm^2. The hardening depth instead increases from 1 μm for Fe and Ck15 to 2 μm for Ck45 and 7 μm for C80. For C105 the hardening depth is again reduced to 2 μm. The total nitrogen take-up as measured with RNRA does not change much with the carbon content (12 at.% mean concentration in the first 300 nm). There is only a slight increase for C80 (to 13 at.%). Thus, the increase of the ε phase with the carbon content is a result of the thermodynamics being changed by carbon [57], but not of a changed nitrogen take-up during the laser nitriding treatment.

4. Comparison

The Mössbauer phase analysis shows for both processes, gas nitriding and laser nitriding, the formation of the ε phase. The difference between the two is the additional formation of the γ phase in the case of the laser nitriding process, whereas the gas nitrided samples show the γ' phase instead. This is easy to understand when looking at the treatment temperatures: 823 K for gas nitriding and up to 4500 K for laser nitriding [59]. The occurrence of the α" phase in the laser nitrided samples is not easy to understand. The fast quenching process should hinder the formation of the complicated structure [60,61]. This may be explained by the transformation of low nitrogen ε into α" [62] or the annealing of the nitrogen martensite [60] during the treatment of neighboring areas.

Although there are great differences in treatment temperature, kinetics and phase formation, the resulting surface hardness is comparable for both processes. Figure 5 compares the microhardness depth profiles of steel C15 after laser nitriding and after gas nitriding (1A). For both cases a surface hardness of about HU 1=5000 N/mm^2 is obtained. Nevertheless, the shape of the profiles is completely different. For laser nitriding, high hardness values are already obtained at the very surface, which then soon drop to the value of the untreated material at about 1 μm. The gas nitrided sample is quite soft at the very surface and the hardness only increases to the maximum at about 0.5 μm. The hardening depth here is in the order of 3-4 μm. In both cases, the ε phase should be responsible for the high hardness (HU(ε) = 7100 N/mm^2 [63]).

5. Conclusion

The effect of oxygen additions to the nitriding atmosphere during gas nitriding and the influence of carbon on laser nitriding of iron have been investigated. It was demonstrated that Mössbauer spectroscopy, especially in its surface sensing modifications, can

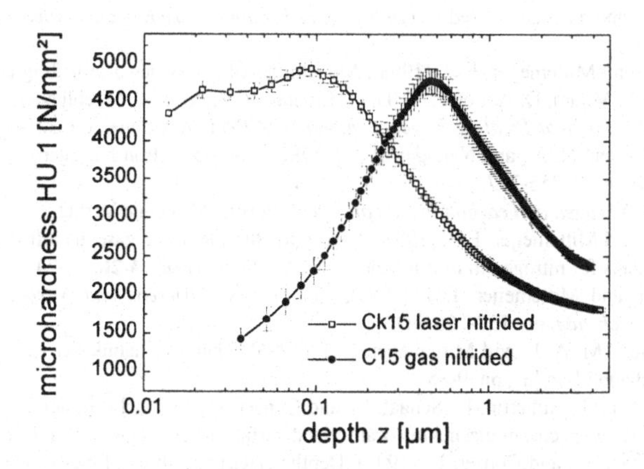

Figure 5. Comparison of the hardness obtained by laser nitriding (□) and gas nitriding (•).

help to resolve unknown mechanisms by an accurate and depth-dependent phase analysis.

An oxygen addition to the nitriding atmosphere proves to be advantageous, especially for the nitriding of alloyed and chromium-containing steels; for higher chromium contents this is even essential for a successful nitriding. The passivation layer existing at the surface of these steels is destroyed by external oxidation during the oxinitriding process. This external oxidation and the internal nitriding occur parallel in the early stages of the oxinitriding. After that, nitriding takes place unhindered.

Laser nitriding leads, within seconds, to nitride formation, resulting in surface layers of high hardness. Although the resulting phase composition is different, the ε phase occurs for both processes, its amount increasing with the carbon content of the laser nitrided steel and also with the oxygen activity for gas nitrided alloyed steel. The achieved hardness values are comparable. For laser nitriding the hardening depth is much larger for C80 as compared to the other steels. This beneficial effect may be due to the eutectoid composition.

Acknowledgment

This work was partially supported by the Deutsche Forschungsgemeinschaft (DFG, grant Scha 632/3). Prof. Dr. H.-J. Spies and his group (TU Bergakademie Freiberg, Institut für Werkstofftechnik) performed the sophisticated gas nitriding treatments. We also acknowledge the kind help of Dr. J. Barnikel and R. Queitsch (ATZ Vilseck) with the laser treatments.

References

1. ASM (1978), *Metals Handbook*. Metals Park, Ohio: American Society for Metals.
2. Mittemeijer, E. J. and Grosch, J. (eds.) (1991), *Berichtsband Nitrieren und Nitrocarburieren 1991*. Wiesbaden: AWT.
3. Mittemeier, E. J. and Grosch, J. (eds.) (1996), *Berichtsband Nitrieren und Nitrocarburieren 1996*. Wiesbaden: AWT.
4. Somers, M. A. J. and Mittemeijer, E. J. (1998), Modelling of the kinetics of nitriding and nitrocarburizing of iron, in D. L. Milam, D. A. Poteet, G. D. Pfaffmann, V. Rudnev, A. Muehlbauer, and W. B. Albert (eds.), *Proc. 17th ASM Heat Treating Society conference*. ASM International, pp. 321-330.
5. Wriedt, H. A., Gokcen, N. A., and Nafziger, R. H. (1987), The Fe-N (Iron-Nitrogen) system, *Bulletin of Alloy Phase Diagrams* **8**, 355-377.
6. Kunze, J. (1990), *Nitrogen and carbon in iron and steel*. Berlin: Akademie Verlag.
7. Somers, M. A. J. and Mittemeijer, E. J. (1995), Layer-growth kinetics on gaseous nitriding: evaluation of diffusion coefficients for nitrogen in iron-nitrides, *Metall. Mater. Trans.* **A 26**, 57-74.
8. Somers, M. A. J. and Mittemeijer, E. J. (1996), *Kinetik des Nitrierens und Nitrocarburierens: Verbindungsschichtwachstum*, pp. 157-166. in [3].
9. Kooi, B. J., Somers, M. A. J., and Mittemeijer, E. J. (1996), Thermodynamik der Fe-N Phasen und das Fe-N Zustandsschaubild, in [3], pp. 9-18.
10. Abada, L., Rixecker, G., Aubertin, F., Schaaf, P., and Gonser, U. (1993), Information depths of Conversion X-ray Mössbauer spectra in plasma nitrocarburized surface layers, *phys. stat. sol (a)* **139**, 181-187.
11. Rixecker, G., Schaaf, P., and Gonser, U. (1993), Depth selective analysis of phases and spin textures in amorphous, nanocrystalline and crystalline ribbons treated with an excimer laser, *J. Phys. D: Appl. Phys.* **26**, 870-879.

12. Niederdrenk, M., Schaaf, P., Lieb, K.-P., and Schulte, O. (1996), Characterization of magnetron-sputtered ε iron-nitride films, *J. Alloys and Compounds* **237**, 81-88.
13. Levin, L., Ginzburg, A., Klinger, L., Werber, T., Katsman, A., and Schaaf, P. (1998), Controlled formation of surface layers by Pack Aluminization, *Surf. Coatings Technology*. In print.
14. Schaaf, P., Illgner, C., Niederdrenk, M., and Lieb, K.-P. (1995), Characterization of sputtered iron-nitride films and laser-nitrided iron, *Hyperfine Interactions* **95**, 199-225.
15. Rissanen, L., Neubauer, M., Lieb, K.-P., and Schaaf, P. (1998), The new cubic iron nitride phase FeN prepared by reactive magnetron sputtering, *J. Alloys and Compounds* **274** (1-2), 74-82.
16. Schaaf, P. (1998), Iron nitrides and laser nitriding of steel, *Hyperfine Interactions* **111**, 113-119.
17. Eckstein, H.-J. and Lerche, W. (1968), Untersuchungen zur Beschleunigung der Nitrierung in der Gasphase, *Neue Hütte* **13**, 210-215.
18. Spies, H.-J. and Bergner, D. (1992), Innere Nitrierung von Eisenwerkstoffen, *Härterei-Technische Mitteilungen* **47**, 346-356.
19. Spies, H.-J. and Höck, K. (1996), Verbindungsschichtfreies Gasnitrieren, *Härterei-Technische Mitteilungen* **51**, 233-237.
20. Spies, H.-J. and Vogt, F. (1997), `Gasoxinitrieren hochlegierter Stähle'. *Härterei-Technische Mitteilun-* **52**, 342-349.
21. Spies, H.-J., Schaaf, P., and Vogt, F. (1998), Einfluß von Sauerstoffzusätzen beim Gasnitrieren auf den strukturellen Aufbau von Nitrierschichten'. *Materialwissenschaft und Werkstofftechnik*. In print.
22. Mittemeijer, E. J. and Somers, M. A. J. (1997), Thermodynamics, kinetics and process control of nitriding'. *Surface Engineering* **13**, 483-497.
23. Hoffmann, R., Mittemeijer, E. J., and Somers, M. A. J. (1994), Die Steuerung von Nitrier- und Nitrocarburierprozessen, *Härterei-Technische Mitteilungen* **49**, 177.
24. Berg, H.-J., Spies, H.-J., and Böhmer, S. (1991), Einsatz eincs Nitriersensors für die präzise Steuerung von Gasnitrierprozessen, *Härterei-Technische Mitteilungen* **46**, 375.
25. Somers, M. A. J. and Mittemeijer, E. J. (1991), Oxidschichtbildung und gleichzeitige Gefügeänderung der Verbindungsschicht, in [2], pp. 152-162.
26. Somers, M. A. J. and Mittemeijer, E. J. (1991), Verbindungsschichtbildung während Gasnitrieren und Gas- und Salzbadnitrocarburieren, in [2], pp. 24-38.
27. Somers, M. A. J. and Mittemeijer, E. J. (1992), Oxidschichtbildung und gleichzeitige Gefügeänderungen der Verbindungsschicht, *Härterei-Technische Mitteilungen* **47**, 169-174.
28. Somers, M. A. J., Kooi, B. J., Sloof, W. G., and Mittemeijer, E. J. (1994), Phase transformations and nitrogen distribution on oxidation of iron-nitride layers, *Materials Science Forum* **154**, 87-96.
29. Schaaf, P., Emmel, A., Illgner, C., Lieb, K.-P., Schubert, E., and Bergmann, H. W. (1995), Laser Nitriding of Iron by Excimer Laser Irradiation in Air and in N_2 gas, *Mater. Sci. Eng. A* **197**, L1-L4.
30. Barnikel, J., Schutte, K., and Bergmann, H. W. (1997), Nitrieren von Aluminiumlegierungen mit UV-Laserstrahlung, *Härterei-Technische Mitteilungen* **52**, 91-93.
31. Illgner, C., Schaaf, P., Lieb, K.-P., Queitsch, R., and Barnikel, J. (1998), Material transport during excimer laser nitriding of iron, *J. Appl. Phys.* **83**, 2907-2914.
32. Landry, F., Neubauer, M., Lieb, K.-P., and Schaaf, P. (1998), Optimised irradiation parameters for laser nitriding of iron and X5CrNiMo18.10, In: B. L. Mordike and F. von Alvensleben (eds.): *Proc. EKLAT 1998, September 1998, Hannover*. DGM Informationsgesellschaft, Frankfurt, 1998. In print.
33. The nitriding potential $K_N = p(NH_3)/p(H_2)^{3/2}$ (sometimes r_N) should have the unit [$bar^{-1/2}$] (or [$(101300\ Pa)^{1/2}$]). Usually the unit [$bar^{-1/2}$] is assumed and then omitted.
34. Hoffmann, F., Hoffmann, R., and Mittemeijer, E. J. (1992), Zur korrekten Nutzung der Nitrierkennzahl, *Härterei-Technische Mitteilungen* **47**, 365.
35. Spijkermann, J. J. (1981), Conversion electron Mössbauer spectroscopy, In: I. J. Gruvermann (ed.): *Mössbauer Effect Methodology, Vol. 7*. New York-London: Plenum Press, pp. 85-96.
36. Tricker, M. J. (1981), Conversion electron Mössbauer spectroscopy and its recent development, In: J. G. Stevens and G. K. Shenoy (eds.): *Mössbauer spectroscopy and its chemical applications*, Advances in Chemistry Series 194. Washington DC: American Chemical Society, Chapt. 3, pp. 63-100.
37. Keisch, B. (1972), A detector for efficient backscattering Mössbauer effect spectroscopy, *Nucl. Instrum. Methods* **104**, 237-240.
38. Fultz, B. T. (1983), Radiation detector, *United States patent 4, 393, 306*.
39. Schaaf, P. and Gonser, U. (1990), Depth analysis by combination of conversion electron, conversion X-ray and γ-ray Mössbauer spectroscopy, *Hyperfine Interactions* **57**, 2101-2104.

40. Schaaf, P., Blaes, L., Welsch, J., Jacoby, H., Aubertin, F., and Gonser, U. (1990), Experience with a toroidal proportional detector for backscattered Mössbauer γ-rays and x-rays, *Hyperfine Interactions* **58**, 2541-2546.
41. Schaaf, P., Krämer, A., Blaes, L., Wagner, G., Aubertin, F., and Gonser, U. (1991), Simultaneous conversion electron, conversion X-ray and transmission Mössbauer spectroscopy, *Nucl. Instr. Meth.* **B 53**, 184-186.
42. Gonser, U., Schaaf, P., and Aubertin, F. (1991), Simultaneous Triple Radiation Mössbauer Spectroscopy (STRMS), *Hyperfine Interactions* **66**, 95-100.
43. Schaaf, P., Wenzel, T., Schemmerling, K., and Lieb, K.-P. (1994), A simple six-input multichannel system for Mössbauer spectroscopy, *Hyperfine Interactions* **92**, 1189-1193.
44. Kündig, W. (1969), A least square fit program, *Nucl. Instrum. Methods* **75**, 336-340.
45. Landry, F. and Schaaf, P. (1998), GöMOSS - PC Fitprogram for the analysis of Mössbauer spectra, unpublished.
46. Gonser, U. and Schaaf, P. (1991), Surface phase analysis by conversion x-ray and conversion electron Mössbauer spectroscopy, *Fresenius J. Anal. Chem.* **341**, 131-135.
47. Wagner, F. E. (1976), Applications of Mössbauer scattering techniques, *J. de Physique* **37**, C6 673-689.
48. Ron, M. (1980), Iron-carbon and iron-nitrogen systems, In: R. L. Cohen (ed.): *Applications of Mössbauer spectroscopy II*. New York: Academic Press, pp. 329-392.
49. Schaaf, P., Illgner, C., Landry, F., and Lieb, K.-P. (1998), Correlation of the microhardness with the nitrogen profiles and the phase composition in the surface of laser-nitrided steel, *Surface and Coatings Technology* **100-101**, 399-402.
50. Illgner, C., Schaaf, P., Lieb, K.-P., Queitsch, R., Schutte, K., and Bergmann, H.-W. (1997), Lateral and Depth Profiles of Nitrogen in Laser Nitrided Iron, *Nucl. Instrum. and Methods* **B 122**, 420-422.
51. Oliver, W. C. and Pharr, G. M. (1992), An improved technique for determining hardness and elastic modulus using load and displacement sensing indentation experiments, *Journal Materials Research* **7**, 1564-1583.
52. Behnke, H.-H. (1993), Bestimmung der Universalhärte und anderer Kennwerte an dünnen Schichten, insbesondere Hartstoffschichten., *Härterei-Technische Mitteilungen* **48**, 3-10.
53. Schaaf, P., Wiesen, S., and Gonser, U. (1992), Mössbauer study of iron-carbides: Cementite $(Fe,M)_3C$ with various manganese and chromium contents, *Acta Metall.* **40**, 373-379.
54. Schaaf, P., Krämer, A., Wiesen, S., and Gonser, U. (1994), Mössbauer study of iron-carbides: mixed carbides M_7C_3 and $M_{23}C_6$, *Acta Metall.* **42**, 3077-3081.
55. Schaaf, P., Landry, F., Neubauer, M., and Lieb, K.-P. (1998), Laser-nitriding investigated with Mössbauer spectroscopy, *Hyperfine Interactions* p. in print..
56. Schaaf, P. (1999), The Mössbauer parameters of ε-carbonitride, in preparation.
57. Kunze, J. (1996), Thermodynamische Gleichgewichte im System Eisen-Stickstoff-Kohlenstoff, *Härterei-Technische Mitteilungen* **51**, 348-355.
58. Horstmann, D. (1985), *Das Zustandsschaubild Eisen-Kohlenstoff*. Düsseldorf: Verlag Stahleisen mbH.
59. Illgner, C. (1997), Untersuchungen zum Lasernitrieren von Eisen, PhD Dissertation, Universität Göttingen, Göttingen.
60. Jack, K. H. (1951), The occurence and crystal structure of alpha"-iron nitride, a new type pf interstitial alloy formed during the tempering of martensite, *Proc. Roy. Soc.* **A 208**, 216.
61. Weber, T., de Wit, L., Saris, F. W., and Schaaf, P. (1996), Search for giant magnetic moment in ion beam synthesized α''-$Fe_{16}N_2$, *Thin Solid Films* **279**, 217-221.
62. Rochegude, P. and Foct, J. (1985), The transformation $\varepsilon \rightarrow \alpha''$ in Iron-Nitrogen solid solutions studied by Mössbauer spectrometry, *phys.stat.sol.(a)* **88**, 137-142.
63. Weber, T., de Wit, L., Saris, F. W., Königer, A., Rauschenbach, B., Wolf, G. K., and Krauss, S. (1995), Hardness and Corrosion resistance of single phase nitride and carbide of iron, *Mater. Sci. Eng.* **A 199**, 205.

PHASE COMPOSITION OF STEEL SURFACES AFTER PLASMA IMMERSION ION IMPLANTATION

Y. JIRÁSKOVÁ[1], O. SCHNEEWEISS[1], V. PEŘINA[2], C. BLAWERT[3] AND B.L. MORDIKE[3]

[1)]*Institute of Physics of Materials, AS CR, Žižkova 22, CZ - 616 62 Brno, Czech Republic*

[2)]*Nuclear Physics Institute, AS CR, 250 68 Řež near Prague, Czech Republic*

[3)]*Institut für Werkstoffkunde und Werkstofftechnik, TU Clausthal, Agricolastr. 6, D-386 78 Clausthal-Zellerfeld, Germany*

Abstract

The formation of non-equilibrium phases at surfaces of austenitic X6CrNiTi1810, ferritic X10CrAl18 and austenitic-ferritic X2CrNiMoN2253 stainless steels modified by nitrogen plasma immersion ion implantation at 300 °C and 400 °C is studied using Conversion Electron Mössbauer Spectroscopy. The nitrogen depth profiles are followed by Rutherford Backscattering Spectroscopy. The results of the phase analysis are related to the surface hardness measured for each treatment.

1. Introduction

Surface processing of steels belongs to the most intensive studies of both basic and applied material research. In order to improve the surface properties, e.g., wear behaviour and hardness, with maintaining the advantageous properties of the bulk material, various methods have been used. The Plasma Immersion Ion Implantation (PI^3) is one of the newest among them. It allows a controlled treatment of stainless steels, keeping the nitrogen in a solid solution and improving the surface properties with only minor effects on the bulk ones.

Within the general effort to clarify the many different aspects involved in the PI^3 of steels, the problems of the phase composition and structure of the implanted layers as well as their thermal evolution appear. A positive feature of PI^3 processing is ascribed to the reproducible formation of a phase called expanded austenite or S-phase at low temperatures. A recent review about expanded austenite, its preparation by different surface treatment techniques and its properties, has been done by Williamson [1].

In this paper the experimental observations of a phase formation in different types of nitrogen modified stainless steels in dependence on temperature of PI^3 treatment are presented and related to measurements of surface microhardness. To reach the aim of

M. Miglierini and D. Petridis (eds.), Mössbauer Spectroscopy in Materials Science, 173–182.

these studies the ^{57}Fe Conversion Electron Mössbauer Spectroscopy (CEMS) was used for the phase analysis and the Rutherford Backscattering Spectroscopy (RBS) applied for determination of nitrogen depth profiles.

2. Experimental

The commercial austenitic X6CrNiTi1810 (X6), ferritic X10CrAl18 (X10) and austenitic-ferritic (duplex) X2CrNiMoN2253 (X2) stainless steels were studied. Their nominal chemical compositions are given in Table 1. The duplex steel X2 consists of a paramagnetic austenite (fcc) and a ferromagnetic ferrite phase (bcc). The changes in the phase composition after nitrogen implantation are compared with the phases formed at the surfaces of austenitic X6 and ferritic X10 steels implanted by nitrogen at the same conditions.

Five millimetres thick discs were cut from the bars (25 mm diameter). The surface of the samples was grinded, polished to a 1 μm diamond mirror finish and cleaned in an alcohol ultrasonic bath prior to PI^3 treatment.

PI^3 was used for nitriding of the samples at 300 °C and 400 °C for 3 hrs. The process parameters are summarised in Table 2. The specimens of 10 mm in diameters for CEMS measurements were prepared by cutting of implanted discs using spark erosion.

TABLE 1. Composition of studied commercial stainless steels in wt. %. The remainder is Fe.

Steel	C	Si	Mn	P	S	Cr	Ni	Ti, Al, Mo, N
X6CrNiTi1810	≤0.08	1	2	0.045	0.03	17÷19	9÷12	Ti: <5x %C≤0.8
X10CrAl18	≤0.12	0.7÷1.4	≤1	0.04	0.03	17÷19		Al: 0.7÷1.2
X2CrNiMoN2253	<0.3	1	2	0.03	0.02	21÷23	4.5÷6.5	Mo: 2.5÷3.5, N: 0.08÷0.2

TABLE 2. Treatment parameters for PI^3 nitriding experiments in pure nitrogen with high voltage pulses of 100 μs and 40 kV.

Temperature (°C)	Time (min)	Pressure (μbar)	Pulse frequency (Hz)	Dose (10^{17}N/cm^2)	Current density (mA/cm^2)
300	180	1.4	72	18	1.06
400	180	1.5	149	26	0.98

The nitrogen depth profiles were obtained by the Rutherford Backscattering Spectroscopy [2-4] using 2 MeV protons and a scattering angle of 170 °. This method is suitable for surface analysis in depth range up to 10 μm. The spectra were evaluated using the GISA 3.991 interactive PC program package with an interactive graphic interface [5].

The Conversion Electron Mössbauer Spectroscopy was used to determine the phase composition of the implanted surfaces. It was considered to be the appropriate method because both the implantation zone and the escape range for conversion electrons are in the same order of 500 nm. A conventional gas flow (94 % He, 6 % CH_4) electron counter was used. The ^{57}Co in a Rh-matrix was used as source. The calibration was performed against the standard pure α-iron foil. Computer processing of measured spectra was carried out using least-square analysis by assuming discrete single-, double- and six-line components with the CONFIT program package [6].

The surface hardness was measured by a Leitz Durimet microhardness tester using a Vickers indenter at a load of 25 g.

3. Results and discussion

3.1. NITROGEN DEPTH PROFILING

An example of nitrogen depth profiles from the surfaces of implanted austenitic X6 and ferritic X10 steels is depicted in figure 1. The distribution of nitrogen is influenced by the implantation up to about 300 - 500 nm from the surface whereas the diffusion into the sample volume prevails in deeper distance from the surface. Both regions are influenced by the temperature used in the PI^3 process and by the structure of the steels as can be seen from a comparison of both depth profiles.

The concentration of nitrogen in the depth up to about 300-500 nm is higher in the specimens implanted at 300 °C although the dose was higher at 400 °C. It means that at 400 °C higher amount of nitrogen diffuses into the specimen volumes. In both steels treated at 300 °C the maximum of nitrogen content is ~30 at. %, in those ones treated at 400 °C it reaches 15-20 at. %.

The influence of the different structures of steels can be clearly seen in the depth profiles. The diffusion zone in the austenitic steel X6 treated at 300 °C is about half of the diffusion zone of the specimen treated at 400 °C while in the case of ferritic steel X10 the depth of diffusion zones for both temperatures of treatment is almost the same. This means that the diffusion of nitrogen through the ferrite (bcc) structure is faster than through the austenitic (fcc) structure. This is in good relation to the diffusion constants for γ- and α-iron given in [7]. The remarkably slower diffusion found for the ferritic specimen X10 treated at 400 °C can be explained by the formation of Cr(Fe)N and the involved trapping of nitrogen as nitride. It can be also supposed that the depth profile is, in addition to the structure and the temperature of treatment, influenced by the density of defects the mobility of which depends on the temperature as well. The nitrogen atoms can be trapped in the defects and their mobility is reduced.

According to the Mössbauer results obtained from the untreated surface of austenitic X6 specimen, the as-prepared surface contains about 47 % of martensitic (bct) phase. This phase arises in the austenitic matrix due to the mechanical treatment of disc surfaces prior to the PI^3 process. The mechanical grinding and polishing introduce a very high content of defects into the surface layers causing an induction of stresses evoking the $\gamma \rightarrow \alpha'$ transformation. Heat treatments, which should simulate an influence

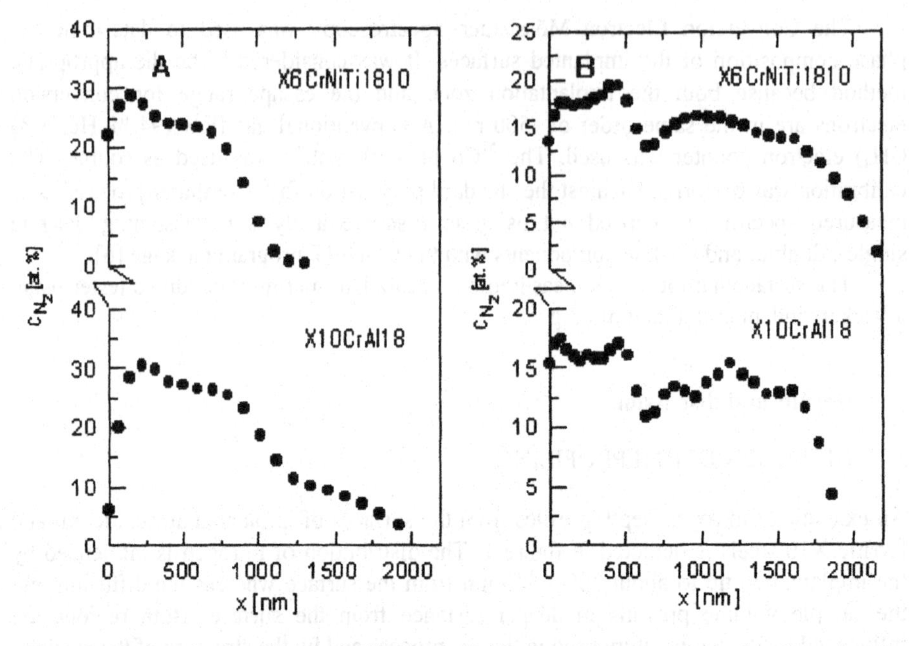

Figure 1. Nitrogen depth profiles obtained by Rutherford Backscattering Spectroscopy from investigated steels implanted at 300 °C (A) and 400 °C (B) using PI^3.

of temperature during PI^3 processing, have shown that the temperature of 300 °C for 3 hours has a minor effect on the defect structure. Only the temperature of 400 °C has evoked the annealing out of stresses and the reversal transformation of martensitic phase into the austenitic one. Similar effect was found in the duplex steel (X2) and it can be supposed in the ferritic one as well, even if it is not so pronounced from the analysis of the Mössbauer results which do not allow to distinguish the ferritic and martensitic phase precisely.

3.2. PHASE ANALYSIS

The CEMS spectra for the investigated specimens X6, X10 and X2 treated at 300 °C (A) and 400 °C (B) are depicted in figure 2. The hyperfine parameters (B-hyperfine magnetic induction, ΔB-the width of distribution of hyperfine inductions, IS-isomer shift and EQ-quadrupole splitting) are summarised in Table 3 for austenitic specimens X6, in Table 4 for ferritic specimens X10 and in Table 5 for duplex specimens X2.

The surfaces of the austenitic specimens X6 implanted at 300 °C and 400 °C reveal a low concentration of oxides FeO and Fe_2O_3, respectively, in dependence on the treatment temperature (Table 3). About 10 % of the ε-$Fe_{3-x}N$ phase with hcp structure was found only in the specimen treated at 300 °C. We suppose that this phase was facilitated by the high content of martensitic phase (47%) in the untreated surface due to the defect structure mentioned above. This phase consists of two six-line components with hyperfine induction values of 17.9 T and 9.7 T which are, according to [8], characteristic for the ε-nitride with x = 0.53. In the austenitic X6 specimen implanted at

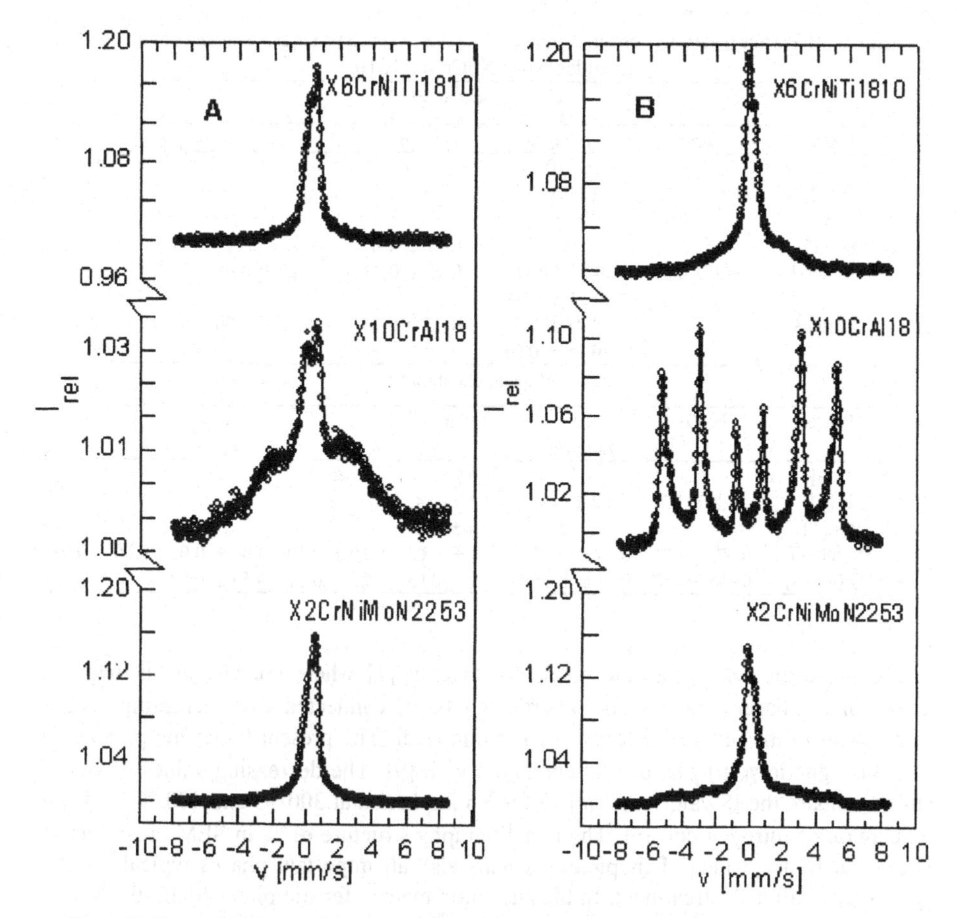

Figure 2. Conversion electron Mössbauer spectra of investigated specimens implanted at 300 °C (A) and 400 °C (B) using PI³.

300 °C about 87 % of surface can be ascribed to the expanded austenite γ_N. Its spectrum consists of a paramagnetic part γ_N(p) (76 %) represented by two doublets and a ferromagnetic part γ_N(m) (11 %) represented by a distribution of hyperfine inductions with the mean value of 11 T which is lower than that found in [1]. The structure of this part is not fully understood up to now and desires further investigations.

The CEMS spectrum of the X6 specimen implanted at 400 °C reveals a change in the magnetic nature of γ_N. The hyperfine parameters demonstrate a transition from a predominantly paramagnetic behaviour in the X6 implanted at 300 °C to a ferromagnetic one in X6 implanted at 400 °C even if the content of nitrogen, according to RBS depth profile (figure 1), is lower.

TABLE 3. Hyperfine parameters obtained from the Mössbauer spectra analysis of surfaces of austenitic stainless steel X6CrNiTi1810.

PI^3 treatment at 300 °C

Phase	FeO	ε-$Fe_{3-x}N$	γ_N (m)	γ_N (p)	
I [%]	3	10	11	28	48
B [T]		17.9 ± 0.3 9.7 ± 0.4	10.5 ± 0.4		
ΔB [T]			9.8 ± 1.0		
IS [mm/s]	0.97± 0.04	0.35 ± 0.04 0.33 ± 0.04	0.05 ± 0.04	0.20 ± 0.01	0.44 ± 0.00
EQ [mm/s]		-0.28 ± 0.04 -0.13 ± 0.04	-0.20 ± 0.03	0.33 ± 0.00	0.23 ± 0.00

PI^3 treatment at 400 °C

Phase	Fe_2O_3	α′-Martensite	γ_N (m)		γ_N (p)	
I [%]	1	1.5	58.5	25	3	11
B [T]	52.8 ± 1.7	33.2 ± 0.4	11.5 ± 0.4			
ΔB [T]			19.4 ± 1.2			
IS [mm/s]	0.67 ± 0.18	0.04 ± 0.05	0.27 ± 0.02	0.10 ± 0.00	0.06 ± 0.02	0.00 ± 0.01
EQ [mm/s]	0.14 ± 0.12	0.01 ± 0.05	-0.09 ± 0.03	0.22 ± 0.01	0.55 ± 0.02	

This result contradicts the result obtained in [1] where the change in magnetic nature of γ_N phase, paramagnetic→ferromagnetic, is connected with increasing content of nitrogen in the implanted layer of austenitic steel. The present hyperfine parameters of paramagnetic γ_N(p) part can be compared with [9]. The decreasing values of isomer shift (compare the IS values in Table 3 for X6 implanted at 300 °C and 400 °C) indicate a decrease of nitrogen content. The metallography structure of γ_N in SEM cross section is seen in figure 3 [11]. This picture reveals also all important phases typical for that type of steel after PI^3 treatment. In the austenitic matrix ferritic plates (delta-ferrite) are

Figure 3. Expanded austenite layer obtained on the surface of austenitic X6CrNiTi1810 steel after PI^3 at 400 °C for 3 hours.

randomly distributed. This phase was not detected by CEMS measurements because of transformation into the expanded austenite phase. The dark areas along the grain boundaries are Cr(Fe)N precipitates.

The surface of the ferritic X10 specimen before implantation consists of a ferrite phase with mean hyperfine induction of 26 T. This phase is similar to the ferrite phase of the duplex steels. After an implantation at 300 °C the expanded austenite γ_N phase consists of only 2 % of paramagnetic part γ_N(p) and of 65 % of ferromagnetic part γ_N(m) (Table 4) with the hyperfine parameters comparable with those obtained for the X6 specimen implanted at the same temperature.

TABLE 4. Hyperfine parameters obtained from the Mössbauer spectra analysis of surfaces of ferritic stainless steel X10CrAl18.

PI³ treatment at 300 °C			
Phase	Ferrite	γ_N (m)	γ_N (p)
I [%]	33	65	2
B [T]	22.5 ± 2.5	11.0 ± 0.6	
ΔB [T]	21.8 ± 2.7	8.9 ± 0.5	
IS [mm/s]	0.08 ± 0.09	0.30 ± 0.03	0.46 ± 0.02
EQ [mm/s]	-0.16 ± 0.05	-0.01 ± 0.03	0.21 ± 0.02
PI³ treatment at 400 °C			
Phase	α'-Martensite /Ferrite		γ
I [%]	97		3
B [T]	29.4 ± 0.9		
ΔB [T]	9.5 ± 0.2		
IS [mm/s]			0.24 ± 0.02
EQ [mm/s]			0.17 ± 0.01

In the specimen X10 treated at 400 °C the paramagnetic component represented by doublet of IS=0.24 mm/s and EQ=0.17 mm/s can be ascribed to the carbonitride and/or Cr(Fe)N precipitates. The surface is formed predominantly by ferromagnetic components with mean value of hyperfine induction of 29 T, which represent the ferrite/martensite phase. It is impossible to distinguish these phases on the basis of hyperfine parameters. The phase analysis does not prove the presence of expanded austenite. This steel contains the lowest concentration of elements stabilising the austenitic structure from all studied specimens. This, together with the nitrogen supersaturation and the high treatment temperature, is the reason for a decomposition of expanded austenite phase or a direct formation of ferrite/martensite phase and Cr(Fe)N precipitates.

The substrate of the duplex X2 specimen before implantation was composed of a dominant ferromagnetic, ferrite/martensite, phase (68 %) and a minority austenitic phase (32 %). After the implantation at 300 °C and 400°C (Table 5) the compositions of the surface layers are similar to the austenitic X6 specimen. The paramagnetic part of expanded austenite γ_N(p) dominates at lower temperature. The higher PI³ treatment temperature shifts the character of implanted surfaces from the magnetic point of view more to the ferromagnetic one and, according to the hyperfine parameters, the minor austenitic phase contains a lower content of nitrogen.

TABLE 5. Hyperfine parameters obtained from the Mössbauer spectra analysis of surfaces of austenitic-ferritic stainless steel X2CrNiMoN2253.

		PI^3 treatment at 300 °C			
Phase	γ_N (m)	γ_N (p)			
I [%]	13.5	25	52.5	6	3
B [T]	14.0 ± 0.1				
ΔB [T]	3.7 ± 0.1				
IS [mm/s]	0.25 ± 0.01	0.23 ± 0.01	0.45 ± 0.00	0.34 ± 0.01	0.29 ± 0.02
EQ [mm/s]	- 0.04 ± 0.01	0.32 ± 0.00	0.21 ± 0.00	0.67 ± 0.03	0.90 ± 0.05
		PI^3 treatment at 400 °C			
Phase	Ferrite/α'-Martensite	γ_N (m)		γ_N (p)	
I [%]	8.7	53.7	25	4.6	8
B [T]	29.4± 0.4	3.3 ± 0.3			
ΔB [T]		25.4 ± 1.2			
IS [mm/s]	-0.04 ± 0.02	0.24 ± 0.02	0.13 ± 0.11	0.02 ± 0.01	-0.06 ± 0.36
EQ [mm/s]	0.06 ± 0.02	-0.12 ± 0.02	0.19 ± 0.11	0.63 ± 0.01	

3.3. MICROHARDNESS

Typical surface hardness values for the different steels after PI^3 treatment at different temperatures are shown in figure 4. Due to the shallow layer thickness the microhardness values are influenced by the substrate material. They were 252 HV for the austenitic, 225 HV for the ferritic, and 275 HV for the duplex austenitic-ferritic steel. Therefore, the measured values do not represent the real implanted surfaces hardness, however, they can be satisfactory used for the comparison of the phase composition changes in the layers.

The microhardness values of the specimens treated at 300 °C are equal in the range of experimental uncertainty. This is in good agreement with the Mössbauer phase analysis revealing a high content of the expanded austenite phase, 67 % in the ferritic X10 steel and ~86 % in austenitic X6 and duplex X2 steels. Here, the microhardness is predominantly determined by this phase.

After the treatment at 400°C the situation is different. The stability of the expanded austenite is determined by the composition of the material and by the temperature. A decomposition of the expanded austenite starts first in the material with the lowest content of austenite stabilising elements which is in this case the ferritic X10 specimen. Here, no expanded austenite phase was found by Mössbauer phase analysis and the high microhardness value of 1130 HV is given by ferrite/martensite phase and Cr(Fe)N precipitates. With increasing Ni content the stability of austenite structure is increased. It means that the duplex X2 specimen with the lower content of Ni (see Table 1) might have started to decompose but the majority phase is still expanded austenite (~79 %) determining the hardness. The hardness and the amount of ferrite/martensite (~ 8.7 %) are higher than in the austenitic X6 specimen which has the highest content of Ni and therefore also the highest stability of all three materials. Here the hardness is only slightly affected by the precipitation of Cr(Fe)N.

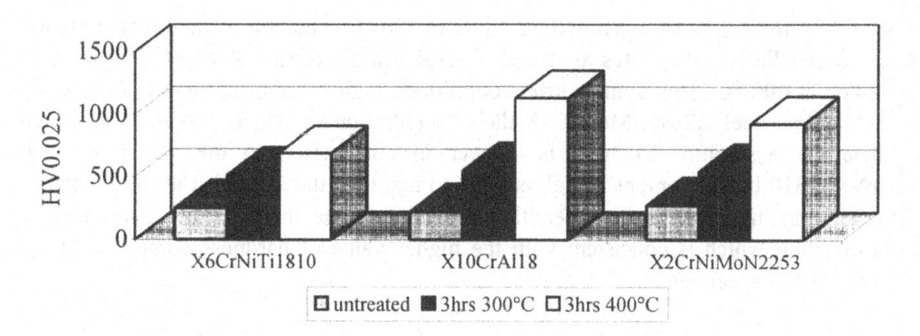

Figure 4. Surface microhardness of austenitic X6CrNiTi180, ferritic X10CrAl18 and duplex X2CrNiMoN2253 steels after different treatment.

4. Conclusions

The Mössbauer phase analysis of studied surfaces of austenitic X6CrNiTi1810, ferritic X10CrAl18 and duplex X2CrNiMoN2253 steels modified by nitrogen plasma immersion ion implantation confirmed the formation of an expanded austenite phase in all type of steels implanted at 300 °C for 3 hours.

This phase was found also in the specimens of the austenitic X6 and the duplex X2 steels implanted at higher temperature of 400 °C for 3 hours. As the Mössbauer observations have shown, this phase changes the magnetic nature in dependence on the temperature of treatment. In specimens treated at 300 °C for 3 hours the concentration of nitrogen according to the RBS depth profiles is higher and the CEMS spectra demonstrate a predominant paramagnetic phase. When the implantation temperature was changed to 400 °C for 3 hours the concentration of nitrogen decreases due to the more pronounced diffusion into the specimen volumes and CEMS spectra of both specimens, X6 and X2, reveal a transition into the predominant magnetic phase. This can be explained by increasing mobility of elements stabilising the austenitic structure at higher temperature and formation of carbonitrides and/or Cr(Fe)N, TiN precipitates. On the other hand, the change of magnetic nature of expanded austenite observed in our work is opposite to that one presented by Williamson [1] for the stainless steel ASI 304 with similar chemical composition. The reason can be found in different conditions of implantation and/or in different depth of investigated surface region (500 nm in our case, 100 nm in Williamson work) where the chemical composition can differ substantially.

The content of expanded austenite together with the composition of studied steel and the temperature of PI^3 treatment determine also the microhardness values. At lower temperature of PI^3 treatment the hardness is determined predominantly by content of expanded austenite which is in all samples high and the hardness values of specimens do not differ substantially. If the content of elements stabilising the austenitic structure is very low or zero then the higher PI^3 treatment temperature and the high nitrogen content evoke a decomposition of expanded austenite or direct formation of ferritic/martensitic phase and Cr(Fe)N precipitates as it was observed in the ferritic X10CrAl18 specimen implanted at 400 °C. This composition conditioned high value of hardness (1130 HV). In the duplex steel X2CrNiMoN2253 the concentration of Ni, as the element which stabilise the austenitic structure, is lower in comparison to the austenitic steel X6CrNiTi1810 but its content is high enough to maintain the expanded austenite phase. However, the higher content of ferritic/martensitic phase indicates the beginning of decomposition which is consistent with the higher value of hardness compared to the X6CrNiTi1810 specimen.

Acknowledgement

This work was partly supported by the Research Grant No. A4032601 of the Academy of Sciences of the Czech Republic, by the Grant Agency of the Czech Republic under project No. 202/97/0444, by the Bundesministerium für Bildung, Wissenschaft, Forschung und Technologie (BMBF) under contract number 13N6345 and by the Deutsche Forschungs Gemeinschaft (DFG) under contract number Mo319/24-1.

References

1. Williamson, D.L., Ozturk, O., Wei, R. and Wilbur, P.J. (1994) Metastable phase formation and enhanced diffusion in fcc alloys under high dose, high flux nitrogen implantation at high and low ion energies, *Surf. Coat. Technol.* **65**, 15-23.
2. Chu, W.-K., Mayer J.W. and Nicolet, A. (1978) *Backscattering Spectrometry*, Academic Press, New York.
3. Tesmer, J.R. and Nastasi, M. (1995) *Handbook of Modern Ion Beam Material Analysis*, Material Res. Soc. Pitsburg, Pennsylvania.
4. Werner, H.W. and Garten R.P.H. (1984) A comparative study of methods for thin-film and surface analysis, *Rep.Prog. Phys.* **47**, 221-344.
5. Saarilahti, J. and Rauhala E. (1992) Interactive personal-computer data analysis of ion backscattering spectra, *Nucl. Instr. Meth. Phys. Res.* **B64**, 734-738.
6. Veselý, V. (1986) FFT-based processing of unresolved spectra with multiple convolutions, *Nucl. Instr. Meth. Phys. Res.* **B18**, 88-100.
7. Adda, Y., and Philibert, J. (1966) *La diffusion dans les solides,* 1. Part, Institut National des Sciences et Techniques Nucléaires, Saclay.
8. Kopcewicz, M., Jagielski, J., Turos A. and Williamson D.L. (1992) Phase transformations in nitrogen-implanted α-iron, *J. Appl. Phys.* **71**, 4217-4226.
9. Génin ,J.-M.R. (1990) Mössbauer spectra analysis of Fe-N austenites, *Script. Metall. Mat.* **24**, 399-402.
10. Wagner, G., Leutenecker, R. and Gonser U. (1990) High resolution CEM-spectrum of nitrogen implanted austenite, *Hyper. Int.* **56**, 1653-1656.
11. Blawert, C., Mordike, B.L., Jirásková, Y. and Schneeweiss O. (in press), Phase and microstructural study of the surfaces layers produced by plasma immersion ion implantation of stainless steel X6CrNiTi1810, *J. Vac. Sci. Technol.*

SPARK ERODED Fe-N POWDERS

S. HAVLÍČEK, Y. JIRÁSKOVÁ AND O. SCHNEEWEISS
Institute of Physics of Materials, ASCR, Žižkova 22,
CZ-61662 Brno, Czech Republic

1. Introduction

When iron is exposed to nitrogen at elevated temperature several FeN phases can be formed even if the equilibrium solubility of nitrogen is limited in both, bcc and fcc, modifications of iron. The methods and the technological parameters used in the nitride processing have to be chosen from the point of view of technical applications.

Two types of iron nitrides, γ'-Fe_4N and, especially, α''-$Fe_{16}N_2$ are of large interest from the magnetic point of view. Both are ferromagnetic at room temperature with the magnetic moment equal or higher than in a pure iron. According to the phase diagram [1] the high content of nitrogen makes their direct production impossible and the special processes must be used. The attempts to prepare pure α''-$Fe_{16}N_2$ nitride, which is more interesting for the magnetic applications, were not successful. Usually a content of 18 - 25% is the maximum reached in α-Fe matrix by usage of conventional method of nitridation, e.g., plasma nitridation, ion implantation followed by heat treatment of α'-martensite or ε-nitrides at temperatures close to the reported metastable existence region for α''-$Fe_{16}N_2$.

Spark erosion method known and widely used for preparation of amorphous, nanocrystalline and crystalline powder materials is one of the unconventional methods not commonly used for production of iron nitrides. It offers new possibilities of enrichment of iron particles by nitrogen atoms and preparation of material with high content of nitrogen. A comparison of conventional and the spark erosion processing is depicted on figure 1.

The main aim of this work was to prepare the iron nitrides with possibly high content of nitrogen by the use of spark erosion of α-iron in the gaseous and liquid ammonia surroundings and to follow the changes in a phase composition of obtained powder under a heat treatment with the special attention to the formation of α''-$Fe_{16}N_2$ phase.

2. Experimental

The powder samples were prepared from pure iron ingot by a spark erosion method. Erosion was carried out using two pure iron rods as electrodes in a gaseous ammonia atmosphere of normal pressure (SEG) and in a liquid ammonia (SEL). The temperature of the liquid ammonia was kept using an ethylacohol bath in region (-80°C to

M. Miglierini and D. Petridis (eds.), Mössbauer Spectroscopy in Materials Science, 183–188.

-50°C). This method yielded powders with particle size about 8μm. Further heat treatment was curried out in vacuum furnace or in a protection $H_2 + N_2/NH_3$ atmosphere.

The phase analysis of powders was done using Mössbauer spectroscopy. The spectra were taken at room temperature in transmission geometry using ^{57}Co in Rh matrix as a source. The calibration was performed against a standard pure α-iron foil giving the magnetic splitting $g_0 = 3.9156$ mm/s corresponding to hyperfine induction of 33 T. Computer processing of measured spectra was carried out using least-square-fit by assuming of discrete single-, double-, and six-line components the hyperfine parameter of which are isomer shift IS, quadrupole splitting EQ and magnetic induction B. (CONFIT program package [3]). The contents of phases were determined according to corre-

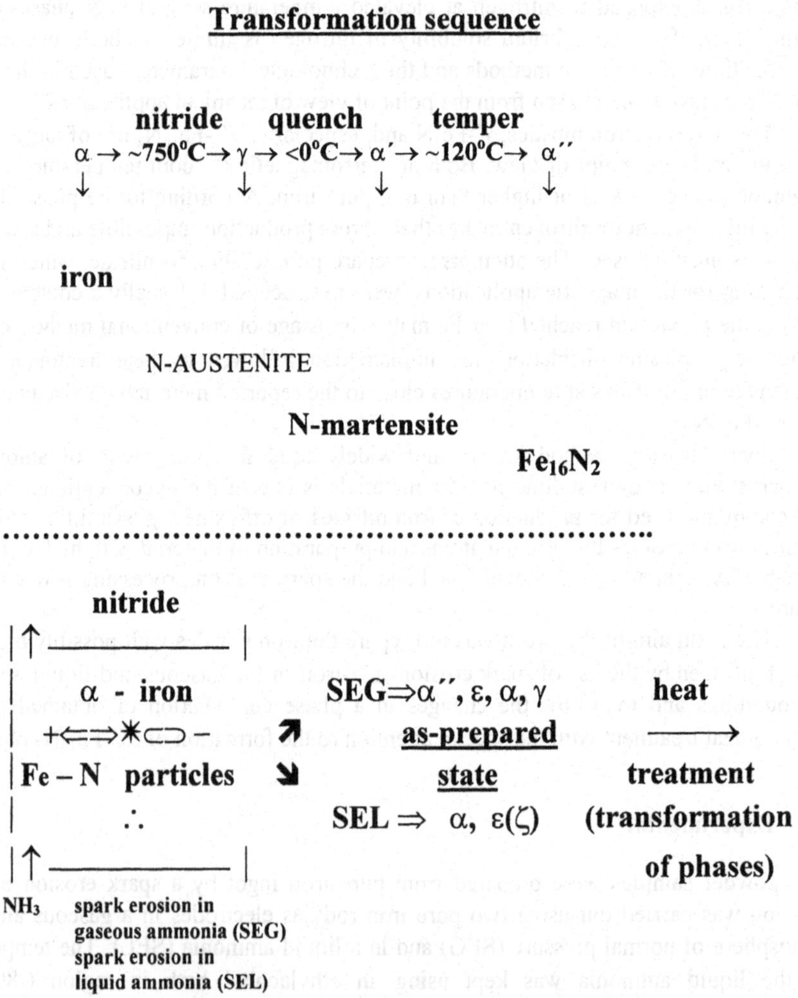

Transformation sequence

nitride quench temper

$\alpha \rightarrow \sim 750°C \rightarrow \gamma \rightarrow <0°C \rightarrow \alpha' \rightarrow \sim 120°C \rightarrow \alpha''$

↓ ↓ ↓ ↓

iron

N-AUSTENITE

N-martensite

$Fe_{16}N_2$

....

nitride

α - iron SEG ⇒ α'', ε, α, γ heat

+⇐⇒✱⊂⊃ – ➚ as-prepared ⟶

Fe – N particles ➘ state treatment

∴ SEL ⇒ α, ε(ζ) (transformation of phases)

NH_3 spark erosion in gaseous ammonia (SEG)

spark erosion in liquid ammonia (SEL)

Figure 1. Comparison of original bulk preparation $\alpha''-Fe_{16}N_2$ phase [2] (top) and method spark erosion of preparation $\alpha''-Fe_{16}N_2$ phase (bottom).

sponding subspectra intensities I. In all phases detected the same values of the Lamb-Mössbauer factor were expected.

3. Results and Discussion

Mössbauer transmission spectrum of the as-prepared SEG sample is shown in upper part of figure 2. The sample prepared in the gaseous ammonium atmosphere (SEG) contains 68 % of ferromagnetic α+α' phases with the mean value of hyperfine induction 33 T. Only 12 % of sample volume is occupied by desired α''- $Fe_{16}N_2$ phase represented by three six-line components with the hyperfine induction 37.7 T (FeIII), 35.3 T (FeII) and 29.5 T (FeI). These values are in good agreement with [4] and correspond to the three different crystallographic Fe sites of a body-centred tetragonal structure having different distances to the nearest neighbour nitrogen atoms. About 17 % of sample volume is occupied by paramagnetic austenitic γ-FeN_x phase represented by single-line component with the isomer shift IS=-0.09 mm/s and double-line component of hyperfine parameters IS=0.18 mm/s and EQ=0.191 mm/s. The relative ratio of intensities of both components r =32/68 (i.e. single-line 32%, double-line 68% of the subspectra of γ-FeN_x phase), indicates a concentration of nitrogen $x \geq 0.5$ in agreement with [5]. The rest of sample volume (3 %) is formed by ferromagnetic ε-$Fe_{3+x}N$ phase.

The paramagnetic $\varepsilon(\zeta)$- Fe_2N phase was found to be dominating (91 %) in the sample prepared in the liquid ammonia (SEL) the spectrum of which is shown in the lower part of figure 2. This phase is represented by double-line component with IS=0.35 mm/s and EQ=0.30 mm/s. The room temperature Mössbauer measurements are not sufficient to distinguish whether it is the ε or ζ phase since the hyperfine parameters of both phases are nearly the same. Some discrepancies also remain in the values of IS and EQ in the literature. Some authors are in good agreement with our result, e.g., [6] but in other papers the hyperfine parameters close to the IS=0.42 mm/s and EQ=0.23 mm/s, e.g., [7,8] are present.

A heat treatment of SEL sample at 150 °C for 90 hrs does not influence the chemical and phase composition substantially as can be seen from a comparison of spectra in figure 3 and phase analysis in table 1. Pronounced change of the Mössbauer spectrum was observed after annealing at 185 °C for 66 hrs. The phase analysis has shown that the content of $\varepsilon(\zeta)$- Fe_2N phase is strongly reduced (22 %) and the new phases α''-$Fe_{16}N_2$, ε-$Fe_{3+x}N$, γ'-Fe_4N and Fe_3O_{4-x} arise. Because of observed progressive oxidation of sample several attempts with gas mixture as protection atmosphere and vacuum treatments of powder SEL as well as pure iron samples were carried out. While the pure iron powder sample has been stable no protect atmosphere was found to prevent oxidation of SEL sample.

The SEG sample was treated at the temperature of 200 °C and the following composition was found: 83% of $\alpha(\alpha')$, 9% of γ'-Fe_4N_{1-x}, and 4% of unresolved paramagnetic phases, probably cubic $\gamma''$and γ''' FeN. Iron oxides were not identified. Partial decomposition of the γ phase and transformation of α'' phase into $\alpha(\alpha') + \gamma'$ due to the annealing can be expected according to the equilibrium phase diagram [1].

The above observation of oxidation are in good agreement with results reported about the oxidation of α-Fe and ε-Fe_2N_{1-x} by Jutte et al. [9]. They found that in an early

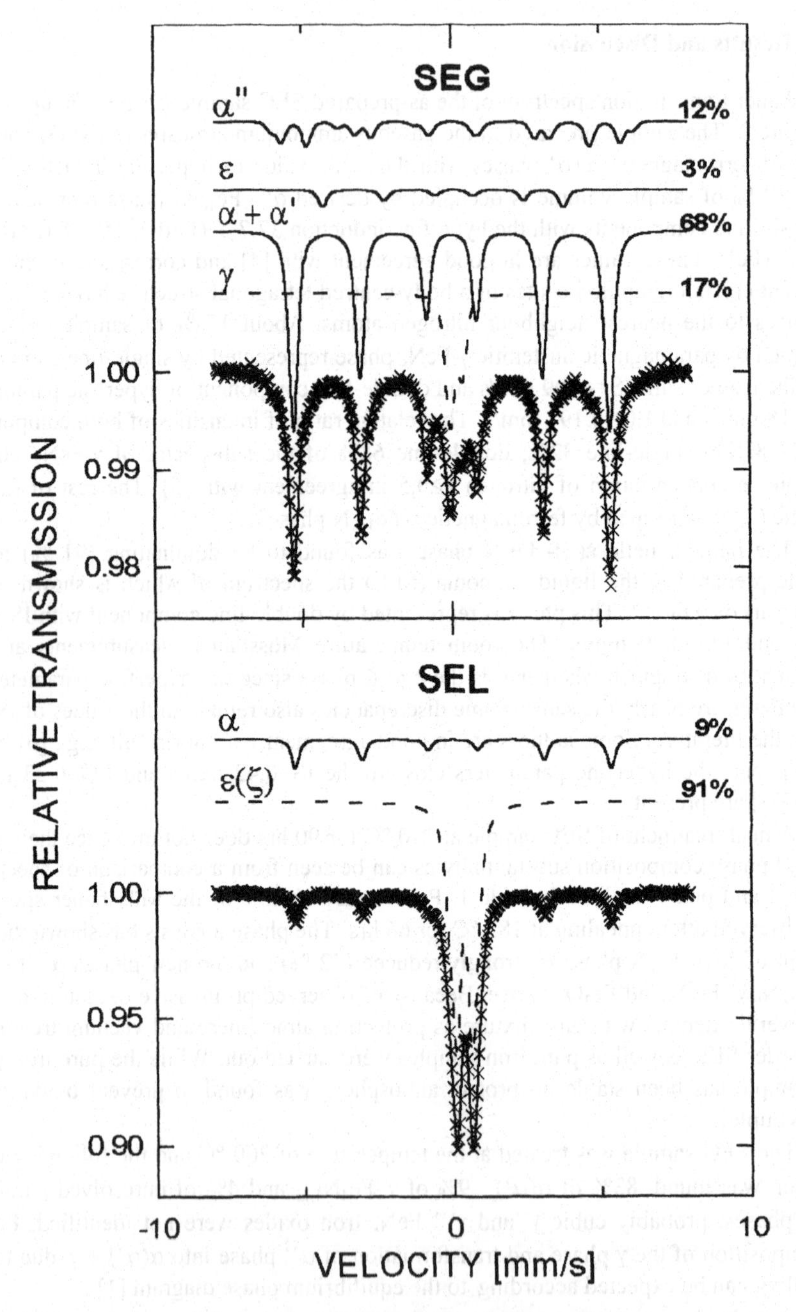

Figure 2. Mössbauer transmission spectra of Fe-N powder samples prepared by spark erosion in the gaseous ammonia (SEG) and in the liquid ammonia (SEL).

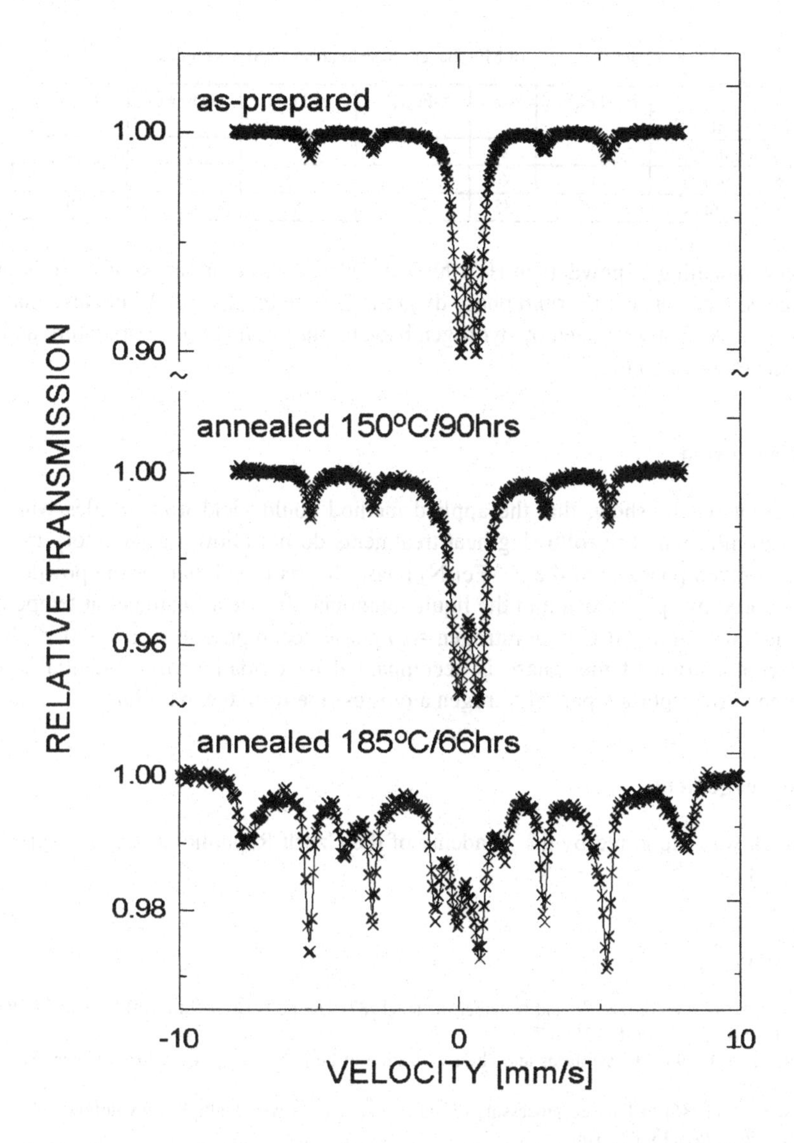

Figure 3. Mössbauer transmission spectra and phase composition of Fe-N particles prepared by spark erosion in liquid ammonia (SEL).

stage of oxidation the oxidation rate of ε-Fe_2N_{1-x} is much larger than that of α-Fe.

This was explained as large effective area in ε-Fe_2N_{1-x} than in α-Fe due to porous structure of ε-Fe_2N_{1-x}. They state that oxidising ε-Fe_2N_{1-x} increases the nitrogen concentration in the remaining nitride volume, because the outward flux of iron cations, necessary for oxide growth and γ'-Fe_4N_{1-x}. In our case formation of the γ' was not observed in SEL sample, where progressive oxidation was observed.

Table 1. Results of Mössbauer phase analysis of SEL samples.

Treatment	α''-$Fe_{16}N_2$	$\alpha(\alpha')$	ε-$Fe_{3+x}N$	γ'-Fe_4N	$\varepsilon(\zeta)$-Fe_2N	Fe_3O_{4-x}
As-prepared		9			91	
150°C/ 90hrs		6			94	
185°C/ 66hrs	7	35	4	6	22	26

The annealing of powders in $H_2 + NH_3$ at 500°C causes formation of γ'-Fe_4N phase as expected from the equilibrium phase diagram. On the other hand it indicates that iron oxides must be firstly reduced by hydrogen back to pure iron at this temperature and it is subsequently nitrided to the γ'.

4. Conclusion

The present results show, that the applied method could yield iron nitrides with high content of nitrogen. The following heat treatments do not allow a satisfactory transformation of given phases into the α''-$Fe_{16}N_2$ phase. It was found that for the powder samples prepared by spark erosion in the liquid ammonia after heat treatment at temperature between 150°C and 180°C high nitrogen $\varepsilon(\zeta)$ phase decompose to α, α'' and γ' phases. The decomposition of this phase is accompanied by oxidation of substantial part of specimens. An important part of nitrogen atoms escape from the material.

Acknowledgement

This work was supported by the Academy of the Czech Republic under the project No. A4032601.

References

1. Wriedt, H. A., Gokcen, N. A. and Nafziger, R. H. (1987) The Fe-N (Iron-Nitrogen) Systém , *Bulletin of Alloy Phase Diagrams* **8,** 355-377.
2. Jack, K. H. (1995) The synthesis and characterization of bulk α'' - Fe_{16} N_2, *J.Alloys Comp.* **222,** 160 - 166.
3. Veselý, V. (1986) FFT-based processing of unresolved spectra with multiple convolutions, *Nucl. Instr. Meth. Phys. Res.* **18,** 88 - 100.
4. Bao, X. and Metzger, R. M. (1994) Synthesis and properties of α''-$Fe_{16}N_2$ in magnetic particles, *J. App. Phys.* **75,** 5870-5872.
5. Nakagawa, H., Nasu, S., Fujii, H., Takahashi, M. and Kamamaru, F. (1991) ^{57}Fe Mössbauer study of FeN_x (x = 0.25 - 0.91) alloys, *Hyperfine Int.* **69,** 455 - 458.
6. Moriya, T., Sumitomo, Y., Ino, H., Fujito, E., F. and Maeda, Y. (1973) Mössbauer Effect in Iron - Nitrogen Alloys and Compounds, *Phys. Soc. Japan* **35,** 1378 - 1385.
7. Foct, J., Dubois, J., M. and Le Caer, G. (1974) Étude par spectrométrie Mössbauer des phases interstitielles, *J. de Physique* **35,** 493 - 458.
8. Firraro, D., Rosso, M., Principi, G. and Frantini, R. (1982) The influence of carbon on nitrogen substitution in iron ε - phases, *J. Mater. Sci.* **17,** 1773 -1778.
9. Jutte, R.H., Kooi, B. J., Somers, M. A. J. and Mittemeijer, E. J. (1997) On the oxidation of α-Fe and ε-Fe_2N_{1-x}: I. Oxidation Kinetics and Microstructural Evolution of the Oxide and Nitride Layers, *Oxidation of Metals* **48,** 87-109.

ION-BEAM IRRADIATED METALLIC SYSTEMS

A Mössbauer Spectroscopy Study

G. PRINCIPI[1] AND A. GUPTA[2]
[1)] *INFM, Unità di Trento, and Settore Materiali - DIM, Università di Padova, via Marzolo 9, 35131 Padova, Italy*
[2)] *IUC Centre, University Campus, Khandwa Road, Indore 452001, India*

1. Introduction: Ion Beam/Solid Interactions

The interaction of energetic ion bems with solid targets has been studied for some decades. It is now well established that two independent processes contribute to the *stopping power*, $S = -dE/N \cdot dx$, the energy loss dE of an ion crossing a layer dx thick of a solid having N atoms per volume unit: (*i*) S_e, the *electronic stopping*, due to electronic excitation and ionisation, and (*ii*) S_n, the *nuclear stopping*, due to elastic collisions with the nuclei of the target atoms [1].

As qualitatively depicted in Figure 1, S_n predominates in the low energy range (up to hundreds of keV), while it is negligible in comparison with S_e (ratio of the order of 10^{-3}) in the high-energy range (tens or hundreds of MeV). Up to the eighties it was believed that only the nuclear stopping process, which directly results in atomic displacements and relocations, was able to give rise to defect accumulation and to a detectable atomic mixing at the interfaces of a layered target [2]. An overwhelming quantity of experimental data has been collected in the keV range, with the attempt of finding new processes to modify the structure of material surfaces (ion implantation) and of thin films (ion-beam mixing). With the advent ion accelerators able to deliver heavy ions at energies up to GeV, drastic topological modifications induced by pure electronic en-

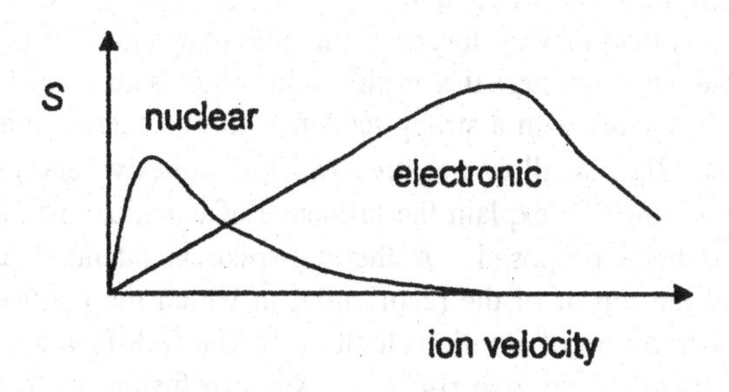

Figure 1. Qualitative representation of nuclear and electronic energy loss of energetic ions in a solid as a function of the ion velocity.

M. Miglierini and D. Petridis (eds.), Mössbauer Spectroscopy in Materials Science, 189–202.

ergy loss has been also evidenced in a variety of materials.

The current analytical techniques for studying ion-beam irradiated materials are Rutherford backscattering spectrometry (RBS) and nuclear reaction analysis (NRA), by which depth profiles of elements are available. With systems containing a Mössbauer isotope as a constituent or as a probe, conversion electron Mössbauer spectroscopy (CEMS) has often provided useful structural local information [3,4]. After a description in outline of the mechanisms of interaction, some recent studies in which Mössbauer spectroscopy has given a determinant contribution will be reviewed.

2. Mechanisms of Interactions

2.1. IN THE keV RANGE

Early work in late fifties on ion-beam interaction with solids was addressed to the direct implantation in semiconductors for damaging and doping effects [1]. Only about twenty years later it was discovered that instead of implanting an ion A into a matrix B, where only a single atom is added to B per incident ion, the ion irradiation of an interface A/B was more effective, giving rise to hundreds or thousands of intermixed atoms per incident ion [5]. Reviews on the mechanisms of ion-beam induced atomic mixing in solids have been given by Cheng [6], Nastasi and Mayer [7], Bolse [8,9] and Kelly and Miotello [10].

The perturbations produced by the ion beam into the specimen are essentially of two classes, resulting from two different mechanisms. In the *collisional* or *ballistic mechanisms* the isotropic recoil mixing, due to primary collisions, and the anisotropic ion cascades, due to secondary collisions, are fast and athermal. The *diffusional mechanisms* are connected with the thermally activated mobility of atoms and defects. Below a certain transition temperature $T_{transition}$, which is higher than room temperature for most metal/metal couples, the mixing process is athermal and is influenced by the incident energy, the masses of incident ion and target atom, and the chemical driving forces. Above $T_{transition}$ the effects of long distance atomic transport become predominant in the mixing process. These effects are diffusionlike processes and are strongly dependent on the defect concentration and mobility. The *radiation-enhanced diffusion* plays a very important role in this temperature regime.

The role of chemical driving forces in the athermal regime it is to be pointed out: very similar collisional characteristics in different systems lead to different degrees of mixing [7]. Binary systems with a strong tendency to mix thermodynamically, i.e. with heath of formation ΔH_{mix} small or negative and low cohesive energy ΔH_{coh}, present a higher amount of mixing. To explain the influence of chemical affinity the diffusion in *thermal-spikes* has been proposed. A thermal-spike is intended as a small volume around the path of the ion or of the recoil atom in which the majority of atoms are in motion with the distribution of atomic velocities of Maxwell-Boltzmann type. The local temperature is high enough to give rise to transient diffusion processes originating the observed intermixing.

The ion beam mixing process may also be classified in terms of duration. In the *prompt stage* (10^{-13} to 10^{-11} s) the incoming energetic ion transfers part of its kinetic energy knocking out target atoms (primary collisions). These atoms recoil and collide with other atoms (secondary collisions), giving rise to higher generations of collisions. These last collisions produce many low-energy recoils, which induce small random displacements (collision cascade). At the end of the collision cascade the *thermal-spike stage* (10^{-11} to 10^{-10} s) can take place, when the velocity of the moving atoms have approximately reached equipartition of kinetic energy, which is dissipated by both lattice and electron conduction. In the *delayed stage* (more than 10^{-9} s) the solid reaches an uniform ambient temperature and the involved processes are of low-energy and thermally activated, giving rise to *radiation enhanced diffusion* mainly determined by vacancy-interstitial recombination.

2.2. IN THE MeV RANGE

Early investigations on the electronic energy loss of ions in solids were mainly restricted to insulating materials, because it was generally admitted that the high mobility of free electrons in metals and semiconductors would reduce the sensitivity to the high energy ion beam, hindering a significant atomic rearrangement in response to electronic excitation. The systematic use of heavy-ion accelerators has then extended the attention also to amorphous and crystalline metals [11,12]. The higher sensitivity observed in amorphous metals was attributed to the lower electron mobility, which allows the energy deposited in the electronic system to be confined long enough to form a transiently heated region, called *latent track*. A transient thermodynamical model, or *thermal spike*, has been successfully applied by Toulemonde and co-workers to experimental data [13-17]. The idea is that the damage produced by heavy-ion high-energy irradiation may result from an increase of the lattice temperature due to a two step process: thermalization of the deposited energy on the electron system via electron-electron interaction and transfer of this energy to the lattice via electron-atom interaction.

3. Selected Examples

3.1. Ag-Fe

This system is of particular interest because its large positive heat of formation (ΔH_{mix} = +42 kJ/mole, which implies the complete immiscibility even in the liquid state) and the mean atomic number $\overline{Z} = 36.5$ favour the Cheng's conditions [6] for the formation of thermal spikes ($\overline{Z} > 20$).

Bilayers and multilayers of this system have been studied by Schaaf and co-workers [18-20] using CEMS, RBS, perturbed angular correlations (PAC), and scanning tunneling microscopy (STM). Figure 2 illustrates some of the results. The broadening in the depth distributions of Ag and Fe after Ar and Xe ions irradiation, as deduced from the RBS analysis, is mainly due to increase in the surface roughness, detected by STM, than to atomic mixing at the interface. The effective mixing as a function of the ion fluence results to be practically zero. At the same time, the combination of PAC and CEMS

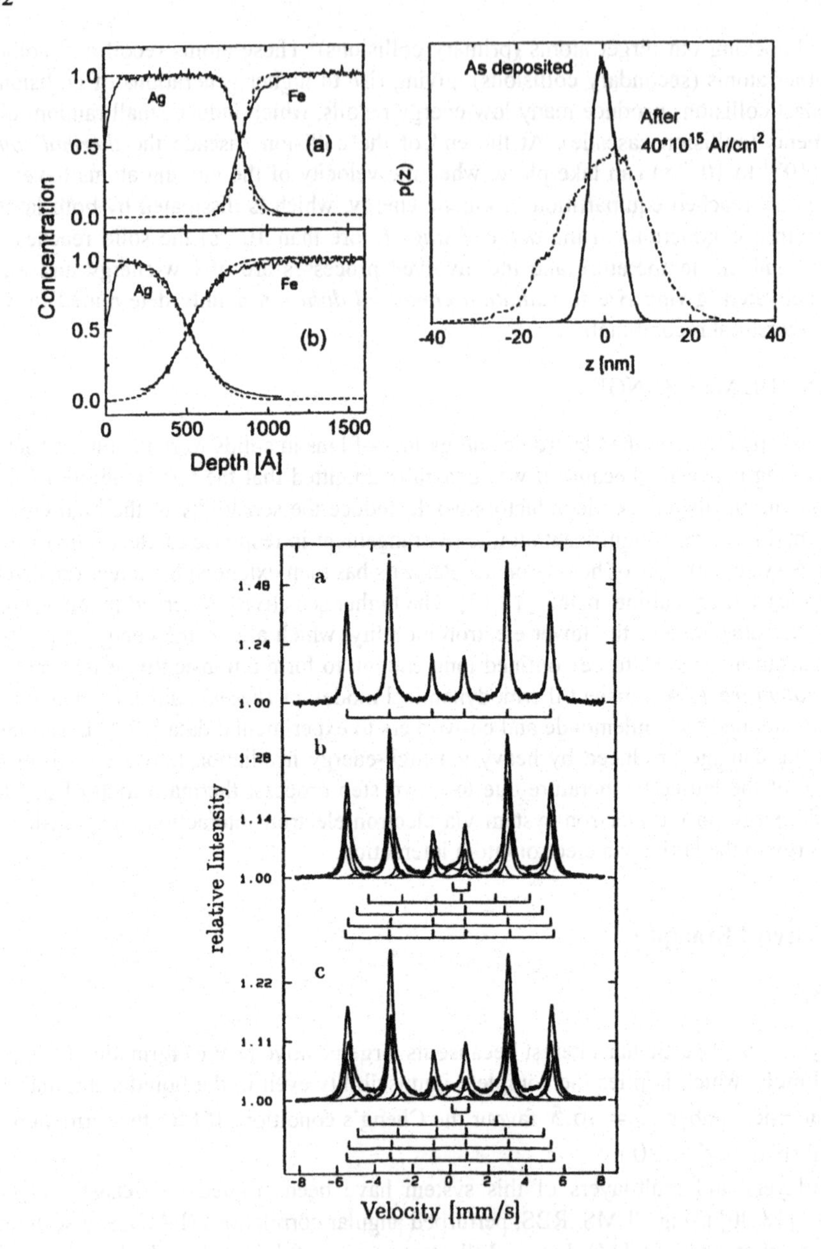

Figure 2. Concentration profiles obtained from the RBS spectra of an Ag/Fe bilayer (a) as deposited and (b) after irradiation at 77 K with 8×10^{15} Xe/cm^2 of 750 keV (top left). Surface roughness as measured with STM of an Ag/Fe bilayer as deposited (solid line) and after irradiation at 77 K with 300 keV Ar at the fluence indicated (dashed line) (top right). CEMS spectra of a ^{57}Fe-enriched Ag/Fe bilayer (a) as deposited; (b) mixed at 77 K with 450 keV Xe ions with a fluence of 2.8×10^{15} Xe/cm^2 and (c) 5.8×10^{15} Xe/cm^2 [Refs. 18-20].

measurements gives evidence that Ag-decorated defects may form as a consequence of long-range ballistic transport due to head-on primary collisions, according to the ballistic model. The CEM spectra of 450 keV Xe irradiated samples exhibit, besides the contribution of pure α-Fe, additional magnetic subspectra attributed to Fe atoms with Ag atoms neighbours and a paramagnetic fraction attributed to Fe atoms in the Ag matrix.

The authors conclude that the ballistic mixing of the Ag/Fe interface, which should occur in any case, is subsequently balanced by a demixing due to a chemically driven relaxation, most probably a thermal spike diffusion process. The large ΔH_{mix}, which would interfere during the thermal spike stage producing segregation of the atomic species relocated by collisional processes, is responsible for the low mixing efficiency observed.

3.2. Fe-Si

3.2.1. *keV range*

This system has a negative ΔH_{mix} and the possibility of amorphization of multilayer specimens by solid state reaction and ion beams in the keV range has been documented by Principi and co-workers [21-23]. The effect of low temperature 100 keV Kr ion irradiation of multilayer of overall composition $Fe_{50}Si_{50}$ is illustrated in the X-ray diffraction (XRD) patterns and in the CEM spectra of Figure 3.

The XRD patterns collected in the reciprocal space range encompassing the Fe(110) and FeSi(211) peaks display a broadening of the Fe(110) peak after irradiation, indicating the occurrence of structural disorder, and absence of the FeSi(211) peak.

The CEM spectrum of as deposited specimen consists of α-Fe and small amounts of low field components attributed to iron atoms at the Fe-Si interface. Irradiated samples exhibit a weak sextet due to unmixed metallic iron, a quadrupole doublet, having parameters intermediate among those of crystalline FeSi and amorphous FeSi, and a

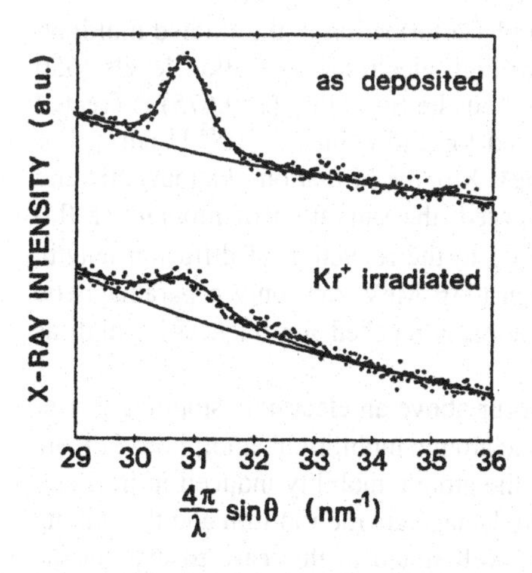

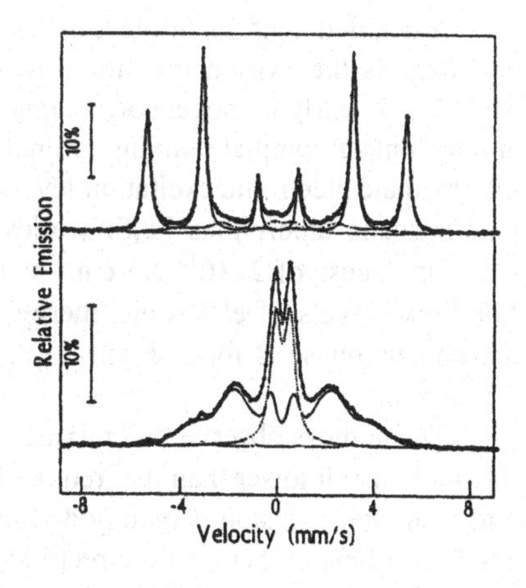

Figure 3. Partial XRD patterns of a Fe-Si multilayer as deposited and irradiated with 100 keV Kr ions to a dose of 1.8×10^{16} Kr/cm^2 (left) and CEMS spectra of the same samples (right) [Ref.22].

broad magnetic component characteristic of an iron-rich amorphous Fe-Si alloy.

This partial amorphization during ion mixing may be understood in terms of two different processes: (*i*) rapid quenching associated with the cooling phase of the thermal spike (evaluated to be of the order of 10^{12} K s^{-1}) which kinetically constrains the formation of the crystalline phase; (*ii*) a polymorphous transformation driven by the difference in the free energy between multilayers and amorphous phase. The authors conclude in favour of process (*ii*), because at low-temperature ion-beam mixing the mass transport occurs mainly as a result of diffusive motion during collision cascade and long-range diffusion cannot take place. The formation of mixed amorphous and crystalline phase structure in the present case may be attributed to: (*i*) the fact that the composition of the system corresponds to that of an intermetallic phase, and (*ii*) that the structural defect and chemical disordering due to irradiation are not able to raise the free energy of the crystalline FeSi compound definitely above that of the amorphous phase.

The above results essentially agree with those reported in more recent papers by Dufour and co-workers [24,25] the mixing process induced by 360 keV Ar ions on (Fe 4.5 nm/Si 3.5 nm)×50 multilayers is initially characterised by a large compositional interfacial profile, with preferential formation of a Si-rich amorphous Fe-Si alloy (the doublet), and by the formation of a not completely homogeneous amorphous alloy at high fluences (based on the exhibition of two peaks in the hyperfine field distribution of the broad magnetic component).

3.2.2. *MeV range*

Dufour and co-workers [24,25] have also studied the electronic excitations by 197 MeV Ar, 736 MeV Kr, 858 MeV Xe and 650 MeV U ions of the same (Fe 4.5 nm/Si 3.5 nm)×50 multilayers mentioned above. The multilayer results insensitive to swift Ar ions, no change being detected in the CEM spectrum with respect to the unirradiated specimen. On the contrary, swift heavy ions are able to induce mixing effects, since the corresponding CEM spectra display increasing changes by increasing the fluence and the ion mass.

A detailed analysis of the kinetics of Fe-Si phase formation in the studied multilayers suggests the existence of three regimes. For high electronic excitation levels (650 MeV U), a nearly homogeneous magnetic amorphous Fe_xSi_{1-x} alloy (x=0.675) is created and an almost complete mixing is finally observed for a low fluence (10^{13} U cm^{-2}). For intermediate electronic excitation levels (858 MeV Xe), the formation of a magnetic and paramagnetic amorphous Fe_xSi_{1-x} alloys is observed, the saturation of mixing for fluences in excess of 2×10^{13} Xe cm^{-2} being ascribed to the reduction of diffusion length. For lower levels of electronic energy deposition (736 MeV Kr), only a paramagnetic amorphous phase is formed and no clear saturation is reached for fluences as high as 10^{14} Kr cm^{-2}.

The authors observe that mixing effects occur above an electronic stopping power threshold much lower than that required to induce atomic mobility in amorphous silicon, and then interdiffusion is mainly dominated by the atomic mobility induced in iron layers. Nevertheless, there is no direct link between damage creation in iron and the mixing process. The thermal spike model seems to be well suited in this case to explain the mixing effects as a result of the lattice heating up.

3.3. Fe-Tb

Perpendicular magnetic anisotropy (PMA) in rare-earth-transition-metal (RE-TM) multilayers is of great interest due to their possible application in magneto-optical recording, and also from the point of view of basic studies in interface magnetism. PMA shows a strong variation with the thickness of RE and TM layers, as well as with the state of the RE/TM and TM/RE interfaces [26,30]. It is generally agreed that the dominant contribution to PMA comes from the single ion anisotropy of the rare earth ions coupled with the anisotropic distribution of TM-RE bonds at the interfaces [26,27,31]. With increasing layer thickness the crystallographic structure of Fe layer changes from amorphous to crystalline bcc structure around a critical thickness d_c of 2.0-2.4 nm [27,28]. For both amorphous and crystalline structure of Fe layer, the PMA sensitively depends on the state of the interface between Fe and Tb layers. It has also been observed that the magnitude of PMA also depends on the thickness of the intermixed layer between Fe and Tb layers [32,33].

3.3.1. *keV range*

In a study by Tosello et al. [34] low temperature irradiation by low dose 100 keV argon ions on PMA of (Fe 6 nm/Tb 0.7 nm)×10 multilayers, the main effect was to increase the width of interfacial region, accompanied by an increase in the average angle ϕ between spin direction and normal to the film plane.

The CEM spectra in Figure 4 have been analysed assuming superposition of two components: (*i*) a sharp sextet, with an internal field of 33 T, which corresponds to α-iron, and (*ii*) a broad sextet, representative of the interface, with an average internal field of about 31 T, depicted on the right side of Figure 5 as a hyperfine field distribution.

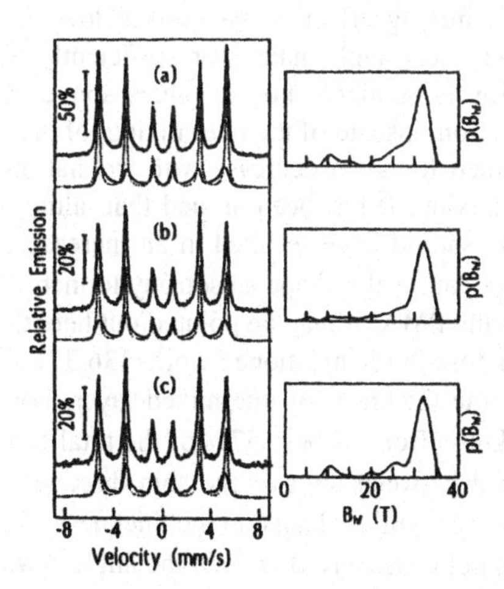

Figure 4. CEM spectra (left) of Fe(6 nm)/Tb(0.7 nm) multilayer (a) as deposited and after irradiation with 100 keV Ar ions to the doses of (b) 2×10^{14} Ar/cm^2 and (c) 6×10^{14} Ar/cm^2. The spectral components are the sharp sextet of a-iron (dashed line) and the broad magnetic pattern of interface iron (full line). The hyperfine field distributions of the broad magnetic pattern are on the right side [Ref. 34].

The magnetic texture of the samples is measured from the relative intensity $X=(4\sin^2\phi)/(1+\cos^2\phi)$ of the second and fifth peaks of the Mössbauer spectrum, where ϕ is the angle between the γ-ray and the average direction of the magnetic moments of iron atoms [35]. Formation of a second phase at the interface is excluded on the basis of present data.

Fitting of shown spectra gives an appreciable increase of ϕ, from 63° to 73°, i.e. a decrease in PMA. In addition, the relative area of the broad component, which can be used for a rough evaluation of the thickness of the interface, increases considerably from 0.4 to 0.5. A main effect of irradiation is of relaxing the stresses of the iron layers as well as those at the interface with terbium: numerous displacements are created by low-energy events in a collision sequence, involving the atoms of the multilayer in the interaction depth. This process induces relaxation of stresses generated during evaporation, which affects the PMA in the region where iron and terbium atoms interact.

3.3.2. *MeV range*

It has been shown by Richomme et al. [36,37] that both mixing and demixing at the interface induced by heavy ion irradiation in the MeV range causes PMA to decrease. The irradiation effects by swift heavy ions are quite complex and depend on the electronic energy loss S_e as well as on the thickness of the Fe and Tb layers. Using a variety of ion species and energies Richomme et al. [36,37] have covered a range of S_e from 15 keV/nm to 56 keV/nm in the study of multilayers with Fe layer thickness both below and above the critical thickness d_c below which the structure of the layer is amorphous. Depending on the values of S_e three distinct types of effects of ion irradiation have been observed. For values of S_e below 17 keV/nm, irradiation results in segregation of Fe and Tb atoms in the interfacial region. The segregation effect is pronounced in multilayers with Fe thickness close to the critical thickness of 2.2 nm. For the case 30 keV/nm $\leq S_e \leq$ 52 keV/nm, a small demixing effect is observed at low dose, while intermixing takes place by increasing the dose and finally, for sufficiently high doses, a homogeneous amorphous Fe-Tb alloy is obtained. For S_e values above 52 keV/nm, the pronounced effect is the amorphization, instead of the intermixing, of the bulk of the Fe layers.

The PMA has been found to decrease with irradiation dose in both cases of demixing as well as mixing. It has been argued that, although the sharpening of interfaces due to demixing should have resulted in an increase in PMA, an increase in the thickness of α-Fe layer causes the shape anisotropy to increase and thus the net effect of the two is to decrease the PMA. It may be pointed out here that the observed variation in PMA with irradiation dose in the mentioned works [36,37] cannot be understood only in terms of a change in the thickness of intermixed layer. For example, the effect of the increase in α-Fe thickness from 53% to 57% of the total iron content due to Kr irradiation causes the angle ϕ between the average spin direction and the film normal to increase from 68^0 to 75^0. A similar change in the thickness of α-Fe produced by either Xe or Pb irradiation does not cause any change in the angle ϕ within the experimental error. Thus, in order to understand the observed variation in PMA, the effects of irradiation other than demixing or mixing should also be considered.

The effect of 80 MeV Si ion irradiation on (Fe 3 nm/Tb 2 nm)×20 multilayers with high PMA has been studied by Gupta et al. [38] using X-ray reflectivity and CEMS. For comparison, the effect of 150 MeV Ag ion irradiation, where some interfacial mixing is expected to occur, has also been studied. These experiments, very recent, will be described in the following with some details.

X-ray reflectivity. In the X-ray reflectivity patterns of Figure 5 of the virgin and of the irradiated specimens the first and second order Bragg peaks due to multilayer periodicity are distinctly visible. X-ray reflectivity patterns contain information about the surface and interface roughness, thickness of individual layers, interdiffusion, etc. [39-42].

For the theoretical fit to the experimental data, the multilayer was considered to be consisting of alternating layers of Fe and Tb with an intermixed layer of composition $Fe_{0.5}Tb_{0.5}$ at each interface. A thin layer of oxide was also assumed at the top. The refractive indices of Fe and Tb layers were taken to be that of bulk materials, while that of Fe-Tb intermixed layer was taken to be the average of Fe and Tb. The following quantities were taken as parameters, (*i*) the thickness of the top oxide layer, (*ii*) the intermixed layer at the interfaces, and (*iii*) the roughness of the interfaces. It may be noted that the above parameters affect different aspects of the reflectivity curve. The interface roughness affects the overall decay rate of the envelope of the curve, while the intermixing at the interface affects the height of the Bragg peaks, specially the second one. The effect of the top oxide layers is to produce a periodic modulation of the reflectivity curve with period depending on the thickness [43]. The theoretical curve which best fits the experimental data is shown as continuous curve in Figure 1. From the corresponding fitted parameters one may deduce that even in the as-deposited film the interface between Fe

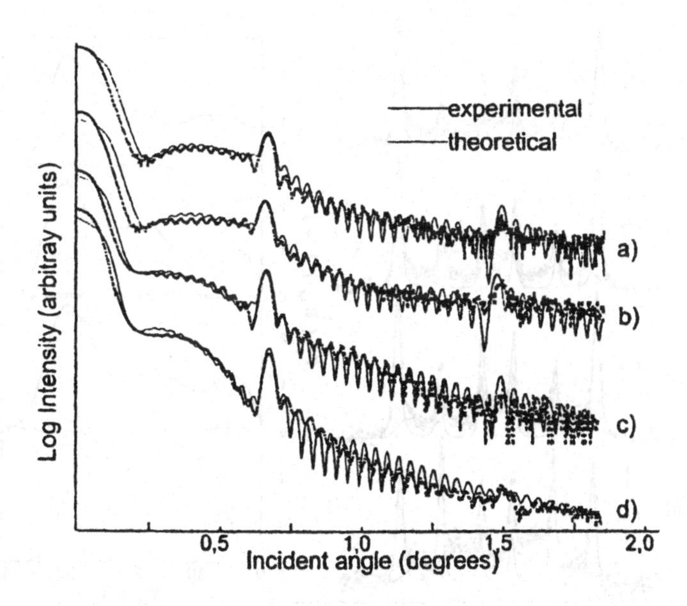

Figure 5. X-ray reflectivity patterns of Fe(3 nm)/Tb(2 nm) multilayers: a) as deposited, and after 80 MeV Si ion-irradiation to a dose of b) 10^{14} ions/cm^2 and c) 10^{15} ions/cm^2. Curve d) corresponds to a further irradiation of specimen c) by 150 MeV Ag ions to a dose of 10^{13} ion/cm^2. The continuous curves represent the best theoretical fit to the experimental data [Ref. 38].

and Tb is not very sharp (about 1.0 nm of interdiffused layer exists at each interface) and that the roughness of interfaces is about 0.2 nm, while that of the top surface is 1.0 nm. The main effect of the irradiation up to the highest Si dose is to increase the interfacial roughness to a value of 0.65 nm. No significant increase in the thickness of the intermixed layer is observed. Ag ion irradiation causes both the interfacial roughness and the intermixed region to increase to 1 nm and 1.2 nm, respectively.

Mössbauer spectroscopy. Figure 6 shows the CEM spectra before and after irradiation. The results of the fitting are summarised in Table 1. A qualitative estimate of thickness of the intermixed layer, as obtained from the Mössbauer measurements, agrees reasonably well with that obtained from the X-ray reflectivity data. From the fitting of Mössbauer data one finds that the area under the broad hyperfine component, which has contributions from Fe atoms at the interface as well as in the intermixed layer, is about 50%. If one neglects the difference in the recoilless fractions of the iron atoms in the bulk α-Fe layer and in the interfacial regions, then 50% of Fe atoms can be assumed as residing in the interfacial region. For perfectly sharp interfaces two monolayers of each of Fe layer will be interfaced with Tb atoms. This will constitute about 10% of the total atoms in a layer. Therefore 50-10=40% of Fe atoms exist in the intermixed layers. On the other hand, from the fitting of X-ray reflectivity data, the thickness of the intermixed layer is about 1.0 nm. Taking the composition of the intermixed layer as $Fe_{0.5}Tb_{0.5}$, this thickness corresponds to about 30% of Fe atoms in the intermixed region, in very good

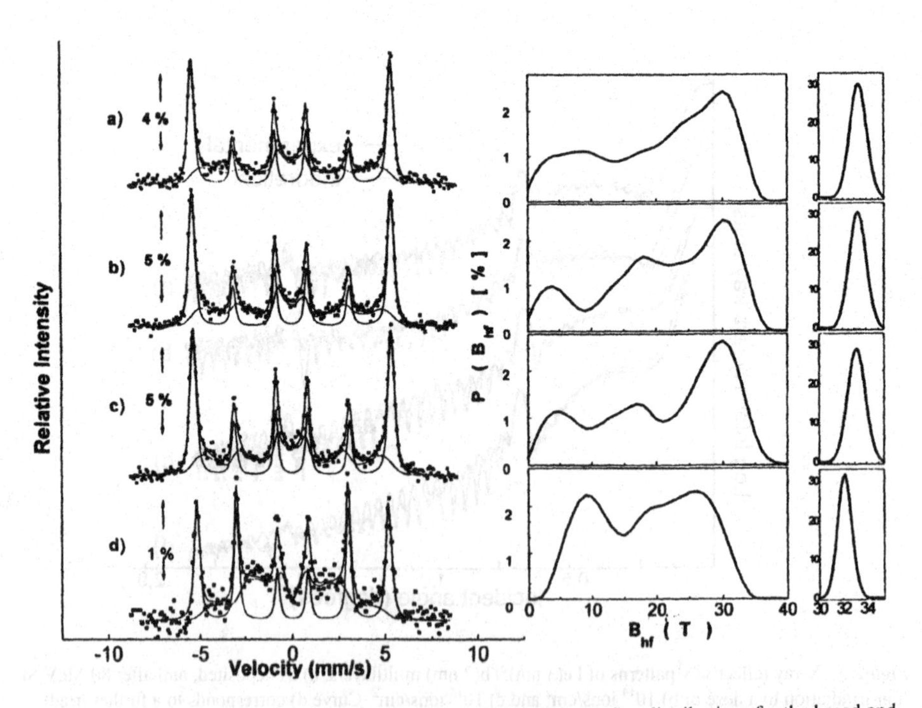

Figure 6. CEM spectra of the same samples of Fig. 5 [Ref. 38]. Hyperfine distributions for the broad and narrow spectral components are also depicted.

agreement with Mössbauer data.

TABLE 1. Fitted parameters of sharp sextet in spectra of Figure 6: average hyperfine field, $<B_{hf}>$, width of the distribution, ΔB_{hf}, PMA calculated from relative intensity of 2nd and 5th lines, ϕ [Ref. 38].

Spectrum	$<B_{hf}>$ (T)	ΔB_{hf} (T)	Relative area (%)	ϕ (degrees)
a)	33.40±0.03	0.61±.07	50±1	33.5±1
b)	33.33±0.03	0.67±.05	53±1	36±1
c)	33.21±0.03	0.74±.05	53±1	40±1
d)	32.14±0.03	0.61±.05	48±1	62±1

From Table 1 it may be noted that irradiation by Si ions causes small but significant changes in the parameters of the sharp hyperfine component: (*i*) decrease in the average field; (*ii*) increase in the width of the field and, (*iii*) increase of relative area. Changes (*i*) and (*ii*) are indicative of creation of some disorder in the bulk of Fe layer, while change (*iii*) provide an evidence of incipient demixing between Fe and Tb layers. Most significant effect of irradiation is observed on the PMA. The angle ϕ increases from 33.5° for the virgin sample to 40° for the sample with highest irradiation dose, indicating a decrease in PMA upon irradiation. Irradiation with 150 MeV Ag ions results in an increase in the mixed region accompanied by a large decrease in PMA. Incorporation of Tb atoms into the lattice of Fe layers shifts to left the corresponding B_{hf}.

Discussion. Both X-ray reflectivity and Mössbauer measurements show that irradiation with Si ions does not induce any additional intermixing at the interface or, rather, a small demixing occurs according to CEMS data analysis. This agrees with the S_e TRIM code calculated value in the present case, which is 6.9 keV/nm in Fe and 4.7 keV/nm in Tb, and then is much below the threshold value of 25 keV/nm for latent track formation or intermixing in Fe, as estimated by Richomme et al. [36,37]. However, the values of S_e achieved here may be sufficient to produce isolated point defects and short range movement of atoms. Such process would be responsible for the creation of disorder in the bulk of Fe and Tb layers, causing the observed changes in the hyperfine field parameters of α-Fe layers, and may also cause the interface roughness to increase.

The X-ray reflectivity measurements show distinct increase in the interface roughness upon irradiation. Figure 7 shows that PMA does not have the almost linear increase with irradiation dose exhibited by the roughness. Assuming a perfectly flat and sharp interface in the beginning, an intermixing at the interface, which is uniform in the x-y plane, would cause a gradient of the iron concentration along the z direction (Figure 8a). However, a surface of constant concentration will still remain flat. On the other hand, segregation of Fe and Tb atoms would modify the topology of the surface of constant concentration and thus would cause the interface roughness to increase (Fig. 8b).

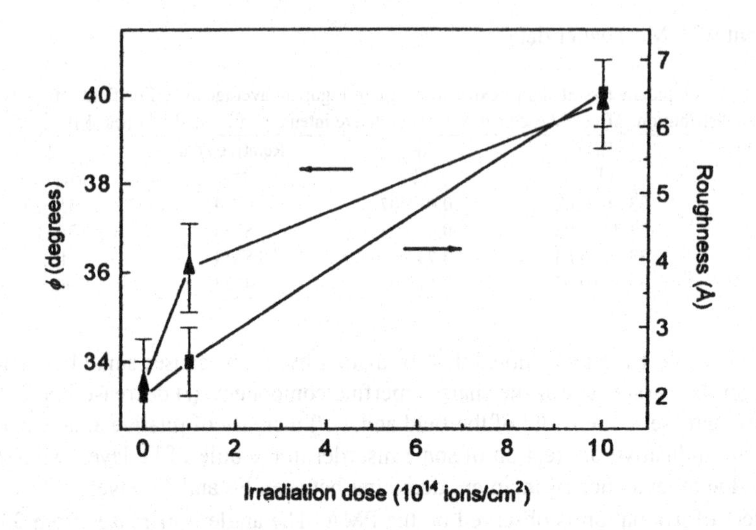

Figure 7. Variation of interfacial roughness and angle ϕ with the dose of 80 MeV Si ions irradiation in Fe-Tb multilayers [Ref. 38].

Magnetic anisotropy of the magnetic layers with thickness t can be phenomenologically written as $K\,t = K_v\,t + 2\,K_s$ [44,45]. The volume term, K_v, contains contributions from shape, magnetocrystalline and magnetoelastic anisotropies, while K_s is the interface contribution. Shape anisotropy would always try to keep atomic spins in the film plane in order to minimise the magnetostatic energy. The magnitude of the shape anisotropy will also depend on the magnetic moment per iron atom and thus on the structure of the Fe layer. Magnetocrystalline anisotropy depends on the crystal structure of the layers, which is small for Fe layers with bcc structure. Magnetoelastic anisotropy has its origin in the internal stresses present in the individual layers. The stresses may

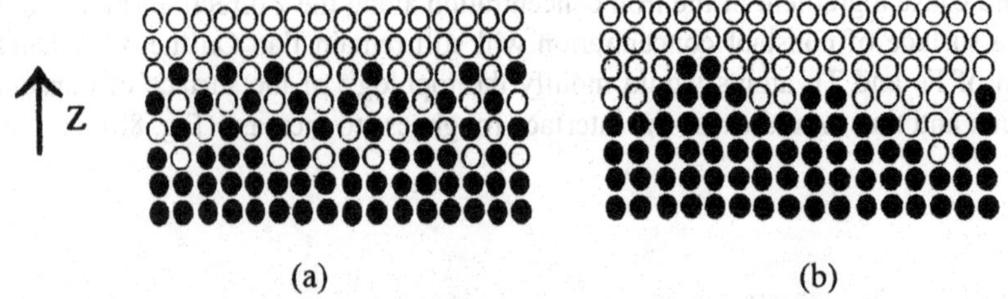

Figure 8. Schematic vision of the distribution of atoms of two species at the interface: a) in the presence of finite interdiffusion; b) after a partial segregation [Ref. 38].

arise by differences in thermal expansion between film and substrate or between the layers. Structural defects may also give rise to internal stresses. Interface anisotropy K_s in RE/TM multilayers has its origin in single ion anisotropy coupled with anisotropic distribution of RE/TM pairs [26].

In the present case the only observed change with irradiation dose at the interfaces is an increase in roughness, without any change in the thickness. This suggests that the decrease of PMA can be attributed at least partly to the increase of interfacial roughness. On the other hand, the atomic rearrangements associated with the electronic energy loss should also cause the relaxation of internal stresses in the films, which would result in a decrease in the magnetoelastic contribution to the anisotropy. Thus, the observed decrease in PMA with Si dose is a result of an increase in the interfacial roughness and stress relaxation in the bulk of the films. The initial faster decrease in PMA, as seen from Figure 7, suggests that most of the structural relaxation has taken place up to an irradiation dose of 10^{14} ions/cm^2 and the subsequent slow decrease in PMA is mainly due to increase in interfacial roughness.

Electronic energy loss S_e associated with 150 MeV Ag ions is just above the threshold for intermixing, as observed by Teillet et al. [37], and thus causes an additional intermixing at the interfaces, as expected. Our data indicate a noticeable increase of interfacial roughness and of thickness of the intermixed region. Both these effects would cause PMA to decrease. It may be noted that an increase in interfacial roughness from 2.5 to 6.5 Å causes an increase of ϕ at the most by ~3.5^0. Therefore, in the Ag ion irradiated specimen the contribution of increased interfacial roughness to the observed increase in ϕ should also be of the same order. The rest of the change in ϕ (18.5^0) should be attributed to increased intermixing at the interfaces. This suggests that the effect of intermixing on PMA is much stronger as compared to that of the interfacial roughness. There are evident analogies with the effect on PMA due to the keV energy irradiation of Fe/Tb multilayers reported by Tosello et al. [34].

4. Conclusions

The currently used analytical tools to investigate ion-beam irradiated systems can be profitably complemented by Mössbauer spectroscopy for its unique ability in giving detailed structural information on the surroundings of resonant atoms. Sometimes the more relevant features are revealed by means of this technique. The few examples reported here are persuasive cases of application.

References

1. Dearnaley, G., Freeman, J.H., Nelson, R.S., and Stephen, J. (1973) *Ion Implantation*, North Holland, Amsterdam.
2. Fischer, B.E. and Spohr, R. (1983), *Rev. Mod. Phys.* **55**, 907.
3. De Waard, H. and Niesen, L. (1989) in G.J. Long (ed.), *Mössbauer Spectroscopy Applied to Inorganic Chemistry*, vol. 2, Plenum Press, New York, pp. 1-119.
4. Principi, G. (1993) in G.J. Long and F. Grandjean (eds.) *Mössbauer Spectroscopy Applied to Magnetism and Materials Science*, vol. 1, Plenum Press, New York, pp. 33-76.
5. Lee, D.H., Hart, R.R., Kiewit, D.A., and Marsh, O.J. (1973), *Phys. Stat. Sol.* **15a**, 645.

6. Cheng, Y.T. (1990), *Mater. Sci. Rep.* **5**, 45.
7. Nastasi, M. and Mayer, J.W. (1994), *Mater. Sci. Eng.* **R12**, 1.
8. Bolse, W. (1994), *Mater. Sci. Eng.* **R12**, 53.
9. Bolse, W. (in press), *Mater. Sci. Eng.* **A**.
10. Kelly, R. and Miotello, A. (1996), *Surf. Coat. Techn.* **83**, 134.
11. Klaumünzer, S., Hou, M.D., and Schumacher, G. (1986), *Phys. Rev. Lett.* **57**, 850.
12. Auduard, A., Balanzat, E., Fuchs, G., Jousset, J.C., Leseur, D., and Thomé, L. (1987), *Europhys. Lett.* **3**, 327.
13. Toulemonde, M., Dufour, C., and Paumier, E. (1992), *Phys. Rev.* **B46**, 14362.
14. Wang, Z.G., Dufour, C., Paumier, E., and Toulemonde, M. (1994), *J. Phys. Cond. Matter* **6**, 6733.
15. Wang, Z.G., Dufour, C., Paumier, E., and Toulemonde, M. (1994), *J. Phys. Cond. Matter* **7**, 2525.
16. Dufour, C., Wang, Z.G., Levalois, M., Marie, P., Paumier, E., Pawlak, F., and Toulemonde, M. (1996), *Nucl. Instr. and Meth.* **B107**, 218.
17. Toulemonde, M., Dufour, C., Wang, Z.G., and Paumier, E. (1996), *Nucl. Instr. and Meth.* **B112**, 26.
18. Neubauer, M., Lieb, K.P., Schaaf, P., and Uhrmacher, M. (1996), *Thin Solid Films* **275**, 69.
19. Neubauer, M., Lieb, K.P., Schaaf, P., and Uhrmacher, M. (1996), *Phys. Rev.* **B53**, 10237.
20. Crespo-Sosa, A., K.P., Schaaf, P., Bolse, W., M., Lieb, Gimbel, M., Geyer, U., and Tosello, C. (1996), *Phys. Rev.* **B53**, 14795.
21. Principi, G., Gupta, A., Gupta, R., Tosello, C., Gratton, L.M., Maddalena, A., and Lo Russo, S. (1991), *Hyp. Inter.* **69**, 627.
22. Gupta, A., Gupta, R., Principi, G., Jannitti, E., Tosello, C., Gratton, L.M., Lo Russo, S., Rigato, V., and Frattini, R. (1992), *Surf. Coat. Techn.* **51**, 451.
23. Gupta, A., Principi, G., Gupta, R., Maddalena, A., Caccavale, F., and Tosello, C. (1992), *Phys. Rev.* **B50**, 2833.
24. Dufour, C., Bruson, A., Marchal, G., George, B., and Mangin, Ph. (1993), *J. Magn. Magn. Mater.* **93**, 545.
25. Bauer, Ph., Dufour, C., Jaouen, C., Marchal, G., Pacaud, J., Grilhe, J., and Jousset, J.C. (1996), *J. Appl. Phys.* **81**, 116.
26. Sato, N., (1986), *J. Appl. Phys.* **59**, 2514.
27. Scholz, B., Brand, R.A., and Keune, W. (1989), *J. Magn. Magn. Matter.* **104**, 1089.
28. Honda, S., Nawate, M., and Sakamato, I. (1996), *J. Appl. Phys.* **79**, 365.
29. Richomme, F., Teillet, J., Fnidiki, A., Auric, P., and Houdy, Ph. (1996), *Phys. Rev.* **B 54**, 416.
30. Tappart, J., Keune, W., and Brand, R.A. (1996), *J. Appl. Phys.* **80**, 4503.
31. Shan , Z. S., Sellmyer, D.J., Jaswal, S.S., Wang, Y.J., and Shen, J.X. (1990), *Phys. Rev.* **B42**, 10446.
32. Findiki, A., Juraszek, J., Teillet, J., Richomme, F., and Lebertois, J.P. (1997), *J. Magn. Magn. Matter.* **165**, 405.
33. Cherifi, K., Dufour, C., Piecuch, M., Bruson, A., Bauer, Ph., Marchal, G., and Mangin, Ph. (1991), *J. Magn. Magn. Matter.* **93**, 609.
34. Tosello, C., Gratton, L.M., Principi, G., Gupta, A., and Gupta, R. (1996) *Surf. Coat. Techn.* **84**, 338.
35. Gonser, U. (1975), in U. Gonser (ed.), *Topics in Applied Physics, Vol.5, Mössbauer Spectroscopy*, Springer-Verlag, Berlin, pp. 1-95.
36. Richomme, F., Findiki, A., Teillet, J. and Toulemonde, M. (1996), *Nucl. Instr. Meth.* **B107**, 374.
37. Teillet, J., Richomme F., and Fnidiki, A. (1997), *Phys. Rev.* **B55**, 11560.
38. Gupta, A., Paul, A., Gupta, R., Avasthi, D.K., and Principi, G. (in press), *J. Phys. Cond. Matter*.
39. Huang, T.C. and Parrish, W. (1992), *Adv. X-ray Analysis* **35**, 137.
40. Parratt, L.G. (1954), *Phys. Rev.*, 359.
41. Underwood, J.H. and Barbee, Jr., T.W (1981), *Appl. Opt.* **20**, 3027.
42. Sinha, S.K., Sirota, E.B., Garoff, S., and Stanley, H.B. (1988), *Phys. Rev.* **B38**, 2297.
43. Stettner, J., Schwalowsky, L., Seeck, O.H., Tolan, M., and Press, W. (1996), *Phys. Rev.* **B53**, 1398.
44. Draaisma, H.J.G., den Broeder, F.J.A., and de Jonge, W.J.M. (1987), *J. Magn. Magn. Mater.* **66**, 351.
45. den Broeder, F.J.A, Hoving, W., and Bloemen, P.J.H. (1991), *J. Magn. Magn. Mater.* **93**, 562.

ELECTRODEPOSITED, ION BEAM MIXED AND BALL MILLED FeCrNi ALLOYS

Comparative Study of Formation and Transformation of Metastable Phases

E. KUZMANN[1], M. EL-SHARIF[2], C.U. CHISHOLM[2], G. PRINCIPI[3], C. TOSELLO[4], K. HAVANCSÁK[1], A. VÉRTES[1], K. NOMURA[5], V.K. GARG[6] AND L. TAKÁCS[7]

[1)] *Department of Nuclear Chemistry, Eötvös University, Budapest, Hungary*

[2)] *Glasgow Caledonian University, Glasgow, U.K.*

[3)] *Department of Materials Science, Padova University, Padova, Italy*

[4)] *Department of Physics, Trento University, Trento, Italy*

[5)] *Graduate School of Engineering, University of Tokyo, Tokyo, Japan*

[6)] *Department of Physics, University of Brasilia, Brasilia-DF, Brazil*

[7)] *Department of Physics, University of Maryland, Baltimore, U.S.A.*

Abstract

Conversion electron and transmission Mössbauer spectroscopy, X-ray diffractometry and electron microprobe analysis were used to perform comparative study of Fe-Cr-Ni electrodeposited, ion beam mixed vacuum deposited and ball milled alloys of same composition around 50%Fe-25%Cr-25%Ni. The main phase of the electroformed Fe-Cr-Ni microcrystalline samples is ferromagnetic contrary to the paramagnetic character of thermally prepared alloys of equivalent composition. Metastable phases have been shown in Fe-Cr-Ni multilayers, consisting of a few atomic layers of iron and nickel as well as chromium, prepared by high vacuum deposition and ion beam or laser irradiation. The main phase of evaporated and ion beam mixed Fe-Cr-Ni films has been found to be ferromagnetic similarly to that observed in the case of electrodeposited alloys. It was shown that transformation of metastable phases into the stable one occurs in high vacuum evaporated and ion beam mixed or laser irradiated Fe-Cr-Ni films due to appropriate heat treatment in vacuum. Transformation of metastable phases in electrodeposited Fe-Cr-Ni alloys was achieved by irradiation with energetic heavy ions at room temperature. Metastable phases, including the ferromagnetic one, were found in Fe-Cr-Ni alloys prepared by ball milling method, too.

M. Miglierini and D. Petridis (eds.), Mössbauer Spectroscopy in Materials Science, 203–214.

1. Introduction

Films of iron-chromium-nickel alloy electrodeposits have gained great importance in recent years because they have considerable wear and corrosion resistance, high strength, good adhesion and attractive appearance. They can therefore be developed as protective coatings for softer and more corrosive substrate materials. They could also provide an economical substitute for stainless steel. The recently developed preparation procedure of iron-chromium-nickel electrodeposits [1,2] can produce thick (1-500 μm) coatings and enhances significantly the practical applicability. The study of the structure of these coatings by X-ray diffraction revealed the microcrystalline or amorphous-like character of the electrodeposits [3,4]. Mössbauer spectroscopy can be especially advantageously applied in these Fe-Ni-Cr alloy coatings because it is an excellent unique tool to study the short range order arrangement and/or the magnetic state of the iron atoms in microcrystalline or amorphous alloys. Previous investigations [3-9] using Mössbauer spectroscopy have shown that the dominant phase of microcrystalline electrochemically deposited Fe-Ni-Cr is ferromagnetic. In contrast the thermally prepared alloys of equivalent metallic composition consisted only of paramagnetic phase.

The aim of the present work was to perform comparative study of the occurrence, short range order and transformation of the metastable ferromagnetic phases in FeNiCr alloys prepared by different non-equilibrium techniques like electrochemical deposition, ion beam mixing or laser beam irradiation of multilayers, and ball milling. For these investigations ^{57}Fe transmission (TM) and conversion electron Mössbauer (CEM) spectroscopy, X-ray diffractometry (XRD) were performed on the differently prepared Fe-Ni-Cr alloys.

2. Experimental

Electrodeposition was carried out using copper cathode substrates of 25x25mm working area and 0.2 mm in thickness using a circulation cell. Table 1 shows the bath composition and Table 2 shows the other electrodeposition conditions.

Although the chromium concentration is almost forty times the iron concentration in the electrolyte, the average concentration of the deposits can be achieved 25% Cr, 25% Ni and 50% Fe (±10%) for suitable deposition conditions. The element concentration in the individual samples were determined by electron microprobe measurements.

This new electrochemical procedure can control the Cr(II) ions generated at the cathode surface during the deposition by maintaining a constant cathode potential and by circulation and agitation of the electrolyte. The final technique involved anodically generated chlorine. The electrolyte flow was maintained from the graphite anode towards the cathode, carrying the dissolved chlorine with reacts with and destroys the Cr (II) in the cathode vicinity.

The heat treatment of the electrodeposits was performed in a quartz tube in a furnace at 500 °C between 5 minutes and 100 hours (in 12 steps) isothermally in flowing pure argon gas atmosphere.

TABLE 1. Bath Composition

$CrCl_3.6H_2O$	0.80 mol/l
$NiCl_2.6H_2O$	0.20 mol/l
$^{57}FeCl_2.4H_2O$	0.02 mol/l
NaCl	0.50 mol/l
NH_4Cl	0.50 mol/l
H_3BO_3	0.15 mol/l
H_2O deionised	500 g
dimethylformamide	500 g

TABLE 2. Electrodeposition Conditions

Cathode potential	-1.8 V (SCE)
Electrolyte temperature	20 °C
Flow rate (circulation cell)	100 ml/min
Anode (40x60 mm)	graphite
Cathode (25x25 mm)	copper
Bulk pH	1.8

The irradiation of electrodeposits was carried out at room temperature with 209 MeV ^{84}Kr ions with different fluences up to $3x10^{14}$ ions/cm^2 at the Laboratory of Nuclear Reactions, JINR, Dubna.

50% Fe-25%Ni-25%Cr overall composition multilayered samples were prepared by physical vapour deposition using e-gun according to the following scheme: SiO_2/Ni(4.6)/^{57}Fe(10)/Cr(5)/Ni(4.6)/57Fe(10)/Cr(5)/Al(2), where the numbers between parenthesis represent the layer thicknesses in nm. The outer thin layer of aluminium was deposited for protective purposes. The deposition was performed at a pressure of about $6x10^{-6}$ Pa onto a 100 nm thick SiO_2 substrate layer thermally grown on a silicon wafer.

The as-deposited multilayers were ion beam mixed with 100 keV xenon ions to a dose of $2x10^{16}$ ions/cm^2, at a current density of 3.5 A/cm^2, keeping the sample temperature at 150 K. Laser irradiation of the multilayered samples were performed with a nanosecond (full width at half maximum of 18 ns) ruby laser (λ=694 nm) under argon protective atmosphere at an energy density of 1.1 J/cm^2. Annealing has been done in a vacuum of 6 x10^{-5} Pa.

The milled samples were prepared by a SPEX 8000 Mixer Mill in a round bottom hardened steel vial with the help of both 8 steel balls of diameter 3/8 inch and 3 balls of diameter 1/2 inch. Samples were prepared with the composition 50% Fe-25%Ni-25%Cr milled for 5 and 15 hours.

X-ray diffractograms were recorded by means of a computer controlled DRON-2 diffractometer using $Co_{K\alpha}$ radiation and β-filter and by Philips equipment in glancing angle geometry with $Cu_{K\alpha}$ radiation.

Conversion electron ^{57}Fe Mössbauer spectra of electrodeposited samples were recorded by a conventional Mössbauer spectrometers using CEM detector at room temperature. Transmission Mössbauer measurements were performed between 20K-295K. A 10^9Bq activity ^{57}Co(Rh) source supplied the γ-rays. Isomer shifts are given relative to the α-iron. The evaluation of Mössbauer spectra was performed by least-square fitting of lines using the MOSSWINN program [10]. The hyperfine field distributions were obtained by a modified Hesse-Rübartsch method.

3. Results and Discussion

3.1. ELECTRODEPOSITS

The XRD results are consistent with the microcrystalline character of electrodeposits [2].

Fig. 1 and Fig. 2 show the transmission and conversion electron Mössbauer spectra of thick and thin electrodeposited samples prepared with different plating parameters resulted in different composition in the range of 40-50% Fe-25-35%Ni-15-25%Cr in comparison with spectra of samples in another composition region.

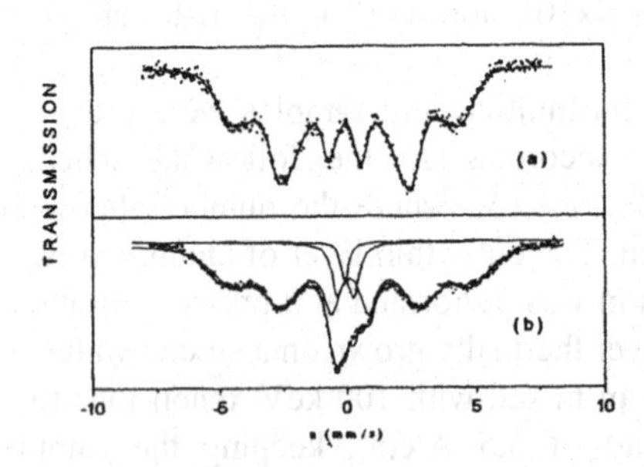

Figure 1. Mössbauer spectra of electrodeposites of 46%Fe-33%Ni-21%Cr (a) and 18%Fe--77%Ni-5%Cr (b).

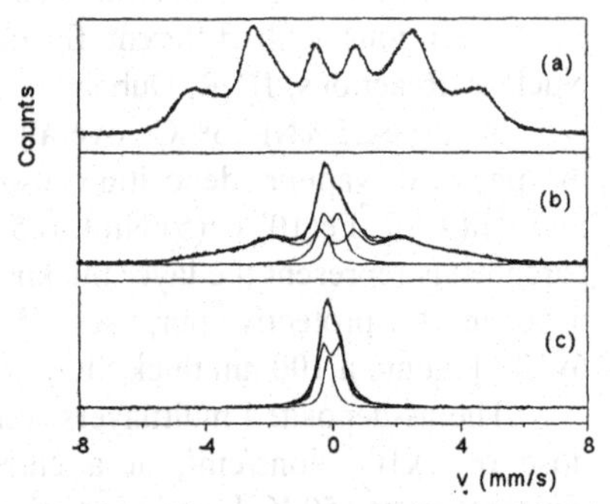

Figure 2. CEM spectra of electrodeposites of 51%Fe-24%Ni-25%Cr (a) and 19%Fe--37Ni-44%Cr (b) and 17%Fe--36Ni- 47%Cr (c).

The typical spectrum of the electrodeposits (Figs. 1a, 2a) exhibit a magnetically split broad six line envelop in which small components are in the paramagnetic region. This is in agreement with those reported earlier for very wide range of compositions [3-11]. However, in the range with smaller than 20% iron content, it is also possible to prepare electrodeposits, by suitable selection of the condition of electrolysis, which spectra reflect considerable paramagnetic components (Figs. 1b, 2b) or the total absence of the magnetic part (Fig. 2c). This latter spectrum was either decomposed into a singlet and a doublet evaluated to a quadrupole splitting distribution involving a Lorentzian singlet. All the other spectra of electrodeposits were decomposed into a singlet, a doublet and a sextet by conventional least-square fitting of Lorentzians. The evaluation of these spectra were also performed by calculating two distributions and a singlet simultaneously, namely, a hyperfine field distribution for the magnetic component and a quadruple splitting distribution for the paramagnetic contribution other than that of the singlet. The details of this method will be published elsewhere. A distribution-pair derived in this way are shown in Fig 3. The results obtained by the two different methods were in satisfactorily correspondence with each other. However, the latter method can derive the quadrupole splitting distribution for the paramagnetic contribution and can also determine more adequately the hyperfine field distribution.

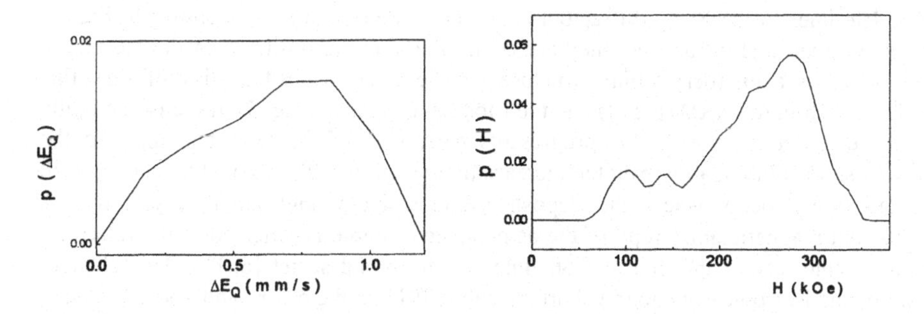

Figure 3. A distribution-pair derived for the electrodeposit which Mössbauer spectrum is shown in Fig. 1a.

For the investigated large number of electrodeposits in the total composition range, the Mössbauer parameters of subcomponents were found in the following ranges: the isomer shift is -0.07 to -0.10 mm/s for the singlet, the average quadrupole splitting is between 0.55-0.85 mm/s and the average isomer shift is between -0.03 to 0.12 mm/s for the doublet, and the average hyperfine field is between 200 and 285 kOe and the average isomer shift is between 0.0 and 0.05 mm/s for the magnetic component. For 50%Fe-25%Ni-25%Cr composition, we have found characteristic values with -0.08 mm/s of the isomer shift of singlet, 0.8 mm/s of the quadrupole splitting of doublet and 255 kOe of the average hyperfine field.

The singlet is associated with the stable paramagnetic phase of the alloy because it is the fingerprint of the thermally prepared alloy (Fig. 10c) with crystalline fcc structure in stable paramagnetic state [3].

The doublet is assigned to a disordered paramagnetic microcrystalline or amorphous Fe-Ni-Cr alloy phase [12] since the deposit showing only paramagnetic spectrum (Fig 2c) was found to be paramagnetic in magnetic test and its XRD reflected a dominant amorphous contribution.

The magnetic part is very similar to the spectrum of an amorphous magnetic alloy. This spectral component can be interpreted as a superposition of a large number of magnetically split subspectra belonging to Fe atoms with different structural micro-environments of the alloying elements due to the amorphous or microcrystalline nature of these deposits. This component can be attributed to a strongly disordered ferromagnetic solid solution Fe-Ni-Cr phase in these electrodeposited alloys [8]. All samples which exhibit magnetic parts in their spectra were found to be ferromagnetic in magnetic tests. The present and earlier results [3-9] of Mössbauer analysis showed that the ferromagnetic phase is dominant in the electrodeposits in a wide range of compositions where the existence of the ferromagnetic state is unexpected considering both the equilibrium phase diagrams and the only paramagnetic state of the thermally prepared alloys.

No contribution of the stable paramagnetic phase was observed in the spectra of elecrtrodeposites prepared under the condition described in the experimental section.

The transformation of metastable phases were tested due to ageing and due to radiation effects.

The lines in the X-ray diffractogram of the as-deposited sample shown in Fig. 4a are very broad and diffuse and they are centred around the positions of the fcc phase. This suggests a microcrystalline structure for the electrodeposited alloy of very fine grain size. Both the XRD (Fig. 4) and the Mössbauer spectra (Fig. 5) of the isothermally aged electrodeposites exhibit a progressive narrowing of the lines belonging to the FeNiCr solid solution alloy with the increase of ageing time. This can be associated with an ordering process where the deposit becomes increasingly more crystalline. A significant magnetic anisotropy of the as-deposited sample is reflected in the relatively small intentensity of the 2nd and 5th lines of the broad sextet [6,.11]. This changes towards the isotropic state after a short annealing [9]. On the other hand, there appeared new reflections in the XRD and new Mössbauer spectral components relating to magnetite, and heamatite indicating the occurrence of a significant oxidation during the ageing. The oxidation process can be explained by the release of oxygen occluded in the deposits. Although, the kinetics of oxide formation has been determined in details, a less knowledge could be obtained about the transformation process of the metastable phases [11].

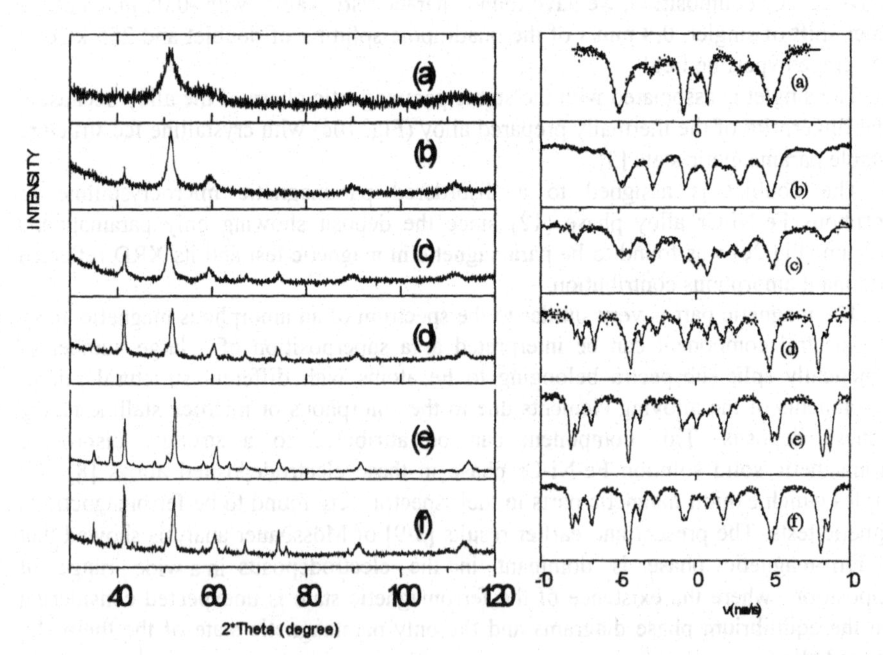

Figure 4. XRD of 49%Fe-46%Ni-5%Cr electrodeposited alloy, as-deposited(a) and aged isothermally at 500 °C in Ar gas atmosphere for 5 min (b), 1h (c), 40h (d), 50h(e) and 100h (f).

Figure 5. Mössbauer spectra of 49%Fe-46%Ni-5%Cr electrodeposited alloy, as-deposited(a) and aged isothermally at 500 °C in Ar gas atmosphere for 5 min (b), 1h (c), 40h (d), 50h(e) and 100h (f).

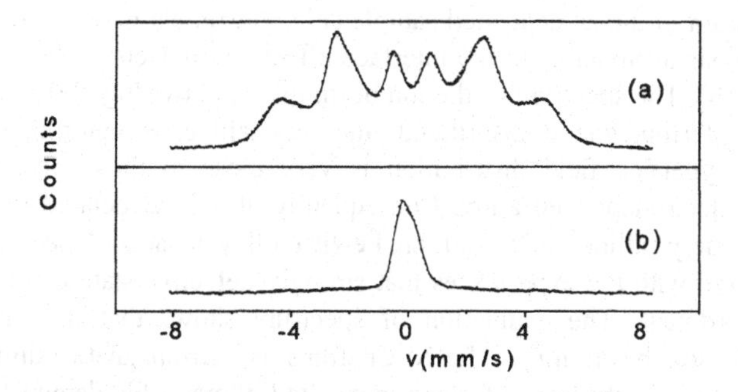

Figure 6. CEM spectra of electrodeposited 51%Fe-24%Ni-25%Cr samples (a) non-irradiated and (b) irradiated with 209 MeV ^{84}Kr ions with a fluence of $3x10^{14}$ ions/cm^2.

When the electrodeposits were irradiated with energetic heavy ions at room temperature, significant changes were found between Mössbauer spectra (Fig. 6) of the irradiated and non-irradiated electrodeposited alloys. These changes reflect the transformation of the metastable ferromagnetic phase to the paramagnetic one. This can be understood as radiation induced phase transformation [12] and can indicate the metastability of the dominant ferromagnetic phase of the electrodeposits.

3.2. ION BEAM MIXED AND LASER IRRADIATED MULTILAYERS

Fig. 7 shows the Mössbauer spectra of as-deposited, ion beam mixed and laser irradiated multilayers as well as their states after ageing at 450 °C, 640 °C and 800 °C for 3 hours.

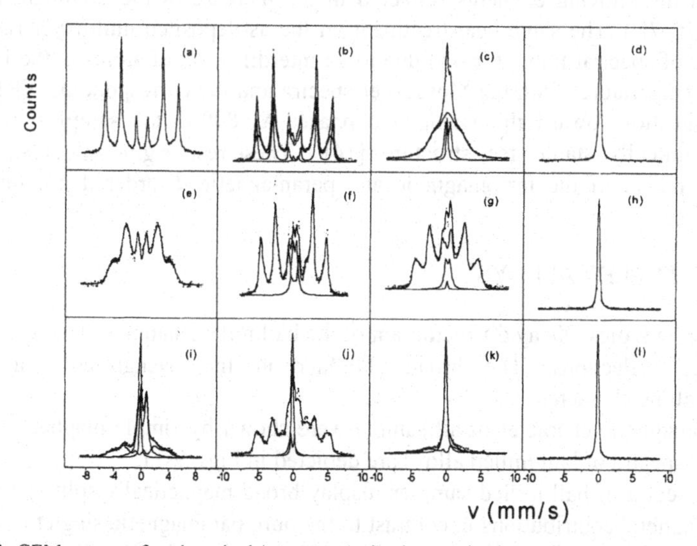

Figure 7. CEM spectra of as-deposited (upper row) , ion beam mixed (middle row) and laser irradiated (lower row) multilayers as well as their states after ageing at 450 °C (second column), 640 °C (third column) and 800 °C (fourth column) for 3 hours

The spectrum of the as-deposited sample (Fig. 7a) reflect mainly α-iron as well as small contributions attributable to the interface effect and defects in the neighbourhood of iron atoms [13]. The spectrum of the ion beam mixed alloy (Fig. 7e) is a broad sextet typical for the ferromagnetic disordered microcrystalline or amorphous alloy. The parameters of hyperfine field distribution is very close to those obtained with the corresponding electrodeposited alloy. Consequently, this is associated with a strongly disordered microcrystalline solid solution FeNiCr alloy in accordance with its X-ray diffractogram and with the experiences that strongly defective state can be induced by the ion bombardment. The evaluation of spectrum shows that the main phase of evaporated and ion beam mixed Fe-Ni-Cr films is ferromagnetic similarly to that observed previously in the case of electrodeposited alloys. This demonstrates that the appearance of the metastable ferromagnetic phase is not restricted only for the electrodeposited alloys. The ion beam mixing of multilayers is recognized as an excellent tool to produce non-equilibrium phase Fe-Ni-Cr alloy.

In the spectrum of laser irradiated multilayer (Fig. 7i) the singlet is attributed to the stable paramagnetic phase, the other paramagnetic component with an average quadrupole splitting of 0.8 mm/s, which well agrees with that of the corresponding component in the electrodeposited sample of the same composition, is assigned to a disordered paramagnetic solid solution, while the magnetic component is associated with a microcrystalline ferromagnetic solid solution, the latter being the dominat phase in the ion beam mixed sample. The considerable occurrence of paramagnetic phases, especially the stable equilibrium phase, in the laser irradiated sample indicates that the heat effect of the laser pulse also plays a roll in the formation of the structure of this alloy.

Annealing of the ion beam mixed and laser beam irradiated films at 450°C induces a solid state diffusion among the parent elements which reduces the coordination of iron atoms with the alloying elements reflected in the increase of the dominant hyperfine field (Fig 7f, 7j). The same heat treatment on the as-deposited multilayer results in a broadening of spectral lines (Fig 7b) due to an interdiffusion of atoms at the interfaces. At higher temperatures both the Mössbauer spectra and the corresponding XRD show a gradual transition toward the stable fcc phase. At 800°C the samples display the thermodynamically stable fcc structure [14]. These results give an evidence for the metastable nature of the ferromagnetic and paramagnetic disordered microcrystalline FeNiCr phases.

3.3. BALL MILLED ALLOYS

Fig. 8 shows a typical X-ray diffractogram of the ball milled samples. The diffractogram exhibit fcc reflections. The broad shoulder in the background can indicate microcrystalline character.

Ferromagnetic character of all sample were shown by simple magnetic tests. The Mössbauer spectra of ball milled alloys are depicted in Figs. 9-11.

The spectra of ball milled samples display broad magnetically split spectrum part and paramagnetic contributions in contrast to the pure paramagnetic singlet of thermally prepared samples with the same composition. The spectra of ball milled samples were evaluated similarly to those of electrodeposited and ion beam mixed multilayered ones.

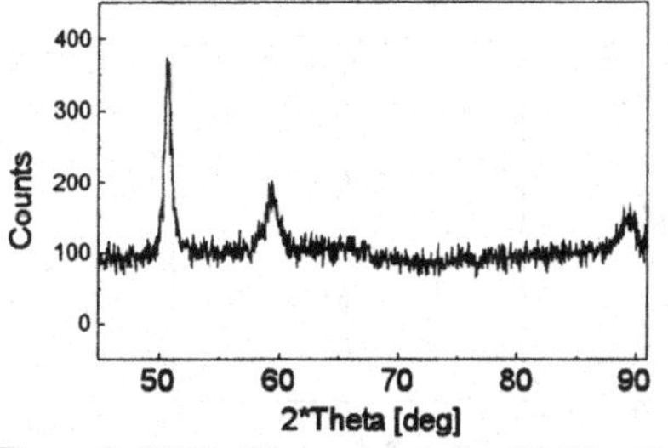

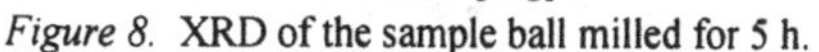

Figure 8. XRD of the sample ball milled for 5 h.

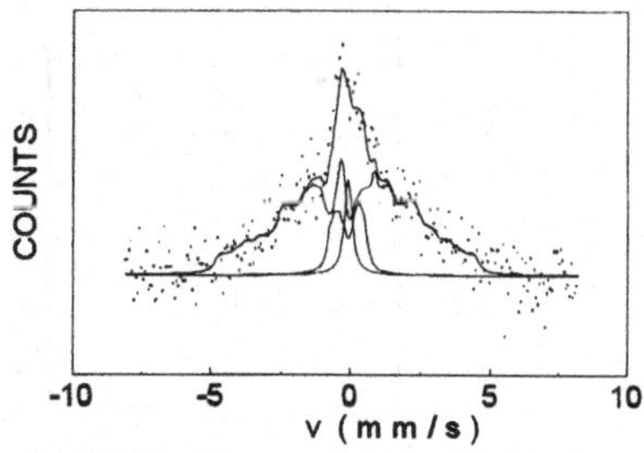

Figure 9. CEMS of the sample ball milled for 5h.

The spectra were either decomposed into a singlet, a doublet and a sextet or evaluated as a combination of a hyperfine field and quadrupole splitting distributions with the presence of a singlet. The isomer shift of the singlet was found to be -0.08 mm/s (at 293K) in agreement with that characteristic for the thermally prepared alloys, consequently, the singlet was assigned to the paramagnetic equilibrium phase in this alloy. The doublet or the quadrupole splitting distribution gave an average quadrupole splitting of 0. 8 mm/s which makes possible to assign this component to a paramagnetic phase of a defective FeNiCr paramagnetic solid solution which can similarly occur in the electrodeposited and mixed multilayered FeNiCr alloys. Similarly to the electrodeposits and to the ion and laser beam mixed multilayers, the magnetically split spectrum part can be attributed to a strongly disordered ferromagnetic solid solution FeNiCr phase with a large variety of iron microenvironments having different distribution of alloying elements in the neighbourhood of iron atoms.

In the samples (especially which was ball milled for 5 hours) the magnetically split component is dominant. This, similarly to those mentioned in the previous cases, reveals that the main phase of this ball milled sample is also ferromagnetic which never occur with alloy prepared thermally with conventional methods and cannot be expected on the

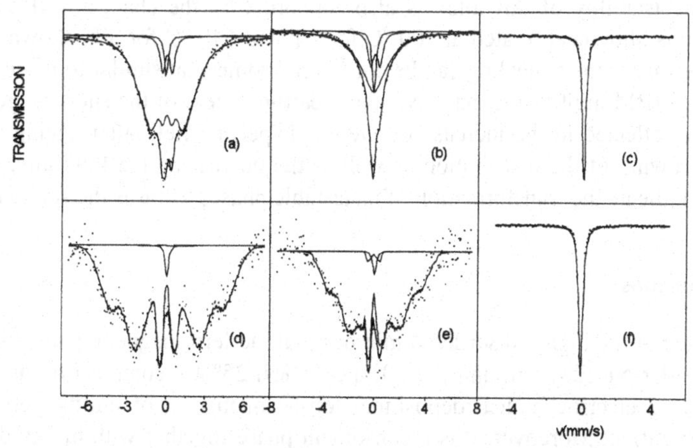

Figure 10. 295K (top) and 78K (bottom) Mössbauer spectra of the samples ball milled for 5 h (a, d), for 15 h (b, e) and the thermally prepared samples of the corresponding composition (c, f).

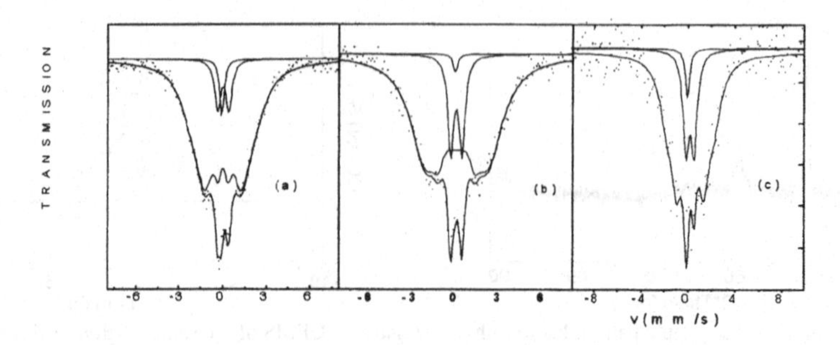

Figure 11. 293 K Mössbauer spectra of sample ball milled for 5 hours (a) and aged at 500°C for 1h (b) as well as at 640°C for 3h (c).

equilibrium phase diagram. This is in good correspondence with that observed with the electodeposited and ion beam mixed multilayered samples. The very broad linewidths of the well resolved magnetically split spectrum recorded at 77 K show a high degree of disorder of the this ball milled alloy. The high disorder is rather expected in the samples prepared by ball milling technique at suitable parameters [15].

The observed feature can also be attributed to the very inhomogenous character of the ball milled alloys.

Significant differences were found among the Mössbauer spectra of samples processed for different milling times. As the milling time increases from 5 to 15 hours a significant increase of paramagnetic components and a narrowing of the magnetic field distribution can be observed. This reflects significant differences in the phase composition and in the degree of short range ordering of samples formed at different milling times. This is in agreement with the character of ball milling that more stable phases can be prepared at longer milling times, consequently a clear indication of the metastability of ferromagnetic phase of alloy prepared with 5 hours milling.

The comparison of the evaluation of CEM and transmission spectra of sample milled for 5 hours indicates that the surface behaves similarly to the bulk.

The metastability of this alloy is also supported by the changes reflected in the spectra, of this alloy heat treated at 500 °C for 1 h and 640 °C for 3 h, shown in Fig. 6. The changes due to the annealing can be considered somewhat similar to those observed with the aged IBM multilayers, namely, some rearrangement of the short range order of alloy can be reflected in the increase of average hyperfine field after ageing at 500 °C and the narrowing of the distribution as well as the decreasing the field after ageing at 640 °C can indicate the transformation of metastable phases towards the stable ones.

4. Conclusions

The existence of the highly disordered microcrystalline ferromagnetic phases have been shown in Fe-Cr-Ni alloys around the 50%Fe-25%Ni-25%Cr composition as dominant phase applying electrochemical deposition, ion-beam mixing of multtilayers and ball milling preparation. Moreover, it is a constituent phase together with highly disordered as well as equilibrium paramagnetic phases in the case of laser beam irradiated multilayers. These metastable phases never occur with alloys prepared thermally with conventional methods and corresponding phases cannot be expected in the ternary Fe-Ni- Cr equilibrium phase diagram.

The metastability of these non-equilibrium phases was also shown.

The electrochemical deposition, the ion beam mixing of multilayers an the ball milling technique as well as partly the laser beam irradiation of multilayers are recognized as excellent methods to produce non-equilibrium FeNiCr alloys.

Mössbauer spectroscopy is a sensitive analytical method to study the metastable phases in FeNiCr alloys prepared by non-equilibrium methods.

Acknowledgements

This work was supported by the Hungarian OTKA Fund (project No 014970), the Hungarian-Italian Science and Technology Program (project No I-17/95), the Hungarian Ministry of Education (project No FKFO 0148/1997), Hungarian Academy of Sciences (project No AKP 9733) and by the Brazilian Council of Scientific Research project No CNPQ VKG-520414/96-9. The help of Drs. R. Gupta, A. Perin and Z. Klencsár is also acknowledged. The irradiation at JINR, Dubna, was financially supported by the Hungarian Academy of Sciences.

References

1. Watson, A., Chisholm, C.U. and El-Sharif, M. (1988) The role of cromium II and VI in the electrodeposition of chromium nickel alloys from trivalent chromium -amide electrolytes, *Trans. IMF* **64**, 149-154.
2. El-Sharif, M., Watson, A. and Chisholm, C.U. (1988) The sustained deposition of thick coatings of chromium/nickel and chromium/nickel/iron alloys and their properties, *Trans. IMF* **66**, 34-41.
3. Kuzmann, E., Czakó-Nagy, I., Vértes, A., Chisholm, C.U., Watson, A., El-Sharif, M., Kerti, J. and Konczos, G. (1989) Mössbauer study of electrodeposited Fe-Ni-Cr alloys, *Hyp. Int.* **45**, 397-402.
4. Watson, A., Kuzmann, E., El-Sharif, M.R., Chisholm C.U. and Czakó-Nagy, I. (1988) The value of Mössbauer spectroscopy in elucidation of the microstructure of electrodeposited alloys, *Trans IMF* **66**, 84-88.
5. Vertes, A., Watson, A., Chisholm, C.U., Czakó-Nagy, I., Kuzmann, E. and El-Sharif, M.R. (1987) A comparative study of Mössbauer spectroscopy and X-ray diffractometry for the elucidation of the microstructure of electrodeposited Fe-Ni-Cr alloys, *Electrochimica Acta* **32**, 1761-767.
6. Kuzmann, E., Vértes, A., Chisholm, C.U., Watson, A., El-Sharif, M. and Anderson, A.M.H. (1990) Mössbauer study of electrochemically deposited Fe-Ni-Cr and Fe-Ni-Cr-P alloys, *Hyp. Int.* **54**, 821-824.
7. Kuzmann, E., Vértes, A., Chibirova, F. Kh., Chisholm, C.U. and El-Sharif, M. (1993) Mössbauer and X-ray diffraction study of electrochemically deposited Fe-Ni-Cr and Fe-Ni-Cr-P alloys, *Nucl. Instr. Meth.* **B76**, 96-98.
8. Kuzmann, E., Vértes, A., Chisholm, C.U. and El-Sharif, M. (1995) Conversion electron Mössbauer study of short range ordering in electrochemically deposited Fe-Ni-Cr alloys, *J. Radional. Nucl. Chem.* **190**, 327-332.
9. Kuzmann, E., Varga, I., Vértes, A., Czakó-Nagy, I. and Chisholm, C.U. (1991) Mössbauer study of electrodeposited Fe-Ni-Cr alloys, *Hyp. Int.* **69**, 459.
10. Klencsár, Z., Kuzmann, E. and Vértes, A. (1996) User-Friendly Software for Mössbauer-Spectrum Analysis, *J. Radional. Nucl. Chem. Lett.* **210**, 105-112.
11. Kuzmann, E., El-Sharif, M.R., Chisholm, C.U. and Vértes, A. (in print) Mössbauer and X-ray diffraction study of isothermal ageing effect at 500°C in electrochemically deposited Fe-Ni-Cr alloys, *J. Radional. Nucl. Chem.*

12. Kuzmann, E., Nomura, K., El-Sharif, M., Chisholm, C.U., Principi, G., Tosello, C., Gupta, R., Havancsák, K. and Vértes, A. (1997) Comparative study of electrodeposited and ion beam mixed Fe-Ni-Cr alloys, *Hyp. Int.* **112**, 175-178.
13. Principi, G., Gupta, R., Tosello, C., Gratton, L.M., Jannitti E., Kuzmann E. and Varga I. (1996) Structural investigation of FeNiCr films prepared by non-equilibrium techniques, *Suppl.Nuovo Cim.* **50**, 707-711.
14. Perin, A., Gupta, R., Principi, G., Tosello, C., Gratton, L.M., Jannitti, E., Kuzmann, E., Klencsar and Z., (n print) Thermal stability of ion-beam mixed and laser-irradiated FeNiCr multilayers, *Surface and Coatings Technology*.
15. Takács, L. (1996) Nanoncrystalline Materials by Mechanical Alloying and their Magnetic Properties, in C. Suryanarayana, J. Singh and F.H. Froes (eds.), *Processing and Properties of Nanocrystalline Materials*, The Minerals, Metals & Materials Society, Warrendale, P.A., pp 453-464.

APPLICATION OF THE MÖSSBAUER EFFECT IN SOME TRIBOLOGICAL PROBLEMS

A.L. KHOLMETSKII[1], V.V. UGLOV[1], V.V. KHODASEVICH[1], V.M. ANISCHIK[1], V.V. PONARYADOV[1] AND M. MASHLAN[2]
[1)]Department of Physics, Belarus State University, 4, F. Skorina Avenue, 220080, Minsk, Belarus
[2)]Department of Experimental Physics, Palacky University, Svobody 26, 771 46 Olomouc, Czech Republic

1. Introduction

It is well-known that different kinds of mechanical treatment of metal surfaces have essential influence on their properties [1]. It has been established that the most intensive structural changes occur in the surface layer with the depth <1 μm. At the same time, the Röntgen methods provide investigation of surface layers with the depth < 5 μm, while electron methods (microscopy, electronography, etc.) allow to look for the surface layers with the depths < 0.03...0.05 μm.

Application of conversion electron Mössbauer spectroscopy (CEMS) in the field of tribology seems to be promising since it deals with the depths of surface layers of about 0.1 μm. In addition, simultaneous measurement of the investigated samples by means of conversion X-ray Mössbauer spectroscopy (XMS) allows to look for the surface layer ~10 μm. The combination of CEMS and XMS could be very useful in a number of tribological problems.

One should notice that the application of Mössbauer spectroscopy in tribology requires in many cases to investigate the samples in wide range of their sizes. The standard CEMS detectors usually do not satisfy this requirement. In order to avoid such restriction we have developed a special CEMS detector for tribological applications [2], which allows to measure the samples of almost arbitrary shape and size.

2. CEMS Methodology for Tribological Applications

At the first sight, the requirement of high efficiency for CEMS detector is in contradiction with the possibility to change the shape and size of a sample. Indeed, in order to provide high efficiency, it is necessary to place the sample inside the detector realizing a 2π or 4π geometry. On the other hand, such a placement inevitably leads to some restrictions on the shape and size of the sample. This is actually true, with the one exception: the case where air of natural composition is used as a working gas in the detector. In such case the detector could have practically infinite working volume. It solves the

M. Miglierini and D. Petridis (eds.), Mössbauer Spectroscopy in Materials Science, 215–226.

first problem (elimination of restriction on admissible sample's size). On the other hand, it is well-known (see, e.g. [3, 4]) that the registration of current pulses in gas detectors with natural air is inconvenient due to a long ion drift time in air (10^{-4} - 10^{-5} s). At the same time, the duration of light pulses accompanying the processes of ionization and excitation of air atoms, is less than 10^{-7} s [5]. The main luminous element, nitrogen, emits photons in the visible and ultra-violet ranges [5] well matched to usual photomultipliers (PM). However, the average amplitude of light pulses produced by low-energy conversion and Auger electrons in air without electric field is small and insufficient for a perfect functioning of the detector (low efficiency, an influence of the dark and radiation [6] background of the PM).

In order to increase the amplitude of the light pulse per single electron one can use a discharge mode of the detector, where each discharge is produced by an initial ionization in the working volume. It implies an application of an external electric field. However, a controlled evolution of such discharges is possible only in a gas of optimal and fixed composition, while in natural air a quenching mechanism is absent. It means there is always a finite probability of self-sustaining discharge development, which prevents from further registration of electrons.

A transformation of microdischarges into self-sustaining discharge could be avoided by an isolating film (transparent to light) placed on a surface of one of the electrodes. However, a new problem arises: the distortion of the electric field in the working volume by charges accumulating on the isolating film. Our calculations show that such a distortion is not practically significant [2].

The above mentioned combination: the use of natural air as working gas, the registration of light pulses from microdischarges in air, and the elimination of self-sustaining discharges in air due to an isolating film on a surface of an electrode - is implemented in the simples way in the following construction of the CEMS detector (fig. 1).

The sample under investigation (S) is placed near the input window of the PM. The sample is irradiated by a collimated tangential beam from the Mössbauer source (MS). The S, PM and MS are placed in a hermetic chamber (HC). The sign of high-voltage on the sample (u_s) is positive and opposite to the sign of high-voltage on photocathode of the PM.

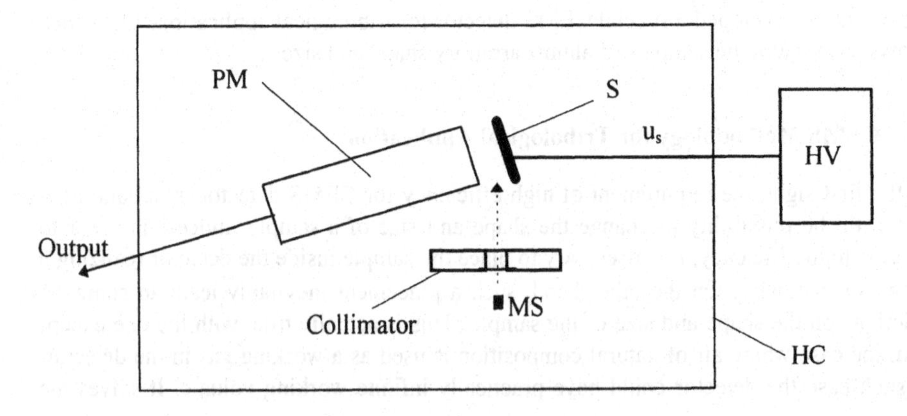

Figure 1. Scheme of air scintillation detector.

The electrons leave the surface of the sample and cause the micro-discharges in the gap between S and PM. The value of the electric field in the gap is determined by the difference of the electric potentials of the sample and the photocathode of PM. A simplicity of the described construction of the detector is provided by the triple function of the PM: its photocathode is one of the electrodes, its glass bulb plays the role of the isolating film between the electrodes, and the PM properly detects the light pulses. We call the construction in fig. 1 as air scintillation detector (ASD).

An optimal functioning of an ASD could be realized by means of variation of the high-voltage supplied to the sample under different pressure in the inner volume of the ASD. It is obvious, that the lower the air pressure p, the higher the selectivity for low-energy electrons (for fixed thickness of the gap l between PM and sample), due to the dependence $dE/dx \sim 1/E$. However, there is an admissible lowest magnitude of p (for fixed electric field), below which the production of microdischarges from natural ions in the air is possible. The theoretical analysis and experiments allowed to find the optimal working parameters of the ASD: p=4.0×10^3 Pa; l=5 mm, the difference between the sample's and the photocathode's electric potentials Δu=2300 kV; the angle between the sample's plane and the gamma-beam axis ϑ=5°. Under these conditions the value of the resonant effect for a Mössbauer spectrum of α-Fe is more than 10% [2].

We notice that in case of CEMS the tangential incidence of the gamma-beam on a surface of a sample provides an increase of the count-rate by$1/\sin\vartheta$ times in comparison with the case of normal incidence of gamma-beam due to a corresponding increase of the path length of the gamma-quanta in the surface layer referring to the maximum escape length of electrons. For the chosen value of $\vartheta = 5°$, $1/\sin\vartheta \approx 10$. Hence, the count-rate of the ASD is several times larger compared to normal incidence used in standard CEMS detectors. In addition, the tangential incidence of gamma-beam on a surface of a sample makes the ASD directly sensitive to structural and magnetic anisotropy of the sample, that could be important for some practical applications.

The developed ASD was used in the time-modulation Mössbauer spectrometer [7]. All CEMS and XMS spectra were measured under room temperature. Processing of obtained Mössbauer spectra was carried out by the NORMOS software developed by R.A. Brand (Ungewandle Physik, University of Duisburg, Germany).

The developed instruments were applied by us for investigation of some topical tribological problems.

3. Polishing and Grinding of the Surfaces of Steels

3.1. POLISHING OF A MAGNETIC STEEL

Investigation of a polishing process is a classical problem of tribology. According to modern viewpoint, this process entails an intensive motion of molecules on a surface of metal so that the state of a surface is close to melting. As a result so called Beilby layer is formed which consists of very small crystals [8]. The measurement of their dimensions is usually impossible by means of Röntgen methods (the so called Röntgen-amorphous state). At the same time, it is well-known that polishing leads to an increase of hardness of a surface of metal.

In order to investigate a process of Beilby layer formation as well as a mechanism providing an increase of its hardness we chose the sample of high-quality constructive steel 45 in a cylindrical form (diameter 18 mm, and height 16 mm) in a state of delivering. The face surfaces of the sample were ground on a grinding machine with a carbide silicon bob. Then one of the face surfaces was treated by hand polishing. At the last state of polishing a diamond slurry with a dimension of 0.25 μm was used.

Mössbauer spectrum of the initial grinding surface of steel is shown in fig. 2a. It represents a superposition of a broad sextet of α-Fe (the line width is of about 0.5 mm/s) and γ-Fe (austenite) (the isomer shift with respect to α-Fe δ=(-0.05±0.03) mm/s). Its concentration is of (11±1.1) %. A presence of austenite in the surface layer of steel after grinding is not surprising. Local heating of a surface under grinding and its fast cooling make possible a transformation of different initial iron magnetic phases into austenite [1, 9]. A distribution of the effective magnetic field H_{ef} for magnetic phases of iron (fig. 3a) shows that the surface of steel 45 consists of ferrite (H_{ef}≈33 T) and martensite with H_{ef}≈26...30 T.

The spectrum of the sample after polishing is present in fig. 2b, and a respective

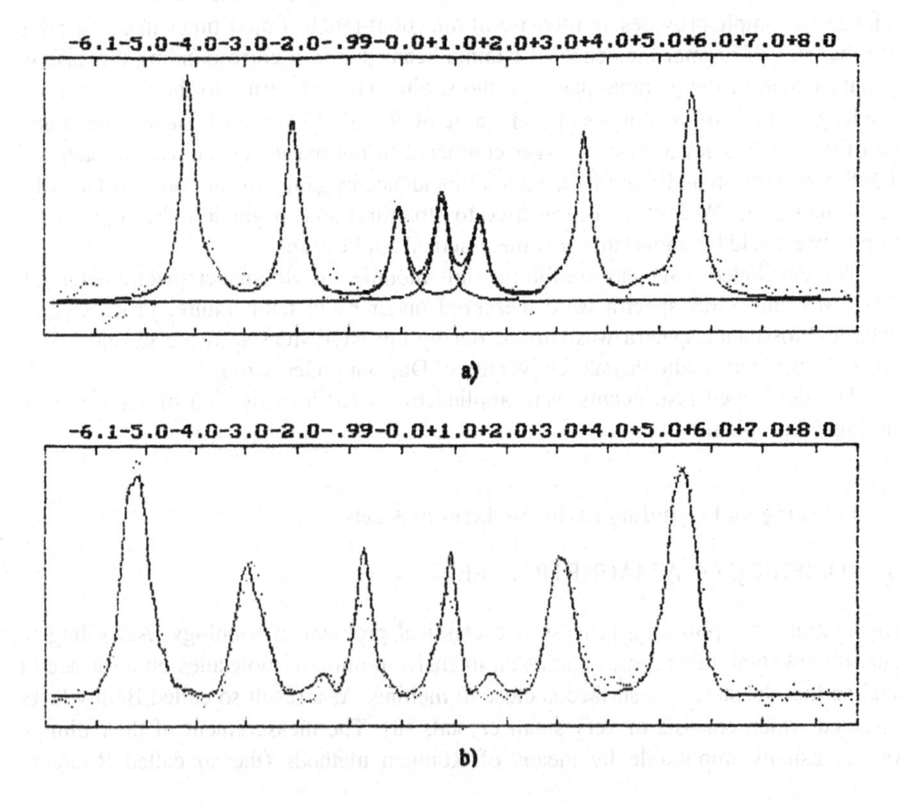

Figure 2. Mössbauer spectra of ground (a) and polished (b) surfaces of steel 45.

distribution of H_{ef} is shown in fig. 3b. One can see a full transformation of austenite into magnetic phases of iron. In addition, one reveals a broadening of H_{ef} distribution in the range 30-35 T and the appearance of additional peaks for H_{ef}=17.0 T and H_{ef}=21.5 T. Such values of H_{ef} correspond to carbides χ-Fe_5C_2 and θ-Fe_3C (cementite), respectively [9]. Total concentration of carbides is of (8.3±1.1)%, that is close to the concentration of austenite in the ground surface of steel. This allows to propose that there is a full transformation of austenite into carbide phases of iron under polishing. This is similar to the phase transformations in a Fe-C system under tempering of steels for the temperature t>300°C [9, 10].

Thus, one may conclude the following.

The process of polishing is accompanied by structural and phase transformations in the surface layers of steels. It is well-known that under polishing the surface atoms have high mobility, and the effects of microplastic deformation and dispergation take place, too. Decomposition of austenite under plastic deformation is a well-known effect [11]. However, we have additionally revealed a formation of carbides after polishing. This allows to propose, that an increase of Beilby layers hardness after polishing occurs not only due to dispersion hardening, but also due to transformation of γ-Fe into carbides.

3.2. POLISHING OF A STAINLESS STEEL

In order to obtain more detailed information about transformation of austenite into magnetic phases of iron, we have investigated a polishing process in a surface of stainless steel 12Cr18Ni9Ti doped by chromium, nickel and titanium. The dimensions of the sample were 40×40×1 mm.

A series of Mössbauer spectra is seen in fig. 4. Only one line shows up in the spectrum obtained from the surface of an unpolished sample (fig. 4a), its isomer shift with respect to α-Fe is equal to δ=(-0.04±0.003) mm/s). After diamond polishing a group of new sextets arises beside the line belonging to stainless steel (fig. 4b). A total concentration of magnetic phases of iron is equal to (53±12)%. It is extremely difficult to identify the individual magnetic phases. A respective broad distribution of p(H_{ef}) (fig. 5a) indicates a formation of solid solution of Fe+Cr+Ti; the observed maxima 21.5 T, 18.0 T and 11.0 T correspond to the carbides θ-Fe_3C and χ-Fe_5C_2. The obtained results are in agreement with the data obtained in [12].

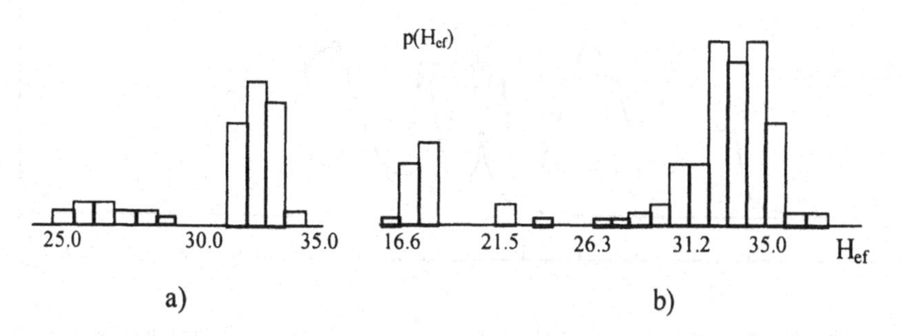

Figure 3. Distributions of H_{ef} for the Mössbauer spectra measured on ground (a) and polished (b) surfaces of the steel 45.

Fig. 4c represents a surprising spectrum obtained after diamond polishing in the presence of external magnetic field. It is known, that a magnetic field under mechanical treatment of metal surfaces influences their properties. The obtained result proves that a modification of surface layers after such treatment should be explained by intensive phase transformations caused by the presence of magnetic field. In particular, we see an essentially larger transformation of paramagnetic Fe into magnetic Fe. A concentration of paramagnetic phase decreases down to (8.7±1.4) %. In a corresponding distribution of H_{ef} one sees an appearance of the maximum H≥30 T, which indicates a forma-

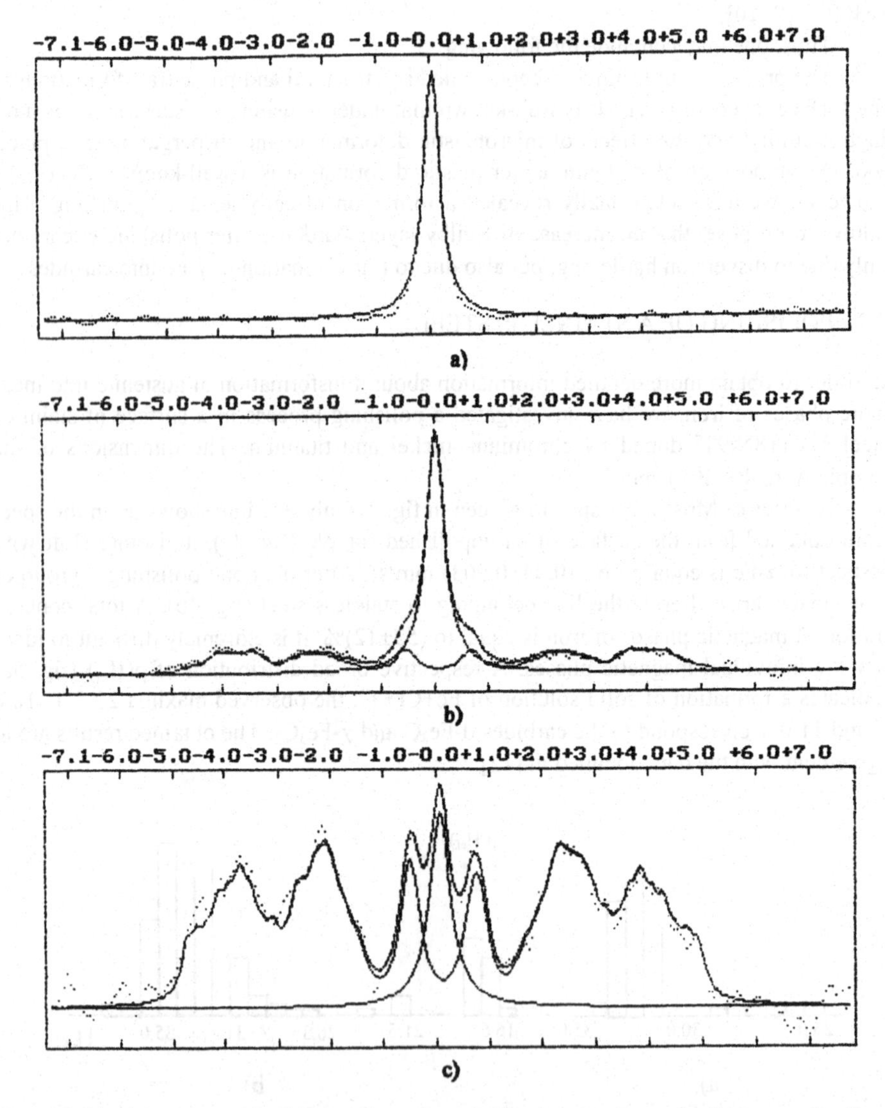

Figure 4. Mössbauer spectra of stainless steel: a - after annealing; b - after diamond polishing; c - after diamond polishing in a presence of magnetic field.

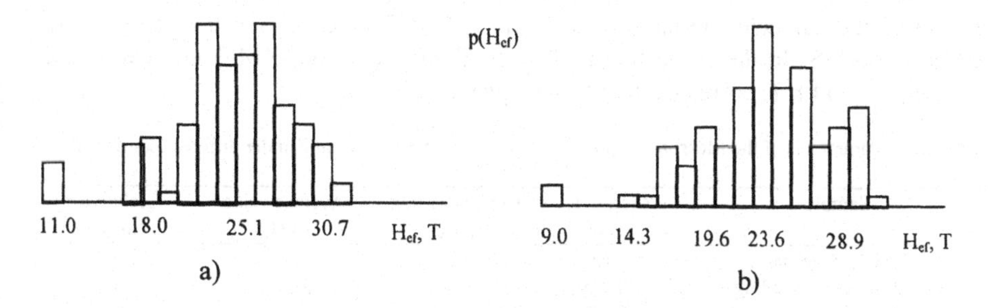

Figure 5. Distributions of H_{ef} for Mössbauer spectra of stainless steel after diamond polishing (a) and after diamond polishing in the presence of magnetic field.

tion of pure ferrite resulting from oxidation of doped elements..

Thus, one can conclude that the presence of magnetic field promotes a γ-Fe decomposition as well as oxidation of doped elements in the surface layer of the stainless steel. A single real effect which could act together with a mechanical (and temperature) influence on a surface is a magnetostriction.

3.3. GRINDING OF A GREY PIG IRON

The effects caused by grinding constitute a separate group of tribological problems. It is known that a grinding and other similar methods of treatment of surfaces do not provide a combination of such properties as hardness, plasticity, wearness, etc. Our investigation of grinding process aimed to clarify an influence of oils on a structure of metal's surfaces as well as an influence of plastic metals (Cu, Ni) added to oils.

In our investigations we chose a grey pig iron GPI 21-40, the dimensions of the sample were 40×20×20 mm. The measurements after different kinds of grinding were carried out by means of CEMS and XMS. Fig. 6 shows the obtained distributions of H_{ef} for different modes of grinding. The principal parameters of the measured Mössbauer spectra are presented in table 1.

One can see that a dry grinding increases the concentration of γ-Fe in the surface layers of pig iron. An application of the oil BNZ-4 prevents a formation of austenite. In this case one may suppose that a temperature of the surface does not reach the temperature of the $\alpha \rightarrow \gamma$ transformation. It is interesting to analyse a behaviour of the average value of H_{ef} and the total square of the spectra f (in relative units) for different kinds of grinding. Dry grinding decreases both these values in the surface layer of 0.1 μm (CEMS measurements) as well as in the surface layer ~10 μm (XMS measurements). It could be explained by a very high density of structural defects generated by dry grinding. Grinding with BNZ-4 decreases a density of structural defects in the surface of pig iron, that leads to an increase of f. On the other hand, one can see that grinding in BNZ-4+Cu+Ni again decreases the value of f under CEMS measurement. It definitely proves a formation of Cu-Ni film on the surface of pig iron which screens the low-energy conversion and Auger electrons. On the contrary, the value f increases under XMS measurement for this grinding variant. It means that plastic particles of Cu and Ni "heal" the structural defects in the surface of pig iron. Moreover, the atoms of Cu and Ni penetrate into α-Fe lattice inducing an increase of $<H_{ef}>$ in both Mössbauer spectra measured by CEMS and XMS. In the respective $p(H_{ef})$ distribution (fig. 6d, XMS) one can see a maximum H_{ef}=34.8 T, corresponding to such penetration.

TABLE 1. Parameters of the Mössbauer spectra of grey pig iron after grinding under different conditions.

Sample	Measure-ment	Concentra-tion of γ-Fe	Average value of H_{ef}	*f*, relative units
GPI 21-4, initial	CEMS	30.2 %	27.1 T	1.00
GPI 21-40, dry grinding	CEMS	43.8 %	26.9 T	0.80
GPI 21-40, grinding in oil	CEMS	2.0 %	26.9 T	0.93
GPI 21-40, grinding in oil + Cu+Ni	CEMS	4.3 %	27.9 T	0.90
GPI 21-40, initial	XMS	<1.0%	29.5 T	1.13
GPI 21-40, dry grinding	XMS	5.5%	29.3 T	1.10
GPI 21-40, grinding in oil	XMS	<1.0%	28.7 T	1.23
GPI 21-40, grinding in oil + Cu+Ni	XMS	<1.0%	29.7 T	1.35

The revealed effects are practically significant since they indicate that the films of Cu+Ni formed during a grinding have better adgesion properties than the surface films of plastic metals formed by standard methods.

4. Investigation of Phase Composition of High-Speed Steel after High Current Density Nitrogen and Boron Implantation

At present the formation of thick (≥ 10 μm) modified layers inside different materials by means of ion-beam technology is of great interest (see, e.g., [13]). The use of high energy (about 1 MeV) ions is very expensive and therefore it is not economically profitable for broad applications. A new ion-bean method of treatment has been developed - low energy high current ion implantation (HCI) of metalloid ions at high temperature [13].

This method allows a formation of surface layers with thickness of several tens of micrometers saturated with nitrogen or boron ions due to quantum diffusion, which enhances the mechanical properties of the material. The profile of elements implanted by this method is very similar to the one of traditional methods of steel treatment - nitriding and boriding, but the enhancement of mechanical properties is greater than after traditional methods. In order to understand such an enhancement of the mechanical properties of steels it is necessary to know the phase composition and structural changes in the implanted layers and their dependence on a mode of implantation. In order to solve this problem we applied both the CEMS and XMS.

The samples used were 2 mm × 15 mm Ø of AISI M2 high-speed steel (0.86% C, 6.0% W, 5.0% Mo, 4.1% Cr, 1.9% V, 0.5% Co, in wt%). The steel samples were subjected to annealing. Finally, they were polished to a mirror finish on one face by dia-

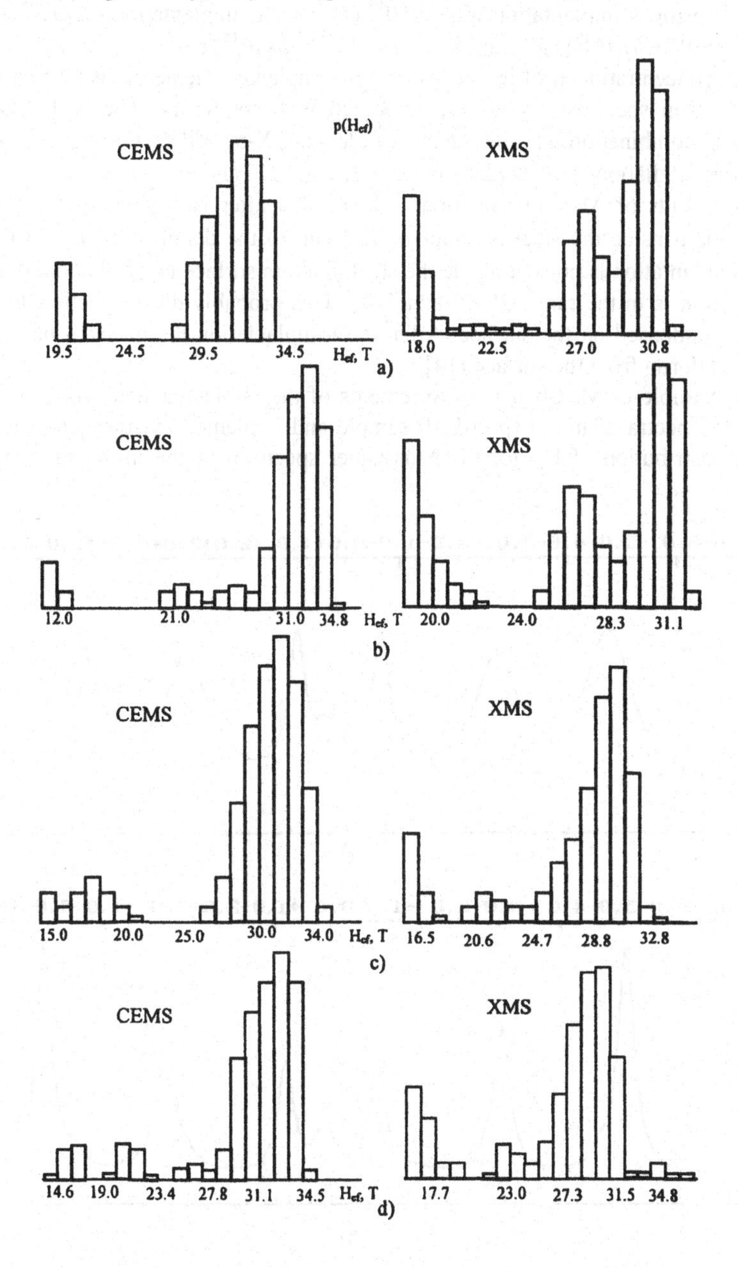

Figure 6. Distributions of H_{ef} for Mössbauer spectra of GPI 21-40 obtained after different kinds of grinding: a- initial sample; b - dry grinding; c - grinding in BNZ-4; d -grinding in BNZ-4+Cu (15 wt%)+Ni (3wt%).

mond slurry. The hardness of the steel after the thermal treatment was 64 HRC.

The samples of steel were treated with a special installation. The procedure consisted of ion implantation of the nitrogen and boron ions at a temperature of (500±5°C). A total dose for N implantation was 7×10^{19} cm^{-2}; for B implantation - 2×10^{18} cm^{-2}; for B+N - $(2\times10^{18} + 7\times10^{19})$ cm^{-2}; for N+B - $(7\times10^{19} + 2\times10^{18})$ cm^{-2}.

The concentration profiles of doped and implanted elements were measured by Auger electron spectroscopy (AES) combined with sputtering. The XMS data were analysed in combination with the results of glancing X-ray diffraction using monochromatic Co K_α radiation. The incident angles were 0.5°, 1° and 2°.

The results of AES-investigations showed that after N-implantation the nitrogen concentration inside the steel is about 8-9 at% up to the depth of 14 μm [14]. Boron after B-implantation is present inside the steel in a thin surface layer of about 0.4 μm, its concentration is in the range 20-40 at% [14]. This probably did not penetrate into the steel but condensed on the surface. After B+N-implantation the nitrogen has sputtered condensed boron from the surface [14].

As example of Mössbauer measurements of the implanted steel M2, fig. 7 shows the CEMS spectra of initial (polished) sample and implanted by nitrogen sample. The obtained distribution of H_{ef} for the Mössbauer spectrum of the initial sample (fig. 8)

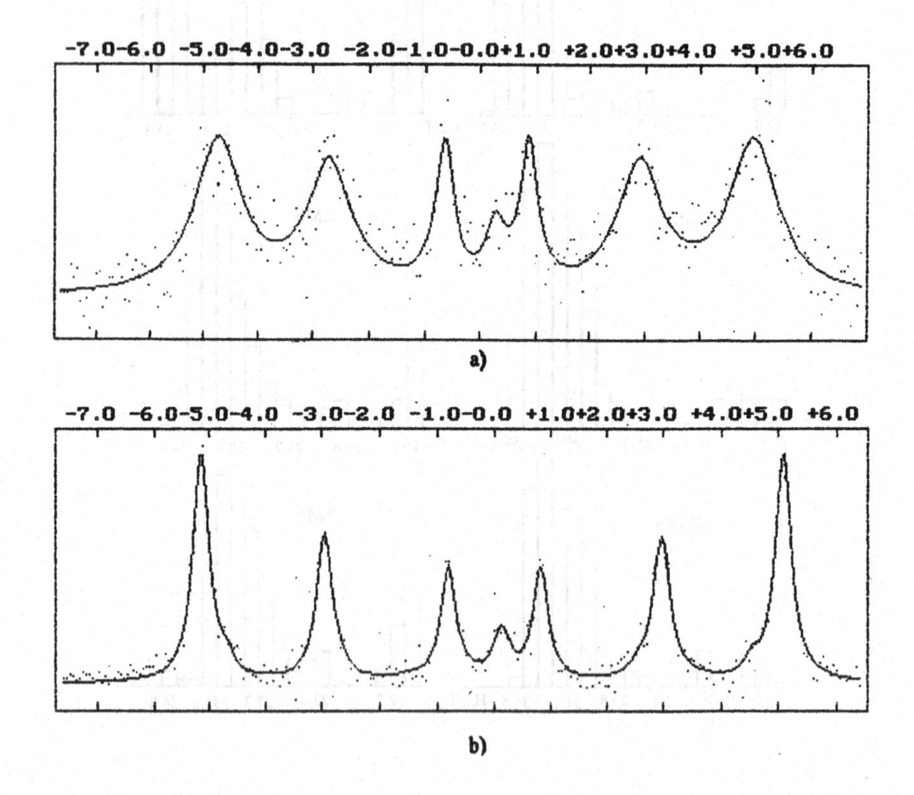

Figure 7. Mössbauer spectra of initial sample (a) and implanted by nitrogen sample (b) of M2 steel.

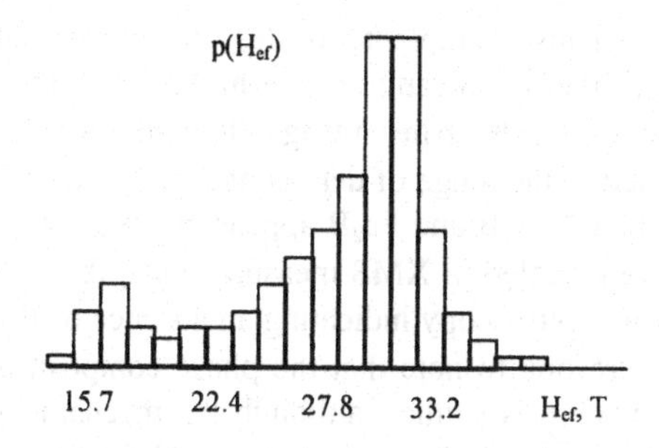

Figure 8. Distribution of H_{ef} for Mössbauer spectrum of initial sample of the M2 steel.

shows that its magnetic phases represent a martensite-carbide mixture. According to Röntgen measurements a dominant carbide is Fe_3W_3C (its concentration is close to 10 wt %). A broad distribution of H_{ef} and a big value of Zeeman lines width (more than 0.7-0.8 mm/s) for the initial sample suggest an existence of a number of non-equivalent sites of iron in the M2 steel.

After high current nitrogen implantation one observes a decrease of concentration of carbides as well as a great narrowing of Zeeman lines (from 0.7-0.8 mm/s for the initial sample to ~0.3 mm/s for implanted sample). In general, the latter effect takes place in more or less extent for all variants of implantation both for CEMS and XMS measurements. It could be partially explained by formation of clusters of nitrogen-doped elements of the steel M2 with appearance of zones of pure ferrite. A narrowing of Mössbauer lines of the implanted steels allows to apply a discrete approach to processing of the spectra obtained. The results of such processing are present in table 2.

TABLE 2. Phase composition of M2 steel implanted by nitrogen and boron ions.

Variant of implantation	Method of measurement	Concentration of martensite	Concentration of austenite	Iron phases formed due to implantation (concentration, %)
Initial sample	CEMS	(82±2) %	(2.7±0.5) %	-
M2←N	CEMS	(88±2) %	(4.5±0.6) %	$Fe_{16}N_2$- (5±1) % ε-Fe_3N - (2.±1) %
M2←N	XMS	(70±2) %	(5.3±0.5) %	$Fe_{16}N_2$- (19±1) % ε-$Fe_{3.2}N$ - (4.±1) %
M2←B	CEMS	(80±1) %	(4.0±0.3) %	Fe_2B - (7.6±0.5) % FeB - (8.5±0.7) %
M2←B	XMS	(85±2) %	(6.2±0.6) %	-
M2←B←N	CEMS	(67±2) %	(4.1±0.3) %	$Fe_{16}N_2$- (13±1) % ε-$Fe_{3.2}N$+Fe_2B - (10±1) % FeB - (5.1±0.3) %
M2←B←N	XMS	(69±4) %	(6±1) %	$Fe_{16}N_2$- (13±1) % ε-$Fe_{3.2}N$ - (9.±1) %
M2←N←B	CEMS	(61±3) %	(11±1) %	$Fe_{16}N_2$- (15±2) % ε-$Fe_{3.2}N$ +Fe_2B - (10±1) % FeB - (3.4±0.4) %
M2←N←B	XMS	(63±3) %	(5.8±0.6) %	$Fe_{16}N_2$- (17±1) % ε-$Fe_{3.2}N$ - (10.±1) %

One can see that all variants of implantation lead to increase of concentration of austenite in comparison with the initial sample. The highest amount of austenite is observed for the variant N+B - (11±1) %. After nitrogen implantation the phases $Fe_{16}N_2$, ε-$Fe_{3.2}N$ and ε-Fe_3N are formed in the range of depths of surface layer of 0.1...10 μm. After boron implantation the phases FeB and Fe_2B appear in the layer with the depth <0.3 μm. No phases of boron are revealed in XMS measurements, that is in agreement with the results of Auger electron spectroscopy indicating an absence of B atoms in the depth more than 0.4 μm. It is interesting to note that the phase composition of the M2 steel after N+B and B+N implantation is similar, although the mechanical properties of the surface layers are essentially different: the best variant is N+B. For the latter case a microhardness of the surface is 1720 kg/mm^2, and the saturated friction coefficient is of 0.72 [14].

Acknowledgement

The authors warmly thanks to Prof. K. Ruebenbauer (Institute of Physics & Computer Science, Pedagogical University, Cracow, Poland) and Dr. O. Schneeweiss (Institute of Physics of Materials, Academy of Sciences of the Czech Republic, Brno) for helpful discussions on the subject. This work was supported by IAEA (Contract No. 8926/R2).

References

1. Akhmatov, A.S. (1963) *Molecular Physics of Friction,* Metallurgiya, Moscow (in Russian).
2. Kholmetskii, A.L., Misevich, O.V., Mashlan, M., Chudakov, V.A., Anashkevich, A.F., and Gurachevskii, V.L. (1997) Air scintillation detector for conversion electrons Mössbauer spectroscopy, *Nuclear Instruments and Methods* **B129**, 110-116.
3. Aoyama, T., Totogawa, M., and Watanabe, T. (1987), *Nuclear Instruments and Methods* **A255**, 524-528.
4. Abramov, A.I., Kazanskii, J.A., and Matusevich, E.S. (1970) *Experimental Methods in Nuclear Physics,* Atomizdat, Moscow (in Russian).
5. Vagin, J.P. (1970) Scintillations in air, *Atomnaya Energia* **28**, 177-182 (in Russian).
6. Anshakov, O.M., Nalibotskii, B.P., Pertsev, A.N., Kholmetskii, A.L., and Chudakov, V.A. (1983) Investigation of photomultipliers noise produced by gamma-quanta with different energies, *Pribory i Technica Experimenta* **5**, 44-51 (in Russian).
7. Kholmetskii, A.L., Mashlan, M., Chudakov, V.A., Misevich, O.V., and Gurachevskii, V.L. (1992) A time-modulation method for Mössbauer spectra registration, *Nuclear Instruments and Methods* **B71,** 461-465.
8. Beilby, G. (1921) *Aggregation and Flow of Solids,* McMillan, London.
9. Litvinov, V.S., Karakishev, S.D., and Ovchinnikov, V.V. (1982) *Nuclear Gamma-Resonance Spectroscopy of Alloys,* Metallurgiya, Moscow (in Russian).
10. Cristofaro, N., and Kaplow, R. (1977), *Metal. Trans.* **A8**, 35-41.
11. Brömer, H. (1994) Hyperfine fields in steels, *Nucleonika* **39**, 69-90.
12. Vertes, C., Vass, G., Kuzmann, E., Romhanyi, K., Lakatos-Varsanyi, M., and Vertes, A. (1994) Conversion electron Mössbauer and XPS study on the effect of polishing of a stainless steel sample, *J. Radioanal. Nucl. Chem. Letters* **186**, 447-454.
13. Williamson, D.L., Ozturk, O., Wei, R., and Wilbur, P.J. (1994) *Surface and Coating Technology* **64**, 15-24.
14. Uglov, V.V., Rusalsky, D.P., Khodasevich, V.V., Kholmetskii, A.L., Wei, R., Vajo, J.J., Rumyanceva, I.N., and Wilbur, P.J. (in press), Modified layer formation by means of high current density nitrogen and boron implantation, *Surface and Coating Technology.*

MÖSSBAUER STUDIES IN MAGNETIC MULTILAYERS

A. SIMOPOULOS
National Center for Scientific Research Demokritos
153 10 Aghia Paraskevi, Athens, Greece

Abstract

Magnetic multilayers have been the subject of extensive studies over the last two decades in the area of Materials Science. Mössbauer spectroscopy has been employed from the very beginning to explore these materials at the atomic level. We present in this review a short introduction to the basic experimental methods used for their characterization, with some emphasis on the Mössbauer technique, and a literature survey on Mössbauer studies and theoretical calculations of magnetic hyperfine interactions. The focus of this presentation is on the enhanced hyperfine fields observed in Fe atomic layers near the interface and their variation in inner layers. As an example of the methodology followed in such investigations, a more detailed overview is given of recent work in Fe/Pt and Fe/Au multilayers.

1. Introduction

Magnetic multilayers have been extensively studied during the last two decades both for their potential in magnetic storage technologies and their scientific interest as new artificial materials in the nanometer scale. They usually consist of a number of repeated bilayers with one of the components being magnetic and the other non-magnetic. Among the primary questions that have been addressed are the perpendicular anisotropy and its relation with the interface microstructure and the physical parameters of the bilayer, the modification of the atomic magnetic moment and the magnetic hyperfine field at the interface and the neighboring atomic layers, and the intralayer and interlayer magnetic interaction.

Mössbauer spectroscopy (MS) has been the technique of choice for many of these studies due to its high sensitivity which enables the determination of hyperfine interactions at the monolayer level. A prerequisite of such studies is of course the presence of Mössbauer isotopes in one or both components of the bilayer. As such, Fe is the most commonly studied metal as the magnetic component of the bilayer but other metals like Sn, Au, Sb and some of the rare earths have been also investigated. Another important asset of MS is its ability to directly determine the orientation of the magnetic moments with respect to the plane of the multilayer using the relative intensity of the $\Delta m=0$ absorption lines. This is perhaps the most important information that this

M. Miglierini and D. Petridis (eds.), Mössbauer Spectroscopy in Materials Science, 227–242.

technique offers for technological applications such as the perpendicular magneto-optical recording.

The subject of magnetic multilayers has attracted the interest of theoretical studies from the early stages of the experimental investigations [1]. Ohnishi et al [2] predicted enhanced magnetic moments at the surface of ferromagnetic Fe, followed by a damped spatial oscillation in the first monolayers below the surface. Several experimental studies have been focused in this direction. Early MS experiments indicated such a behavior and recent detailed work in multilayers (MLs) of Fe with 4d and 5d transition metals have verified the original theoretical predictions.

We will present in this review some experimental details relevant to the studies of magnetic multilayers (chapter 2) and after a brief review on experimental and theoretical studies of Fe/transition metal multilayers (chapter 3) we will give a detailed account of recent studies of Fe/Pt and Fe/Au MLs (chapter 4) as an example of the ability of MS to detect hyperfine interactions in these nanostructured systems at the atomic layer level.

2. Experimental Details

2.1. METHODS OF PREPARATION

Thin metallic films and multilayers are usually prepared with the techniques of molecular beam epitaxy (MBE), magnetic sputtering and electron gun evaporation. The technique of choice depends on many factors, and in the case of MLs an important factor is the misfit parameter of the two constituent metals. Extended reviews on epitaxial growth are given in references [3] and [4]. It should be emphasized here that the quality of the MLs is of prime importance for studies of physical parameters at the atomic level.

2.2. STRUCTURAL CHARACTERIZATION

Structural characterization can be done with a variety of methods, some of them applied in situ, in the preparation chamber in conditions of ultra high vacuum. The most powerful methods for the surface characterization are high- and low-energy electron diffraction (RHEED and LEED) and Auger electron spectroscopy. A concise review of these techniques is given in reference [5]. X-ray diffraction (XRD) and Mössbauer spectroscopy are good sources of information regarding the structure of the interfaces and the internal layers.

XRD is the first technique to be applied for the characterization of MLs. The multilayer (often called a superlattice), apart from the modified unit cell parameters of the individual components of the bilayer, posses an extra periodic parameter which is the width of the bilayer. Since this parameter varies from a few Å to a few tens of Å it offers extra reflections for XRD. As a result, satellite peaks appear in addition to the main Bragg peak which allow the determination of the actual width of the bilayer. Individual peaks appear also, for the same reason, in the low angle region of the XRD spectra (fig.1).

Transmission electron microscopy (TEM) is the most powerful technique for the determination of the microstructure of thin films. Cross section TEM is the technique of choice for the epitaxial growth characterization of MLs.

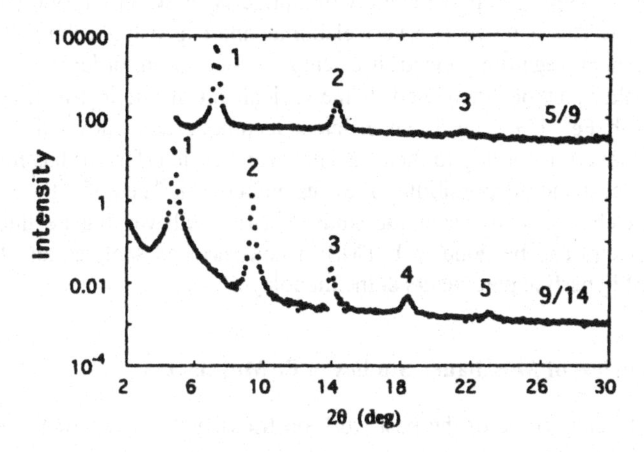

Figure 1. X-ray diffraction in the low angle region for the 5Å Fe/9 Å Pt (top) and 9 Å Fe/14 Å Pt (bottom) multilayer samples.

2.3. MÖSSBAUER TECHNIQUES

Conversion electron (CEMS) and transmission (TMS) Mössbauer spectroscopy have been used for MLs studies. There are advantages and disadvantages in both methods.

CEMS is the appropriate technique for surface studies. Its main advantage is that the detection of conversion electrons is limited to depths of 2-3 thousands Angstroms and all the information is pertinent to surface and subsurface phenomena. In the case of MLs this range is the usual thickness of the sample. The technique suffers from low counting rates and requires strong sources and/or enriched samples in the Mössbauer isotope. Another advantage of this technique is that it is not limited by the substrate of the ML. The main drawback is the limitation to room temperature experiments. There are a few examples of low temperature CEMS experiments with special counters built in the cryostat but the effort and the cost to built such setups are discouraging factors. An interesting variation of this technique is the DCEMS with depth selectivity of electrons by using electron spectrometers. The latest development of DCEMS is the „orange" spectrometer with high luminosity and energy resolution [6].

The main advantage of TMS is its accessibility in any Mössbauer lab and the convenience for temperature variation experiments. It suffers from small absorption efficiency in comparison to CEMS. Apart from isotope enrichment of the samples this drawback can be overcome by stacking together a number (10-20) of film samples. A typical thickness of ML is 1-2 thousand Angstroms and, depending on the individual thickness of the Fe layer, a stack of 10 samples allows an effect of 0.1-1%. The main limitation of TMS is the substrate for the preparation of the films. They must be thin (a few microns) and free of iron. Since each particular multilayer is limited to special substrates for epitaxial growth, this limitation should be considered seriously in planning a TMS experiment with MLs.

For studies of hyperfine parameters with atomic layer sensitivity the technique of choice is DCEMS. This is however a very elaborate and expensive technique and needs further development regarding electron energy resolution in order to achieve this sensitivity. Another method often used, is the enrichment of one to two monolayers in the Fe layer with Fe^{57} [7]. This is an efficient technique and allows the probing of individual monolayers according to their distance from the interface. It has however the drawback of the cost and the possibility of diffusion between ^{56}Fe and ^{57}Fe. As it will be demonstrated in chapter 4, a systematic study of MLs with varying Fe thickness and good quality spectra can be done with TMS in non-enriched samples and lead to the determination of hyperfine parameters at the monolayer level.

3. A Brief Review of Mössbauer Studies in Fe Multilayers

We will present below some of the past work on Mössbauer studies of Fe multilayers. This review is by no means exhaustive and rather serves the purpose of giving typical examples of relevant studies. For convenience we will present these data in alphabetical order of the counterpart element of the Fe/X bilayer.

3.1. Fe/Ag

This system is, together with Fe/Cu, the most extensively studied. In one of the pioneer Mössbauer works on MLs, Tyson et al [8] studied epitaxial films of Fe(110) on Ag(111). They performed TMS measurements at RT and 4.2K using probe layers enriched in ^{57}Fe. In a film consisting of 30 monolayers of iron they measured magnetic hyperfine fields which, at 4.2K, were slightly enhanced (~2%), with respect to the bulk value at the interface while they were less (~1%) than the bulk value at RT. Korecki and Gradmann [9] verified this result by performing CEMS measurements in the temperature region 80K...300K in a film consisting of 21 Fe(110) monolayers grown on W(110) and covered by Ag. They used the probe technique by introducing an ^{57}Fe monolayer (ml) below the Ag surface. Similar results were reported by Koon et al [10] who performed CEMS at RT and 15K in Fe/Ag (100) single crystal MLs with Fe layer thickness of 1.0, 2.4 and 5.5 mls. The 5.5 ml film displayed a spin reorientation from the plane at RT, to an angle out of the plane at 15K.

Ohnishi et al [1,11], using the all-electron full-potential linearized augmented plane wave method (FLAPW), calculated the magnetic moment and the magnetic hyperfine field B_{hf} for each monolayer in a free standing Fe (100) layer. The results showed a spatial oscillation of the B_{hf} near the Fe surface with a reduction by ~ 25% in the surface ml and a similar enhancement in the second ml. They attributed this behaviour to Friedel oscillations. The calculation shows that these oscillations disappear when the Fe surface is covered by Ag. Similar behavior was predicted by Freeman and Fu [12] but with an amplitude half of that predicted by Ohnishi et al [11]. Such an oscillation was verified experimentally [9] although the maximum enhancement was marginal (2-3%). We will discuss further the subject of oscillation in the next chapter on other systems where such phenomena appear more pronounced.

3.2. Fe/Au

Przybilski et al [13] studied the influence of noble metal coating on the magnetic properties of thin Fe films. They performed CEMS measurements at 80 K and RT on 21 mls of Fe grown on W and covered by Cu, Ag and Au respectively. They introduced one probe monolayer at the interface with the noble metal. The B_{hf} showed an increase in going from Cu to Ag to Au. A 7% increase with respect to the bulk value was observed in the case of Au at 80 K.

Shinjo and his group performed a number of studies in Fe/Au MLs using ^{57}Fe, ^{197}Au [14] and ^{119}Sn [15] Mössbauer probes. The samples were prepared on Ag buffer layers and deposited on polyamide (kapton) films. The growth orientation was along the (111) direction. The ^{57}Fe TMS displayed a ~3% enhancement of B_{hf} at RT for an 8 Å Fe layer and an orientation of the magnetic moment which depended on the thickness of the counterpart Au layer. The ^{197}Au TMS displayed transferred magnetic hyperfine interactions in the first 2 mls at the interface with Fe. This magnetic perturbation of the Au layer was verified by ^{119}Sn probes introduced in the Au layer.

Recent results on a series of Fe/Au MLs will be presented in the next chapter.

3.3. Fe/Cr

Zukrowski et al [16] studied Cr/Fe/Cr trilayers grown on W(110) substrates by using CEMS and the probe monolayer technique. The results indicate an oscillation of B_{hf} up to the 4th ml from the interface. The effect of the antiferromagnetic nature of Cr is also displayed for measurements below the Nèel temperature.

The same system grown on GaAs(100) was studied by Klinkhammen et al [17] using the same technique. The analysis of the spectra led to a structural model for the determination of roughness parameters at the interface.

3.4. Fe/Cu

In one of the pioneering works of Mössbauer spectroscopy in MLs Keune et al [18] performed TMS studies on Fe/Cu MLs with an 18 Å Fe thickness and 1000 Å Cu thickness. The films were grown on a 2000 Å single crystalline Cu along the (001) direction. The results demonstrated the formation of low spin fcc Fe similar to the γ-phase. The same group performed CEMS measurements in the temperature range 80K...300K on two samples with 3 and 7 ^{57}Fe mls respectively grown on (001) single crystalline Cu [19]. The results were interpreted with the formation of an fct-like high-spin (ferromagnetic) phase and a low-spin (antiferromagnetic) phase. The formation of the two phases depended on the thickness of the Fe layer and the preparation temperature. Keavney et al [20] arrived at similar conclusions by studying with TMS $Fe/Cu_{1-x}Au_x$ MLs grown on NaCl(001). The Fe thickness was 6.5 mls with the two interface mls composed of natural Fe and the interior mls enriched in ^{57}Fe while the

$Cu_{1-x}Au_x$ layer was 40 mls. By changing the composition of the CuAu alloy they were able to change the lattice constant of the ML from a=3.606 Å (x=0) to 3.704 Å (x=0.2). The results demonstrated the coexistence of the low- and high-spin phases with the abundance of the latter increasing with x.

3.5. Fe/Ni

Colombo et al [21] studied by both CEMS and TMS Fe/Ni MLs with Fe thickness of 17 Å and Ni thickness of 200 Å . They used the probe technique to enrich the interface Fe mls and the interior separately. The spectra were interpreted with an interface component with the direction of the spins at 39° to the normal to the plane, and an interior layer component with the spins at an angle of 45°.

3.6. Fe/Pd

Boulefeld et al [22] have performed CEMS studies on a series of Fe/Pd MLs with natural abundance Fe grown on graphite and sapphire [1120] substrates. The samples consisted of the same number (n) of mls for the two constituents (symmetric MLs) which varied from 1 to 44. The data indicate enhanced B_{hf} values for n>2 (~10%) and the average hyperfine field displays an oscillation behavior up to the 10th Fe ml. The authors interpreted their data as arising from interdiffusion in the first 2-3 Fe mls at the interface.

The same system was studied by Kisters et al [23] in a 40 Å Fe(100) film grown on 400 Å Cr and covered by 15 Å of Pd. By using a 2 ml probe technique they scanned 8 Fe mls below the Fe/Pd interface. The results showed a ~12% enhancement of B_{hf} at the 1st ml below the interface with a damped oscillation which faded out at the 8th ml. They interpreted these data by a superposition of a short range exponential distance dependence and an RKKY like oscillating term, the former reflecting an exponentially decreasing exchange interaction due to 3d-4d hybridization.

3.7. Fe/Pt

Brand et al. [24] studied by TMS a series of $Fe_{1-x}Co_x/Pt$ (x: 0-0.9) grown along (111) on kapton . Perpendicular magnetic anisotropy was found to increase for increasing x and decreasing thickness of the FeCo layer. For x=0 they observed a ~5% enhancement of B_{hf} at 4.2 K for a 5 Å Fe/20 Å Pt system. We have published recently a detailed study on a series of Fe/Pt MLs [25]. An account of the Mössbauer results of this study will be given in the next chapter.

3.8. Fe/Sb

One of the first Mössbauer studies in MLs was performed by the group of Shinjo [26] on Fe/Sb MLs by employing ^{57}Fe and ^{121}Sb TMS. Using the ^{57}Fe ml probe technique they showed the „head-tail" effect where the 1st Fe ml grown on Sb was non-magnetic while the last Fe ml below the Sb layer was magnetic. The ^{121}Sb data revealed transferred magnetic hyperfine interactions in the Sb layer.

3.9. Fe/V

Jaggi et al [27] have presented detailed CEMS data on symmetric Fe/V MLs with individual layer thickness ranging from 3 to 11 mls. They analyzed the spectra with a simple model based on a concentration profile with only one free parameter, the diffusion length. This analysis was rather satisfactory and gave for all samples a diffusion length of ~1 ml. They verified this model by analyzing probe ml data of Shinjo et al [28] in a similar system. All B_{hf} values are equal or smaller than the bulk iron value. Self consistent band-structure calculations on Fe/V MLS [29] predict an oscillatory behavior of the magnetic moment of the Fe monolayer as a function of the distance from the interface. Such an oscillation is not supported by the model analysis of the Fe/V CEMS data.

4. Mössbauer Studies of Fe/Pt and Fe/Au Multilayers

We will present in this chapter recent CEMS and TMS measurements on a series of Fe/Pt and Fe/Au MLs. This presentation is more detailed than in the review given in the previous chapter and it serves the purpose of an example of the methodology followed for Mössbauer studies of magnetic multilayers. Some of these data have been published already [25], and the reader can refer to this paper for further details.

4.1. MATERIALS

The multilayer samples were prepared using magnetron sputter deposition in a deposition chamber which was cryogenically pumped to a base pressure of 1.3×10^{-5} Pa. Clean polished Si and mica films were used as substrates. They remained at RT during deposition. The sputter deposition rates, between 0.2 and 0.5 Å per sec, were monitored using calibrated quartz crystals. These crystals also provide the individual component thicknesses of the bilayers. The multilayer films were grown to a total thickness of ~2000 Å consisting of N bilayers where N varied from 24 to 135, depending on the thickness of the bilayer. 8 samples of Fe/Pt and 4 samples of Fe/Au MLs were prepared under the same sputtering conditions. The Pt and Au thicknesses were kept constant at 9 and 13 Å respectively. Samples of Fe/Pt with constant Fe thickness and varying Pt thickness were also prepared in order to study the effect of the nonmagnetic component in the magnetic properties of the ML. The thickness of the Fe component increased in a step-wise manner in order to have samples with 1, 2, 3 etc Fe mls and a constant Pt(Au) thickness. The samples will be referred from now on with two numbers X/Y, the first giving the Fe and the second the Pt(Au) thickness in Å.

Si samples were used for CEMS and mica samples for TMS measurements after removing the mica substrate. This removal is essential since mica has 10 times more iron than the multilayer. Mica removal can be done easily with a good quality scotch tape. Prior wetting of the samples makes removal easier. In the case of Fe/Au MLs this procedure was not 100% successful and a few mica layers were left in the samples, as witnessed by a paramagnetic component in the spectra.

4.2. STRUCTURAL CHARACTERIZATION

X-ray diffraction (XRD), transmission electron microscopy (TEM), x-ray absorption spectroscopy (XAS), extended x-ray-absorption fine structure (EXAFS) and Rutherford back scattering (RBS) have been employed for structural characterization.

XRD, in-house and synchrotron facilities, show fcc-Pt(Au)-[111] structure with sharp low and high angle superlattice satellites (Figure 1). In the high-angle XRD spectra a complete quantitative characterization was performed by multidimensional optimization of a stochastic model, where the structural parameters of the superlattice were refined [30]. Table I contains the average d spacings of the individual Fe and Pt(Au) layers, derived from the structural refinement procedure. We notice an increase of d_{Fe} as the Fe layer thickness decreases.

TEM and high resolution imaging reveal the multilayer film morphology and lattice structure. The plan-view electron diffraction patterns demonstrate the polycrystalline in-plane nature of the film with a 5-8° mosaic of the lattice planes perpendicular to the textured {111} growth direction. The film growth structure imaged in cross section is found to be formed by densely packed columns with an average in-plane size of 300 Å.

Fe-K-edge XAS measurements [31] reveal the local symmetry of iron. The results indicate an fcc-like symmetry of the Fe layer for Fe thickness less than ~8 Å which gradually transforms to bcc symmetry for larger thicknesses.

4.3. MAGNETIC CHARACTERIZATION

SQUID magnetometry was used for magnetic characterization. Figure 2 displays the temperature variation of the saturation magnetization for a number of Fe/Pt MLs. This variation is faster than that displayed in bulk iron with a slope increasing as the Fe layer thickness decreases and the Pt thickness increases. From the temperature variation of the magnetization we determined the spin-wave stiffness constant b from the Bloch equation:

$$M_s(T)=M_0(1-bT^{3/2}) \tag{1}$$

It should be noted that the magnetization data for sample 3/9 closely follow the $T^{3/2}$ dependence while the data of sample 3/19 and 3/39 are fitted well by the linear temperature expression. The b parameter values are considerably larger than those of bulk Fe for small Fe layer thicknesses, approaching the α-Fe value as the Fe thickness increases. The opposite effect is observed when the Pt thickness increases with constant Fe layer thickness.

The magnetic moment per Fe atom was calculated from the saturation magnetization at 4.2 K and the areal density of Fe atoms determined from the RBS data. These values are listed in Table I. Although a large error (~10%) is estimated for these calculations, an enhancement of the magnetic moment, which decreases as the Fe interplanar spacing d_{Fe} increases, is evident from these values.

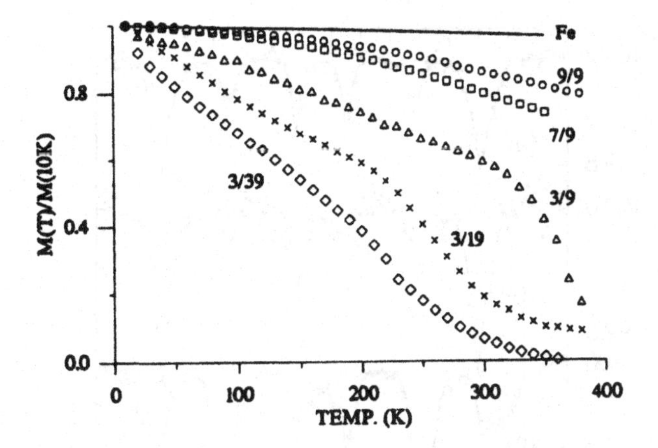

Figure 2. Temperature variation of the saturation magnetization of Fe/Pt multilayers. The corresponding curve for bulk α-Fe is included for comparison.

4.4. MÖSSBAUER RESULTS AND ANALYSIS

CEMS spectra were taken at RT on samples grown on Si, and TMS spectra were taken at 4.2 K on samples grown on mica after removing the substrate. CEMS spectra taken at the same temperature on samples with different substrates were identical. For TMS measurements, the films were cut in 1x1 cm squares and 10-15 squares were stacked together. All isomer shift (i.s.) values refer to α- Fe at RT.

4.4.1. Fe/Pt multilayers

Figure 3 displays some typical TMS spectra taken at 4.2 K for Fe/Pt MLs. The thinnest Fe layer samples (3/9 and 3/19) display magnetic hyperfine spectra with spin orientation out of the plane of the film as indicated from the intensity of the $\Delta m=0$ lines. The direction of B_{hf} with the normal to the plane is at $\Theta = 39^{\circ}$ for the 3/9 sample and 20° for the 3/19 sample. This result implies reduction of the exchange interaction between the magnetic layers as the thickness of the Pt layer increases and is consistent with the linear temperature variation of the magnetization for the 3/19 sample (Fig.2), indicating a two-dimensional character for this sample. Spectra for the 3/39 sample were not possible since the signal to noise ratio is too small. The angle Θ increases as the Fe layer thickness t_{Fe} increases and becomes 68° for the sample 5/9 and 90° for samples with $t_{Fe}>5$ Å and $t_{Pt}=9$ Å. The 3/19 sample displays a paramagnetic doublet at RT with a small quadrupole splitting ($e^2qQ/2=0.28$ mm/s) while magnetic hyperfine structure appears at ~ 240 K in coexistence with a paramagnetic component indicating superparamagnetic effects.

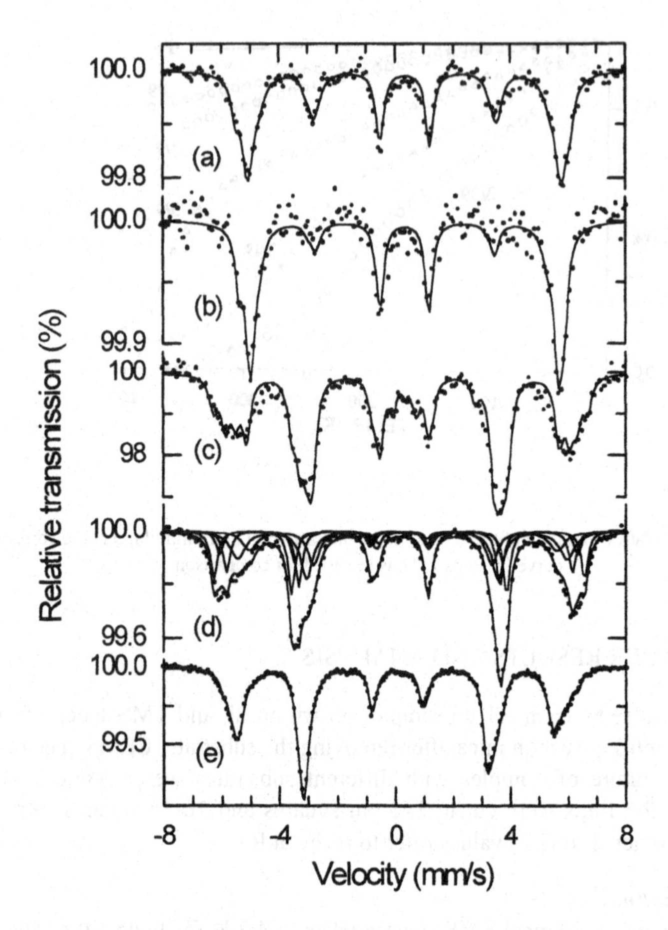

Figure 3. Mössbauer spectra of Fe/Pt MLs at 4.2 K. (a) 3/9 (b) 3/19 (c)7/9 (d) 9/9 and (e) 25/9. Individual components are shown for sample 9/9.

The spectra of the 3/9 and 3/19 samples were analyzed with 3 magnetic components with B_{hf} values of 362(352), 339(333)and 314(310) kOe and isomer shift (i.s) values of 0.39(0.36), 0.45(0.45) and 0.45(0.40) mm/s where the number in parenthesis refer to the 3/19 samples. The i.s. values are larger than the bulk Fe value (0.10 mm/s at 4.2K) indicating the influence of the neighbor Pt atoms. Figure 4 depicts the structural model we have adopted for the microstructure of the MLs and used for the analysis of the Mössbauer spectra. The 3 Å Fe layer thickness samples corresponds to 1-2 mls of Fe. According to Fig.4 the three components of the fit account for the structural components S_0, S_{Fe} and S_{Pt} in the order of decreasing B_{hf} and increasing i.s. values which is in agreement with the increasing number of nearest neighbor (nn) Fe atoms. The relative abundance of these components is 18%, 70% and 12%, indicating considerable diffusion at the interface.

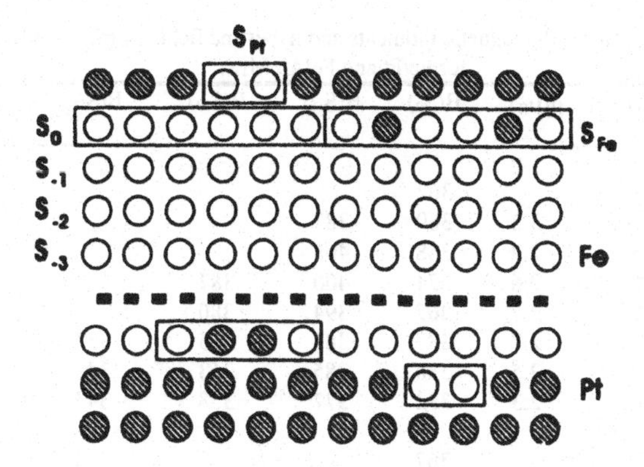

Figure 4. Assumed structural model of the Fe/Pt structure used in the analysis of Mössbauer spectra. Open circles denote Fe atoms and shaded circles Pt atoms.

Spectral structure is shown also for the thicker Fe layer samples. As shown in Fig. 3 this structure is reduced as t_{Fe} increases. We fitted these spectra by adding one component for each two extra Fe mls in the corresponding sample, since each two extra mls are symmetric with respect to their morphology and their distance from the two opposite interfaces. Thus, an S_{-1} component was added to samples 5/9 and 7/9, S_{-1} and S_{-2} to samples 9/9 and 12/9 and so on up to component S_{-5} for the thicker samples (25/9, 38/9, 60/9). The fits were constrained to the instrumental linewidth (0.28 mm/s) for all the components and to relative intensities of the interior ml components calculated from the relative abundance of the corresponding ml in a sample of Fe thickness t_{Fe}. The same constrains were used for the RT spectra. B_{hf} values of the S_i components are given in Table I.

The most striking result from the above analysis is the appearance of magnetic components with B_{hf} values of ~400 kOe, namely ~18% higher than the bulk value. It should be noted that these values are considerably higher than the values reported for FePt alloys which are in the range of 350-370 kOe. These latter values are close to the values assigned to the interface components which indeed can be considered as microscopic FePt alloys. Following the work of Kisters et al [23] on Fe/Pd we have plotted in Figure 5 the hyperfine fields assigned to each Fe ml averaged over all the Fe/Pt samples we measured at LHe, as a function of the distance of the corresponding ml from the interface. The results show the first period of a damped oscillation in close similarity with the Fe/Pd data (also plotted for comparison in the same figure). This oscillation is also present in the RT data. The striking similarity should perhaps be expected, since both the Pd and Pt counterparts of the Fe bilayer are both „nearly magnetic" metals. We have also plotted in Fig. 5 values calculated by Freeman and Fu [12] for a free standing nine mls Fe slab based on the FLAPW method. As we can see from this figure there is a close agreement between theoretical and experimental values, despite the fact that the Fe counterpart for the theoretical result is vacuum.

TABLE I. Interpanar distances d, magnetic moments and hyperfine fields $B_{hf}(S_i)$ in kOe, measured at 4.2 K for Fe/Pt and Fe/Au MLs.

Sample	d_{Fe}(Å)	$\mu(\mu_B)$	$B(S_0)$	$B(S_{-1})$	$B(S_{-2})$	$B(S_{-3})$	$B(S_{-4})$	$B(S_{-5})$
Fe/Pt								
3/9	2.209	-	362					
5/9	2.194	3.2	369	387				
7/9	2.195	2.7	388	408				
9/9	2.166	2.6	374	400	387			
12/9	2.128	2.7	367	394	380			
25/9	2.083	2.7	337	383	365	352	335	343
38/9	2.052	2.5	338	385	363	345	337	345
60/9	2.040	2.3	332	377	358	343	341	340
Fe/Au								
3/13	2.182	-	367					
10/13	2.098	2.3	367	385	369			
18/13	2.065	-	359	379	358	334	357	348
32/13	2.046	2.2	344	377	360	342	357	340
Bulk Fe(110)	2.03	2.2						340

Recently Wu et al [32] have determined magnetic moments and hyperfine fields for an Fe/Pt bilayer with 5 Fe and 4 Pt mls. In this calculation they also employ the FLAWP technique but they let the system relax first (zero total force on the atoms). They obtained B_{hf} values of 275, 328 and 315 kOe for the S_0, S_{-1} and S_{-2} mls respectively, ~80 kOe smaller than the experimentally determined values. The interplanar distances, as determined by the relaxation calculation, were 1.71 and 1.73 Å respectively, i.e. much smaller than the experimentally determined values for the corresponding sample 9/5 (Table I). When they used larger interplanar values they got B_{hf} values which were close to the experimental data, namely 356, 408 and 402 kOe respectively [33]. This

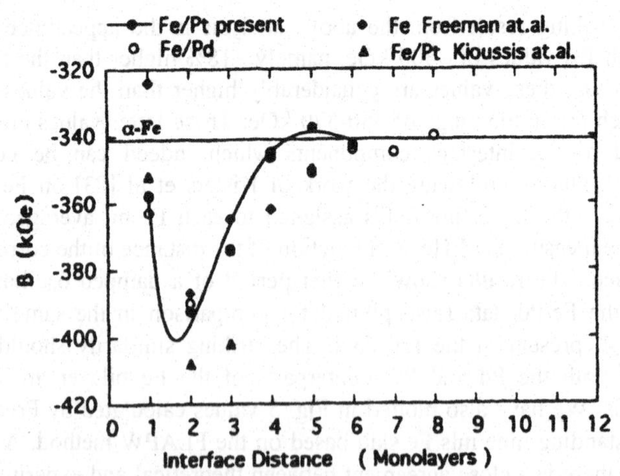

Figure 5. Variation of the hyperfine field B_{hf} assigned to each Fe ml with the distance from the interface. Experimental data are for Fe/Pt and for Fe/Pd taken from references [23] and [25] respectively. The solid line represents least square fit to the Fe/Pt data. Diamond and triangle symbols represent theoretical calculations.

sensitivity of the hyperfine field calculation to the interplanar spacings implies that the local spin density is substantially affected by these distances. An inspection of Table I shows a monotonic increase of the B_{hf} value for the S_0, S_{-1} and S_{-2} components with d_{Fe}.

Figure 6 shows that this variation is linear for the S_{-1} component. We have included in this figure the Fe/Au data (vide infra) and the calculated value by Wu et al [32]. The results of this figure indicate an ecumenical character of this relation.

4.4.2. Fe/Au multilayers.

Figure 7 displays TMS spectra at 4.2K for four Fe/Au multilayers with t_{Fe} varying from 3 to 28 Å and t_{Au} being constant at 13 Å. The thinnest Fe layer sample is paramagnetic at RT, although a faint magnetic component can be discerned, and at LN the spectrum is fully magnetically split indicating superparamagnetic behavior with a blocking temperature above ~100 K. The thicker Fe layer spectra display an asymmetric broadening which decreases as t_{Fe} increases. The evidence of spectral structure is further corroborated by the fact that lines 2 and 5 are unequal in height and width. Such an asymmetry can be achieved only with components displaying a distribution in both B_{hf} and i.s. values. Following the discussion above for the analysis of the Fe/Pt spectra we have analyzed the Fe/Au spectra in a similar manner. The corresponding B_{hf} values are included in Table I.

We should notice here that in the analysis employed for both the Fe/Pt and Fe/Au MLs the isomer shift varies systematically from a value close to the corresponding alloy value for the interface Fe ml to a value close to the bulk Fe value for the S_{-4} and more inner layers. The i.s. values for the thinnest Fe layer sample (3/13) at 4.2 K are 0.41, 0.60 and 0.80 mm/s for the S_0, S_{Fe} and S_{Au} components respectively, indicating that the interface is an FeAu alloy, as in the case of the Fe/Pt thinnest samples. The spin orientation for the 3/13 ML is 28° with respect to the vertical to the plane, i.e. in between the values we obtained for the 3/9 and 3/19 Fe/Pt samples.

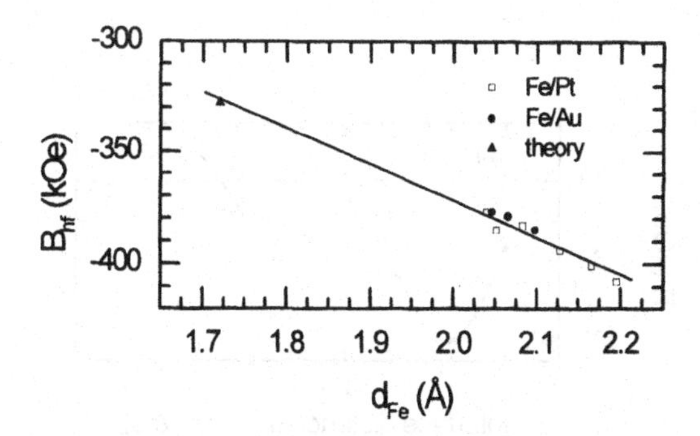

Figure 6. Variation of the hyperfine field B_{hf} of the 1st atomic layer under the interface with the average lattice spacing d_{Fe} of the Fe layer for the Fe/Pt and Fe/Au samples. A value from the calculation of ref. [32] is included.

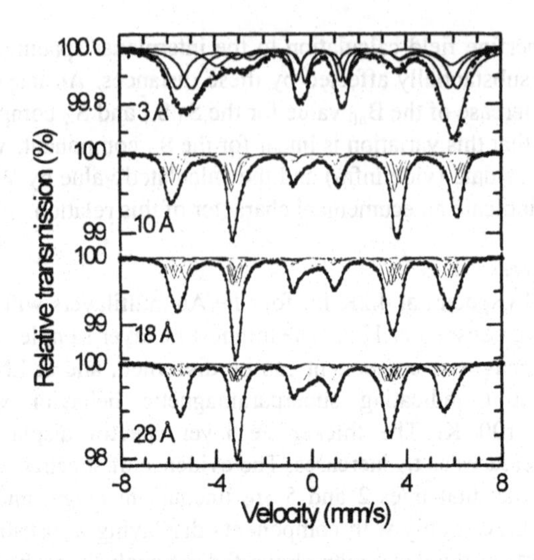

Figure 7. Mössbauer spectra of xÅ Fe/13Å Au with x equal (from above) to 3, 10, 18 and 28 Å.

A plot of the average B_{hf} values for each ml as a function of its distance from the interface (Fig.8) again displays the Friedel oscillations observed for the Fe/Pd and Fe/Pt MLs (Fig.5). The maximum again appears for the S_{-1} ml (380 kOe).

An oscillatory behavior of B_{hf} also appears in first-principles calculations in Fe/Au MLs by Guo and Ebert [34]. It is worth noting here that unlike Pd and Pt which are „nearly magnetic" metals, Au is a non-magnetic metal with a filled 5d shell. Thus this oscillatory behavior seems to be of a more general character. The linear variation of $B_{hf}(S_{-1})$ with d_{Fe} depicted in figure 6 suggests that at least one important factor for the observed oscillation is the variation of the interplanar spacing t_{Fe} as we move from the interface to the interior of the Fe layer component. Finally, we should mention in this discussion, that Friedel oscillations have been detected recently by perturbed angular correlation spectroscopy in bcc-Fe/Co multilayers [35].

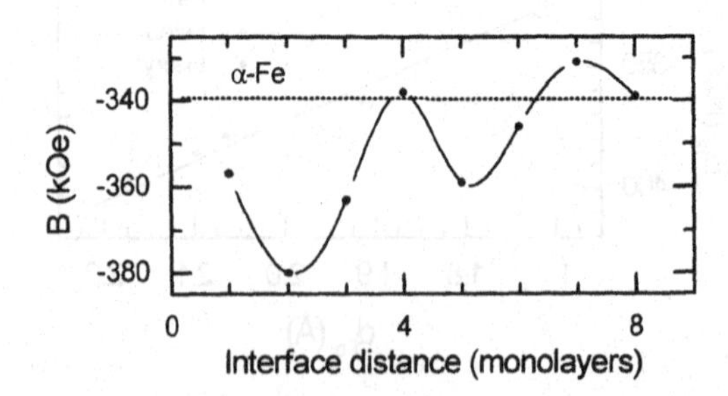

Figure 8. Variation of the hyperfine field B_{hf} assigned to each Fe monolayer with the distance from the interface for Fe/Au multilayers. The solid line is a fit of a damped oscillation function to the data.

5. Conclusions

We have presented in this paper a short review on Mössbauer studies in magnetic multilayers. We ended the review with a more detailed exposure of recent studies of Fe/Pt and Fe/Au multilayers. The emphasis in this presentation was put on the determination of hyperfine parameters at the monolayer level using the Mössbauer technique. The following are the main conclusions of the present review:

1. Both CEMS and TMS are suitable Mössbauer techniques for investigations of multilayers at the monolayer level. The 1-2 ml probe technique is ideal for such studies provided that interdiffusion between ^{56}Fe and ^{57}Fe is eliminated during preparation. The alternative technique of using samples with natural abundance of ^{57}Fe can also give results at the monolayer level if a multicomponent analysis of the spectra is done in a consistent way. This can be achieved if a number of samples with different number of Fe monolayers is studied.
2. Early theoretical works predict an enhancement of the magnetic moment and the hyperfine field at the mls close to the interface. These parameters are influenced by Friedel oscillations as we move to the interior of the Fe layer. Such oscillations were observed originally in Fe/Ag MLs and recently, more pronounced, in Fe/Pd, Fe/Pt and Fe/Au MLs.
3. Combined structural and Mössbauer analysis on Fe/Pt and Fe/Au data indicate that the B_{hf} values for each Fe monolayer are affected by the interplanar spacing of this particular ml in a universal way. Whether this is the only factor producing the observed damped oscillations is a subject for further theoretical and experimental investigations.

Acknowledgements

The author wishes to acknowledge collaboration in studies of magnetic multilayers with Prof. Thomas Tsakalakos, Dr Eamon Devlin, Dr Constantin Chassapis, Dr Alan Jankowski and Prof. Mark Croft. The author is also thankful to Professor Nikos Kioussis for the many valuable discussions on theoretical aspects of magnetic multilayers and Dr George Kallias for his assistance in the typesetting. The support of the Greek General Secretariat for R and D through the PENED program is also acknowledged.

References

1. Freeman, A. J.(1980) Symposium on surfaces and magnetism, *J. Magn. Magn. Mater.* **15-18,** 1070-3.
2. Ohnishi, S., Weinert, M. and Freeman, A.J. (1984) Interface magnetism in metals, *Phys. Rev.* B**30**, 36-43.
3. Bauer, E. and van der Merwe, J. H. (1986) Structure and growth of crystalline superstructures, *Phys. Rev.* B **33,** 3567-71.
4. Markov, I. and Stoyanov S. (1987), *Contep. Phys.* **28**, 267.
5. Gradmann, U. (1993), Magnetism in ultrathin transition metal films, in K. H. J. Buschow (ed.), Handbook of Magnetic Materials, North Holland, Amsterdam, pp.1-96.
6. Stahl, B. and Kankeleit E. (1997), A high luminosity UHV orange type magnetic spectrometer developed for depth selective Mössbauer spectrometry, *Nucl. Instr. Meth. Phys. Res. B* **122**,149-161.

7. Korecki, J. and Gradmann, U. (1983), In situ Mössbauer analysis of hyperfine interactions near Fe(100) surfaces and interfaces, *Phys. Rev. Lett.* **59**, 2491-4.
8. Tyson. J. et al (1981), *J. Appl. Phys.* **52**, 2487.
9. Korecki, J and Gradmann, U.(1986), Spatial oscillation of magnetic hyperfine field near the free Fe(100)-surface, *Europhys. Lett.* **2**, 651-7.
10. Koon, N.C. et al. (1987), Direct evidence for perpendicular spin orientations and enhanced hyperfine fields in ultratin Fe(100) films on Ag(100), *Phys. Rev. Lett.* **59**, 5483-6.
11. Ohnishi, S., Weinert, M. and Freeman, A. J.(1983), *Phys. Rev.* B **28**, 6741.
12. Freeman, A. J. and Fu, C. L.(1987), Strongly enhanced 2D magnetism at surfaces and interfaces, *J. Appl. Phys.* **61**, 3356-61.
13. Przybilski, M., Gradmann, U. and Kropp, K. (1990), Influence of coating materials on magnetic properties of thin iron films, *Hyperf. Inter.* **57**, 2045.
14. Kobayashi, Y. et al (1995) ^{197}Au and ^{57}Fe Mössbauer study of Au/Fe multilayers, Internatinal Conference on the Applications of the Mössbauer Effect, I. Ortalli (Ed.), Italian Physical Society, Bologna, pp. 619-622.
15. Emoto, T., Hosoito, N. and Shinjo, T. (1997), Magnetic polarization of conduction electrons at Au layers in Fe/Au multilayers by ^{119}Sn Mössbauer spectroscopy, J. Phys. Soc. Japan **66**, 803-8.
16. Zukrowski, J. et al (1996), Mössbauer spectroscopy of Cr(110)/Fe(110)/Cr(110) sandwiches, *J. Magn. Magn. Mat.* **145**, 57-66.
17. Klinkhammer, F. et al (1996), Interface roughness in Fe(100)/Cr film structures studied by CEMS, *J. Magn. Magn. Mat.* **161**, 49-56.
18. Keune, W. et al (1977), Antiferromagnetism of fcc Fe thin films, *J. Magn. Magn. Mat.* **6**, 192-95.
19. Ellerbroke, R. D. et al (1995), Mössbauer Effect study of magnetism and structure of fcc-like Fe(001) films on Cu(001), *Phys. Rev. Lett.* **74**, 3053-56.
20. Keavney, D. J. et al (1995), Site- specific Mössbauer Evidence of structure-induced magnetic phase transition in fcc Fe(100) thin films, *Phys. Rev. Lett.* **74**, 4531-4.
21. Colombo, E. et al (1992), Static magnetization direction in fcc (111) Fe/Ni multilayers, *J. Magn. Magn. Mat.* **104-107**, 1857-8.
22. Boulefeld, A., Emrick, R. M. and Falko, C. M. (1991), Magnetism of Fe/Pd superlattices, *Phys. Rev.* B **43**, 13152-8.
23. Kisters, G. et al (1994) , CEMS interface study of Fe(100)/Pd film structures, *Hyper. Inter.* **92**, 285-9.
24. Brand, R. A. et al (1993), Local magnetic properties of Pt/$Fe_{1-x}Co_x$ multilayers studied by Mössbauer spectroscopy, *J. Magn. Magn. Mat.* **126**, 248-50.
25. Simopoulos, A. et al (1996), Structure and enhanced magnetization in Fe/Pt multilayers, *Phys. Rev.* B **54**, 9931-41.
26. Shinjo, T. et al (1983), Interface magnetism of Fe-Sb multilayered films with artificial superstructure from ^{57}Fe and ^{121}Sb Mössbauer , Neutron diffraction and FMR experiments, *J. Phys. Soc. Japan* **52**, 3154-62.
27. Jaggi, N. K. et al (1985), Mössbauer spectroscopy study of composition modulated [110] Fe-V films, *J. Magn. Magn. Mat.* **49**, 1-13.
28. Shinjo, T. et al (1984), *J. de Phys. Colloq.* **C5**, C5-361.
29. Hamada, N., Terakura K. and Yanase A.(1983),Calculation of magnetic moment distributions in multilayered films, *J. Magn. Magn. Mat.* **35**, 7-8.
30. Chassapis, C. S. and Tsakalakos, T. (1997), Multidimentional optimization of a stochastic model for X-ray diffraction from superlattices, *Computer Physics Communications* **99**, 163-7.
31. Croft, M. et al (1997), Fe-fcc layer stabilization in [111]-textured Fe/Pt multilayers, *Nanostructured Materials* **9**, 413-22.
32. Wu, R. Chen, L. and Kioussis, N. (1996), Structural and magnetic properties of fcc Pt/Fe(111) multilayers, *J. Appl. Phys.* **79**, 4783-7.
33. Kioussis, N., private communications.
34. Guo, G. Y. and Ebert, H. (1996), First-principles study of the magnetic hyperfine field in Fe and Co multilayers, *Phys. Rev.* B **53,** 2492-2502.
35. Swinnen, B. et al (1997), Oscillation of the Fe and Co magnetic moments near the sharp (1-10) Fe/Co interface, *Phys. Rev. Lett.* **78,** 362-5.

MÖSSBAUER SPECTROMETRY APPLIED TO IRON-BASED NANOCRYSTALLINE ALLOYS I.

High Temperature Studies

JEAN-MARC GRENÈCHE[1] AND MARCEL MIGLIERINI[2]

[1)] *Laboratoire de Physique de l'Etat Condensé, UPRESA CNRS 6087, Université du Maine, Faculté des Sciences, 72085 Le Mans Cedex 9, France*

[2)] *Department of Nuclear Physics and Technology, Slovak University of Technology, Ilkovičova 3, SK 812 19 Bratislava, Slovakia*

Abstract

Nanocrystalline alloys which are obtained after subsequent annealing of the amorphous precursor consists of nanocrystalline grains embedded within a residual amorphous matrix. Thus, they exhibit a two-phase magnetic behaviour which is strongly dependent on the volumetric fraction of the crystalline phase at elevated temperatures. We emphasise the high efficiency of ^{57}Fe Mössbauer spectrometry which is able to elucidate the different kinds of Fe atoms, particularly the crystalline phase, the amorphous residual matrix, and the interface zone between crystalline grains and the amorphous phase and which reveals these different magnetic high temperature behaviours in iron-based nanocrystalline alloys. The fitting procedures involving either distributions of hyperfine magnetic fields or both distributions of hyperfine magnetic fields and of quadrupolar splitting are first described and then discussed when analysing temperature dependencies of Mössbauer spectra, particularly in the vicinity as well as above the Curie point of the residual amorphous phase. Superparamagnetic and penetrating magnetisation effects are discussed on the basis of different contributions of the magnetic energy. Part II., the following paper concentrates on the distributions of hyperfine fields.

1. Mössbauer Spectrometry and Nanocrystalline Alloys

Mössbauer spectrometry which is complementary to diffraction techniques, is often proving decisive in the area of material science, although it is limited both to some isotopes and to solid materials or frozen solutions [1-3]. Indeed, its local behaviour enables to be an atomic scale sensitive tool while its time of measurement allows to investigate relaxation phenomena and dynamic effects. It is important to emphasise that ^{57}Fe is a valuable isotope which can be encountered in different kinds of materials

M. Miglierini and D. Petridis (eds.), Mössbauer Spectroscopy in Materials Science, 243–256.

studied in physics and/or chemistry ranging from metallurgy to archaeology via mineralogy and biology. In addition, ^{57}Fe Mössbauer spectrometry is experimentally easy to use over a wide temperature range, thanks to its high recoilless factor value. One remembers that the relevant hyperfine parameters are the isomer shift (IS), the quadrupolar splitting (Δ), the hyperfine field (H_{hyp}), the quadrupolar shift (ε), the linewidth at half height (Γ), the asymmetry parameter (η) and in the case of monocrystalline samples, the angles which define the directions of γ-beam and of hyperfine field with respect to the electric field tensor axes.

In crystalline compounds, the spectra exhibit sharp lorentzian lines (Fig. 1a) and the determination of hyperfine parameters and their temperature dependencies as well, lead to the number and relative proportions of iron sites, their electronic states, in addition to information on their respective structural symmetry and magnetic character. In non crystalline systems, the spectra which consist of broadened non-lorentzian lines (Fig. 1b) in agreement with the atomic disordered arrangement, have to be reproduced by means of either discrete distributions or continuous distributions of hyperfine parameters (Fig. 1c - solid curves): their mean values and their standard deviations are useful to characterize the amorphous state and/or the heterogeneous nature of structural and magnetic arrangements.

There exists numerous fitting methods but the refinement of hyperfine parameters requires great attention because of several artifacts induced by the overlapping of lines and/or the magnetic or quadrupolar texture in the case of anisotropic samples during the fitting procedure [4, 5]. As the time scale given by the Larmor precession time is about 10^{-8} s in the case of ^{57}Fe, one also expects to observe the blocked state of a superparamagnetic transition in most iron-based particle systems which exhibit magnetic fluctuations, within the usual temperature range (0 - 400K). Further high field experiments provide information on the magnetic moment arrangements and on the

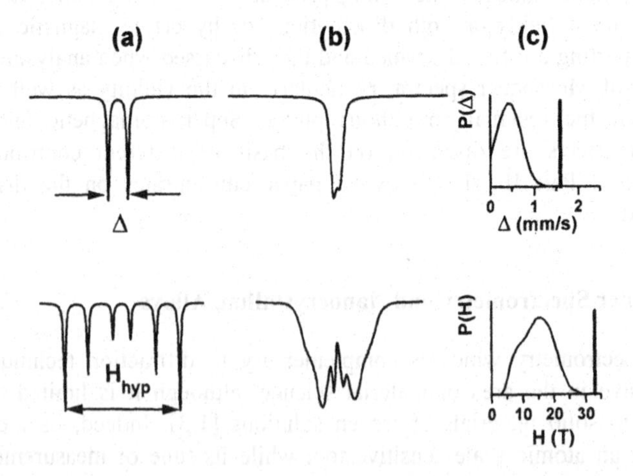

Figure 1. Model Mössbauer spectra of crystalline (a) and non-crystalline (b) system. Single discrete values and distributions of the most relevant hyperfine parameter as derived from the spectra (a) and (b) are plotted in (c) by thick vertical lines and by solid curves, respectively. The doublet-like spectra in the upper row indicate paramagnetic, whereas those at the bottom (sextets) magnetic ordering of the resonant atoms.

relaxation phenomena [6, 7]. Consequently, ^{57}Fe Mössbauer spectrometry can be effectively applied to investigate both the structural and magnetic properties of iron-based nanocrystalline alloys which consist of an assembly of nanocrystalline grains embedded in a residual amorphous matrix.

Since their discovery by Yoshizawa *et al.* [8, 9], the nanocrystalline alloys were widely investigated: thus, these new materials are very attractive because they exhibit unusual two-phase structural and magnetic behaviours and are promising for industrial applications due to their excellent soft magnetic properties combining high saturation magnetic flux density with high permeability. The magnetic coupling between crystalline grains via the intergranular phase, the suppression of the effective magnetic anisotropy and the decrease of the magnetostriction originate these soft magnetic properties [10-13]. Indeed, on the basis of the random anisotropy model earlier proposed by Alben *et al.* [14], Herzer concluded that the effective magnetocrystalline anisotropy was averaged out by exchange interaction when the grain size is lower than 20 nm. Consequently, the occurrence of larger grains drastically increases the magnetocrystalline anisotropy, leading to a degradation of soft magnetic properties.

Nanocrystalline alloys result from a subsequent annealing under controlled atmosphere to prevent oxidation, applied to their amorphous precursor alloys which are prepared either by melt-spinning or sputtering techniques. The appropriate amorphous alloys are characterised by two exothermic peaks, corresponding to the primary and to the secondary crystallization at which the onset of nanocrystallization and the complete amorphous-crystalline transformation occur, respectively. Nanometre size crystalline grains (typically 10-15 nm in diameter) precipitate randomly distributed within the amorphous residual matrix when the annealing temperature is comprised between these two crystallization temperatures and the volumetric fraction of the crystalline phase increases with increasing annealing temperature and with increasing annealing time. From the experimental point of view, the as-quenched ribbons prepared by the melt-spinning technique are annealed under vacuum or controlled atmosphere to prevent from oxidation.

Two families of nanocrystalline soft magnetic alloys can be distinguished at present:

(i) the FINEMET-type deriving from the pioneering alloy with the nominal composition of $Fe_{73.5}Cu_1Nb_3B_9Si_{13.5}$. They are constituted of Fe-metalloid based alloys (70-80 at. %) with addition of small amounts of Cu and Nb atoms which accelerate the nucleation and limit the growth of FeSi crystalline grains, respectively [8, 9, 15],

(ii) the NANOPERM-type resulting from FeMB alloys (M = Zr, Hf, Nb; 85-92 at. % Fe) and eventual addition of Cu, Cr, Ti, Ta and/or metalloids as Si. Annealing treatment favours the emergence of bcc-Fe crystalline grains [16-18].

The magnetic performances of these typical FINEMET and NANOPERM alloys are essentially characterized by saturation magnetisation B_s and coercive field H_c comprised within the ranges 1.2-1.4 T and 1.5-1.7 T, 0.5-1.0 T and 5-10 A/m, respectively. Because they exhibit performances slightly lower than those of usual FeSi microcrystalline alloys, nanocrystalline alloys display as highly challenging materials in the field of microelectronics and magnetic devices [17, 18]. Besides these promising technological aspects, the fundamental interest is essentially due to their two-phase structural and magnetic behaviours.

In the following section, we report general features concerning the structural aspects of nanocrystalline alloys and the contribution of Mössbauer spectrometry to describe the microstructure of these alloys. The third section is devoted to the Mössbauer studies performed at elevated temperatures, *i.e.* above the Curie temperature of the amorphous phase and below the temperature at which the amorphous-crystalline transformation occurs: their interpretation is discussed in terms of intergrains, intragrains magnetic interactions, and interactions between grains and the intergranular phase. Because of extremely different magnetic behaviours, we consider first two ideal situations with low and high volumetric fraction of crystalline phase. But, some aspects of intermediate fraction of crystalline phase which are now in progress, are only briefly reported.

Detailed discussion on the distributions of hyperfine fields in the amorphous residual phase and the interfacial regions is presented in Part II., the following paper [19].

2. Structural Aspects

The structure of nanocrystalline alloys can be first examined by means of diffraction techniques and transmission electron microscopy: they reveal unambiguously the presence of FeSi or bcc-Fe crystalline grains, in FINEMET- and NANOPERM-type alloys, respectively. In the case of FINEMET-type alloys, the Si content is estimated from the value of the lattice parameter, assuming neither lattice distortions, nor stresses. One generally observes by means of transmission electron microscopy the presence of equiaxial crystalline grains randomly distributed in the residual amorphous matrix whereas the size and the shape of the grains remain rather homogeneous. The analysis of X-ray patterns may give an estimate of the mean size of crystalline grains and of the volumetric fraction of the crystalline phase.

The distribution of atoms in as-quenched and nanocrystalline alloys can be analysed by atom probe field ion microscopy (APFIM) technique [18, 20]: in FeZrB nanocrystalline alloys, it is concluded that (i) bcc-Fe grains contain higher Fe content than the nominal one of the amorphous precursor, and (ii) a very small amount of Zr and B atoms (less than 1 at. %) as impurity, (iii) the remaining amorphous matrix contains lower Fe and higher Zr and B concentrations as in the as-quenched state, and (iv) a significant enrichment of Zr and B elements in the remaining amorphous phase occurs close to the crystalline grains, that probably favours the achievement of the bcc-Fe nanostructure.

In Si-containing nanocrystalline FINEMET alloys, the complexity of the hyperfine structure of both low and high temperature Mössbauer spectra prevents from an immediate analysis, as illustrated in Fig. 2. Indeed, the bcc-FeSi crystalline phase with a DO_3 superstructure exhibits different non equivalent iron sites which give rise to several magnetic sextets, in addition to a broad line sextet due to the remaining amorphous phase.

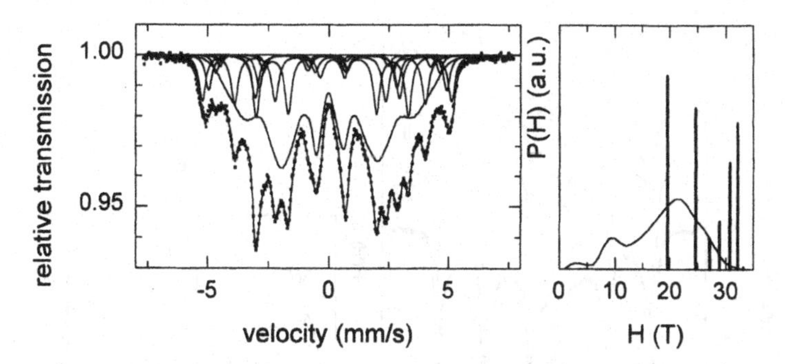

Figure 2. Room temperature ^{57}Fe Mössbauer spectrum of a $Fe_{73.5}Cu_1Nb_3Si_{13.5}B_9$ nanocrystalline alloy (left) with its spectral components and the corresponding distribution of hyperfine fields P(H) (right). Discrete H-values of the particular crystalline components are plotted by thick vertical lines.

Different fitting procedures were proposed in literature to derive the hyperfine data of both the crystalline and the amorphous phases: the number of sextets attributed to the crystalline phase was taken between 4 to 7, while the contribution due to the amorphous phase was described by means of distributions of hyperfine fields with different constrained profiles, according to the authors (see [21] and references therein). Due to a lack of spectral resolution, it is important to emphasise that the uncertainties of both hyperfine parameters and relative proportions of crystalline and amorphous phases remain high. Consequently, the fitting model has to be applied to Mössbauer spectra recorded at different temperatures, particularly at higher temperatures because the resolution of the hyperfine structure is improved. Thus, the validity of the fitting model first requires a physical evolution of the different hyperfine parameters versus temperature (as example, the isomer shift should linearly decrease with increasing temperature). By comparing the temperature dependence of the hyperfine data with those characteristic of microcrystalline iron-silicon alloys, one can estimate the silicon content within the crystalline grains. In addition, the relative contents of crystalline and amorphous have to be temperature independent or slightly temperature dependent consistently with their respective f factor temperature dependences. Finally, the silicon content and the volumetric fraction of the crystalline phase (Fe at.%) can be compared with those obtained by X-ray, transmission electron microscopy and static magnetic measurements. But, the structural nature of the intergranular phase cannot be accurately described [21].

The Mössbauer spectra taken from NANOPERM alloys exhibit better defined hyperfine structure. As shown in Fig. 3, the low temperature Mössbauer spectra first consist of (i) a sextet with sharp lorentzian lines, the intensities of which strongly depend on the volumetric fraction of the crystalline phase, and (ii) a sextet with broad asymmetrical non lorentzian lines. The spectra can be *a priori* described on the basis of two components attributed to (i) a crystalline phase and (ii) an amorphous residual matrix, respectively. But a more detailed analysis gives rise to an additional magnetic component, the lines of which are located at the internal wings of the outermost lines: this latter component is attributed to an interface zone which is defined as the intermediate region between the core of the crystalline grains and the residual

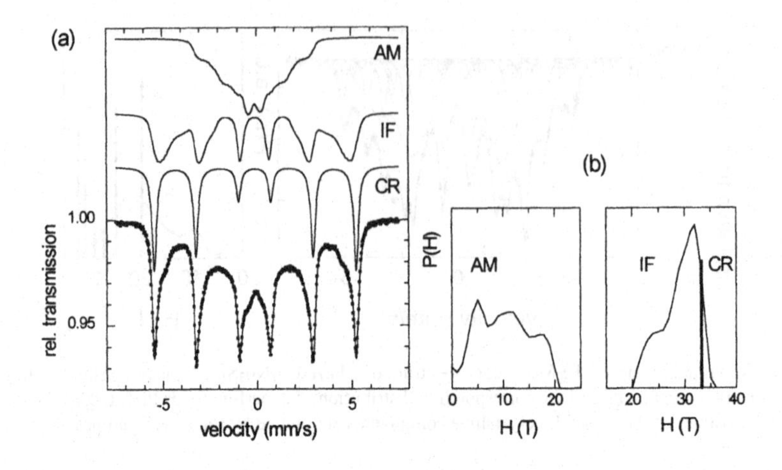

Figure 3. Mössbauer spectrum (a) with the partial subspectra and the corresponding P(H) distributions (b) for the $Fe_{80}Mo_7Cu_1B_{12}$ (470° C/1h) nanocrystal.

amorphous phase as sketched schematically in Fig. 4 [22-26].

The fitting procedure which has been developed in conjunction with the expected topology, is reported in more details [21, 25, 26]. The upper and lower limits of the hyperfine field distributions of both interface and amorphous components, the correlation factor between isomer shift and hyperfine field distributions characteristic of the amorphous remainder, and the magnetic texture of crystalline and amorphous contributions remain the crucial parameters. They can be refined by applying the fitting model to spectra recorded at different temperatures. It is important to emphasise that this fitting procedure has been successfully applied to different systems studied in a wide range of temperatures [25-29]. Because the volumetric fractions of the crystalline phase, the interface and the amorphous residual phase are proportional to the respective areas

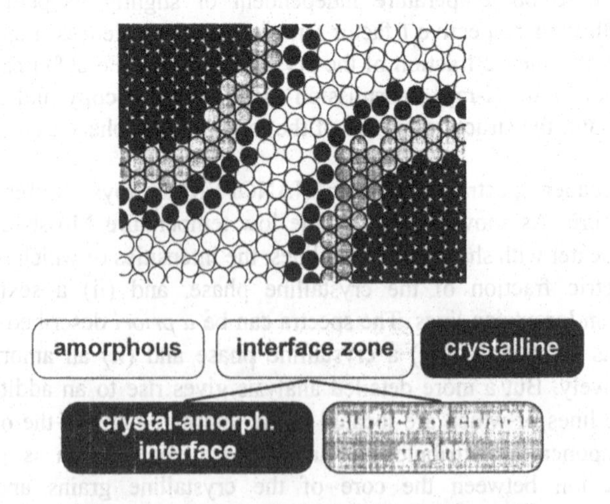

Figure 4. Schematic structural model of a nanocrystal.

of corresponding components (S_{cr}, S_{int}, and S_a), assuming the same value of the recoil-free fractions of each component, one can estimate the thickness δ of the interface on the basis of a simple geometrical model based on spherical grains with radius R, via $(S_{cr}+S_{int})/S_{cr} = (1+ \delta/R)^3$. The values of the thickness are generally comprised between 0.5-0.8 nm, consistent with 2-3 atomic layers [30].

3. High Temperature Mössbauer Studies

When the measurement temperature increases and becomes higher than the expected Curie temperature of the residual amorphous phase T_C(am), the hyperfine structure drastically changes: the broad line magnetic sextet tends to collapse to a paramagnetic doublet. It is important to emphasise that the evolution of Mössbauer spectra strongly depends on the volumetric fraction of the crystalline phase, which governs the nature and the strength of intergrain magnetic interactions and of exchange magnetic interactions within the crystalline grains and the residual amorphous matrix.

Indeed, the nanocrystalline alloys can be described as a two-phase magnetic system where the crystalline grains form an assembly of single domain crystalline grains embedded within an amorphous matrix which is either ferromagnetic or paramagnetic. The intrinsic magnetic energy of a grain results from (i) its magnetocrystalline anisotropy, (ii) its shape anisotropy and (iii) the exchange RKKY interactions; because of its small size, (iv) the surface magnetic anisotropy becomes important together with (v) the thermal energy. Extrinsic contributions are due to (vi) the magnetic dipolar interactions between grains, (vii) the exchange RKKY interactions between grains and between a grain and the amorphous matrix, and (viii) the magnetoelastic anisotropy term originating from internal and external stresses. Details are given in [30].

When the temperature remains lower than (T_C(am)), exchange interactions (iii) and (vii) and those intrinsic of the amorphous residual phase are prevailing: they originate ferromagnetic arrangements in both crystalline and residual amorphous phases, whatever the volumetric fraction of the crystalline phase is. The present analysis is consistent with static magnetic measurements.

3.1. LOW CRYSTALLINE CONTENT

Several experimental features occur when the temperature increases. The first case is illustrated in Fig. 5. One observes that the broad line sextet component attributed to the amorphous phase collapses to an asymmetrical broad paramagnetic doublet, whereas the two magnetic components attributed to the crystalline grains and to the interface remain as magnetic sextets. Such a situation is typical for a rather low volumetric fraction of the crystalline phase, lower than 20 at. % corresponding to an intergrain distance higher than 10 nanometres.

The fitting procedure previously mentioned can be applied, except that the paramagnetic component has to be reproduced by means of a distribution of quadrupolar splitting $P(\Delta)$ linearly correlated to that of isomer shift. The correlation does not strongly differ from that encountered in as-quenched amorphous alloys. The mean hyperfine field of the amorphous residual phase decreases with increasing temperature, according to the magnetisation temperature dependence. The Curie temperature value characteristic of the amorphous residual phase can be extrapolated and remains comparable to that of the amorphous precursor, in agreement with magnetic measurements. The small increase is consistent with the slight change of the nominal composition which occurs at this stage of crystallization.

Because the large spectral absorption attributed to the amorphous phase is splitted into two lines at high temperatures, the two other magnetic sextets appear weakly resolved. One has to take care of the refinement of hyperfine parameters during the fitting procedure: the lower is the volumetric fraction of the crystalline phase, the higher the difficulty to refine hyperfine parameters. Contrary to low temperature range where the hyperfine field temperature dependence is in fair agreement with the expected one for a bcc-Fe crystalline bulk, a reduction might be observed at high temperatures together with a broadening of the sextet lines. In addition, the component assigned to the interface exhibits expanded broad lines, preventing from an accurate refinement of hyperfine parameters.

All these experimental features are consistent with the presence of magnetic fluctuations, *i.e.* a superparamagnetic state. Indeed, in the case of low volumetric

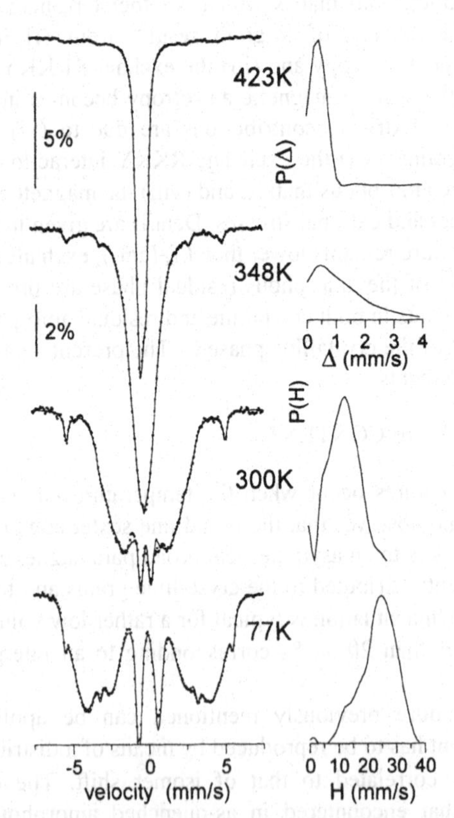

Figure 5. Mössbauer spectra taken at the indicated temperatures (left) and corresponding hyperfine field distributions P(H) (right) for a $Fe_{86.5}Zr_{6.5}Cu_1B_6$ alloy annealed at 500° C/1h.

fraction of the crystalline phase, the crystalline grains are magnetically independent: they can be consequently considered as an assembly of non-interacting single domain particles. Thus, the main contribution to the magnetic energy originates from the thermal energy which overcomes effective magnetic anisotropy contribution. Consequently, it favours fluctuations of magnetisation among easy directions of each particle, that progressively occur when the temperature increases.

Such a feature was observed on FeCrCuNbBSi nanocrystalline alloys with a fraction of crystalline phase estimated at 14%: the temperature dependence of the hyperfine field corresponding to the crystalline Fe-Si grains displays a kink, the temperature of which was close to the Curie point of the amorphous phase, and decreases more rapidly than that of the bulk bcc-Fe at high temperatures [31, 32]. The presence of the kink is attributed both to the suppression of the exchange interactions within the residual amorphous phase and to the exchange interactions between the residual amorphous phase and the crystalline grains, and a sudden occurrence of the fluctuations of magnetisation due to the important thermal contribution [32]. The present interpretation is well supported by static magnetic measurements including magnetisation and coercive field high temperature dependence (see [33] and references therein). High temperature in-field Mössbauer experiments have to be performed in order to confirm the presence of superparamagnetic effects.

The superparamagnetic temperature range depends (i) on the relaxation times of the magnetisation of the crystalline grains, *i.e.* their size and shape and their distributions, and (ii) on the nature of intergranular phase, acting as a transmitting medium of magnetic dipolar interactions between crystalline grains which increase the blocking temperatures. An estimate of the strength of the intergranular magnetic coupling is given by comparing the Curie temperature of the crystalline phase to that of the bulk FeSi alloy: one finds in FeCr based nanocrystalline alloys $\Delta Tc \approx 190K$, corresponding to $\approx 2.6\ 10^{-21}J$ [32]. Modelling of the magnetisation temperature dependence is in progress in the case of two-phase magnetic systems.

3.2. HIGH CRYSTALLINE CONTENT

The second scenario concerns nanocrystalline alloys which are characterized by a high volumetric fraction of the crystalline phase higher than 50% corresponding to intergrain distances lower than 2nm. The main features are the following as illustrated in Fig. 6: the broad line sextet due to the amorphous phase progressively turns into both a quadrupolar doublet and a magnetic sextet, with asymmetrical and broad lines, whereas the outer part of the spectra always consists of two magnetic components due to the crystalline grains and the interface. Such a case is opposite to the previous one: the two latter magnetic components which are well resolved can be accurately fitted, contrary to the central part because of the presence of mixed quadrupolar and low field magnetic contributions.

The magnetic sextet attributed to the crystalline grains exhibits sharp lorentzian lines even at elevated temperatures, whereas the temperature dependence of the hyperfine field fairly agrees with that of a bcc-Fe bulk for temperatures below 600K [34]. The component assigned to the interface exhibits broad and asymmetrical lines consistent with the structural and/or magnetic disorder of the neighbouring Fe nuclei.

The hyperfine field characteristic of the interface is lower than that of the crystalline grains and the reduction, which is almost constant whatever the temperature is, depends on the constituents of the nanocrystalline alloys. One finds 3,5-4T [26], 4T [27], and 4-5T [28] in the case of Zr, Ti, and Nb containing NANOPERM alloys, respectively.

Concerning the third contribution attributed to the residual amorphous phase, the fitting procedure first consists of considering only a distribution of hyperfine fields: one observes the emergence of a low field peak when the temperature increases and the presence of remaining high field components. It demonstrates clearly that Fe nuclei located in the residual amorphous phase exhibit different neighbourhoods, consistent with a chemical non-homogeneity. This point is supported by recent atom-probe field ion microscopy observations which clearly evidence, in the case of FeZrB nanocrystalline alloys, for an enrichment in Zr and B at the vicinity of the crystalline grains [17, 20] and will be discussed to more detail in Part II, the following paper [19].

The mean hyperfine field progressively decreases when the temperature increases and remains small at the highest temperatures. Such a situation strongly differs from that observed in the case of low volumetric fraction of crystalline phase, which exhibits a sharp transition. As illustrated in Fig. 7, two different linear behaviours are generally

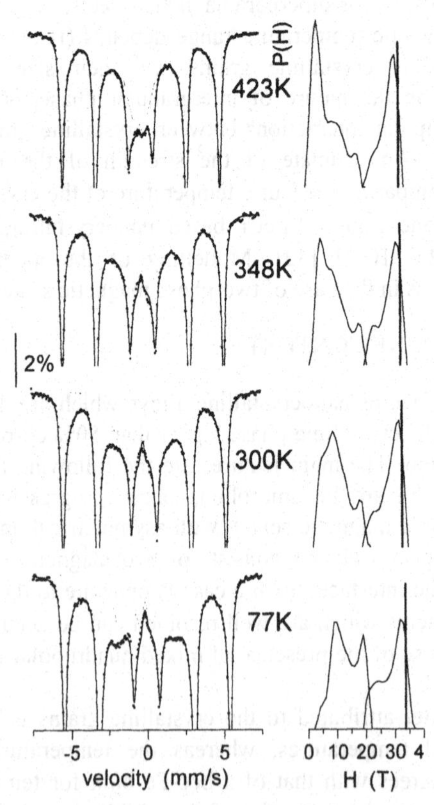

Figure 6. Mössbauer spectra taken at the indicated temperatures (left) and corresponding hyperfine field distributions P(H) (right) for a $Fe_{86.5}Zr_{6.5}Cu_1B_6$ alloy annealed at 600° C/1h.

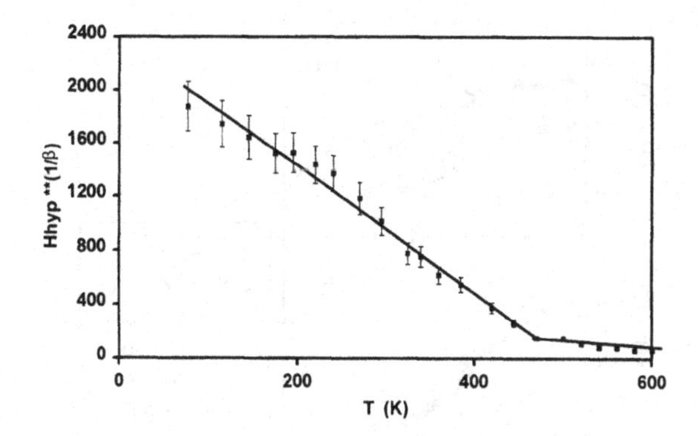

Figure 7. Temperature dependence of the mean value of the hyperfine field characteristic of the residual amorphous phase in reduced coordinates [34].

evidenced when the mean value of the hyperfine field is plotted in reduced coordinates $H_{hyp}^{1/\beta}$ *versus* T, where β represents an empirical fit parameter taken equal to 0.36 [30]. The low temperature part is unambiguously attributed to the residual amorphous phase and one can estimate its Curie temperature, which is strongly higher than that of the as-quenched amorphous phase, because of the strong change of nominal composition. This implies a presence of rather high magnetic hyperfine fields at the Fe sites at temperatures extremely higher than the Curie temperature. Consequently, an assumption can be made that strong exchange interactions between crystalline grains, which are rather close to each other, are penetrating through the residual amorphous matrix even in the paramagnetic state with subsequent induction of hyperfine fields at the Fe sites.

One can apply another fitting procedure which involves the presence of both magnetic sextet and quadrupolar doublet. The first stage consists in recording paramagnetic Mössbauer spectra of the as-quenched amorphous alloy at several temperatures: they exhibit an asymmetrical doublet with broad lines, whatever the nominal composition is. Then, one establishes the isomer shift temperature dependence and the fitting model combining distribution of quadrupolar splitting linearly correlated to that of the isomer shift. Assuming that the quadrupolar structure of the amorphous residual phase slightly differs from that of the as-quenched alloy, Mössbauer spectra of nanocrystalline alloys can be described by means of a theoretical paramagnetic spectrum (its shape is calculated from the combined distributions previously established, taking into account the recording temperature for the isomer shift) superimposed to a magnetic sextet based on a free distribution of hyperfine fields. Such a method gives rise to a temperature dependence of the quadrupolar component [28], which increases with increasing temperature, that is consistent with the previous description.

3.3. INTERMEDIATE CRYSTALLINE CONTENT

The microstructure of nanocrystalline alloys characterized by intermediate volumetric fraction of crystalline phase consists of both interacting and non interacting crystalline grains, according to their mutual distance. Thus, one expects at high

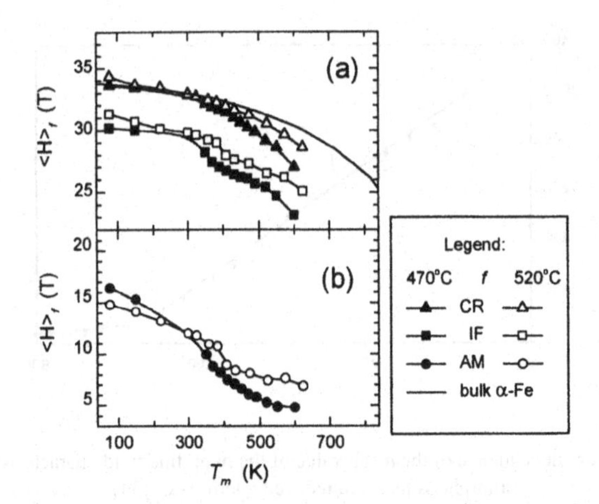

Figure 8. Comparison of the hyperfine field temperature dependencies of bcc-Fe crystalline grains in $Fe_{80}Nb_7Cu_1B_{12}$ nanocrystalline alloys (see the legend) and bcc-Fe bulk phase (solid line) [35].

temperature a combination of phenomena,as described in the two previous sections.The contribution attributed to crystalline grains have to result from components with hyperfine field value typical for bcc α-Fe due to interacting grains, and with lower hyperfine fields due to non-interacting grains, because of fluctuation effects. The amorphous matrix has also to exhibit both paramagnetic and magnetic components, but the overlapping of hyperfine structure prevents *a priori* from an accurate modelling of the interface. This description is consistent with that reported in Fig. 8. It is worth to note that the usual situation may differ from a pure combination of the ideal situations: one has to consider the nature of constituents and the homogeneity of amorphous precursor.

4. Conclusions

Mössbauer spectrometry is a suitable tool for investigating the structural and the magnetic properties of nanocrystalline alloys, particularly NANOPERM alloys. Nevertheless, great attention has to be paid to the fitting procedure because of the complexity of the spectra and of the heterogeneous nature of these two-phase materials. The fitting model which reveals three different iron-based zones, *i.e.* the crystalline phase, the interface and the amorphous remainder, has to be supported by the temperature dependence of the hyperfine structure which is strongly related to the volumetric fraction of the crystalline phase, *i.e.* both annealing temperature and time, and to the nature of amorphous precursor (constituents and amorphous homogeneity). The thickness of the interface is generally estimated at 2-3 atomic layers. The interface behaves as a spin glass like [30] and plays an important role in transmitting magnetic interactions between grains, together with the amorphous remainder. On the basis of present Mössbauer data and of hyperfine field distributions obtained at both low and high temperatures [19], the microstructure of nanocrystalline alloys can be modelled

[36] in conjunction with APFIM results [18]. It is important to emphasise that Mössbauer spectrometry might contribute in understanding the mechanism of nucleation and growth of grains in nanocrystalline alloys.

Acknowledgement

This work was partially supported by the Slovak Ministry of Education by a French-Slovak co-operation grant, and by the SGA grant # 1/5103/98. Dr N. Randrianantoandro (Le Mans) and Dr. A. Ślawska-Waniewska (Warsaw) are gratefully acknowledged for fruitful discussions.

References

1. G.J. Long (ed.), (1984, 1987, 1989), in *Mössbauer Spectroscopy Applied to Inorganic Chemistry Vol 1, 2, and 3*, Plenum Press, New-York.
2. G.J. Long and F. Grandjean (eds.), (1993, 1996), in *Mössbauer Spectroscopy Applied to Magnetism and Materials Science Vol 1, 2*, Plenum Press, New-York.
3. I. Ortalli (ed.), (1996), in *Proceedings of ICAME: International Conference on the Applications of the Mössbauer Effect*, Italian Physical Society, Bologna.
4. Longworth, G. (1987), in G.J. Long (ed.), *Mössbauer Spectroscopy Applied to Inorganic Chemistry Vol. 2*, Plenum Press, New-York, p. 289.
5. Campbell, S.J. and Aubertin, F. (1989), in G.J. Long and F. Grandjean (eds.), *Mössbauer Spectroscopy Applied to Inorganic Chemistry Vol. 3*, Plenum Press, New-York, p. 183.
6. Topsoe, H., Dumesic, J.A., and Morup, S. (1980), in R.L. Cohen (ed.), *Applications of Mössbauer Spectroscopy*, Vol 2, Academic Press, New-York, p. 55.
7. Dormann, J.L., Fiorani, D., and Tronc, E. (1997), in I. Prigogine and Stuart A Rice (eds.), *Advances in Chemical Physics Vol XCVIII*, John Wiley & Sons, Inc., p. 283.
8. Yoshizawa, Y., Oguma, S., and Yamauchi, K. (1988) New-Fe-based soft magnetic alloys composed of ultrafine grain structure, *J. Appl. Phys.* **64**, 6044-6046.
9. Yoshizawa, Y. and Yamauchi, K. (1990) Fe-based soft magnetic alloys composed of ultrafine grain structure, *Mat. Trans. JIM* **31**, 307-314.
10. Herzer, G. (1989) Grain structure and magnetism of nanocrystalline ferromagnets, *IEEE Trans. Magn.* **25**, 2327-3329.
11. Herzer, G. (1990) Grain size dependence of coercivity and permeability in nanocrystalline ferromagnets, *IEEE Trans. Magn.* **26**, 1397-1400.
12. Herzer, G. (1991) Magnetization process in nanocrystalline ferromagnets, *Mater. Sci. Eng.* **A 133**, 1-5.
13. Herzer, G. (1993) Nanocrystalline soft magnetic materials, *Phys. Scr.* **T49**, 307-314.
14. Alben, R., Becker, J.J., and Chi, M.C. (1978) Random anisotropy in amorphous ferromagnets, *J. Appl. Phys.* **49**, 1653-1658.
15. Yamauchi, K. and Yoshizawa, Y. (1995) Recent development of nanocrystalline soft magnetic alloys, *Nanostructured Materials* **6**, 247-254.
16. Suzuki, K., Kataoka, N., Inoue, A., Makino, A., and Masumoto, T. (1990) High saturation magnetization and soft magnetic properties of FeZrB alloys with ultrafine structure, *Mat. Trans. JIM* **31**, 743-746.
17. Makino, A., Inoue, A., and Masumoto, T. (1995) Nanocrystalline soft magnetic Fe-M-B (M=Zr, Hf, Nb) alloys produced by crystallization of amorphous phase, *Mat. Trans. JIM* **36**, 924-938.
18. Inoue, A., Takeuchi, A., Makino, A., and Masumoto, T. (1996) Soft and hard magnetic properties of nanocrystalline Fe-M-B (M-Zr, Nd) base alloys containing intergranular amorphous phase, *Sci. Rep. RITU* **A42**, 143-156.
19. Miglierini, M. and Grenèche, J.M. (1999) Mössbauer spectrometry applied to iron-based nanocrystalline alloys II. Hyperfine fields of amorphous and interfacial regions, in M. Miglierini and D. Petridis (eds)., *Mössbauer Spectroscopy in Materials Science*, Kluwer Academic Publishers, Dordrecht, pp. 257-272.

20. Hono, K, Zhang, Y., Inoue, A., and Saturai, T. (1997) APFIM studies on nanocrystallization of amorphous alloys, *Mater. Sci. Eng.* **A226-228**, 498-502.
21. Grenèche, J.M. (1997) Nanocrystalline iron-based alloys investigated by Mössbauer spectrometry, *Hyp. Int.* **110**, 81-91.
22. Grenèche, J.M. and Ślawska-Waniewska, A. (1997) Interface effects in $Fe_{89}Zr_7B_4$ nanocrystalline alloy followed by Mössbauer spectroscopy, *Mat. Sci. Eng.* **A226-228**, 526-530.
23. Ślawska-Waniewska, A., Roig, A., Molins, E., Grenèche, J.M., and Zuberek, R. (1997) Surface effects in Fe-based nanocrystalline alloys, *Appl. Phys.* **81**, 4652-4654.
24. Ślawska-Waniewska, A., Brzózka, K., and Grenèche, J.M. (1997) Surface effects in Fe-based nanocrystalline alloys, *Acta Physica Polonica* **91**, 229-232.
25. Miglierini, M. and Grenèche, J.M. (1997) Mössbauer spectrometry of Fe(Cu)MB-type nanocrystalline alloys: I. The fitting model for the Mössbauer spectra, *J. Phys.: Condens. Matter* **9**, 2303-2319.
26. Miglierini, M. and Grenèche, J.M. (1997) Mössbauer Spectrometry of Fe(Cu)MB-type nanocrystalline alloys: II. Topography of hyperfine ineractions in Fe(Cu)ZrB alloys, *J. Phys.: Condens. Matter* **9**, 2321-2347.
27. Miglierini, M. and Grenèche, J.M. (1997) Hyperfine interactions in amorphous and nanocrystalline $Fe_{80}Ti_7Cu_1B_{12}$ alloy, *Czech. J. Phys.* **47**, 507-512.
28. Miglierini, M., Škorvánek, I., and Grenèche, J.M. (1998) Microstructure and hyperfine interactions of the $Fe_{73.5}Nb_{4.5}Cr_5CuB_{16}$ nanocrystalline alloys: Mössbauer effect temperature measurements, *J. Phys.: Condens. Matter* **10**, 3159-3176.
29. Brzózka, K., Ślawska-Waniewska, A., Grenèche, J.M., Jezuita, K., and Gawronski, M. (in press) Structure and magnetic hyperfine properties of the amorphous and nanocrystalline FeZrB(Cu) alloys, *Mol. Physics*.
30. Ślawska-Waniewska, A. and Grenèche, J.M. (1997) Magnetic interfaces in Fe-based nanocrystalline alloys determined by Mössbauer spectrometry, *Phys. Rev.* **B 56**, R8491-R8494.
31. Randrianantoandro, N., Ślawska-Waniewska, A., and Grenèche, J.M. (1997) Magnetic properties of nanocrystallized Fe-Cr amorphous alloys, *J. Phys.: Condens. Matter* **9**, 10485-10500.
32. Randrianantoandro, N., Ślawska-Waniewska, A., and Grenèche, J.M. (1997) Magnetic interactions of nanocrystallized Fe-Cr amorphous alloys, *Phys. Rev.* **B 56**, 10797-10800.
33. Grenèche, J.M. and Ślawska-Waniewska, A. (1998) Soft nanocrystalline alloys, in J. Rivas and M.A. Lopez-Quintela (eds.), *Non-Crystalline and Nanoscale Materials*, World Scientific, Singapore, p. 233.
34. Grenèche, J.M., Randrianantoandro, N., Ślawska-Waniewska, A., and Miglierini, M. (1998) Magnetic hyperfine properties in FeZrB-type nanocrystalline metallic alloys, *Hyp. Int.* **113**, 279-285.
35. Miglierini, M. and Grenèche, M. (in press) Temperature dependence of amorphous and interface phases in the $Fe_{80}Nb_7Cu_1B_{12}$ nanocrystalline alloy, *Hyp Int.*
36. Ślawska-Waniewska, A. and Grenèche, J.M. (in press) Structural and magnetic interface properties of nanocrystalline alloys, *Acta Physica Slovaca*.

MÖSSBAUER SPECTROMETRY APPLIED TO IRON-BASED NANOCRYSTALLINE ALLOYS II.

Hyperfine Fields of Amorphous and Interfacial Regions

MARCEL MIGLIERINI[1] AND JEAN-MARC GRENÈCHE[2]
[1)]*Department of Nuclear Physics and Technology, Slovak University of Technology, Ilkovičova 3, SK 812 19 Bratislava, Slovakia*
[2)]*Laboratoire de Physique de l'Etat Condensé, UPRESA CNRS 6087, Université du Maine, Faculté des Sciences, 72085 Le Mans Cedex 9, France*

Abstract

During the crystallization of some amorphous alloys, Fe atoms segregate from the residual amorphous matrix into bcc–Fe crystalline grains. Consequently, the chemical and/or topological short-range order of amorphous remainder is being changed which gives rise to regions depleted in Fe atoms. On the other hand, the nearest surroundings of these Fe atoms are enriched in other constituent elements. As a result, the hyperfine fields experienced by Fe atoms are lowered inside these regions. Considerable fraction of iron is even located in non-magnetic sites. Hyperfine field distributions (HFD) provide information about the structure and magnetic states of atoms located in different structural positions. The position and intensity of HFD peaks reflect the respective short-range orders, and their areas give relative fraction of the corresponding hyperfine interactions. Detailed analysis of three-dimensional HFD mappings provides a support for the interpretations presented in Part I, the previous paper.

1. Disorder in Nanocrystalline Alloys

The nanocrystalline structure, which is created by a suitable heat treatment of amorphous alloys with appropriate composition, substantially improves the magnetic properties of FINEMET and NANOPERM materials [1]. They behave as excellent soft ferromagnets thus competing with other conventional, crystalline or amorphous, materials. The magnetic softening observed is essentially due to presence of nano-sized grains which induce the suppression of the local magneto-crystalline anisotropy by exchange interactions [2]. Nanocrystalline alloys obtained by partial crystallization thus consist of nanocrystalline grains forming the crystalline phase (CR) embedded in a disordered integranular phase. In the latter, the following structurally different atomic sites can be distinguished: (i) the amorphous residual phase (AM) which represents the

M. Miglierini and D. Petridis (eds.), Mössbauer Spectroscopy in Materials Science, 257–272.

remainder of the original amorphous precursor depleted to atoms from which the respective crystallites are created, and (ii) the so-called interface zone (IF) assigned to atoms located on the surface of nanocrystals as well as in their immediate vicinity from the side of AM. Magnetic interactions among CR, AM, and IF and their development with temperature were thoroughly discussed in Part I, the previous paper [3]. Here, we will focus on the intergranular phases (AM and IF) which depict structural disorder having a consequent influence on hyperfine fields.

Nanocrystalline alloys are, due to their interesting magnetic properties, promising candidates for many a technical applications [4]. Structural properties of nanocrystalline alloys (size and form of the grains, homogeneity of the matrix) are studied by X-ray diffraction (XRD), electron diffraction (ED), and transmission electron microscopy (TEM) [5]. XRD and TEM are used to determine the average particle size and the distribution of the particle size, respectively, or to identify the crystalline phases. Structurally different regions can also identified by ED.

The XRD patterns consist of broad and diffuse intensity distributions attributed to the retained amorphous phase and some well-defined but broadened Bragg peaks assigned to the emerging crystalline grains (Fig. 1a). The crystalline phase(s) can be characterized by a structure refinement of X-ray diffractograms using common methods whereas the average grain size may be estimated using the Scherrer equation. However, precise quantitative estimation of the relative fractions of various phases remains difficult particularly in the case of low volume fraction of the crystalline phase. The major problem arises from absorption effects, in addition to residual stresses, texture effects, and instrumental effects. Moreover, it is not possible to distinguish between IF and AM regions since both exhibit structural disorder giving rise to the same broadened component in the diffractograms.

From the reflections in ED and TEM images, a presence of crystallites can be

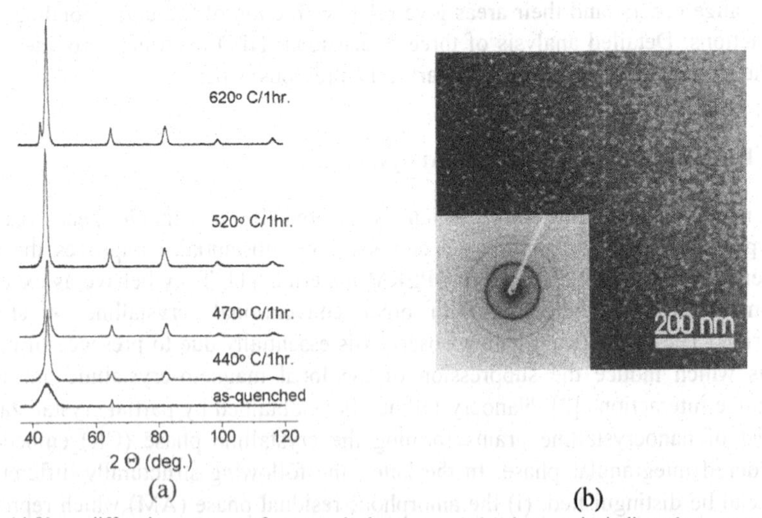

(a) (b)

Figure 1. (a) X-ray diffraction patterns of as-quenched and annealed (1 hour at the indicated temperature) $Fe_{80}Ti_7B_{12}Cu_1$ ribbons. (b) TEM and ED images of the same sample annealed at 520° C.

confirmed (Fig. 1b). An estimation of volumetric ratio of the CR and disordered phase is also possible from TEM images. However, the contribution stemming from atoms located in the IF phase is again hidden.

Classical magnetic measurements provide information about macroscopic magnetic properties (coercivity, magnetization, *etc.*). The data obtained from macroscopic magnetic measurements reflect simultaneously the contributions both from the crystalline and non-crystalline phases present in the sample and in this respect they are connected with the structural arrangements. Annealing results in an increase of T_C due to the precipitation of bcc-Fe which is related to a change in the composition of the residual amorphous phase (Fig. 2a) [6]. A kink is observed (Fig. 2b) in temperature dependence of magnetization when the residual amorphous matrix goes from ferromagnetic to paramagnetic state. Above this temperature range, ferromagnetic Fe-rich nanocrystals exist within a paramagnetic amorphous residue [7].

Like in diffraction methods, direct conclusions about structural arrangements and their influence on hyperfine interactions, which are, consequently, demonstrated in the macroscopic properties, cannot be derived. To understand fully the nature and origin of these unique properties, specific tools for structural and/or magnetic studies on an atomic scale, comprising nuclear probe methods, should be employed [8]. The ^{57}Fe Mössbauer spectrometry can be effectively utilised in these investigations because of its local probe character: it enables to reveal different types of hyperfine interactions as well as to give the crystalline-amorphous ratio and to provide information about structural arrangements. Through scanning the nearest surroundings by means of Mössbauer probe atoms it is possible to reveal *where*, *when*, and *what* ordering is found in the samples studied [9]. Hence, the information on structural and/or magnetic order (disorder) is simultaneously provided via distributions of hyperfine magnetic fields.

The following section briefly summarises the methods of Mössbauer spectra analyses of nanocrystalline alloys. We point out the main advantages and drawbacks of the fitting models used so far and describe our present approach in which independent

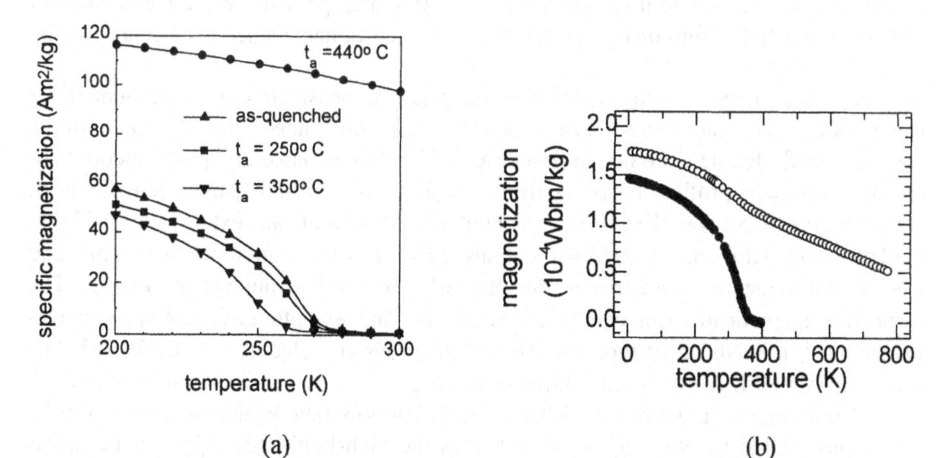

Figure 2. Magnetization *versus* temperature for: (a) $Fe_{80}Mo_7B_{12}Cu_1$ in the as-quenched state and after 1 hour annealing at 250° C, 350° C, and 440° C (after [6]), and (b) $Fe_{73.5}Nb_{4.5}Cr_5CuB_{16}$ alloy in the as-quenched state (●) and after annealing at 550° C for 1 hour (O) (after [7]).

blocks of hyperfine field distributions (HFD) are employed. In the third section, examples of HFDs encountered in iron-based nanocrystalline alloys are given. They are presented in a form of three-dimensional mappings (3-D HFD) which instantaneously illustrate the relation between short-range order (SRO) and the magnetic structure. In this way, they provide a support and an evidence for the ideas presented in Part I, the previous paper [3] concerning the magnetization effects of the nanocrystals. Finally, using the 3-D HFDs a topography of hyperfine interactions is proposed in Section 4.

2. Mössbauer Spectra Analysis

2.1. FINEMET NANOCRYSTALLINE ALLOYS

During crystallization of the FINEMET-type alloys, Fe-Si ultra-fine grains are created with Fe atoms occupying different crystalline sites. Depending on the Si content, the Fe-Si phase exhibits a bcc and/or DO_3 structural arrangement. Alternatively, Fe-Si solid solution is formed. In any case, the resulting Mössbauer spectra are complex, showing multiple narrow lines superimposed on a broadened feature. The former represents structurally different crystallographic sites of the particular CR phase(s) whereas the latter is ascribed to a residual AM phase. An example of a common Mössbauer spectrum can be found in Fig. 2 of Part I [3].

As for the CR phase(s), the problem, in general, is not the line shape (often taken to be Lorentzian) but the number of individual CR sites represented by sextets. However, a variety of approaches was introduced to reconstruct the AM phase. Closer discussion on this topic was published in [10-12], here we only briefly sumnarize some representative examples.

The simplest procedures use a single sextet with large line widths to fit the intergranular phase. The following profiles were applied: Lorentz [13], pseudo-Lorentz [14], and Voigt [15]. Such fitting models are relatively easy to use but the information obtained is limited only to the relative CR/AM ratio, and, perhaps, to the mean value of AM hyperfine field. Both quantities are, however, questionable since the fits are usually unsatisfactory.

Non-equivalent Fe sites within the intergranular phase should be accounted for using HFDs. The latter are implemented via: (i) constrained profiles taken to be Gaussian [16], double Gaussian asymmetrical [17], "skew Gaussian" [18] functions, or (ii) unconstrained profiles resulting from distributions of discrete hyperfine fields of certain number (to cover H-values up to about 35T) of Lorentzian sextets [10, 19-22]. In the latter case, HFDs are derived without any *a priori* assumption about their shape and can be, subsequently, used to describe the SRO of residual amorphous matrix. The respective magnetic ordering can be also revealed. Time evolution of the AM phase was investigated in FeNbCuSiB-type nanocrystalline alloys annealed at 550° C [10,19]. The achieved 3-D HFD mappings are illustrated in Fig. 3. Two magnetically non equivalent sites of iron atoms can be distinguished in AM. The low-field peaks are ascribed to Fe atoms surrounded by Nb, Cu, and B, whereas the high-field ones represent Fe atoms with Si and B in their nearest neighbourhoods. Changes in topological and chemical

SRO are demonstrated by variations in relative intensities and positions of the peaks as a function of annealing time and composition.

Nevertheless, presence of Si within the CR phase leads to complex Mössbauer spectra in which it is practically impossible to identify any interfacial component, *i.e.* the IF phase. Aiming to shed some light on this particular structural component a special interest was recently given to silicon-free nanocrystalline alloys which exhibit only bcc-Fe grains in the first step of crystallization. Since bcc-Fe is in fact one of calibration absorbers for ^{57}Fe Mössbauer spectrometry, its hyperfine parameters are well known and favour detail interpretation of such, relatively simple, Mössbauer spectra.

2.2. NANOPERM NANOCRYSTALLINE ALLOYS

Unlike in Si-containing materials, where the FeSi CR phase yields Mössbauer spectra with many components, NANOPERM-type alloys exhibit only one sextet of Lorentzian lines superimposed on a broadened feature [3]. This is a substantial simplification as far as fitting of the CR phase is concerned. A dilemma still persists what approach to take in the evaluation of the broadened spectral features ascribed to AM and IF regions.

In the most widely studied FeZr(Cu)B nanocrystalline alloys, the intergranular phase is represented by a HFD obtained according to (ii) in Section 2.1. Whatever the fitting procedure (for details see [11, 12]), the HFDs consist of two peaks associated with two magnetically different kinds of Fe atoms. The high-field peaks were thought to be due to Fe_3B [23] or Fe-B [24] crystalline phases, but no experimental support for this assignment was provided from e.g. diffraction techniques. They are assigned also to Fe_3Zr crystalline phase [25] in $Fe_{81}Zr_7B_{12}$ and $Fe_{79}Zr_7B_{12}Cu_2$ nanocrystalline alloys based on X-ray diffraction data which are, nevertheless, not convincing for the latter composition. The AM was described in [26] by a HFD in addition to a quadrupolar doublet associated with non-magnetic Fe phase without any further explanation.

Recently, some attempts have been made to search for IF, too. Using high temperature measurements ($T_m > T_C(AM)$), a small magnetically split subspectrum was

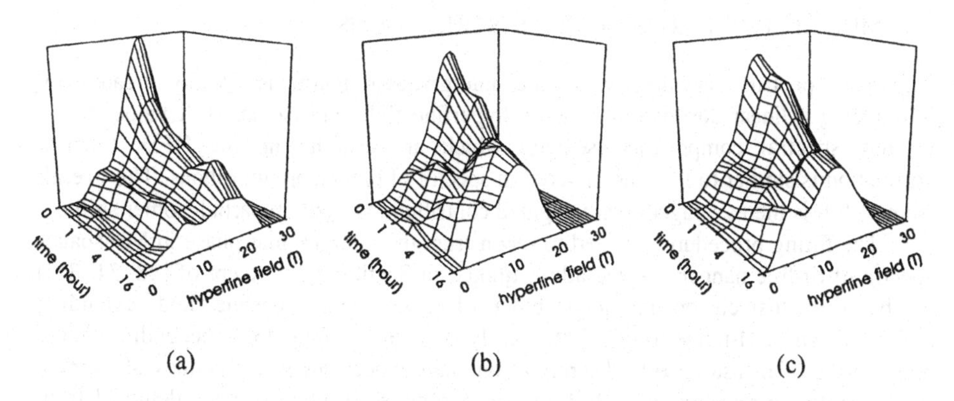

Figure 3. 3-D HFDs (second dimension is time of annealing) of the residual amorphous phases in: (a) $Fe_{73.5}Nb_3Cu_1Si_{13.5}B_9$, (b) $Fe_{70.5}Nb_{4.5}Cu_1Si_{16}B_8$,, and (c) $Fe_{72}Nb_{4.5}Cu_1Si_{13.5}B_9$ nanocrystalline alloys annealed at 550° C.

found by the help of subtraction method in Mössbauer spectra of FeZrCuB nanocrystalline alloys in addition to a central paramagnetic component [27]. The latter was associated with the residual AM and the former is supposed to constitute the so-called interface originating from substantial Zr content on the surface of bcc nanocrystalline grains [27]. This component could not be identified by measurements below T_C(AM) since it strongly overlaps with the amorphous subspectra [27]. There is, however, another component present (its fitting parameters are not given) which creates sort of "shoulder" to the CR sextet. It is interpreted as bcc-Fe containing about 2 at.% Zr and limited amount of B dissolved in the bcc structure [27]. These investigations [27] reply to our previously published results [11, 28]. Both interpretations would be in agreement if the "shoulder" component was assigned to what we call the IF zone, and the distributed magnetically split component was treated as that part of AM which shows higher hyperfine fields due to magnetic interactions among the nanocrystals (see Section 3.2. in Part I [3]). The concept of interface zone being equal to the "shoulder" component was, however, rejected in [29], stating that significant difference would be expected between temperature dependencies of the IF and CR hyperfine fields which is not the case (see e.g. [28]). Again, no details about the fitting results are provided (as e.g., line widths of the "shoulder" component) which might help to verify the conclusions proposed in [27, 29]. Both crystalline and "shoulder" components are assigned to a nanocrystalline bcc Fe phase with 2-3 at. % Zr but also some amount of B dissolved in it [29]. However, the results of atom probe field ion microscopy [30] give no evidence for Zr inside the grains. Moreover, the Mössbauer effect results on FeMCuB- (M = Mo, Nb, and Ti) [31] and FeNbCrCuB-type [7] nanocrystalline alloys show the same IF behaviour even though no Zr atoms are contained in the studied nanocrystalline alloys.

The interfacial regions were identified in FeZrB- and FeZr(Cu)B-type nanocrystalline alloys by other research groups using simple histogram methods for the HFD refinement [32 - 34]. They were also detected in FeNbCrCuB-type alloy [35] where, however, more elaborated fitting procedure was already applied following the concept described in the next section.

2.3. MULTIPLE HYPERFINE FIELD DISTRIBUTIONS

Apparent spectrum asymmetry of the intergranular phase is usually treated by introducing a linear correlation between hyperfine field and isomer shift values to all discrete spectral components (sextets) which are establishing the HFD. Such a correlation is similar to that encountered in spectra of precursor amorphous. The overall isomer shift is then averaged over weighted contributions from all partial isomer shifts.

The fitting procedures applied to reconstruct the intergranular phase in Mössbauer spectra of Si-free nanocrystalline alloys quoted in Section 2.2. (except of [28, 31, 35]) are based exclusively on one single block of sextets with hyperfine fields extending across the whole H-range from 0 T (or nearly zero) up to about 35 T (depending on the temperature of measurement). Therefore, one can expect unphysical values of average isomer shifts corresponding to AM and, in particular, IF phase when calculated from such a fitting procedure. Moreover, a significant line overlap should be considered. The second and fifth lines of sextets with high hyperfine fields contribute to the theoretical curve in lower H-range where another sextets having these low H-values are used. This yields artificial peaks in HFDs which result from overlap between the intermediate and outer sextet lines with different hyperfine fields. Such situation was recognised in [34] where non-physical peaks arose from a single-block HFD fitting.

Whereas all reported HFDs are bimodal in shape, and taking into account also the above mentioned restrictions, we have suggested [11] to decompose the single block HFD into two parts containing independent distributions of hyperfine fields without any constrained profiles. Consequently, the succeeding structural positions of Fe atoms can be distinguished (see also Fig. 4 in Part I [3]): (i) atoms located in the bulk of CR phase; (ii) Fe atoms which constitute the surface of nanocrystals; (iii) Fe atoms situated in the nanocrystal-to-amorphous interface which originate from AM but are positioned between CR grains; and, finally, (iv) a disordered AM arrangement which does not have any direct contact with nanocrystals. The atoms (ii) and (iii) combine into what we call an interface zone - IF.

Atoms (i) are bcc-type ordered and can be analysed by a Lorentzian sextet - CR phase. Atoms (iv) are representing the residual amorphous matrix AM, thus, a distribution of hyperfine parameters must be employed. The IF phase also depicts structural disorder because of perturbed positions of Fe atoms (ii) located at the surface of the nanocrystals (order-disorder boundary) and because of the amorphous nature of atoms (iii). Subsequently, another distribution of hyperfine fields should be applied.

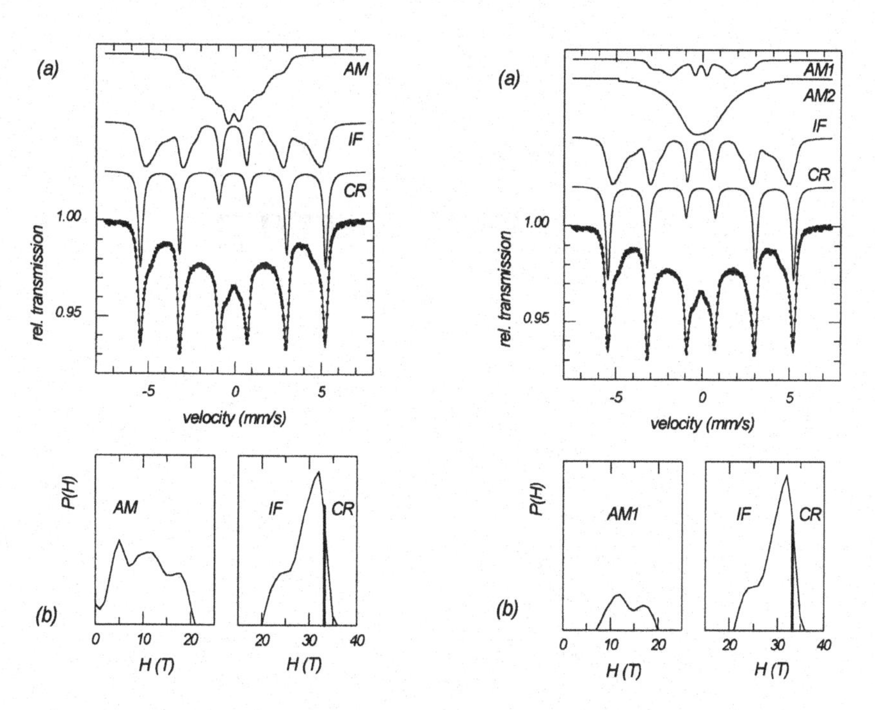

Figure 4. Room temperature Mössbauer spectra of $Fe_{80}Mo_7Cu_1B_{12}$ nanocrystalline alloy (a) and corresponding HFDs (b). The AM remainder is fitted by: (left) one HFD - AM, and (right) by a HFD - AM1 and a quadrupolar doublet - AM2.

The AM block can be eventually decomposed into another two components: a distribution of hyperfine fields resulting from a magnetic subspectrum and a quadrupolar doublet or a distribution of quadrupolar splitting values which represent a paramagnetic component in the amorphous residual phase. Which situation applies depends primarily on the measuring temperature and/or the volumetric content of crystallites as well as on the original composition of the alloy [3]. Examples of both fitting approaches are illustrated in Fig. 4 for the same sample [36].

3. Hyperfine Fields of Amorphous and Interfacial Regions

3.1. TEMPERATURE OF ANNEALING

Temperature of annealing, t_a, controls the amount of nanocrystallites which are demonstrated in the corresponding Mössbauer spectra by rising importance of the CR

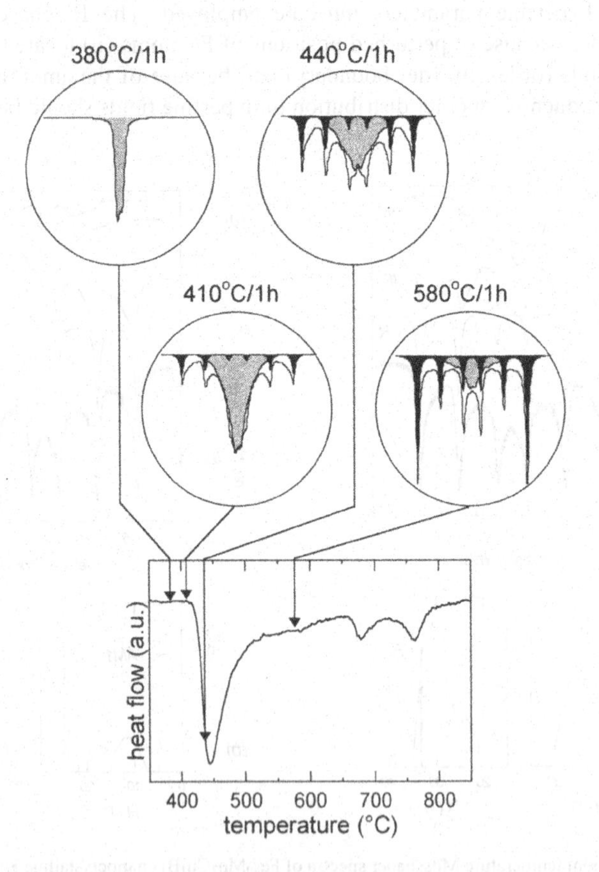

Figure 5. Room temperature Mössbauer spectra of $Fe_{80}Mo_7Cu_1B_{12}$ alloy (CR - black, AM - grey) taken after heat treatments at the indicated temperatures. The corresponding DSC curve is given below.

sextet on account of the AM distributed component. Fig. 5 shows some Mössbauer spectra of a $Fe_{80}Mo_7Cu_1B_{12}$ alloy annealed at t_a indicated on the differential scanning calorimetry curve (DSC). Magnetic states of iron atoms are significantly affected by already small fraction of the CR phase (black). The originally paramagnetic sample (t_a = 380° C) shows distribution of quadrupole splitting values. After some bcc-Fe nanocrystallites have emerged ($t_a \geq 410°$ C), magnetic interactions are observed in the retained amorphous phase (grey). They are demonstrated by apparent broadening of the grey spectral components in Fig. 5 and originate from ferromagnetic interactions among the grains which are penetrating into the AM phase [37-40]. For the sake of clarity, IF are not shown in Fig. 5.

During the crystallization at elevated t_a, Fe atoms segregate from the residual amorphous matrix into bcc-Fe crystalline grains. Consequently, the AM chemical and/or topological SRO is changing which gives rise to regions depleted in Fe atoms. On the other hand, the nearest surroundings of these Fe atoms are enriched in other constituent elements (Mo, Cu, and B in this case). As a result, the magnetic moments of Fe atoms are lowered inside these regions and considerable fraction of Fe is located in non-magnetic sites. These low-field values are indeed electric-quadrupolar interactions and should be treated accordingly by quadrupole doublets (compare the right-hand side of Fig. 4). To allow a thorough view on the evolution of hyperfine interactions within the nanocrystalline material as a whole, in the following we will restrict ourselves only to distributions of hyperfine fields not forgetting about the origin of low-field tails.

Figure 6 illustrates distributions of hyperfine fields within AM and IF regions of FeMCuB-type nanocrystals using t_a as the second dimension in 3-D mappings. No major distinctions are observed in IF-HFDs regardless the composition. This suggests a similarity of SROs which is in accordance with the structural model applied for fitting. The IF-HFDs are assigned to the atoms located on the surface of CR grains, *i.e.* in close contact with those of bulk. The small reduction in field values is due to the proximity of the outer layer (experiencing low fields) which is in close contact with the amorphous phase, and due to frustration originated from the symmetry breaking. On the other hand, variety of nearest surroundings can be seen in AM-HFDs depending on t_a and/or composition. They originate from (i) creation of CR as a function of t_a which, subsequently, influences the composition of AM, and (ii) different original compositions of the as-quenched precursors. Diversity of magnetic arrangements is demonstrated by apparent peaks in HFDs. It is noteworthy that e.g. for M = Mo the prevailing non-magnetic state of AM (pronounced peak at low fields for low t_a) progressively disperses at higher t_a and magnetic regions are observed. The latter are due to changes in chemical SRO but they are also induced by ferromagnetic exchange coupling among CR grains. The M = Nb and Ti alloys are ferromagnetic in the as-quenched state [41] and after the nanocrystallization magnetically distinct regions are clearly seen in their AM phases (Figs. 6a and b).

An example of 3-D mappings obtained from Mössbauer spectra of $Fe_{80}Mo_7Cu_1B_{12}$ nanocrystalline alloys analysed by quadrupole doublets is shown in Fig. 7. AM was fitted by distributions of sextets giving the resulting HFDs. Rising importance of magnetically ordered regions at higher t_a is better seen than in Fig. 6a.

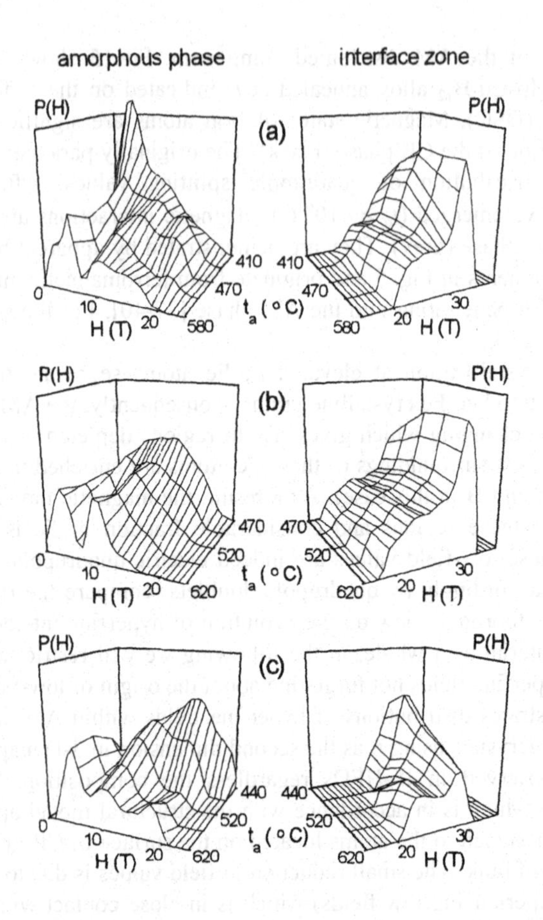

Figure 6. 3-D HFD mappings (second dimension is temperature of annealing t_a) for AM (left) and IF (right) regions in $Fe_{80}M_7Cu_1B_{12}$ nanocrystalline alloys at room temperature: M = Mo (a), Nb (b), and Ti (c).

3.2. TEMPERATURE OF MEASUREMENT

Temperature evolution of hyperfine fields in AM and IF can be better viewed by the help of 3-D mappings using the temperature of measurement T_m as a second dimension. Fe(Cu)ZrB-system annealed at different t_a was thoroughly studied in wide temperature range and the results can be found in [28]. Aiming to reveal the behaviour of the IF zone, other compositions of Si-free nanocrystalline alloys were also studied in wide temperature range. Recent papers report on the research performed upon $Fe_{73.5}Nb_{4.5}Cr_5CuB_{16}$ [7] and $Fe_{80}Nb_7Cu_1B_{12}$ [42, 43] alloys.

By varying temperature of measurement, it is possible to separate one from another the AM and IF contributions to the overall Mössbauer spectrum. This (i) simplifies the fitting procedure, and (ii) opens a new approach to the investigation of hyperfine fields because both components are better resolved. For a full benefit, one must choose appropriate volumetric fraction of CR (*i.e.* annealing treatment) and/or

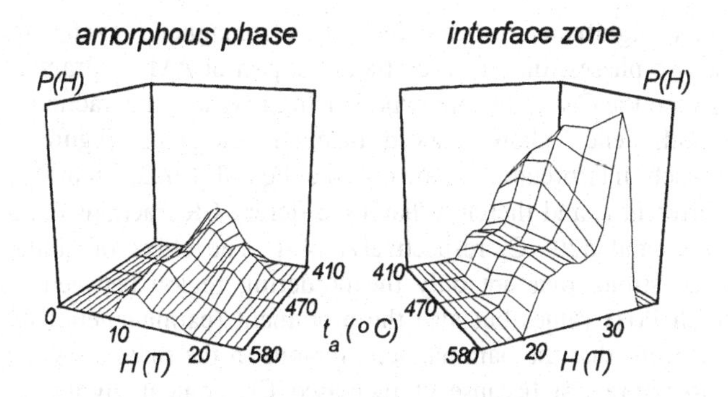

Figure 7. 3-D HFD mappings (second dimension is temperature of annealing t_a) for AM (left) and IF (right) regions in $Fe_{80}Mo_7Cu_1B_{12}$ nanocrystalline alloy obtained from a fitting with doublets and sextets in AM (after [36]).

sample composition (with T_C(AM) close to room temperature). Under such circumstances, it is possible to study either superparamagnetic behaviour of bcc-Fe nanocrystals (low CR fraction) or spreading of ferromagnetic exchange interactions (high CR content) without the risk of further crystallization at elevated temperatures.

Figure 8 introduces 3-D HFD mappings for two alloys having different T_C(AM). Consequently, the residual AM phase exhibits paramagnetic features (prominent low-field HFD peak) at very distinct T_m. It should be noted that in this figure, both AM- and IF-HFDs are plotted in one 3-D mapping together in order to visualise the overlap of hyperfine fields between both regions. This overlap is seen also in Figs. 6 and 7 but might be overlooked in separate 3-D images. Moreover, when plotted together, it is emphasised that the division into two (or more) HFDs is only for the sake of fitting procedure to ensure physically acceptable values of hyperfine parameters or perhaps to ease the interpretation. In reality, hyperfine interactions propagate continuously between structurally different regions and no artificial division is plausible. Thus, while the AM and IF contributions in Fig. 8a are well separated, they are showing non-zero probability values P(H) at about 20T in Fig. 8b.

The „smeared-out" border between AM and IF (*i.e.*, non-zero P(H) values)

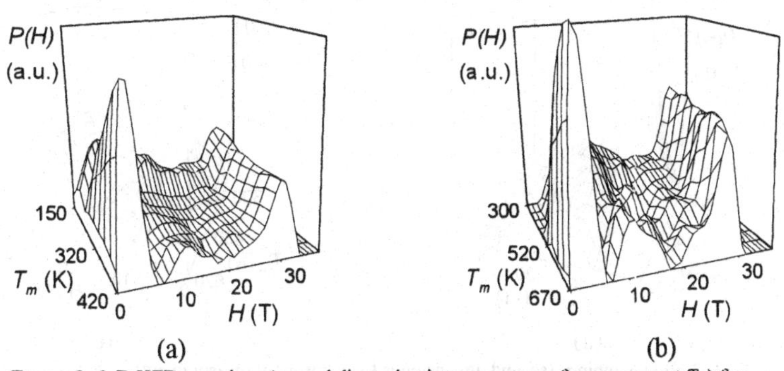

Figure 8. 3-D HFD mappings (second dimension is temperature of measurement T_m) for: (a) $Fe_{73.5}Nb_{74.5}Cr_5Cu_1B_{16}$ annealed at t_a = 550° C and (b) $Fe_{80}Ti_7Cu_1B_{12}$ annealed at t_a = 470° C.

indicates rather significant propagation of ferromagnetic exchange interactions between these two phases. Indeed, even though a part of AM is paramagnetic, regions located in between the CR grains are experiencing magnetic interactions acting among nanocrystals and, hence, show induced magnetic moments. Figure 9 provides an evidence for such interpretation showing the $Fe_{80}Nb_7Cu_1B_{12}$ nanocrystalline alloy annealed at different t_a and therefore having different CR fractions (ca 18% and 30% for t_a = 470° C and 520° C, respectively) [43]. The effect of „induced magnetic moments" is so strong that not only the bordering region between AM and IF is elevated to high P(H) values but also the prominent (paramagnetic) peak is notably reduced. This means that the paramagnetic regions in the same alloy are shrinking at the same temperature just because of increased CR content giving rise to stronger inter-grain magnetic interactions.

3.3. EXTERNAL MAGNETIC FIELDS

The usage of external magnetic fields is twofold: (i) it simplifies the evaluation procedure of Mössbauer spectra, and (ii) tends to equalize the magnetic texture of the studied alloys. Small magnetic fields oriented parallel with the plane of a ribbon-shaped sample (about 0.05T is usually enough) align the net magnetic moment into the absorber's plane in case of pure ferromagnet. Thus, intensities of magnetically split spectrum components (sextets) will be well defined, namely 3:4:1:1:4:3. When, on the other hand, a perpendicular orientation of external magnetic field is used, the $\Delta m = 0$ transitions disappear. Subsequently, a fixed Mössbauer sextet line intensity ratio of 3:0:1:1:0:3 can be incorporated into the fitting procedure. The situation becomes more complex when the magnetic arrangement slightly differs from a pure ferromagnet: the occurrence of canted moments leads to deviate from ideal situation, as previously described.

Examples of both approaches are illustrated in Fig. 10a and b, respectively. It should be noted that in the latter case, high intensity of external magnetic field must be

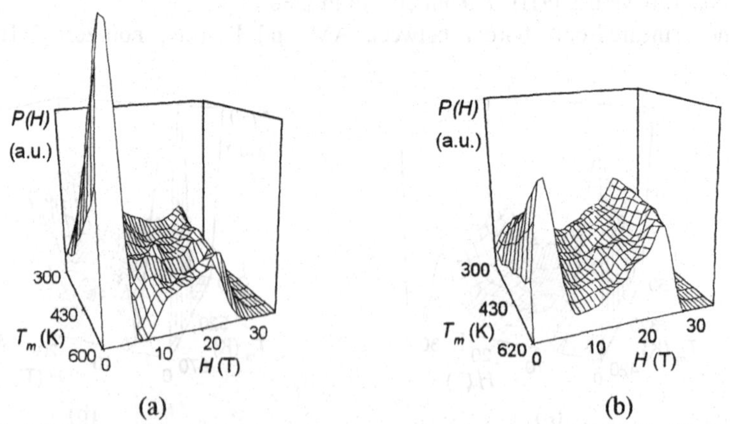

Figure 9. 3-D HFD mappings (second dimension is temperature of measurement T_m) for $Fe_{80}Nb_7Cu_1B_{12}$ annealed at: (a) t_a = 470° C and (b) t_a = 520° C.

used to overcome the demagnetization and to turn all particular magnetic moments out of the ribbon plane. This can be achieved only in fields higher than several teslas. In the presented example, a superconducting magnet yielding $H_{ext} = 4T$ was employed keeping both source and sample at the indicated temperatures. The dominating central part of Mössbauer spectra in Fig. 10b indicates that paramagnetic regions occur in AM even at 65K which was not observed to such extent in out-of-field experiments. At 250K, the contribution of magnetic regions is more suppressed and the separation of non-magnetic and magnetic spectral components is more evident.

4. Topography of Hyperfine Interactions

Employing HFDs obtained from Mössbauer spectra, we are able to distinguish among different structural arrangements in nanocrystalline alloys. At the same time, hyperfine interactions describe structural positions from the point of view of their magnetic states. This is basically achieved via separate distributions of hyperfine magnetic fields for AM and IF phases. The diagnostic potential of Mössbauer spectrometry can be further enhanced by decomposition of HFDs into Gaussian components [36, 44] which provides quantitative assessment of miscellaneous hyperfine interactions observed in the

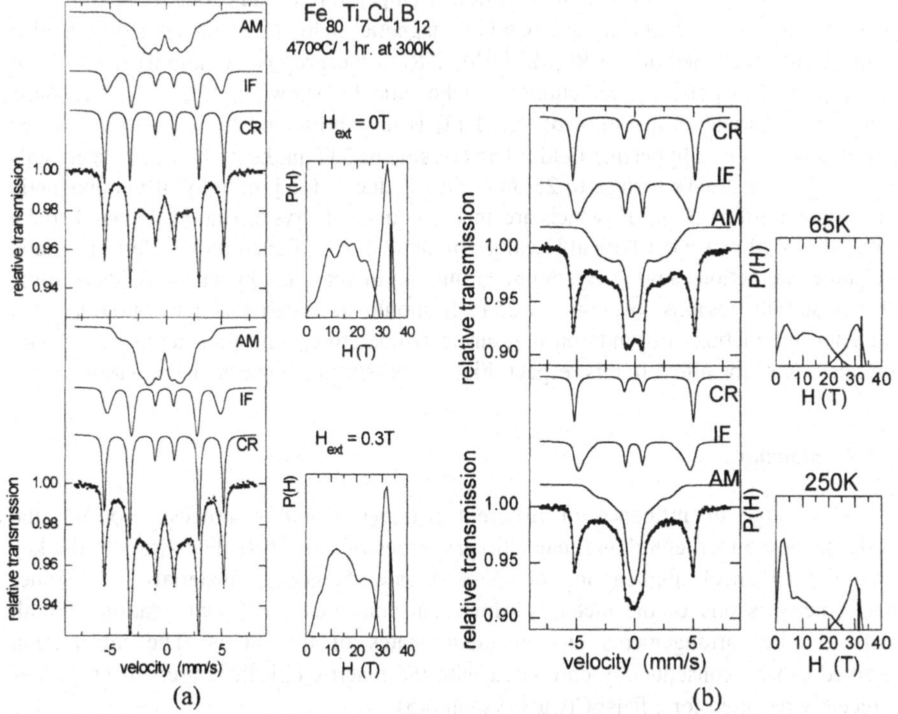

Figure 10. Mössbauer spectra (left columns) and corresponding HFDs (right) for: (a) $Fe_{80}Ti_7Cu_1B_{12}$ annealed at $t_a = 470°$ C and taken at 300K in parallel magnetic field $H_{ext} = 0.3T$, and (b) $Fe_{73.5}Nb_{74.5}Cr_5Cu_1B_{16}$, $t_a = 550°$ C, taken at 65, and 250K in a perpendicular magnetic field $H_{ext} = 4T$.

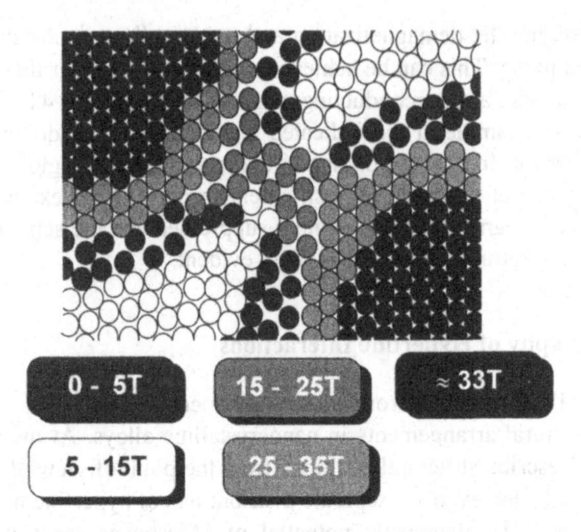

Figure 11. Topography of hyperfine interactions (compare with Fig. 4 in Part I [3]).

disordered intergranular phase.

Based on the evaluation of Mössbauer spectra of Si-free nanocrystalline alloys taken under different experimental conditions (temperature of annealing, temperature of measurement, time of annealing, external magnetic fields, ...) and considering also results from other methods (APFIM, TEM, XRD), we propose a simplified model of topography of hyperfine interactions as schematically shown in Fig. 11. The same atomic arrangement as in Fig. 4 of Part I [3] is now presented from the viewpoint of spatial occurrence of hyperfine fields. The H-value of 33T in the bulk of bcc-Fe crystals decreases at the grain's surface to 25-30T. This is due to the proximity of paramagnetic amorphous regions (0 - 5T) which are in the course of crystallization depleted to Fe atoms. The AM phase has typical H-values of about 5-15T. Penetration of ferromagnetic exchange interactions among crystalline grains can increase the hyperfine fields of some inter-crystalline regions to about 15-25T. It should be noted that the topography of hyperfine interactions depends on measuring temperature, annealing temperature, and composition of the alloy. In this respect, Fig. 11 illustrates one particular situation.

5. Conclusions

The knowledge of influence of different parameters and/or processes during the crystallization on structural and magnetic properties of nanocrystalline alloys is the key issue for technical applications of these advanced modern materials. Mössbauer spectrometry is one of the methods which can contribute to the elucidation of both microstructural arrangements and magnetic states of the atoms. The information obtained can be consequently correlated with the macroscopically observed properties as recently reported for a FeNbCrCuB system [45].

Acknowledgement

This work was partially supported by the Slovak Ministry of Education by a French-Slovak co-operation grant, and by the SGA grant # 1/5103/98.

References

1. Herzer, G. (1992) Nanocrystalline soft magnetic materials, *J. Magn. Magn. Mater.* **112**, 258-262.
2. Herzer, G. and Warlimont, H. (1992) Nanocrystalline soft magnetic materials by partial cristallization of amorphous alloys, *NanoStructured Materials* **1**, 263-268.
3. Grenèche, J.M. and Miglierini, M. (1999) Mössbauer spectrometry applied to iron-based nanocrystalline alloys I.: High Temperature Studies, in M. Miglierini and D. Petridis (eds)., *Mössbauer Spectroscopy in Materials Science*, Kluwer Academic Publishers, Dordrecht, pp. 243-256.
4. Hernando, A., Vázquez, M., and Páramo, D. (1998) Applications of amorphous and nanocrystalline magnetic materials as sensing elements, *Mater. Sci. Forum* **269-272**, 1033-1042.
5. Miglierini, M., Kopcewicz, M., Idzikowski, B., Horváth, Z.E., Grabias, A., Škorvánek, I., Dłużewski, P., and Daróczi, Cs.S. (in press) Structure, hyperfine interactions and magnetic behavior of amorphous and nanocrystalline $Fe_{80}M_7B_{12}Cu_1$ (M = Mo, Nb, Ti) alloys, *J. Appl. Phys.*
6. Idzikowski, B., Baszyński, J., Škorvánek, I., Müller, K.H., and Eckert, D. (1998) Microstructure and magnetic properties of amorphous and nanocrystalline $Fe_{80}M_7B_{12}Cu_1$ (M = Nb, Ti or Mo) alloys, *J. Magn. Magn. Mater.* **177-181**, 941-942.
7. Miglierini, M., Škorvánek, I., and Grenèche, J.M. (1998) Microstructure and hyperfine interactions of the $Fe_{73.5}Nb_{4.5}Cr_5CuB_{16}$ nanocrystalline alloys: Mössbauer effect temperature measurements, *J. Phys.: Condens. Matter* **10**, 3159-3176.
8. Würschum, R. (1995) Nuclear spectroscopy of nanocrystalline metals and alloys, *NanoStructured Materials* **6**, 93-104.
9. Gonser, U., Limbach, T., and Aubertin, F. (1988) Mössbauer spectroscopy in amorphous metals: failures and successes, *J. Non-Cryst. Solids* **106**, 395-398.
10. Miglierini, M. (1994) Mössbauer-effect study of the hyperfine field distributions in the residual amorphous phase of Fe-Cu-Nb-Si-B nanocrystalline alloys, *J. Phys.: Condens. Matter* **6**, 1431-1438.
11. Miglierini, M. and Grenèche, J.M. (1997) Mössbauer spectrometry of Fe(Cu)MB-type nanocrystalline alloys: I. The fitting model for the Mössbauer spectra, *J. Phys.: Condens. Matter* **9**, 2303-2319.
12. Grenèche, J.M. (1997) Nanocrystalline iron-based alloys investigated by Mössbauer spectrometry, *Hyperfine Interactions* **110**, 81-91.
13. Jiang J., Aubertin, F., Gonser, U., and Hilzinger, H.R. (1991) Mössbauer spectroscopy and X-ray diffraction studies of the crystallization in the amorphous $Fe_{73.5}Cu_1Nb_3Si_{13.5}B_9$, *Z. Metallk.* **82**, 698-702.
14. Rixecker, G., Schaaf, P., and Gonser, U. (1992) Crystallization behaviour of amorphous $Fe_{73.5}Cu_1Nb_3Si_{13.5}B_9$, *J. Phys.: Condens. Matter* **4**, 10295-10310.
15. Gupta, A., Bhagat, N., and Principi, G. (1995) Mössbauer study of magnetic interactions in nanocrystalline $Fe_{73.5}Cu_1Nb_3Si_{16.5}B_6$, *J. Phys.: Condens. Matter* **7**, 2237-2248.
16. Pradell, T., Clavaguera, N., Zhu, J., and Clavaguera-Mora, M.T. (1995) A Mössbauer study of the nanocrystallization process in $Fe_{73.5}CuNb_3Si_{17.5}B_5$ alloy, *J. Phys.: Condens. Matter* **7**, 4129-4143.
17. Zemčík, T. (1993) Phase analysis of amorphous and nanocrystalline FeCuNbSiB alloys by ^{57}Fe Mössbauer spectroscopy, *Key Eng. Mater.* **81-83**, 261-266.
18. Knobel, M, Sato Turtelli, R., and Rechenberg, H.R. (1992) Compositional evolution and magnetic properties of nanocrystalline $Fe_{73.5}Cu_1Nb_3Si_{13.5}B_9$, *J. Appl. Phys.* **71**, 6008-6012.
19. Miglierini, M. (1996) Mössbauer study of nanocrystalline alloys: Hyperfine field distributions, *Hyperfine Interactions (C)* **1**, 254-257.
20. Randrianantoandro, N., Grenèche, J.M., Jędryka, E., Ślawska-Waniewska, A., and Lachowicz, H.K. (1995) Nanocrystallized Fe-based metglasses investigated by Mössbauer spectrometry, *Mater. Sci. Forum* **179-181**, 545-550.
21. Randrianantoandro, N., Ślawska-Waniewska, A., and Grenèche, J.M. (1997) Magnetic interactions of nanocrystallized Fe-Cr amorphous alloys, *Phys. Rev. B* **56**, 10797-10800.

22. Borrego, J.M., Peña Rodríguez, V.A., and Conde, A. (1997) Mössbauer study of the nanocrystallization of the amorphous system $Fe_{73.5}Si_{13.5}B_9Cu_1Nb_1X_2$ with X = Nb, Mo, V and Zr, *Hyperfine Interactions* **110**, 1-6.
23. Gorría, P., Orúe, I., Plazaola, F., Fernández-Gubieda, M.L., and Barandiarán, J.M. (1993) Magnetic and Mössbauer study of amorphous and nanocrystalline $Fe_{86}Zr_7Cu_1B_6$ alloys, *IEEE Trans. Magn.* **29**, 2682-2684.
24. Orúe, I., Gorría, P., Plazaola, F., Fernández-Gubieda, M.L., and Barandiarán, J.M. (1994) Temperature dependence of the Mössbauer spectra of amorphous and nanocrystallized $Fe_{86}Zr_7Cu_1B_6$, *Hyperfine Interactions* **94**, 2199-2205.
25. Kopcewicz, M., Grabias, A., Nowicki, P., and Williamson, D.L. (1996) Mössbauer and x-ray study of the structure and magnetic properties of amorphous and nanocrystalline $Fe_{81}Zr_7B_{12}$ and $Fe_{79}Zr_7B_{12}Cu_2$ alloys, *J. Appl. Phys.* **79**, 993-1003.
26. Gómez-Polo, C., Holzer, D., Multinger, M., Navarro, I., Agudo, P., Hernando, A. Vázquez, M., Sassik, H., and Grössinger, R. (1996) Giant magnetic hardening of a Fe-Zr-B-Cu amorphous alloy during the first stages of nanocrystallization, *Phys. Rev. B* **53**, 3392-3397.
27. Kemény, T., Balogh, J., Farkas, I., Kaptás, D., Kiss, L.F., Pusztai, T., Tóth, L., and Vincze, I. (1998) Inter-garin coupling in nasnocrystalline soft magnets, *J. Phys.: Condens. Matter* **10**, L221-L227.
28. Miglierini, M. and Grenèche, J.M. (1997) Mössbauer spectrometry of Fe(Cu)MB-type nanocrystalline alloys: II. The topography of hyperfine interactions in Fe(Cu)ZrB alloys, *J. Phys.: Condens. Matter* **9**, 2321-2347.
29. Vincze, I., Kemény, T., Kaptás, D., Kiss, L.F., and Balogh, J. (1998) Nanostructures, disordered ferromagnetism and spin glasses, *Hyperfine Interactions* **113**, 123-134.
30. Makino, A., Inoue, A., and Masumoto, T. (1995) Nanocrystalline soft magnetic Fe-M-B (M=Zr, Hf, Nb) alloys produced by crystallization of amorphous phase, *Mat. Trans. JIM* **36**, 924-938;
Hono, K, Zhang, Y., Inoue, A., and Saturai, T. (1997) APFIM studies on nanocrystallization of amorphous alloys, *Mater. Sci. Eng.* **A226-228**, 498-502.
31. Miglierini, M. and Grenèche, J.M. (1998) Methodology of interfacial regions in FeMCuB-type nanocrystals, *Hyperfine Interactions* **113**, 375-382.
32. Grabias, A. and Kopcewicz, M. (1998) Crystallization of the amorphous $Fe_{81}Zr_7B_{12}$ alloy induced by short time annealing, *Mater. Sci. Forum* **269-272**, 725-730.
33. Kopcewicz, M., Grabias, A., and Nowicki, P. (1997) Comparison of surface and bulk crystallization of amorphous $Fe_{81}Zr_7B_{12}$ and $Fe_{79}Zr_7B_{12}Cu_2$ alloys, *Mater. Sci. Eng. A* **226-228**, 515.
34. Brzózka, K., Ślawska-Waniewska, A., Nowicki, P., and Jezuita, K. (1997) Hyperfien magnetic fields in FeZrB(Cu) alloys, *Mater. Sci. Eng. A* **226-228**, 654-658.
35. Kopcewicz, M., Grabias, A., and Škorvánek, I. (1998) Study of the nanocrystalline $Fe_{73.5}Nb_{4.5}Cr_5Cu_1B_{16}$ alloy by the radio-frequency-Mössbauer technique, *J. Appl. Phys.* **83**, 935-940.
36. Miglierini, M. (1998) Mössbauer spectrometry in FeMCuB-type nanocrystalline alloys: I. The case of M = Mo, *J. Electrical Eng.* **49**, 21-27.
37. Hernando, A. and Kulik, T. (1994) Exchange interactions through amorphous paramagnetic layers in ferromagnetic nanocrystals, *Phys. Rev.* **B49**, 7064-7067.
38. Navarro, I., Ortuño, M., and Hernando, A. (1996) Ferromagnetic interactions in nanostructured systems with two different Curie temperatures, *Phys. Rev.* **B53**, 11656-11660.
39. Ślawska-Waniewska, A. and Grenèche, J.M. (1997) Magnetic properties of interface in soft magnetic nanocrystalline alloys, *Phys. Rev.* **B 56**, R 8491-8494.
40. Grenèche, J.M., Randrianantoandro, N., Ślawska-Waniewska, A., and Miglierini, M., (1998) Magnetic hyperfine properties in FeZrB-type nanocrystalline metallic alloys, *Hyperfine Interactions* **113**, 279-285.
41. Miglierini, M. (1998) Mössbauer spectrometry in FeMCuB-type nanocrystalline alloys: II. The case of M = Nb, and Ti, *J. Electrical Eng.* **49**, 57-63.
42. Miglierini, M. and Grenèche, M. (in press) Hyperfine fields of amorphous residual and interface phases in FeMCuB nanocrystalline alloys: a Mössbauer effect study, *Hyperfine Interactions.*
43. Miglierini, M. and Grenèche, M. (submitted) Temperature dependence of amorphous and interface phases in the $Fe_{80}Nb_7Cu_1B_{12}$ nanocrystalline alloy, *Hyperfine Interactions.*
44. Miglierini, M. and Grenèche, J.M. (1997) Hyperfine interactions in amorphous and nanocrystalline $Fe_{80}Ti_7Cu_1B_{12}$ alloy, *Czech. J. Phys.* **47**, 507-512.
45. Škorvánek, I., Miglierini, M., and Duhaj, P. (1997) Magnetism and Mössbauer spectroscopy in nanocrystalline FeNbCrCuB alloys, *Mater. Sci. Forum* **235-238**, 771-776.

RADIATION DAMAGE OF NANOCRYSTALLINE MATERIALS

J. SITEK AND J. DEGMOVÁ
Department of Nuclear Physics and Technology, Slovak University of Technology, Ilkovičova 3, 812 19 Bratislava, Slovakia

Abstract

Phenomenological description of mechanism of radiation damage of nanocrystalline materials involves elastic and non-elastic interaction of neutrons with atoms of constituent elements. As a consequence, the atom displacement and formation of new isotopes occurs. Transmission Mössbauer spectroscopy was used to study changes induced by irradiation of amorphous and nanocrystalline Fe-based metallic alloy Fe-Nb-Cu-Si-B and Fe-Zr-B, respectively. In as-cast sample, neutrons weakly modified their structure. In the case of nanocrystalline sample, crystalline component was amorphised and disordered structure of the amorphous rest was modified. Structural changes depend on the state of crystallisation before irradiation and on the total neutron fluence. Rearrangement of atoms due to irradiation caused changes of volumetric ratio of crystalline and amorphous part and average value of the hyperfine field and orientation of the net magnetic moment.

1. Introduction

In the recent years, iron-based alloys were developed with an ultrafine grain structure embedded in an amorphous matrix revealing excellent soft magnetic properties [1]. The most prominent examples are nanocrystalline Fe-Cu-Nb-Si-B and Fe-Zr-B alloys, respectively [2, 3]. Several recent investigations have dealt with the influence of the radiation damage on magnetic properties of metallic glasses [4, 5]. Exchange interactions and the Curie temperature, hyperfine field distribution and parameters of magnetisation processes are governed by changes in a short-range order due to neutron irradiation [6, 7]. The present study was undertaken in order to investigate the influence of neutron irradiation on structural and magnetic properties of nanocrystalline alloys. Phenomenological description of mechanism of radiation damage of nanocrystalline materials involves elastic and non-elastic interaction of neutrons with elements of the alloys. Using transmission ^{57}Fe Mössbauer spectroscopy, we focused our interest on the changes in the orientation of net magnetic moment, in the magnetic hyperfine field and volumetric fraction of crystalline, amorphous, and interface components.

M. Miglierini and D. Petridis (eds.), Mössbauer Spectroscopy in Materials Science, 273–282.

2. Radiation Damage by Neutrons

In the irradiation of amorphous metallic alloys by neutrons, interaction of neutrons with the atoms of constituent elements of the alloy leads to rearrangement of the amorphous structure [8]. Nanocrystalline alloys consist of crystalline component in the nanoscale, embedded in an amorphous matrix. Irradiation of such alloys by neutrons leads to redistribution of the atoms in the amorphous matrix, to disturbance of regular atomic ordering of the crystal lattice and to atom exchange between amorphous and crystalline component [9]. These processes are accompanied by damage of the material structure in two types:

(i) displacement damage-production of atoms shifted out of their original position. In this case, energy is transferred to the nucleus by collision of the neutron with the atomic nucleus, which energy is greater than the threshold energy, E_d.

(ii) capture of neutron in the atomic nucleus with a consequent decay producing particles or photons. The recoil connected with the emission of the decay products can also lead to atom displacement.

Elastic collisions of neutrons with the atomic nucleus produce the primary-knock on atoms (PKA), which, together with the recoil atoms, displaces atoms on their path through the material until their kinetic energy is exhausted. In the first stage of the collision, cascade chains of isolated interstitial atoms and vacancies are formed along the PKA trajectory in crystalline part and redistribution of atoms in the amorphous part occurs (Figure 1). This first stage involves an instantaneous and spontaneous recombination of closed pairs. In the final part of the PKA pathway, mean distance between the collisions becomes comparable with the interatomic distance. As soon as the transferred energy between atoms during collision decreases below E_d, it is dissipated in the form of increased thermal vibrations. In this moment, a considerable amount of energy is concentrated in the very short time (10^{-3} s) in a very small volume (2-5 nm).

The local temperature can reach the melting point in a very short time period. This leads to a collective atomic rearrangement in this region and to formation of random and irregular clusters of vacancies. This disturbed region is called a depleted zone (Figure 1). Relocation occurs at random in the amorphous as well as in crystalline parts. Displacement of this damage zone can achieve a value of approximately 20 nm. The thermal energy is dispersed to the surrounding crystalline lattice and the amorphous matrix. This developed phase leads to the thermal equilibrium and creation of defect distribution in the damage region.

The mean value of number PKA events could be calculated from the cross-section of elastic collisions, and from the neutrons energy spectrum.

Mean energy of PKA, E_a, obtained from neutrons with energy E_n is given by the simplified relationship [10].

$$E_a = \frac{4}{M} E_n \tag{1}$$

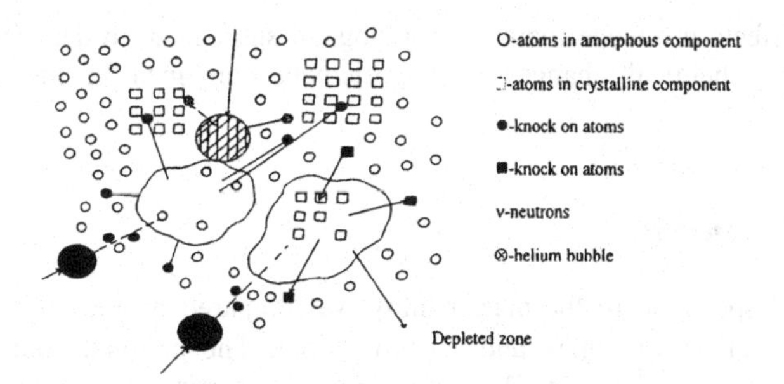

Figure 1. Schematic representation of the radiation damage.

Where M is the mass of the atoms.

The number of pairs $n(E_a)$, created by PKA is given according to Kinchin and Pease model [11] by the relationship:

$$n(E) = \frac{E_a}{2E_d} \tag{2}$$

From (2), one can derive

$$n(E) = \begin{cases} 0 & E<E_T \\ 1 & E_d<E<2E_d \\ E_a/2E_d & 2E_d=<E_a \end{cases} \tag{3}$$

The mean value of E_d is approximately 30 eV for most metallic materials [10]. Using (2), the minimum energy is 60 eV for creation of knockout atoms. Using relationship (1) for most frequently used constituent element iron (M=56), the energy of neutrons for PKA process is about 700 eV.

Non-elastic interactions of neutrons with nuclei (i.e. captures) lead to formation of new isotopes. Neutron capture can occur with the thermal neutrons. Atoms of the inert gases are most important, especially of He. The most frequent is the capture of thermal neutron by ^{10}B, which produces high energy α and Li particles. Especially when higher neutron fluencies are used, the α particles agglomerate and form the helium bubbles. Recoil atoms as decay products also participate on the atom displacement. The recoil atoms have as a rule higher energy than PKA. The contribution of the recoil atoms to the radiation damage increases with increasing energy of the incident neutrons.

As mentioned earlier, mechanism of radiation damage by neutron of nanocrystalline materials is dependent on the constituent elements. Each of them has different cross-section to the thermal and fast neutrons. From the point of view of interaction of thermal neutrons, boron has about 4000 times higher cross-section than all other elements. This means that boron will mainly contribute to radiation damage due to decay products and other elements by the primary knock on atoms.

Summarising the above facts we can say that during neutron irradiation, the radiation damage originates from two different sources of atom displacements.

(i) displacements of recoil atoms caused by high energy α (1,4 MeV) and Li particles (0,9 MeV) as products of nuclear reaction $^{10}B(n,\alpha)^{7}Li$+2,3 MeV

(ii) displacement of atoms caused by fast neutrons.

Both of these effects influence mainly a topological short-range order. The effects of radiation-induced chemical changes of the element contained in the samples are negligible.

3. Experimental Details

A ribbon-shaped specimen of the master alloy was prepared by the method of planar flow casting (about 25 μm thick and 10 mm wide). The nominal composition was $Fe_{73,5}Cu_1Nb_3Si_{13,5}B_9$ and $Fe_{87,5}Zr_{6,5}B_6$. Annealing was carried out in vacuum at the temperature 550°C for 1 hour and 8 hours. Neutron irradiation was carried out in a nuclear pile by neutrons with the total fluencies of 10^{16} up to 10^{19} n/cm^2. Mössbauer spectrometer with a ^{57}Co(Rh) source was used in transmission geometry. Mössbauer spectra were evaluated by the NORMOS [12] and MOSFIT [13] programs, which allow simultaneous treatment of crystalline, residual amorphous matrix and interface by means of single individual lines and a distribution of hyperfine components.

4. Results and Discussion

4.1 AS-CAST ALLOYS

Mössbauer spectra corresponding to the FeNbCuSiB sample taken in as-cast state before and after irradiation with the fluence of 10^{19}n/cm^2 are shown in Figure 2a,b respectively.

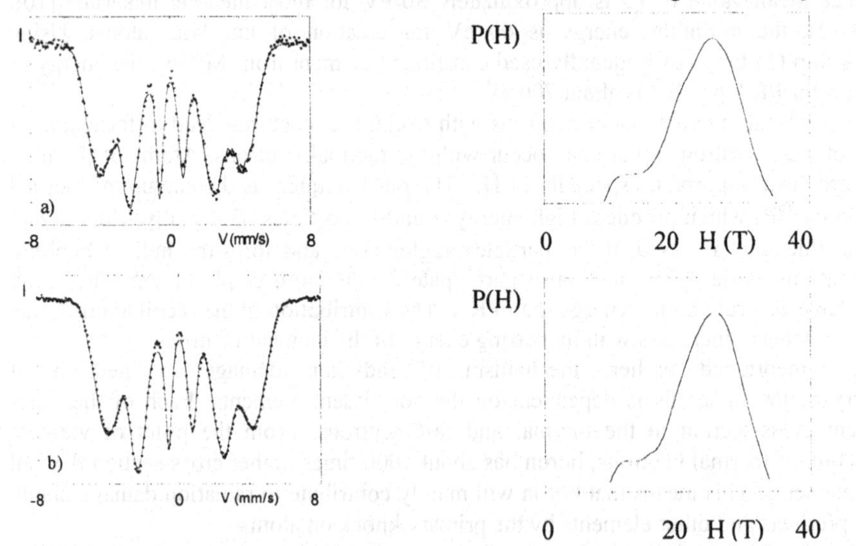

Figure 2. Mössbauer spectra of the FeNbCuSiB sample in the as-cast state before (a) and after (b) irradiation with the fluence of 10^{19} n/cm^2.

Effects of neutron irradiation were studied through changes in the hyperfine field distribution, P(H), and through parameter D_{23} that represents the ratio of the 2nd and 3rd

line areas in the ferromagnetic sextet. D_{23} can be used to determine the relative orientation between the gamma rays and the net magnetic moment.

TABLE 1. Mössbauer parameter of the as cast alloys (<H>-average hyperfine field)

Sample	Parameter	Fluence	
		(10^0 n/cm^2)	(10^{19} n/cm^2)
FeNbCuSiB	D_{23}	2,34	2,62
	<H> (T)	23,38	23,71
FeZrB	D_{23}	2,81	0,77
	<H> (T)	23,40	22,68

An increase of line intensity after irradiation was observed (TABLE 1). This indicates that the net magnetic moment turns into the ribbon plane. The observed changes are supposed to be a consequence of a reorientation of spins in a vicinity of stress centres as a result of atoms mixing after irradiation. Another effect which should be taken into consideration is the surface crystallisation which causes compressive stress in the bulk of the specimen acting in the ribbon plane due to higher density of the crystalline phases with respect to the amorphous ones. The increase of the mean value of the hyperfine field was also observed towards higher neutron fluencies (TABLE 1). After fitting the P(H) distribution with three Gaussians, we found that Gaussian corresponding to the highest field increases with increasing neutron fluencies, relative to the lower one. This phenomenon reflects in a higher value of the average magnetic field. This could be a consequence of the surface crystallisation.

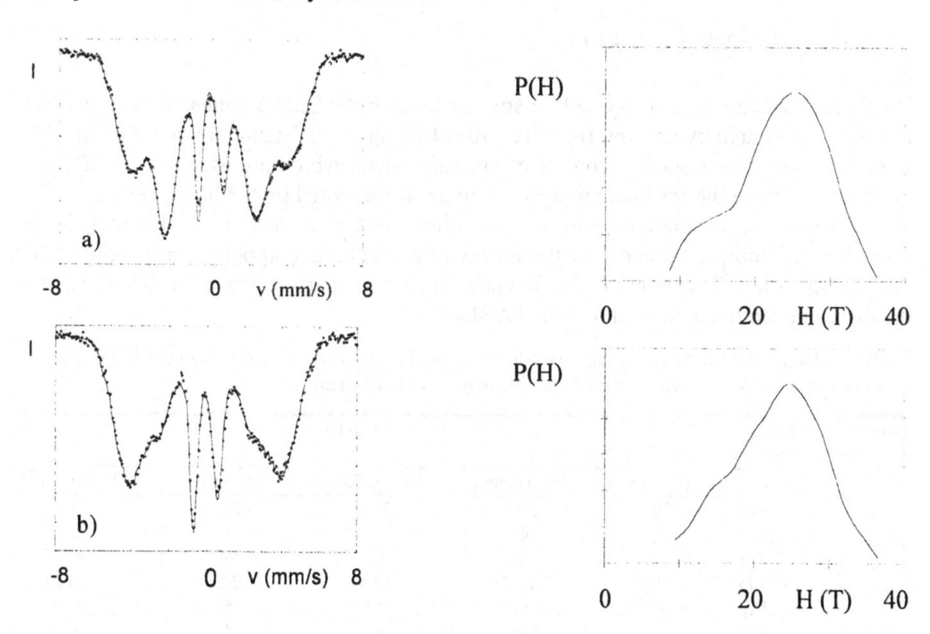

Figure 3. Mössbauer spectra of the as-cast FeZrB alloy before (a) and after (b) irradiation with the fluence of 10^{19}n/cm^2.

Mössbauer spectra of as-cast FeZrB alloy are depicted in Figure 3a, b. The distribution of the hyperfine field P(H) shows a bimodal shape. The P(H) curves obtained were separated into two components, namely the low field and the high field parts. Bimodal structure of the hyperfine field distribution can be attributed to local environment in

(i) Fe-poor regions (high-field part) in which Fe alloys are partially co-ordinated with Zr and B.

(ii) Fe-rich regions (low-field part) in which Fe atoms are mainly surrounded by other Fe atoms with a certain distribution of the Fe-Fe nearest neighbours' distances.

As was pointed out in papers [14, 15], FeZrB amorphous materials exhibit the exchange magnetic frustration. The substitution of Fe atoms by the non-magnetic element (B) reduces the number of antiferromagnetically coupled Fe-Fe pairs and, consequently, the degree of the magnetic frustration. On irradiation by neutrons due to elastic scattering the primary and secondary atom displacements are created. Distribution of energy to the neighbouring atoms leads to structural changes. From the point of view of radiation damage, the most important element is boron, as mentioned above. During neutron irradiation, a part of boron is reduced from the Fe-B co-ordination. Consequently, more free Fe atoms are created, contributing to Fe-Fe co-ordination. This phenomenon causes an increase of low field component in P(H) distribution, decrease of the mean value of internal magnetic field (TABLE 1) and a slight decrease of the second and the fifth lines intensity of Mössbauer spectrum implies a tendency of the net magnetic moment to turn out of the ribbon plane after neutron irradiation, which means that internal stress was removed by irradiation.

Summarising the radiation damage of both samples, the mean short-range order around iron sites in as-cast ribbons is modified with neutron irradiation.

4.2 NANOCRYSTALLINE ALLOYS

Mössbauer spectra of nanocrystalline specimens of FeNbCuSiB annealed during 1 hour at 550°C, irradiated by neutron fluencies from 10^{16} up to 10^{19} n/cm^2 are depicted in Figure 4. Mössbauer spectra consist of six subsextets, which are ascribed to DO_3-FeSi alloy as well as to the residual amorphous phase, represented by the broad sextet.

During the irradiation, part of crystalline phase is destroyed. Fe and Si are dispersed in amorphous part and the atoms in amorphous component are redistributed due to irradiation. The amount of volumetric crystalline part decreases at expense of that residual amorphous part, as shown in TABLE 2.

TABLE 2. Mössbauer parameters of nanocrystalline FeNbCuSiB alloy (A1-annealed 1 hour at 550°C. A8-annealed 8 hours at 550°C, <H>-average hyperfine field, A-Relative area)

Sample	Parameter	Fluence				
		10^0 (n/cm^2)	10^{16} (n/cm^2)	10^{17} (n/cm^2)	10^{18} (n/cm^2)	10^{19} (n/cm^2)
A1	<H> (T)	20,3	20,6	21,3	21,0	21,5
	A_{cr} (%)	45	45	40	40	35
	A_{am} (%)	55	55	60	60	65
A8	<H> (T)	19,6	19,5	19,4	20,1	20,2
	A_{cr} (%)	61	61	53	51	50
	A_{am} (%)	39	39	47	49	50

Crystalline phase contain vacancies or interstitial atoms from other constituent element. Their presence reduces the values of internal magnetic fields in all subsextets. Comparing

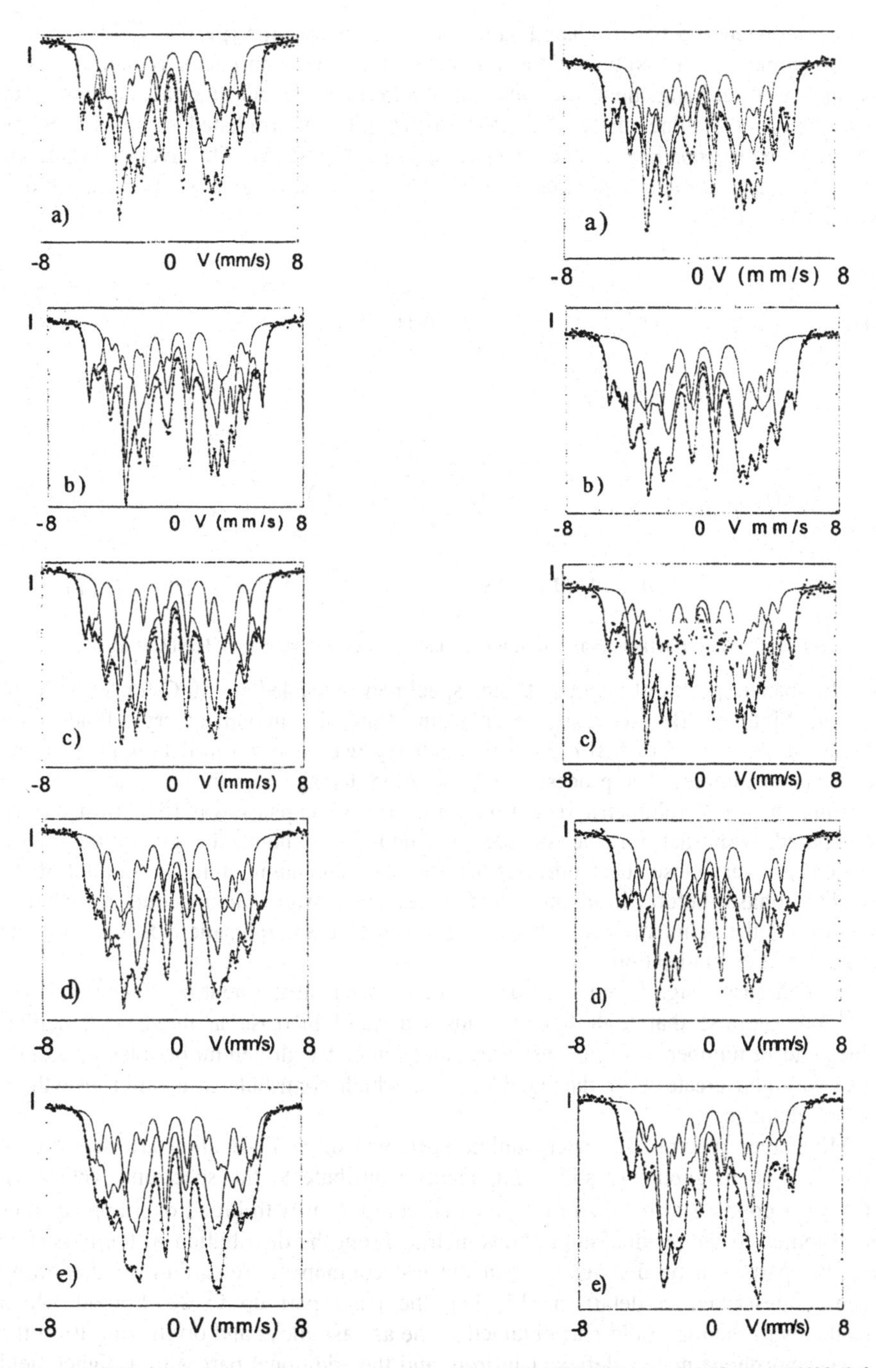

Figure 4. Mössbauer spectra of nanocrystalline specimens of FeNbCuSiB annealed 1 hour at 550°C (left column) and 8 hours at 550°C (right column) (a) non-irradiated samples (b) samples irradiated by neutrons with fluence of 10^{16} n/cm^2 (c) 10^{17} n/cm^2 (d) 10^{18} n/cm^2 (e) 10^{19} n/cm^2.

the non-irradiated and the irradiated samples, the decrease of hyperfine field is in the range of about 0,5 T. New regions are created in the amorphous remainder as a consequence of rearrangement of atoms and of additional Fe and Si atoms coming from the destroyed crystalline phase. The P(H) distribution of irradiated samples exhibits many peaks corresponding to the created regions (Figure 5). The average value of hyperfine magnetic field increases which reflects the new state of the amorphous remainder.

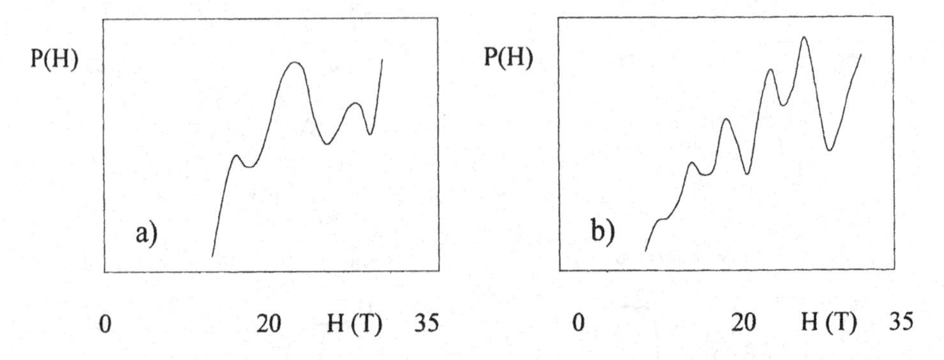

Figure 5. P(H) distribution of (a) non-irradiated and (b) irradiated samples of FeNbCuSiB alloy.

Mössbauer spectra of nanocrystalline specimens annealed at 550°C for 8 hours are shown in. Figure 4. By increasing of annealing time, the amount of crystalline phase increases at about 16% (TABLE 2). This tendency was reported in details [16] on the non-irradiated samples. The process of the radiation damage runs in the same way as in the former case. The volumetric reduction of the crystalline phase was 10%. If this value is compared with that for the sample annealed for 1 hour, the crystalline phase reductions are very close. This indicates that the alloy containing a higher amount of the crystalline phase before irradiation lost a relative lower amount due to neutron irradiation at the same fluence. This also implies that amorphisation depends on the initial stage of crystallisation.

In both cases, significant structural changes occur at the neutron fluence of 10^{19} n/cm^2. We suppose that high fluence causes a rapid increase in number of helium bubbles and in number of vacancies. Vacancies interact with helium bubbles in a more diffuse way and create large damaged regions, which contribute to a rapid growth of structural damage.

Mössbauer spectra of nanocrystalline specimen of Fe-Zr-B alloy are depicted in Figure 6. Two magnetically split components contribute to the spectrum. The sharp sextet corresponding to α-Fe and the smeared component with broad overlapping lines was attributed to the residual amorphous matrix. From the distribution of the hyperfine field, two parts can be distinguished in the last component. According to the fitting procedure described in detail in [17, 18], the main part up to 20 T field which corresponds to the high field part obtained in the as-cast alloys and originating from the residual amorphous matrix deficient in iron, and the additional part with a higher field than 20 T. They originate from Fe atoms placed at the crystalline-amorphous interface. Nevertheless, it is important to emphasise that in the case of these nanocrystalline alloys, an accurate estimate of both the crystalline fraction, the (mean) values of the hyperfine parameters characteristic of the different phases, and the mean orientation of iron magnetic moments can not be obtained from a single Mössbauer spectrum, but from a

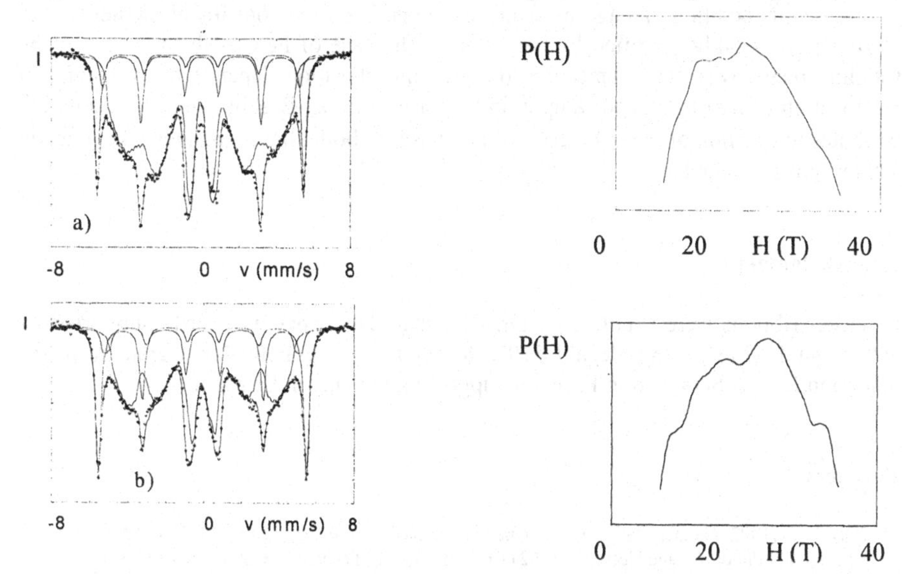

Figure 6. Mössbauer spectra of (a) as cast and (b) nanocrystalline specimen of FeZrB alloy recorded at 77K.

series of spectra recorded at various temperatures. According to a complete study which is described in detail elsewhere [19], one can conclude that this interface has a tendency to increase with increasing of the crystalline component. Main relevant Mössbauer parameters of non-irradiated and irradiated samples are shown in TABLE 3.

TABLE 3. Mössbauer parameters of the alloy $Fe_{87.5}Zr_{6.5}B_6$ after annealing at 550°C for 1 hour S_0 and after neutron irradiation S_{19} (H-hyperfine field, A-relative area, c-crystalline part, a-amorphous part, i-interface)

Sample	Hc (T)	Ac (%)	Hi (T)	Ai (%)	Ha (T)	Aa (%)
S_0	34,2	19	32,5	5	23,8	76
S_{19}	34,1	24	31,6	10	19,1	66

After neutron irradiation, we observed a slight increase of crystalline fraction and interface. We supposed that process of iron separation runs in two ways. The residual amorphous component contains sufficient amount of Fe-B pairs. The reaction of neutron capture by boron produced more free iron atoms contributing to the crystalline component. We can not exclude that a part of the crystalline iron was damaged, however, the process of growth of the crystalline iron due to irradiation is prevailing. The fluence of 10^{19} n/cm^2 is still not sufficiently high to amorphise the complete crystalline structure. The value of hyperfine field of the crystalline fraction decreased slightly, which indicates a partial damage of the crystalline structure of α-Fe. The average hyperfine field of interface decreased, which corresponds to its growth and confirms that iron was co-ordinated by a similar mechanism in the amorphous matrix. The average hyperfine field of the amorphous remainder decreased significantly which indicates that the part of boron is reduced from Fe-B coordination and free iron atoms contribute to the crystalline fraction and interface.

5. Conclusion

The process of radiation damage by neutrons depends on the constituent elements. As-cast alloys are weakly modified by neutrons. In the case of Fe-Cu-Nb-Si-B alloys, the crystalline phase was partly amorphised. The amorphisation depends on the stage of crystallisation of non-irradiated sample. In the case of Fe-Zr-B alloy, neutron irradiation contributes to creation of the α-Fe crystalline phase. In both alloys, changes take place in the amorphous remainder.

Acknowledgement

The master ribbons were supplied by Dr. P. Duhaj. This work was partly supported by grants Vega 1/4286/97 and Vega 1/5103/98 and J. D. is grateful for a grant from the Foundation Robert Schuman in Paris for supporting her stay in Le Mans.

References

1. Herzer, G. (1992) Nanocrystalline soft magnetic materials, *J. Mag. Mag. Mat.* **112**, 258-262.
2. Hampel, H., Pundt, A., and Hesse, J. (1992) Crystallisation of FeCuNbSiB: structure and kinetics examinated by x-ray diffraction and Mössbauer effect spectroscopy, *J. Phys.: Condens. Matter* **4**, 3195-3214.
3. Navarro, I., Hernando, A., Vazquez, M., and Seong-Cho, Y. (1995) Mössbauer spectroscopy in nanocrystalline $Fe_{80}Zr_7B_4Cu_1$, *J. Mag. Mag. Mat.* **145**, 313-318.
4. Shimansky, F. P., Gerling, R., and Wagner, R. (1988) Irradiation -induced defects in amorphous $Fe_{40}N_{40}P_{20}$ 97, *Mat. Sc. Eng.***97**, 173-176.
5. Miglierini, M. and Sitek, J. (1990) Neutron irradiation of metallic glasses and Mössbauer spectroscopy, *Eng. Mat.* **40&41**, 281-285.
6. Miglierini, M., Nasu, S., and Sitek, J. (1992) Influence of neutron irradiation on ferromagnetic metallic glasses, *Hyperfine Interaction* **70**, 885-888.
7. Mihálik, M., Zentko, A., and Macko, L. (1991) Radiation-induced changes of the Curie temperature of Fe-Ni-Cr-Mo-Si-B glass metals, *Acta Phys. Slov.* **40**, 315-322.
8. Miglierini, M., Škorvánek, I., Nasu, S., and Sitek, J. (1992) Neutron irradiation effects on magnetic properties of Fe-based ferromagnetic metallic glasses, *Materials Transaction, JIM* **33**, 327-336.
9. Sitek, J., Seberini, M., Lipka, J., Tóth, I., and Degmová, J. (1997) *Rapidly Quenched and Metastable Materials*, pp. 179-182, Elvice Science Publishers, Amsterdam.
10. Bečvář, J. (1981) *Jaderne elektrárny*, SNTL Alfa, Prague.
11. Kautský, J. and Kočík, J. (1994) *Radiation Damage of Structural Materials*, Academia, Prague.
12. Brand, R. A. (1989) NORMOS program 1989 version, unpublished.
13. Teillet, J. and Varret, F. MOSFIT program unpublished.
14. Brzozka, K., Slawska-Wanievska, A., Nowicki, P., and Jezuita, K. (1997) Hyperfine magnetic field in FeZrB(Cu) alloys, *Mat. Sc. Eng.* **A226-228**, 654-658.
15. Slawska-Wanievska, A. (1998) Interface magnetism in Fe-based nanocrystalline alloys, *J. of Phys.* **8**, 11-18.
16. Miglierini, M. (1994) Mössbauer effect study of the hyperfine field distribution in the residual amorphous phase of Fe-Cu-Nb-Si-B nanocrystalline alloys, *J. Phys. Condens. Matter.* **6**, 1431-1438.
17. Miglierini, M. and Greneche, J. M. (1997) Mössbauer spectrometry of Fe(Cu)MB-type nanocrystalline alloys: I. The fitting model for the Mössbauer spectra, *J. Phys.: Condens. Matter* **9**, 2303-2319.
18. Miglierini, M. and Greneche, J. M. (1997) Mössbauer spectrometry of Fe(Cu)MB-type nanocrystalline alloys: II. The topografy of hyperfine interactions in Fe(Cu)ZrB alloys, *J. Phys.: Condens. Matter* **9**, 2321-2347.
19. Degmová, J., Sitek, J.,and Greneche, J. M., to be published.

DISORDERED NANOCRYSTALLINE Fe-Sn ALLOYS

^{57}Fe and ^{119}Sn Mössbauer Spectroscopy Study

E.P. YELSUKOV[1], E.V. VORONINA[1], G.N. KONYGIN[1],
S.K. GODOVIKOV[2] AND V.M. FOMIN[1]
[1)] Physical-Technical Institute of UrB RAS, 132, Kirov Str., Izhevsk, 426001, Russia
[2)] Moscow State University, Institute of Nuclear Physics, Moscow, 119899, Russia

1. Introduction

The disordered crystalline Fe-Sn alloys are very suitable model objects for studying the mechanisms responsible for the magnetic properties of disordered metal-metalloid systems. A lot of experimental data on the disordered crystalline and amorphous Fe alloys with C, Al, Si, P and Sn allow us to clarify such points as the effect of the type and concentration of the s-p element, topological and chemical disorder, local atomic structure. The nanocrystalline Fe-Sn alloys disordered by mechanical grinding are of special interest for Mössbauer spectroscopy because of the possibility of analysis of the local atomic environment of both resonant nuclei: ^{57}Fe and ^{119}Sn. Earlier, the concentration dependences of the magnetic properties of microcrystalline and amorphous Fe-Al, Fe-Si, Fe-P alloys were explained in terms of the local atomic environment parameters [1-3]. To give a consistent phenomenological description of the hyperfine interaction parameters in terms of the local atomic structure for a series of non-ordered Fe-Sn alloys is the object of this work.

2. Experimental

The nanocrystalline samples $Fe_{1-x}Sn_x$ ($3.2 \le x \le 50$ at. % of Sn) were obtained by mechanical activation in a planetary ball mill. The grain sizes of the powders were within 4-10 nm, while the particle size was 4μm. All the samples were monophase crystalline. The conditions of preparation, testing by X-ray diffraction, magnetic measurements are described in detail in [4].

The Mössbauer spectra (MS) of ^{57}Fe and ^{119}Sn nuclei were taken at 77 K and 14 K using ^{57}Co(Cr) and ^{119}Sn ($CaSnO_3$) sources. The spectra were processed using the regular Tikhonov's algorithm and Levenberg-Marquardt least-squares method taking into account the nearest neighbourhood.

M. Miglierini and D. Petridis (eds.), Mössbauer Spectroscopy in Materials Science, 283–290.

3. Results and Discussion

The ^{57}Fe and ^{119}Sn MS are shown in Fig.1a,b. The shapes of the spectra and P(H) functions are characteristic for the alloys with chemical disorder. Both MS and P(H) functions are seen to broaden uniformly as the Sn concentration increases. It should be specially noted, that there is a similarity of P(H) functions of the ground alloys obtained in this work to those of the thin amorphous Fe-Sn films of close concentrations. Besides, in the P(H) functions we did not find any pronounced components of the equilibrium stoichiometric phases Fe_3Sn, Fe_5Sn_3, Fe_3Sn_2, FeSn and $FeSn_2$ with the well known Mössbauer parameters [5-10]. The results of the processing are the average hyperfine magnetic field at ^{57}Fe $\overline{\mathbf{H}}_{\mathbf{Fe}}(\mathbf{x})$ and at ^{119}Sn $\overline{\mathbf{H}}_{\mathbf{Sn}}(\mathbf{x})$ and the local hyperfine magnetic field, $\mathbf{H}_k^{\mathbf{Fe}}(\mathbf{x})$ presented in Figs. 2a,b and Fig.3, respectively. The results confirmed the fact, mentioned earlier by Trumpy et al. [10], that at small Sn concentrations $\mathbf{H}_k^{\mathbf{Fe}}(\mathbf{x})$, k = 0 increases from H = 337 to 354 kOe. We consider this to be the effect of the coordination spheres more distant than the first one. The increase in the isomer shift of the component corresponding to Fe nuclei without Sn neighbours, $\delta_0^{Fe}(x)$, (Fig. 4a) along with the rise of the ^{119}Sn average isomer shift, $\delta^{Sn}(x)$, confirm this interpretation.

The $\mathbf{H}_k^{\mathbf{Fe}}$ values averaged over the concentration range x ≥ 10 at. % of Sn (Fig.5a) are seen to agree well with the data known for solid solutions [10] and intermetallic compounds [5, 6, 8]. The average concentration value $\Delta H = \mathbf{H}_{k+1}^{\mathbf{Fe}} - \mathbf{H}_k^{\mathbf{Fe}}$ =

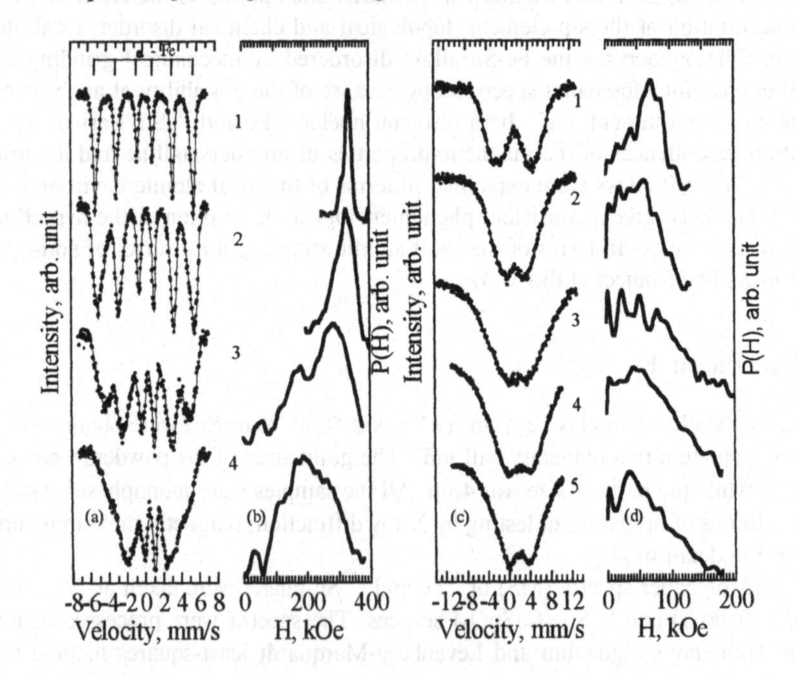

Figure 1. The Mössbauer spectra and the distribution functions of hyperfine magnetic field P(H) of $Fe_{100-x}Sn_x$ alloys at ^{57}Fe (a, b), x = 3.2(1), 14.6(2), 32.7(3), 42.0(4) at.% of Sn and at ^{119}Sn (c, d), x = 7.6(1), 14.6(2), 36.4(3), 46(4), 49.5(5) at. % of Sn.

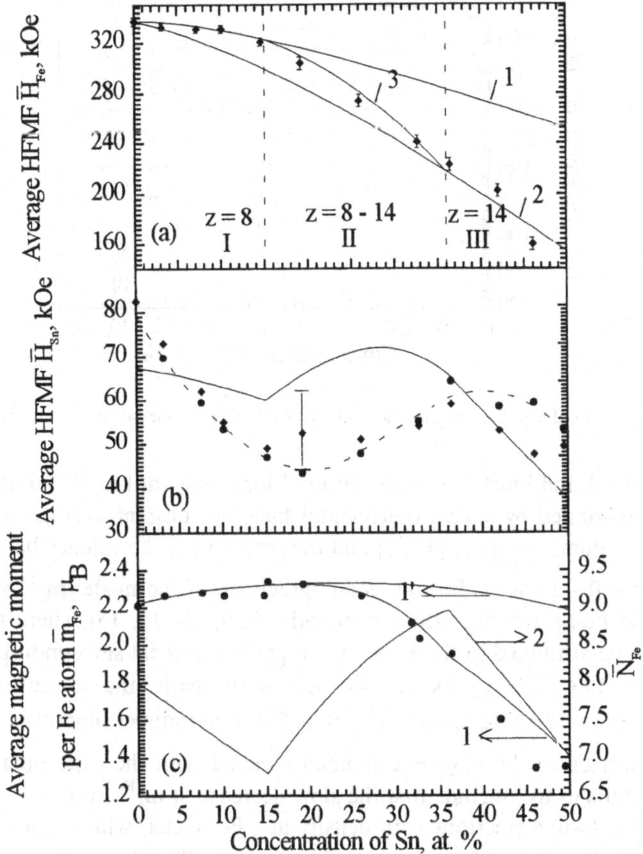

Figure 2. The concentration dependences: (a) - average hyperfine magnetic field $\overline{\mathbf{H}}_{\mathbf{Fe}}$ at ^{57}Fe nuclei, ♦ - this work, solid lines - the calculation according to (1) for z = 8 -curve 1, z = 14 -curve 2 and assuming a random distribution of coordination numbers in range from 8 to 14 and radom atomic distribution for a given z (15-36 at. % of Sn) -curve 3; (b) - average hyperfine magnetic field $\overline{\mathbf{H}}_{\mathbf{Sn}}$ at ^{119}Sn nuclei, ● -14 K, ♦ -77 K, dashed line - the polynomial approximation of the experimental data, solid lines - the calculation according to (5); (c) - average magnetic $\overline{m}_{Fe}$ per Fe atom for z=8 - curve 1, for z = 8 (0÷15at. % of Sn), z = 8÷14 (15÷36 at. % of Sn) and z = 14 (36÷50 at. % of Sn) - curve 1', aveage number of Fe neighbours of Sn atom $\overline{N}_{Fe}$ - curve 2.

24 kOe (k = 0 - 3) coincides with the magnitude of the decrease in the local hyperfine magnetic field $\mathbf{H}_k^{Fe}$ upon substitution of a Sn atom for one of the Fe atoms in the Fe nearest neighbourhood [11]. The $\mathbf{H}_k^{Fe}$ dependence can be approximated by two linear parts with an inflection point at k = 4. From the extrapolation $\mathbf{H}_k^{Fe}$ toward higher *k* we estimate the number of the nearest Sn atoms in the Fe atom neihbourhood, at which $\mathbf{H}_k^{Fe}$ vanishes. It is close to 11.

Starting from the local hyperfine fields $\mathbf{H}_k^{Fe}$, a model of the local magnetic moments $\mathbf{m}_k^{Fe}$ was proposed [12] (Fig. 5b). Its main features are: the unchanged values of

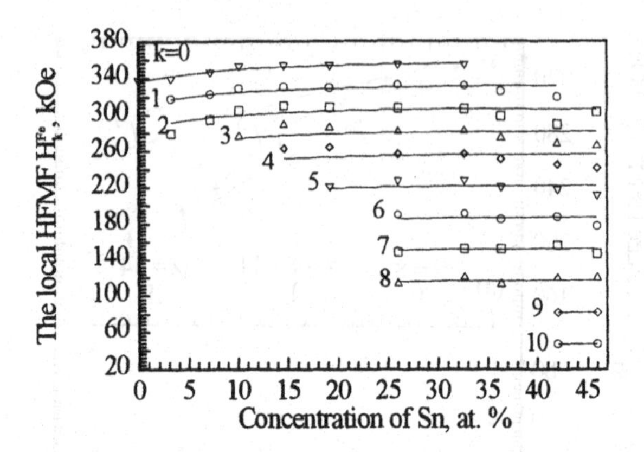

Figure 3. The local hyperfine magnetic field H_k at ^{57}Fe. Solid line - the calculation according to (1).

$\mathbf{m_k}^{Fe}$ for $k = 0$ - 4 and linearly decreasing $\mathbf{m_k}^{Fe}$ for $k > 4$ until $\mathbf{m_k}^{Fe} = 0$ at $k = 11$. The model is corroborated by such experimental facts as constant average magnetic moment of the Fe atom, $\overline{\mathbf{m}}_{\mathbf{Fe}}(\mathbf{x})$, [4, 12] and the structure of the valence band [13] in the alloys with x = 0 - 25 at. % Sn; the good agreement of the model $\mathbf{m_k}^{Fe}$ with the data known for the Fe-Sn intermetallic compounds [5, 9, 10, 14, 15] where the magnetic moments can be attributed to the Fe atoms in particular local surroundings. The value $\mathbf{H_k}^{Fe}/\mathbf{m_k}^{Fe}$ = 113 kOe/μ_B (k ≥ 4) according with the experimental value $\overline{\mathbf{H}}_{\mathbf{Fe}}(\mathbf{x})/\overline{\mathbf{m}}_{\mathbf{Fe}}(\mathbf{x})$ = 120 kOe/μ_B (x = 25 - 50 at.%) is an indirect support of the model.

A comparison of the magnetic moments model with the experimental $\mathbf{H_k}^{Fe}$ dependence permits us to conclude that the slow decrease of $\mathbf{m_k}^{Fe}$ for k = 0 - 4 is due to a decrease in the 4s-like electrons spin density at ^{57}Fe nuclei, which conforms with the concentration dependence of the isomer shifts $\overline{\delta}_{Fe}(\mathbf{x})$ (Fig. 4b).

From a discrete analysis of the ^{57}Fe spectra we obtained the probabilities $\mathbf{P_k}$ of various local surroundings of the Fe atom. The $\mathbf{P_k}$ values for the alloys in the concentration interval x < 15 - 20 at. % of Sn are in agreement with the random atomic distribution in *bcc* lattice with coordination number z = 8. For the alloys with x > 32 at.

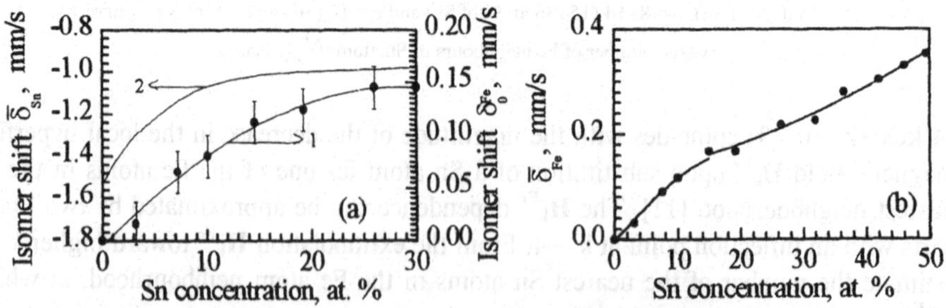

Figure 4. (a) - Concentration dependences of isomer shift $\delta_0^{Fe}(x)$ at ^{57}Fe nuclei for Fe atoms having no Sn atoms in the nearest neighbourhood (curve 1) and average isomer shift at ^{119}Sn $\overline{\delta}_{Sn}(\mathbf{x})$ (curve 2); (b) - the average isomer shift at ^{57}Fe $\overline{\delta}_{Fe}(\mathbf{x})$.

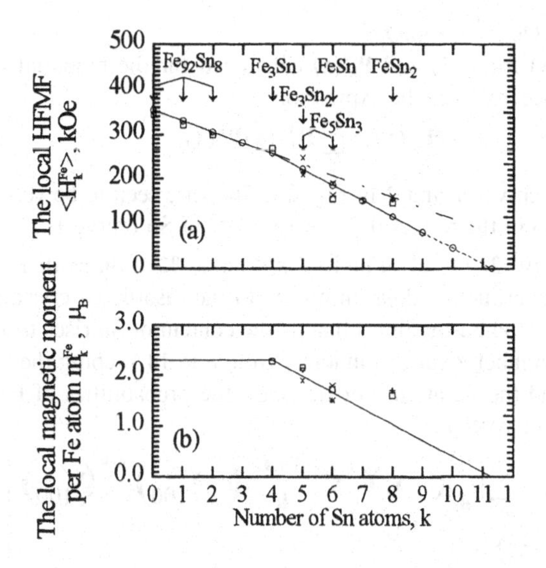

Figure 5. (a) - The local hyperfine field averaged over Sn concentration $<\mathbf{H}_k^{Fe}>$ (T_{meas} = 77 K)in comparison with published data, × -[5], Δ -[6],], • -[8], + -[9], □ -[10], * -[14], , ∇ -[15]; (b) - model for local magnetic moments $\mathbf{m}_k^{Fe}$ at Fe atoms.

% of Sn the experimental $\mathbf{P}_k$ agree with the $\mathbf{P}_k^z$ calculated for the lattice with coordination number z = 14. As it was shown earlier, in the high - concentration alloys with the Sn content 32 < x < 50 at. % the hexagonal structures with coordination numbers z = 10 - 15 are dominant for both intermetallic compounds [7, 10] and metastable alloys [4]. Thus, it is reasonable to assume that the atomic arrangement with z > 8 forms in the disordered $Fe_{1-x}Sn_x$ alloys with x > 32 at. %.

To describe the concentration dependence $\overline{\mathbf{H}}_{Fe}(x)$ for the disordered crystalline Fe-Sn alloys the following expression for the local hyperfine magnetic fields $\mathbf{H}_k^{Fe}$ was written:

$$\left.\begin{aligned} &\mathbf{H}_k^{Fe}(x) = \mathbf{H}_{Fe}\left\{1+\alpha\left[1-\exp\left(-x/x_s\right)\right]\right\} - kh^1, && k < k' \\ &\mathbf{H}_k^{Fe}(x) = \mathbf{H}_{Fe}\left\{1+\alpha\left[1-\exp\left(-x/x_s\right)\right]\right\} - kh^1 - (k-k')h^2, && k' < k \le k_{cr} \\ &\mathbf{H}_k^{Fe}(x) = 0, && k > k_{cr} \end{aligned}\right\} \quad (1)$$

where $\mathbf{H}_{Fe}$ is the hyperfine field at Fe nuclei in α-Fe, $\mathbf{H}_{Fe}$ = 337 kOe at 77K; k' = 4; k_{cr} = 11; h^1 is the contribution to the ^{57}Fe hyperfine magnetic field per one Sn atom in the nearest neighbourhood of the Fe atom, originating from the changes of the 4s-electron spin density at the ^{57}Fe nuclei; h^2 is the contribution to the hyperfine field arising from the changes in the magnetic moment of the resonance atom; α - the scale factor. The best fitting of the experimental values of $\mathbf{H}_k^{Fe}(x)$ is reached with α = 0.06, x_s = 10 at.

% of Sn, h^1 = 25 kOe, h^2 = 35 kOe.

Using $\mathbf{H}_k^{Fe}(x)$ from (1) and $\mathbf{P}_k^z(x)$ calculated for the binomial distribution with z = 8 and z = 14 according to the expression:

$$\overline{\mathbf{H}}_{Fe}(x) = \sum_k \mathbf{H}_k^{Fe}(x)\mathbf{P}_k^z(x) \tag{2}$$

we calculated the curves 1 and 2 in Fig. 2a. They are seen to be very close to the experimental $\overline{\mathbf{H}}_{Fe}(x)$ in the ranges 0 < x < 15 at. % of Sn (curve 1, z = 8) and 36 < x < 50 at. % of Sn (curve 2, z = 14). For 15 < x < 36 at. % of Sn neither of the curves corresponds to the experimental data. Similarly to the disordered crystalline Fe-Si alloys [1], we suppose for the Fe-Sn alloys that as the content of Sn rises from 15 to 36 at. % the coordination number z varies randomly from 8 to 14 keeping the random Sn atoms distribution around the Fe atom. For this case the probabilities of Fe atom local surroundings should be written:

$$\left.\begin{aligned} \mathbf{P}_k^{8-14}(x) &= \sum_{n=0}^{6}\binom{6}{n}(1-x_r)^{6-n}x_r^n\binom{8+n}{k}\left(1-x/100\right)^{8+n-k}\left(x/100\right)^k \\ x_r &= (x-x_1)/(x_2-x_1) \end{aligned}\right\} \tag{3}$$

where x_1 = 15 and x_2 = 36 at. % of Sn. As it is seen, curve 3 (Fig. 2(a)) calculated for this concentration range describes quantitatively the experimental data $\overline{\mathbf{H}}_{Fe}(x)$.

The satisfactory quantitative agreement between model and experimental $\overline{\mathbf{H}}_{Fe}(x)$ allows us to use the same approach for analyzing the concentration dependence of the average magnetic moment per Fe atom $\overline{\mathbf{m}}_{Fe}(x)$ in the disordered Fe-Sn alloys presented in Fig. 2(c). Putting $\mathbf{m}_k^{Fe}$ calculated by the formula similar to (1) and $\mathbf{P}_k^z$ for z= 8 into expression:

$$\overline{\mathbf{m}}_{Fe}(x) = \sum_k \mathbf{m}_k^{Fe}(x)\mathbf{P}_k^z(x) \tag{4}$$

we obtain curve 1 (Fig. 2c) for z = 8, which conforms with experimental data for low Sn concentration x < 20 at. %. For higher Sn concentrations, the calculated magnetic moments predicted by curve 1 exceed significantly the experimental data. Then, in the interval 15 < x < 36 at. % of Sn we take the $\mathbf{P}_k^{8-14}$ values from (3) and at x > 36 at. % - $\mathbf{P}_8^{14}$ for z = 14.The result of calculation is shown in Fig. 2c, curve 2. This curve is seen to agree with experimental $\overline{\mathbf{m}}_{Fe}(x)$ much better than curve 1, although for 36 < x < 50 at. % of Sn the experimental values are repeatedly lower.

The comparison of the average magnetic moment of Fe atom, $\overline{\mathbf{m}}_{Fe}(x)$, with the average ^{57}Fe hyperfine magnetic field (HFMF), $\overline{\mathbf{H}}_{Fe}(x)$, shows their proportional change in the interval 25 - 50 at. % Sn, while there is no proportionality at all between the average ^{119}Sn HFMF $\overline{\mathbf{H}}_{Sn}(x)$ (Fig.2(b)) and $\overline{\mathbf{m}}_{Fe}(x)$. One can see even a slightly pronounced maximum at x = 35 ÷ 40 at. % Sn in dashed line 1 (Fig. 2b), which is the polynomial approximation of the experimental data. However, according to [16, 17] the magnetic moment of surrounding atoms gives rise to the hyperfine magnetic field at ^{119}Sn. Thus, it is unclear point: $\overline{\mathbf{H}}_{Sn}(x)$ does not depend on the $\overline{\mathbf{m}}_{Fe}(x)$ reduction with the Sn content increase. The attempts to process ^{119}Sn MS in a discrete model failed. This difficulty was marked earlier in [18]. Nevertheless, in [10] it has been shown that Sn^{119} HFMF in equilibrium FeSn -intermetallic compounds can be explained in terms of the average magnetic moment of Fe atom, $\overline{m}_{Fe}$, and the number of the nearest neighbouring Fe atoms of the Sn atom. We tried to describe $\overline{H}_{Sn}(x)$ in a similar way in terms of the local atomic structure parameters. That is, for the system studied we write $\overline{H}_{Sn}(x)$ taking into account the average number of nearest neighbouring Fe atoms,

$$\begin{aligned} \overline{H}_{Sn}(x) &= A\overline{m}_{Fe}(x)\overline{N}_{Fe}(x), \\ \overline{N}_{Fe}(x) &= z(1-x) \end{aligned} \tag{5}$$

where A is the proportional coefficient 3.8 kOe/μ_B, according to [10], $\overline{N}_{Fe}(x)$ is calculated through the coordination numbers z of Sn atom

We suppose that the same peculiarities of the local atomic arrangement used for the description of $\overline{H}_{Fe}(x)$ and $\overline{m}_{Fe}(x)$ are valid for the Sn atom surrounding, namely, z = 8 at 0 < x < 15 at. %, z = 14 at 36 < x < 50 at. % of Sn, the random set of z from 8 to 14 is realized at 15 < x < 36 at. % of Sn and is expressed by the formula:

$$\begin{aligned} \overline{z}(x) &= \sum_{z=8}^{14} P_z(x) z; \\ P_z(x) &= \binom{6}{z-8}(1-x_r)^{6-(z-8)} x_r^{z-8}; \\ x_r &= (x-x_1)\big/(x_2-x_1) \end{aligned} \tag{6}$$

In this case the average number of the nearest neighbouring Fe atoms of the Sn atom, $\overline{N}_{Fe}(x)$, (Fig. 2c curve 3) increases significantly in the 15 ÷ 36 at. % Sn interval. The dependence $\overline{H}_{Sn}(x)$ calculated according to (5) with $\overline{m}_{Fe}(x)$ - Fig. 2(c), curve 2, is given in Fig. 2(b), curve 2. The qualitative agreement of the model and experimental $\overline{H}_{Sn}(x)$ data argues that in the disordered crystalline alloys Fe-Sn in the interval 15 < x < 36 at. % of Sn there is the local atomic structure characterized by the increase of Fe atoms in Sn atoms number nearest surrounding, that occurs due to the increase of the coordination number.

4. Conclusion

The analysis of Mössbauer spectroscopy and magnetic measurements data for nanocrystalline disordered Fe - Sn alloys showed that the concentration dependence of the average hyperfine magnetic field and magnetic moment $\overline{H}_{Fe}(x)$, $\overline{H}_{Sn}(x)$, $\overline{m}_{Fe}(x)$ can be described in terms of local atomic structure parameters. The model of the local magnetic moment per Fe atom based on the experimental dependence of the local hyperfine magnetic fields H_k^{Fe} was proposed. Satisfactory quantitative, for

$\overline{\mathbf{H}}_{Fe}(\mathbf{x})$, $\overline{\mathbf{m}}_{Fe}$ (x), and qualitative, for $\overline{\mathbf{H}}_{Sn}(\mathbf{x})$, accordance was obtained in the supposition that the change of the local atomic structure with the Sn content increase consists in the change of the coordination number: in the interval $0 < x < 15$ at. % of Sn the coordination number of *bcc* lattice z = 8, in $15 < x < 36$ at. % of Sn z is varying randomly from 8 to 14, in $36 < x < 50$ at. % of Sn z=14.

Acknowledgment

The financial support of the Russian Fund of Fundamental Research (grant 97 -02 - 16270) is gratefully acknowledged.

References

1. Elsukov, E.P., Konygin, G.N., Barinov, V.A. and Voronina, E.V. (1992) Local atomic environment parameters and magnetic properties of disordered crystalline and amorphous iron-silicon alloys, *J. Phys.: Condens. Matter* **4**, 7597 - 7606.
2. Yelsukov, E.P., Voronina, E.V. and Barinov, V.A. (1992) Mössbauer study of magnetic properties formation in disordered Fe - Al alloys, *J. Magn. Magn. Mater.* **115**, 271.
3. Elsukov, E.P., Vorobyov, Yu.N. and Trubachev, A.V. (1991) Local atomic structure and hyperfine interactions in electrodeposited $Fe_{100-x}P_x$ ($1.8 \le x \le 45$) alloys, *Phys. Stat. Solidi(a)* **127**, 215.
4. Yelsukov, E.P., Voronina, E.V., Konygin, G.N., Barinov, V.A., Godovikov, S.K., Dorofeev, G.A. and Zagainov, A.V. (1997) Structure and magnetic properties of $Fe_{100-x}Sn_x$ ($3.2 < x < 62$) alloys obtained by mechanical milling, *J. Magn. Magn. Mater.* **166**, 334 -348.
5. Yamamoto, H. (1966) Mössbauer effect measurements of intermetallic compounds in iron -tin system: Fe_5Sn_3 and FeSn, *J. Phys. Soc. Japan* **21**, No. 6, 1058 -1062.
6. Nikolaev, V. I., Shtcherbina, Yu. I. and Yakimov, S. S. (1963) Temperature investigation of Mössbauer spectra at ^{57}Fe and ^{119}Sn nulear in ferromagnetic compound $FeSn_2$, *Pisma v ZhETF* **45**, 1277 -1280.
7. Malaman, B., Roques, B., Courtois, A. and Protas, J. (1976) Structure Cristalline du Stannure de Fer Fe_3Sn_2, *Acta Cryst.* **B 32**, 1348 -1351.
8. Le Caër, G., Malaman, B., Venturini, G.,Fruchart, D. and Roques, B. (1985) A Mössbauer study of $FeSn_2$, *J. Phys. F: Met. Phys.* **15**, 1813 -1827.
9. Le Caër, G., Malaman, B. and Roques, B. (1978) Mössbauer effect study of Fe_3Sn_2, *J. Phys. F: Metal. Phys.* **2**, 323 -336.
10. Trumpy, G., Both, E., Djega -Mariadassou, C. and Lecocq, P. (1970) Mössbauer -effect studies of iron -tin alloys, *Phys. Rev. B* **2**, 3477 -3490.
11. Vincze, I. and Aldred, A. T. (1974) Mössbauer measurements in iron -base alloys with nontransition elements, *Phys. Rev. B* **9**, 3845 -3853.
12. Yelsukov, E.P., Voronina, E.V., Fomin, V.M. and Konygin, G.N. (1998) Mossbauer study of local atomic environment effects on magnetic properties in disordered nanocrystalline and amorphous $Fe_{100-x}Sn_x$ ($0<x<50$ at.%), *Phys. Met. Matallogr.* **85**, No 3, 307-314.
13. Kanunnikova, O. M., Gil'mutdinov, F. Z. and Yelsukov, E. P. (1996) Fotoelectronnoe issledovanie poroshkov $Fe_{1-x}Sn_x$ (The XPS investigation of powders $Fe_{1-x}Sn_x$), *Perspectivnye materialy (Advance Materials)* **6**, 71-74.
14. Yamaguchi, K. and Watanabe, H. (1967) Neutron diffraction study of FeSn. *J. Phys. Soc. Jpn* **22**, 1210-3.
15. Venturini, G., Malaman, B., Le Caër, G. and Fruchart, D. (1987) Low -temperature magnetic structure of $FeSn_2$, *Phys. Rev. B* **35**, No. 13, 7038 -7045
16. Goldanskii, V.I. and R.H. Herber (1968) *Chemical Applications of Mössbauer Spectroscopy*, Academic Press.
17. Stearns, M.B. and Norbeck, J.M. (1979) Hyperfine fields at nonmagnetic atoms in metallic ferromagnets, *Phys. Rev. B* **20** 3739.
18. Cranshaw, T.E. (1987) The interaction between ^{119}Sn atoms and impurity atoms of s-p elements of the fourth period dissolved in iron, *J. Phys. F: Met. Phys.* **17**, 1645.

IRON NANOPARTICLES IN X AND Y ZEOLITES PREPARED BY REDUCTION WITH NaN_3

K. LÁZÁR[1], L.F. KISS[2], S. PRONIER[3], G. ONYESTYÁK[4]
AND H.K. BEYER[4]

[1)] *Institute of Isotope and Surface Chemistry, Budapest, P.O.B. 77, H-1525, Hungary*
[2)] *Institute of Solid State Physics and Optics, Budapest, P.O.B. 49, H-1525, Hungary*
[3)] *LACCO CNRS, UMR 6503, Université de Poitiers, Poitiers, F-86022, Cedex, France*
[4)] *Institute of Chemistry, Budapest, P.O.B. 17, H-1525, Hungary*

Abstract

Reduction of extra-framework iron ions to the zerovalent state using sodium vapour generated by thermal decomposition of sodium azide in mixtures with Fe-X and Fe-Y is reported. The resulting products were analysed by in situ Mössbauer spectroscopy, DC magnetometry, X-ray diffraction, and transmission electron microscopy measurements. A part of the iron ions was reduced to the zerovalent state in the samples. Metallic iron was formed in two size ranges upon reduction: metallic particles in the 2 - 7 nm range and small clusters of around 1 nm. Simultaneous recrystallization of the zeolite framework was also observed.

1. Introduction

Various possibilities have been studied for preparing and stabilizing zerovalent metallic iron particles in cages of zeolite framework. One approach is to decompose zerovalent iron compounds while maintaining the zerovalent state upon decomposition. Among these studies, decomposition of iron carbonyls adsorbed in the cages of zeolites was investigated; in Na-Y the formation of iron clusters and larger particles upon evacuation was reported [1,2]. Another possible route for carrying out the decomposition is in a microwave plasma; in this way small iron particles were obtained from ferrocene [3].

Another approach for preparing metallic particles is by reduction of the extra-framework ions by a strong reducing agent, eg. sodium vapour. A proportion of the exchanged iron was reduced using sodium vapour at 1066 K [4], or by prolonged interaction at 673 K [5] in X and Y zeolites.

Instead of the direct evaporation, decomposition of sodium azide, NaN_3 can also be suggested for generating sodium vapour "in situ". The source of sodium can be more

M. Miglierini and D. Petridis (eds.), Mössbauer Spectroscopy in Materials Science, 291–298.

evenly distributed by simple mechanical mixing prior to the thermal decomposition, and in this way more effective reduction can be expected above 670 K in the reaction:

$$2\ Fe^{3+} + 6\ NaN_3 \rightarrow 2\ Fe^0 + 6\ Na^+ + 9\ N_2\uparrow \quad (1)$$

It should, however, be added that the decomposition of NaN_3 is a more complicated process: Na_4^{3+}, Na_6^{5+} and Na_x^0 are formed [6] and these clusters are assumed to perform the reduction of extra-framework iron ions.

In the present contribution, results of reduction of extra-framework iron ions by Eq. (1) are reported for Fe-X and Fe-Y zeolites. The products obtained have been analysed by in situ Mössbauer spectroscopy, DC magnetometry (SQUID), X-ray diffraction (XRD), transmission electron microscopy (TEM), and selected area electron diffraction measurements.

2. Experimental

2.1. PREPARATION

Fe-X and Fe-Y samples were prepared by aqueous ion exchange in 0.1 N $FeSO_4$ solutions under nitrogen atmosphere, and partial exchange of sodium for iron was performed. The samples were subsequently dried in air. The compositions of the elemental cells were $Na_{38.5}Fe_{16.9}[Al_{81.5}Si_{110.5}O_{384}]\ .\ 191\ H_2O$ for Fe-X, and $Na_{18.1}Fe_{16.3}[Al_{54}Si_{138}O_{384}]\ .\ 255\ H_2O$ for Fe-Y.

Before further treatments the adsorbed water was removed by overnight drying at 623 K. The zeolite was then mixed with NaN_3 in an agate mortar under argon atmosphere in a dry box. NaN_3 was used in 30 % excess (related to Na/Fe ratio of 3). Then a pellet was pressed and placed into the in situ Mössbauer cell as fast as possible (ca. 5 min). The mixture was evacuated in the next step at 550 K (10^{-1} Pa) for 20 min to remove the remaining traces of adsorbed water. Finally, the decomposition of sodium azide and the simultaneous reduction of iron ions was carried out by heating the mixture to 800 K for 30 min in a stream of nitrogen. It should be noted that all samples used for other characterizations were prepared in the Mössbauer cell, thus a certain extent of oxidation of the samples might have taken place during the sample transfer due to exposure to air.

2.2. CHARACTERIZATION

In situ 300 and 77 K Mössbauer spectra were obtained by a conventional spectrometer operated in constant acceleration mode. Isomer shift values are related to metallic α-iron, the accuracy of the positional data is ca. ± 0.03 mm/s. Lorentzian line shapes were assumed and no positional parameters were constrained in fitting the spectra.

DC magnetization measurements were performed in a Quantum design MPMS-5S SQUID magnetometer in the 0 - 5 Tesla external magnetic field range and the 5 - 300 K temperature interval. Samples prepared for magnetic measurements were immediately coated with paraffin and their measurements started within a few hours after removal from the preparation cell. For TEM imaging, electron diffraction and EDX observations a CM120 Philips TEM microscope was used - capable for 120 kV accelerating voltage and equipped with LaB_6 filament. (Samples for TEM measurements were kept in air for several weeks before measurements, thus considerable oxidation might be expected).

X-ray diffractograms were recorded with a Philips PW 1200 powder diffractometer equipped with a graphite monochromator using the K_α radiation of copper.

3. Results

3.1. MÖSSBAUER SPECTROSCOPY

300 and 77 K in situ Mössbauer spectra are shown in Figure 1. Each spectrum contains a contribution of larger particles of metallic iron displaying a magnetic sextet. The major contribution in the spectra originates from various doublets of Fe^{2+} components. The Fe-X sample was the same as used for the DC magnetization measurements, thus slight oxidation is also reflected in the spectra (appearance of Fe^{3+} component) due to the brief exposure to air. Decomposition of the spectra also reveals the presence of a singlet component around the 0.0 mm/s isomer shift (IS) value (Table 1).

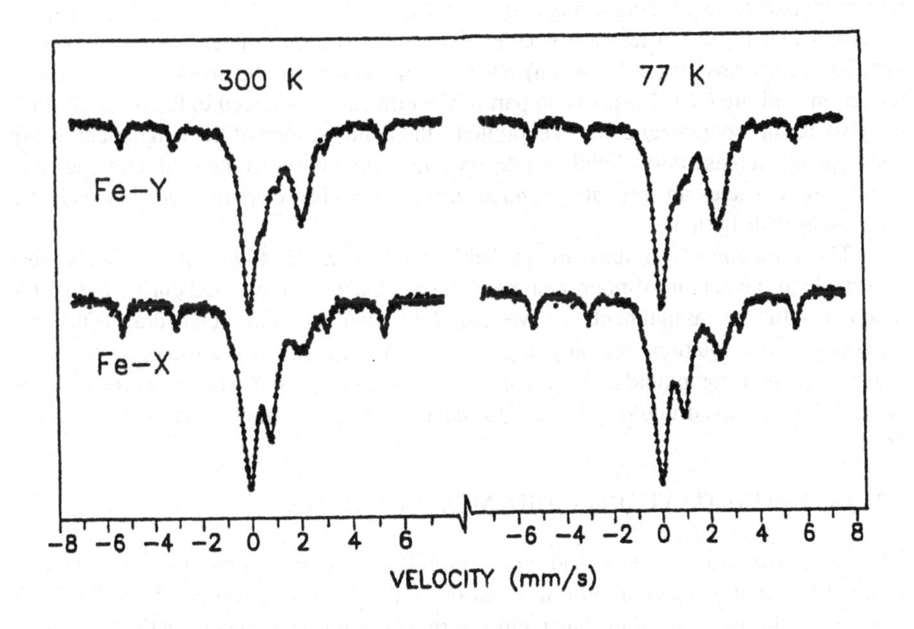

Figure 1. 300 and 77 K in situ Mössbauer spectra of the Fe-X + NaN_3 (bottom) and the Fe-Y + NaN_3 (top) samples.

TABLE 1. Mössbauer parameters obtained from 300 and 77 K in situ spectra (IS: isomer shift related to metallic α-iron, mm/s; QS: quadrupole splitting, mm/s; RI: relative intensity, %).

		300 K			77 K		
Sample	Comp.	IS	QS	RI	IS	QS	RI
Fe-Y	$Fe^{0,a}$	0.0	-	11	0.11	-	10
	$Fe^{0,b}$	-0.05	-	22	-0.02	-	13
	Fe^{2+}_{Td}	0.89	0.52	11	0.97	0.49	10
	Fe^{2+}_{Oh}	1.05	2.62	11	1.22	2.88	9
	Fe^{2+}_{Oh}	1.16	1.77	45	1.23	2.14	57
Fe-X[c]	$Fe^{0,a}$	0.01	-	14	0.10	-	16
	$Fe^{0,b}$	-0.08	-	24	-0.02	-	19
	Fe^{3+}	0.20	1.13	14	0.31	1.19	16
	Fe^{2+}_{Td}	0.79	0.35	21	0.94	0.45	16
	Fe^{2+}_{Oh}	1.10	1.99	26	1.25	2.35	33

[a] Magnetically split component (33 T at 300 K, 34 T at 77 K)
[b] Singlet component (without magnetic splitting)
[c] Sample measured simultaneously with SQUID

The singlet component most probably originated from small iron clusters. The threshold size for exhibiting a magnetic sextet is ca. 4 nm at 77 K [7]; thus, it can be supposed that these clusters are smaller than this threshold value. The other part of metallic iron forms larger (> 4 nm) particles. As shown by the spectra, the extent of reduction is limited, ca. 1/3rd - 1/4th part of the iron can be reduced to the metallic state. Only weak superparamagnetism is exhibited - the contributions of the magnetically split components increase only slightly on decreasing temperature, so the characteristic sizes of metallic particles are probably centred around two characteristic sizes, instead of a continuous distribution.

The remaining iron ions are probably located in the tight smaller pores, less accessible to the sodium vapour, and the Fe^{2+} state has been preserved during the 800 K treatment. In the assignments of various Fe^{2+} doublets the octahedral (Oh) and tetrahedral (Td) notations are only approximate. A description of various coordination states in related compounds, in iron silicate minerals, and a list of the corresponding isomer shift and quadrupole splitting (QS) data can be found in the literature, eg. in ref. [8].

3.2. DC MAGNETIZATION (SQUID) MEASUREMENTS

DC magnetizations of Fe-X and Fe-Y samples were measured, the two samples exhibited similar behaviour. For this reason, only the results obtained on the Fe-X sample are discussed in detail. The temperature dependences of magnetization in various (0.1 and 5 T) fields are shown in Figure 2. Two important features can be deduced. First, the change is monotonic in both cases, i.e. there is no characteristic blocking temperature in the temperature range investigated. The second feature is the expressed increase of the magnetization at low temperatures. (Note the different scaling for the left and right ordinates in Figure 2.) Thus, the presence of (at least two) ferromagnetic components can be suggested.

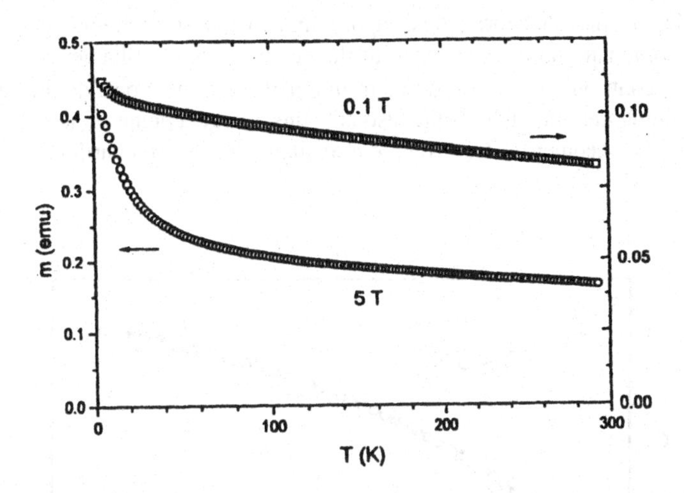

Figure 2. Change of magnetization vs. temperature in the Fe-X sample in 0.1 T (right ordinate) and in 5 T external fields (left ordinate). The magnetization is given in relative units (total moment).

To analyse the ferromagnetic feature in further detail magnetization was also measured as a function of field strength at various temperatures (5, 77 and 293 K). It was found that the magnetization reached at constant fields depended strongly on the temperature, and there was a difference between 293 K (0.16 emu) and 77 K (0.22 emu) values obtained at the highest field strength. The highest magnetization value was found by cooling the sample to 5 K (0.4 emu at 5 T). The curves at 293 and 77 K can be fitted by single Langevin functions by assuming a single characteristic magnetic moment value. In contrast, the 5 K trace cannot be described satisfactorily by this assumption.

Significant improvement can be obtained in the fitting of the 5 K curve by assuming the simultaneous presence of different moments. Excellent agreement with the experimental data can be found by combining two Langevin functions with two characteristic moments and including a small linear (paramagnetic) contribution. Almost equal volume ratios can be attributed to a 10 μ_B component (by a modest rise in the magnetization by increasing the external field) and to a 300 μ_B component (with a sharp, almost step-wise increase). The decomposition is shown in Figure 3.

As for the 77 K and 293 K measurements it is worth mentioning that the 10 μ_B component cannot be saturated even in 5 T field at these temperatures, thus the contribution from this component is negligible. This provides an explanation for the previously mentioned observations, too. In other words, the magnetization curves can be described satisfactorily by assuming only a single (larger) moment, and the decrease of the net magnetization at constant fields at higher temperatures (77 and 293 K) can also be interpreted.

It is assumed that the presence of components with different moments is in agreement with the interpretation of Mössbauer spectra, where the presence of small clusters (contribution in a singlet component at the IS 0.0 mm/s part) and a contribution from larger particles (> 4 nm, reflected in the magnetically split component) was suggested. The 10 μ_B value corresponds approximately to five Fe atoms - this cluster can easily be accomodated in the supercage of the zeolite structure (diameter 1.2 nm). The other, 300 μ_B contribution may originate from a particle located outside the cages. (It is worth mentioning here that the zeolite also contains a large amount of sodium ions, thus only a part of the "geometric" free volume is available for the iron particles.)

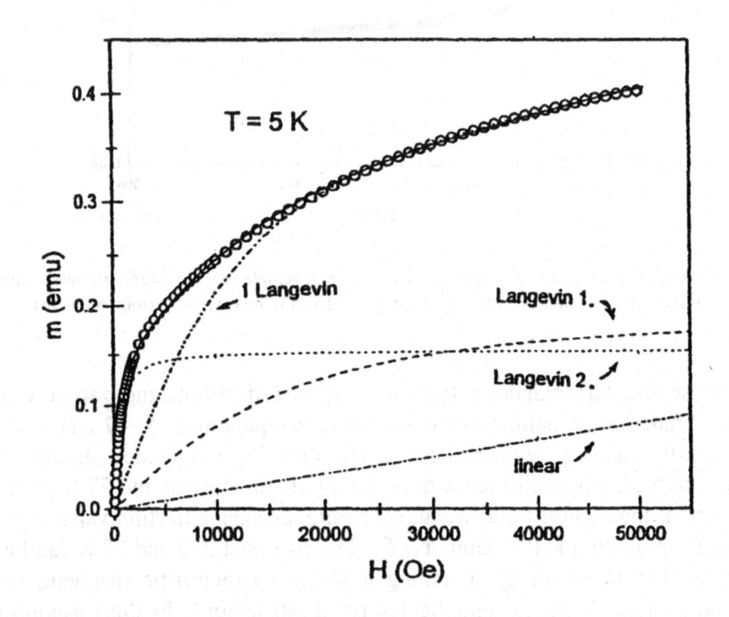

Figure 3. Magnetization of the Fe-X + NaN_3 sample at 5 K, and its decomposition to Langevin functions. (o measured, - fitted: sum of Langevin 1. + Langevin 2. + linear)

3.3. X-RAY DIFFRACTION MEASUREMENTS

The diffractogram of the treated Fe-X + NaN_3 mixture revealed partial recrystallization: peaks of non-specified aluminium silicate phase(s) were observed with typical reflections at $2\Theta = 21.1^\circ$ and at 34.8°. In the Y zeolite samples the original crystallinity was essentially preserved. Reflections of metallic iron (or iron oxide) crystals were not observed in any diffractogram. It should be noted that the lower size limit for the detection of iron crystals is ca. 6 nm, i.e. the presence of larger iron particles in noticeable amounts cannot be expected.

3.4. TRANSMISSION ELECTRON MICROSCOPY AND ELECTRON DIFFRACTION

Various parts of an Fe-X + NaN_3 sample were studied (after several weeks stay in air following the sample preparation). Small agglomerates were detected in a large number distributed evenly in each area with their size ranging mostly in the 2 - 7 nm region. EDAX analysis proved a high iron content for these particles. Selected area electron diffractograms were also obtained on exceptionally large (12 and 36 nm) particles. Characteristic spots were found at d= 2.1 and 1.5 Å for the [200] and [220] reflections of the FeO (wustite) on the first particle, and a diffractogram combined from reflections on [111] (4.9 Å), [220] (3.0 Å) and [311] (2.53 Å) planes of Fe_3O_4 (magnetite) was obtained on the larger, 36 nm particle.

For the interpretation it should be considered that secondary oxidation probably took place during the stay between the preparation and measurement. (Primary in situ Mössbauer data did not reveal the presence of any iron oxide phase.)

Further, it should be mentioned that the lower threshold size for the appearence of particles was in the 1.5 - 2.0 nm range under the given experimental conditions. Thus, detection of the smaller clusters cannot be expected in these TEM studies.

4. Discussion

The combination of results clearly proves that iron is reduced to the metallic state. Only a part (1/3rd - 1/4th) of the total amount can be reduced. Iron particles are found in two size ranges: larger particles prevailing in the 2 - 7 nm range and small clusters located probably inside the supercages. The presence of particles in the 2 - 7 nm size range is clearly demonstrated by the direct TEM measurements, and by the appearance of the magnetic sextet in the Mössbauer spectra (for the > 4 nm particles) and by the decomposition of the Langevin equation for DC magnetization at 5 K. Small amount of these particles is present above the 6 nm, since no characteristic diffraction lines of iron (or its oxides) were found in the X-ray diffractograms. The lower (2 nm) limit is also in accordance with the magnetization measurements at 5 K, the 300 μ_B contribution may result from iron particles with an estimated characteristic size around 2 nm. The 2 - 7 nm particles cannot be accomodated in the structural cages of X and Y zeolites since this size range is too large. Thus, to accomodate these, partial destruction of the framework is necessary. In fact, as was mentioned, recrystallization of the framework was revealed by XRD, thus the 2 - 7 nm iron particles can be accomodated along phase and grain boundaries.

The presence of smaller clusters is also evidenced from the above-mentioned Mössbauer data (singlet component at IS = 0.0 mm/s) and from DC magnetization (10 μ_B component in the Langevin function). It might be mentioned here that chemisorption measurements were also performed on the Na-Y + NaN_3 sample, and it was found that only ca. 1/10th of the metallic iron is easily accessible to gas molecules. This observation also supports the assumption that a number of the particles are embedded in the recrystallized structure [9].

5. Conclusions

The possibility of reducing the extra-framework iron ions with sodium vapour generated by in situ thermal decomposition of sodium azide in mechanical mixtures with Fe-X and Fe-Y zeolites (ca. 5 wt % iron) was investigated. It was found that 1/3rd - 1/4th of the iron can be reduced to the metallic state. Iron particles were formed in two size ranges; 2 - 7 nm particles, and small clusters, accommodated most probably in the supercages (estimated size < 1 nm). The thermal decompositon of azide and reduction were accompanied by a certain degree of recrystallization of the original zeolite lattice.

Acknowledgements

The financial support provided by the Commission of the European Communities in the framework of Chemistry Action COST D5 (PECO project No. CIPE CT 92 6107) and by a grant from the Hungarian National Science Research Fund (OTKA T021131) are gratefully acknowledged. The authors are indebted to L.K. Varga for the fruitful discussions.

References

1. Ballivet-Tkatchenko, D. and Tkatchenko, I. (1979) Small particles in zeolites as selective catalysts for the hydrocondensation of carbon monoxide, *J. Molecular Catalysis* **13**, 1-10.
2. Ziethen, H.M., Doppler, G. and Trautwein, A.X. (1988) Formation and characterization of very small iron clusters in zeolites, *Hyperfine Interactions* **42**, 1109-1112.
3. Suib, S.L. and Zhang, Z. (1988) Structure-sensitive reactions of cyclopropane with cobalt and iron zeolite catalysts, in: W.H. Flank, Th.E. Whyte (eds.) *Perspectives in Molecular Sieve Science*, Am. Chem. Soc., Symp. Ser., 368. A.C.S. Washington, pp. 569-578.
4. Lee, J.B. (1981) Reduction of supported iron catalysts studied by Mössbauer spectroscopy, *J. Catalysis* **68**, 27-32.
5. Schmidt, F., Gunsser, W. and Adolph, J. (1977) Formation of iron clusters in zeolites with different supercage sizes, in: J.R. Katzer (ed.) *Molecular Sieves II.*, Am. Chem. Soc., Symp. Ser., 40. A.C.S. Washington, pp. 291-301.
6. Brock, M., Edwards, C., Förster, H. and Schröder, M. (1994) Decomposition of sodium azide on faujasites of different Si/Al ratios, *Studies in Surface Science and Catalysis* **84**, 1515-1522.
7. Clausen, B.S., Topsþe, H. and Mþrup, S. (1989) Preparation and properties of small silica-supported iron catalyst particles, *Applied Catalysis* **48**, 327-340.
8. Burns, R.G. (1994) Mineral Mössbauer spectroscopy: Correlations between chemical shift and quadrupole splitting parameters, *Hyperfine Interactions* **91**, 739-745.
9. Beyer, H.K., Onyestyák, G., Jönsson, B.J., Matusek, K. and Lázár, K. (1998) Reduction of iron ions to the metallic state in X and Y zeolites by sodium azide, in: M.M.J. Treacy, B. Marcus, J.B. Higgins and M.E. Bisher (eds.) *Proceedings of the 12th International Zeolite Conference, Baltimore*, Materials Research Society, in press.

QUASIELASTIC MÖSSBAUER SCATTERING IN STABLE ICOSAHEDRAL Al-Cu-Fe QUASICRYSTALS

R.A. BRAND[1], J. VOSS[1] AND Y. CALVAYRAC[2]
[1)] *Department of Physics, Universität Duisburg, D-47048 Duisburg, Germany*
[2)] *C.E.C.M./C.N.R.S., 15 rue G. Urbain, F-94407 Vitry Cédex, France*

Abstract

Quasicrystals can show a new type of localised defect termed a "phason". These defects are important for the understanding of both the structure and the atomic motion (diffusion). These defects lead to short-ranged atomic motion without involving vacancies and confined in space. We have studied icosahedral $Al_{62}Cu_{25.5}{}^{57}Fe_{12.5}$ using quasielastic Mössbauer spectroscopy for $T \leq 1100$ K. We find that the Debye Waller factor $f(T)$ of the central elastic (quadrupole-split) line decreases abruptly below the Debye result $f_D(T)$ for $T > 600$ K. But there is an additional quasielastic signal which correlates with the appearance of this anomaly. This is due to atomic motion of the Fe atoms from phason jumps. The electric field gradient shows a different dependence on temperature, decreasing only slowly up to about 800~K, but then much faster. These results give the temperatures where phason motion (≥ 600 K), and then vacancy motion (≥ 800 K), strongly affect the Mössbauer spectrum.

1. Introduction

Quasicrystals with icosahedral point symmetry were discovered in 1982 (and published in 1984 [1]). The structure, lattice dynamics and atomic motion in quasicrystals (QC) are all currently the subject of much debate. Several good surveys of quasicrystal physics are given in [2, 3, 4]. Quasicrystals can show a new type of localised defect termed a "phason". Phasons can lead to discontinuous atomic jumps not involving vacancies and restricted to a finite volume. The existence of such phason jumps (phason modes) and their temperature dependence is a key issue in understanding the structure of stable quasicrystals (see for example the perfect QC model of Katz and Gratias [5], as well as the random tiling model [6]). Kalugin and Katz [8] have hypothesised that phason modes are involved in a new diffusion mechanism not involving vacancies. Especially interesting is the case of the *stable* icosahedral (i-) quasicrystal i-AlCuFe [7]. The phase diagram of this system has recently been extensively studied [9]. It is known that the composition $Al_{62}Cu_{25.5}Fe_{12.5}$ produces a stable icosahedral structure over the widest range of temperature, up to about 1100 K. Our previous extensive room

M. Miglierini and D. Petridis (eds.), Mössbauer Spectroscopy in Materials Science, 299–306.

temperature Mössbauer (and NMR) investigations of the icosahedral and approximant phases are reported on in [10].

It is known from time-of-flight quasielastic neutron scattering experiments that Cu in quasicrystalline AlCuFe is involved in a phason jump process at high temperatures $T \geq 600$ K on a time scale τ of ($\hbar$/250 μeV) [11]. The temperature dependence of this effect was found to be very unusual, indicating a type of "assisted-jump" process involving both Al and Cu atoms. The iron dynamics could not at first be observed, but Coddens et al. [12] were able to show that at high temperatures, a quasielastic line appears in the Mössbauer spectrum at large velocities (ca. 200 mm/s). This was interpreted as signalling the Fe dynamics, which found to be on the time scale of only ($\hbar$/2 μeV). This vast difference has remained a puzzle, but it was later confirmed by quasielastic neutron scattering [13]. It is then crucial to know the temperature dependence of the iron dynamics in order to correlate this with the dynamics of the other elements in i-AlCuFe quasicrystals. Therefore we have started an investigation of i-AlCuFe with high temperature Mössbauer spectroscopy. We have used both natural as well as enriched iron samples.

2. Results

We have studied icosahedral $Al_{62}Cu_{25.5}Fe_{12.5}$ using quasielastic Mössbauer spectroscopy for $T \leq 1100$ K. The samples used in this study have already been described in [10] and in [14]. These are all single phase. The composition used here is known to be stable up to 1100K. The sample powder was mixed with a small amount of high purity BN powder to better distribute the grains in the oven sample holder. This consists of two BeO wafers pressed together in a tantalum clamp. This was mounted in a simple Mössbauer oven with a dynamic vacuum of about 10^{-6} Torr. For the high-velocity spectra, the drive was calibrated using a laser interferometer [15]. X-ray spectra have been taken both before and after the high temperature runs to check for additional phases.

2.1. RESULTS FOR THE NATURAL IRON SAMPLES

The samples with natural iron were used to determine the temperature dependence of the main hyperfine parameters: linewidth $\Gamma(T)$ FWHM; centre lineshift $\delta(T)$; effective electric field gradient (EFG) splitting $\Delta E_Q(T)$; relative resonant area $A_{rel}(T)$ (normalised to that at room temperature). Spectra below room temperature were taken in a conventional bath cryostat down to 4.2 K. The results for $\Gamma(T)$ and $\delta(T)$ are shown in Figure 1, (a) and (b). The natural logarithm of $A_{rel}(T)$ and $\Delta E_Q(T)$ are given in Figure 2 (a) and (b).

The linewidth $\Gamma(T)$ decreases approximately linearly with temperature. In Figure 1 (a), this has been shown including the 95% confidence bounds. The centre lineshift (solid points in Figure 1 (b)) has been fitted using a conventional Debye-Waller model for the second order Doppler shift. It is found that the Debye temperature from $\delta(T)$ is $\Theta_D^{sod} \approx 580 \pm 6$ K. More importantly, it is found that this simple model *adequately fits the data over the whole investigated temperature range.* However, the results for the relative resonant area are quite different. The results (open points in Figure 2 (a)) could be fitted for temperatures below ca. 600 K with a Debye-Waller model, as shown as a solid line ($\Theta_D^{\ln(A)} \approx 550 \pm 50$ K). However, above this temperature, they deviate strongly from this model in such a way that no Debye-Waller fit is possible over the whole range. The dot dashed line for the high temperature slope and the arrow at 600 K are given only as guides for the eye. The missing area for temperatures above ca. 600 K will be shown below to be associated with the quasielastic component appearing at these temperatures.

The EFG splitting is given in Figure 2 (b) as a function of the temperature to the 3/2-power. The reason for this is that there is a slight temperature dependence which shows this behaviour, so that the low temperature section is thus linear in this figure. The origin of this effect has been discussed before [18]. It is of electronic origin, and found in almost all noncubic metals, with a slope of about 3/2 and a coefficient of about 10^{-5} $K^{-3/2}$ usually being found. Since this "usual" temperature dependence is not the subject addressed here, this will not be further discussed, but the extrapolation of a 3/2-power law used to denote the expected behaviour. For temperatures above about 800 K,

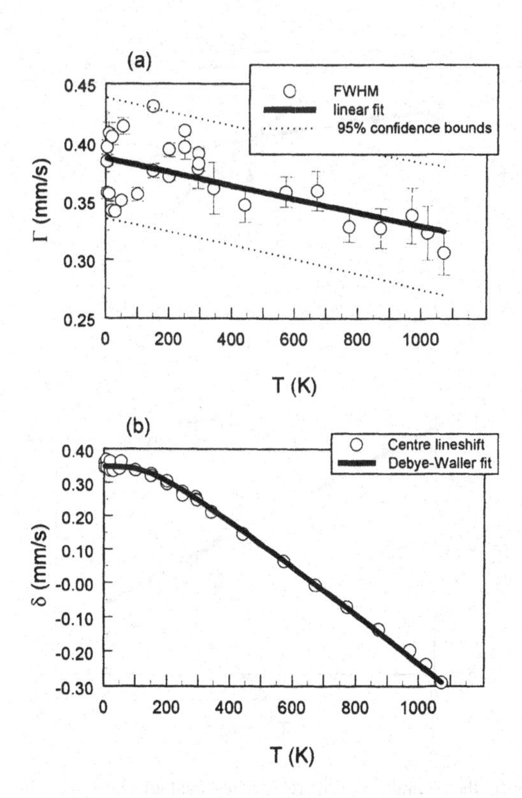

Figure 1. Results for 8a) the linewidth $\Gamma(T)$ and (b) the centre lineshift $\delta(T)$ for $Al_{62}Cu_{25.5}Fe_{12.5}$ in the temperature range up to 1100 K. (a) shows a linear fit including the 95% confidence limits. (b) shows a Debye-Waller fit for the second-order Doppler effect.

this power behaviour is no longer found, and the effective $\Delta E_Q(T)$ decreases much faster. It is interesting that this effect occurs at a temperature definitely *above* that of the anomaly in $\ln(A_{rel}(T))$ which is at about 600 K (these temperatures are both given as arrows in Figure 2 (b)).

2.2. RESULTS FOR THE ENRICHED IRON SAMPLES

Similar spectra were taken for a sample with identical composition, but with 100\% ^{57}Fe enrichment. These samples were studied as a function of temperature in the same range as before. Spectra were taken both in the normal and in several different but high velocity ranges (up to ca. 100 mm/s using a special drive and calibrating with a Laser interferometer [15]). The purpose of these experiments was to detect any quasielastic contribution in the background, as previously seen by Coddens [12]. These samples were made as thick as possible in order to increase the signal to noise ratio of any broad quasielastic component. They were made with about 20 mg/cm^2 powder, so that the Mössbauer thickness parameter at room temperature t_{ab} was about 50. The thickness effects could be treated using the standard transmission integral. However, technical

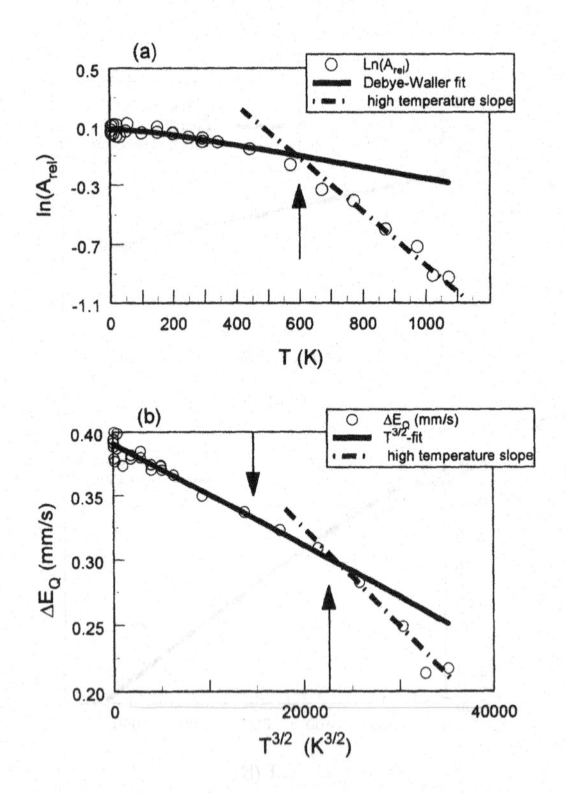

Figure 2. Results for (a) the natural log of the relative resonant area $\ln(A_{rel}(T))$ and (b) the EFG splitting $\Delta E_Q(T)$ for $Al_{62}Cu_{25.5}Fe_{12.5}$ in the temperature range up to 1100 K. (a) shows as well a Debye-Waller fit to the data below 600 K (arrow), and a line indicating the high temperature slope. (b) shows a fit to $T^{3/2}$ up to 800 K (upper arrow) as well as a lower arrow at 600 K.

difficulties caused by a slight non-linearity in the counting efficiency have precluded an exact evaluation of these spectra. They were studied at high Doppler velocities at temperatures between room temperature and 1100~K. They were studied as well at each temperature with a normal velocity scale in order to re-determine the hyperfine parameters of the central quadrupole doublet seen in the natural iron samples. The hyperfine properties obtained from these standard spectra agreed with those of the natural iron samples.

Figure 3 (upper) shows a spectrum taken at room temperature, but at a large velocity scale. The subspectrum labelled elastic is that taken from the spectrum at smaller maximum velocities, and fully agrees with this spectrum as well. The lower spectrum was taken at T = 1090 K, where anomalous values were obtained for the resonant area and EFG splitting in the samples with normal iron. The presence of a new broad contribution is evident: the wings of the spectrum are much too broad to be explained by the subspectrum labelled elastic (again, taken from the low velocity spectrum at the same temperature). It is thus clear that for spectra taken at low maximum velocities, this will appear as an anomalous loss of relative area (relative to the measured background counting rate). The spectra at high velocity have been fit using two subspectra. The first subspectrum is the elastic one. The second is a new broad

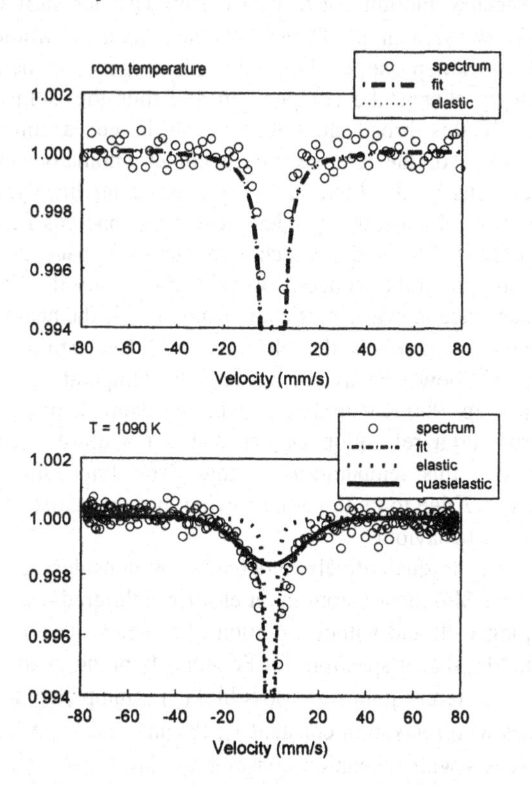

Figure 3. High velocity Mössbauer spectra for $Al_{62}Cu_{25.5}{}^{57}Fe_{12.5}$ at (upper) room temperature and (lower) T = 1090 K. The subspectra denoted elastic have been taken from spectra taken at smaller velocities (2 mm/s). The quasielestic contribution to the high temperature spectrum is evident.

component with a linewidth much larger than the usual Γ. This new component is absent for spectra below about 600 K. Although the amplitude of this component increases rapidly with temperature, it can be directly observed only in a small temperature range. This is because the effective linewidth increases rapidly as well, so that this component is quickly lost again in the background, even for these high maximum velocities. We interpret this new component as arising from quasielastic processes. We shall see below that the additional linewidth τ is given by a combination of the relaxation rates of the iron atoms with and without rotation of the EFG principal axes. This subspectrum is labelled quasielastic.

3. Discussion

The main points from this study are given by the following:

We find that the Debye Waller factor $f(T)$ of the central elastic (and quadrupole-split) line decreases abruptly below the Debye result $f_D(T)$ for $T > 600$ K. But at high velocities, there is an additional quasielastic signal which correlates with the appearance of this anomaly. This is due to atomic motion of the Fe atoms from phason jumps as well as possible vacancy motion. The centre shift (isomer shift plus second order Doppler effect) $\delta(T)$ shows a simple Debye-behaviour over the whole range, excluding the appearance of any new phase, or the partial inhomogeneous melting of the sample (which would strongly change the composition and thus the isomer shift [10] of the remaining solid). If this broad quasielastic line is not accounted for, then the temperature dependence of the relative area becomes anomalous, which has lead to erroneous interpretations in the literature [16] concerning the dynamics of clusters. There are two cases which we can consider. The first concerns Fe motion by phason dynamics. This would lead to motion which is restricted in space, as in the case of cage diffusion. The second possibility concerns vacancy motion itself. This has no reason to be restricted in space since the vacancies come from outside the nearest neighbour shell.

A second important result is that the measured EFG splitting $\Delta E_Q(T)$ deviates below an empirical $T^{3/2}$-power behaviour but only for temperatures above about 800 K (a limit of 800 to perhaps 900 K would agree with our data). Thus in the region of about 600 to 800 K, iron atom relaxation occurs with no additional change in the EFG splitting, while above this temperature it does. The linewidth $\Gamma(T)$ is a smooth decreasing function of T due to motional narrowing, but the statistics is not sufficient to determine precisely its behaviour.

These results can be qualitatively understood by considering a simple model for atomic hopping of the Mössbauer atom in an electric field gradient. Two cases need to be specified: hopping with and without rotation of the EFG tensor. Litterst et al. [17] have calculated the Mössbauer spectrum for Fe atoms hopping in an octahedral cage. In this calculation, two different jumps are possible: (a) a jump (length $\sqrt{2}r$) between two neighbouring apices with relaxation constant τ_1; (b) an axial jump (length $2r$) along the principal axis direction with relaxation constant τ_2. Case (a) includes rotation of the EFG principal axis and thus leads to a collapse in the EFG splitting at large τ_1. Case (b) does not involve such a rotation of the EFG principal axes (no change in EFG

Hamiltonian) and so does not lead to a change in the effective splitting. Both effects lead to a broad quasielastic background which is the signature for atomic motion.

It is then a simple matter to apply these results to our experiments. For an atom in a quasicrystal, a possible phason jump direction is given by surplus space around the atom which however does not constitute a vacancy (missing atom) but does lead to a type of local double well potential. By jumping, the atom leaves one atomic surface and joins another neighbouring one (for a discussion of atomic surfaces and phasons in quasicrystals, see [2 - 4]). Because of this surplus space, we assume that the local EFG principal axis points in the direction of a possible phason jump. Thus a phason jump is interpreted as a cage jump with no change in the EFG principal axis. (Important is only the fact that the EFG tensor remains the same. It seems very reasonable that the local EFG principal axis points in such a direction.)

We have seen that at temperatures above about 600 K, the quasielastic component increases strongly in amplitude, leading to the linear decrease in the logarithm of the relative area $\ln(A_{rel}(T))$. This decrease is in addition to the usual linear decrease at these temperatures due to the phonons. There is however at first no anomalous change in the EFG splitting $\Delta E_Q(T)$. These two facts signal the fact that the phason-dynamics now include the iron atoms to a degree which can affect the spectrum (when the relaxation time τ_2 in the model of Litterst [17] becomes on the order of the Mössbauer lifetime τ = 98 nsec). For temperatures above 800 K, there is in addition an anomalous decrease in the effective EFG splitting. This is the signal that vacancies are now playing a role. The relaxation is now dominated by τ_1 in the model of Litterst. In fact the exact description will then also include vacancy motion around the iron atom, as well as vacancies leaving the nearest neighbour shell. Both lead to a rotation of the EFG tensor principal axes, and thus to a change in the effective EFG splitting.

The energy range of the phason jumps as found from the linewidth of the quasielastic component is ~ ($\hbar/2$ μeV) in the region of T ~ 780 to 1100 K (but increasing with T), in agreement with a preliminary Mössbauer measurement [12]. This energy scale (two orders of magnitude slower), as well as the temperature dependence of the quasielastic signal indicates that Fe does move via phason jumps (as well as by vacancy motion) but is not involved in the complicated „assisted jump“ processes of Cu and Al.

It is interesting to note that roughly the same temperature for the cross-over from phason-jump to vacancy dynamics was found for macroscopic diffusion studies. Blüher et al. [19] have reported that both Pd and Au show novel behaviour of the radio-tracer diffusion coefficients, with different Arrhenius laws below and above 750 K (given in the paper is 450° C). They conclude that for temperatures below this cross-over, diffusion occurs by a new phason-assisted mode, as was proposed by Kalugin and Katz. Above this cross-over, diffusion occurs by the usual vacancy mechanism [8].

In summary, we report on high temperature Mössbauer studies of icosahedral $Al_{62}Cu_{25.5}Fe_{12.5}$ including both natural iron as well as enriched iron samples. We have shown that the high temperature properties indicate relaxation effects. These involve both phason-jumps as well as vacancy diffusion for iron atoms. There is a cross-over near 800 K from the first (dominating below) to the second mechanism (dominating above).

Acknowledgements

The financial aide of the Deutsche Forschungsgemeinshaft: Schwerpunksprogramm Quasikristalle is gratefully acknowledged.

References

1. Shechtman, D., Blech, I., Gratias, D., and Cahnet, J.W. (1984), *Phys. Rev. Lett.* **53**, 1951.
2. Ch. Janot and R. Mosseri (eds.) (1995) *Proceedings of the 5th International Conference on Quasicrystals*, World Scientific, Singapore.
3. F. Hippert and D. Gratias (eds.) (1994) *Lectures on Quasicrystals*, Les Editions de Physique, Paris.
4. S. Takeuchi and T. Fujiwara (eds.) (1998) *Proceedings of the 6th International Conference on Quasicrystals*, World Scientific, Singapore.
5. Katz, A. and Gratias, D. in [2] p. 164; in [3], p. 187; (1993), *J. Non-Crystalline Solids* **153-154**, 187.
6. Widom, M., Deng, D.P., and Henley, C.L. (1989), *Phys. Rev. Lettr.* **63**, 310; Elser, V. in [4], p. 19; Henley, C.L. in [4] p. 27.
7. Tsai, A.P., Inoue, A., and Masumoto, T. (1987), *Jpn. J. Appl. Phys.* **26**, L1505.
8. Kalugin, P.A. and Katz, A. (1993), *Europhys. Lettr.* **21**, 921.
9. Calvayrac, Y., Quivy, A., Bessiere, M., Lefebre, S., Cornier-Quiquandon, M., and Gratias, D. (1990), *J. Phys. France* **51**, 417; Cornier-Quiquandon, M., Quivy, A., Lefebvre, S., Elkaim, E., Heger, G., Katz, A., and Gratias, D. (1991), *Phys. Rev. B* **44**, 2071; Quiquandon, M. et al. (1996), *J. Phys.: Condens. Matter* **8**, 2487.
10. Hippert, F. et al. (1994), *J. Phys.: Condens. Matter* **6**, 11189; Hippert, F. et al. in [2] pp. 464-71; Quivy, A. et al. (1996), *J. Phys.: Condens. Matter* **8**, 4223.
11. Coddens, G., Bellissent, R., Calvayrac, Y., and Ambroise, J.P. (1993), *Europhys. Lettr.* **21**, 921.
12. Coddens, G. et al. (1995), *J. Phys. I (France)* **5**, 771.
13. Coddens, G. et al. (1997), *Phys. Rev. Lettr.* **78**, 4209.
14. Cornier-Quiquandon, M., Cornier-Quiquandon, M., Bellissent, R., Calvayrac, Y., Cahn, J.W., Gratias, D., and Moser, B. (1993), *J. Non-Cryst. Solids* **153-154**, 10.
15. Produced by WissEl GmbH, Starnberg Germany.
16. Janot, C. et al. (1993), *Phys. Rev. Lettr.* **71**, 871.
17. Litterst, F.J., Gorobchenko, V.D., and Kalvius, G.M. (1983), *Hyperfine Interactions* **14**, 21 (which however contains several typographical errors).
18. Jena, P. (1976), *Phys. Rev. Lettr.* **36**, 418; see as well Christiansen, J., Huebes, P., Keitel, R., Klinger, W., Loeffler, W., Sandner, W., and Witthuhn, W. (1976), *Z. Physik B* **24**, 177; Nishiyama, K., Dimmling, F., Kornrumpf, Th., and Riegel, D. (1976), *Phys. Rev. Lettr.* **37**, 357.
19. Blüher, R., Scharwaechter, P., Frank, W., and Kronmüller, H. (1998), *Phys. Rev. Lettr.* **80**, 1014.

QUASI-ELASTIC PROCESSES STUDIED BY METHODS SENSITIVE TO THE MOMENTUM AND ENERGY TRANSFER

Quasi-Elastic Processes

K. RUEBENBAUER
Mössbauer Spectroscopy Laboratory
Institute of Physics and Computer Science, Pedagogical University
PL-03-084 Cracow, ul. Podchorazych 2, Poland

The paper is aimed at the brief review of currently available microscopic methods being sensitive to the energy and momentum transfer, and precise enough to see quasi-elasticity at energies comparable with the hyperfine Hamiltonian. Such methods are invaluable tools, while looking at the diffusivity in a condensed matter at the elementary level provided the momentum transfer is high enough to distinguish atomic jumps, and the energy resolution is sufficient to see diffusivity in solids. Advantages and disadvantages of quasi-elastic Mössbauer spectroscopy (QMS), quasi-elastic neutron scattering (QNS), and Rayleigh scattering of the Mössbauer radiation (RSMS) are reviewed with the attention paid to the application of the synchrotron radiation (SR) for the purpose.

1. Introduction

The fullest information about the atomic diffusive motions in the condensed phase can be obtained by observing atomic correlation functions in the temporal and spatial domains *simultaneously* or in the equivalent energy-momentum space [1]. Hence, one requires a momentum transfer of the same order as for the X-ray diffraction at the energy resolution typical for the hyperfine structure spectroscopy. Few methods are able to meet such demanding conditions, i.e., the emission or absorption quasi-elastic Mőssbauer spectroscopy (QMS) [2], incoherent quasi-elastic scattering of cold neutrons (QNS) [3] and Rayleigh (Bragg) scattering of the Mőssbauer radiation (RSMR) [4]. Neutron spectrometers operating in the momentum transfer range accessible to coherent scattering lack sufficient energy resolution except for extremely large unit cells (jump vectors). QMS is practically limited to the 14.4 keV line of ^{57}Fe, and hence, it allows to study diffusivity of Fe and indirectly Co (emission spectroscopy). On the other hand, QNS has barely sufficient energy resolution, and almost insufficient momentum transfers. RSMR is seldom possible with the favorite 14.4 keV line of ^{57}Fe due to the low brilliance of the ^{57}Co sources except for the amorphous targets. On the other hand, RSMR was quite successful

M. Miglierini and D. Petridis (eds.), Mössbauer Spectroscopy in Materials Science, 307–322.

for the 46.5 keV line of ^{183}W assuring sufficient momentum transfer, albeit at the energy resolution comparable to QNS [4].

A recent advent of the very brilliant synchrotron radiation (SR) sources allows to perform relatively easy time-domain Mőssbauer spectroscopy [5]. One has to note, that again the best line for the purpose is 14.4 keV line of ^{57}Fe. One can look for Fe diffusivity in either coherent forward or coherent nuclear Bragg scattering. An incoherent scattering channel can be used as well, albeit the precision of measurements is significantly inferior as compared to the coherent channel in contrast to the measurements of true inelasticity (scattering with phonons emission or absorption). A diffusivity of Fe atoms has been already observed in the forward direction in a complete agreement with the absorption QMS [6].

The extreme brilliance of SR sources allows for RSMR on non-resonant targets (presently in a time-domain) with the favorite 14.4 keV line of ^{57}Fe applying preferably a time-domain Doppler modulated γ-ray interferometer [7]. Such a device has been already applied successfully to the amorphous target [7]. Hence, one can apply it to a crystalline non-resonant target (presently, single crystal) set to Bragg conditions extending QMS to almost all elements except the lightest and the resonant isotope. Due to the fact that Rayleigh scattering is coherent, one makes observations in the coherent (Bragg) channel for crystalline targets. Therefore, the above method is sensitive solely to jumps between various Bravais sub-lattices [8]. On the other hand, it allows to see jumps of the all atoms diffusing within the specimen [8] even in a better way than for QNS.

Table 1 summarizes currently available methods, while Fig. 1 shows accessible ranges in the energy-momentum space.

Few other methods could see diffusivity indirectly, e.g., perturbed angular correlations [9]. A nuclear magnetic resonance is extremely sensitive on the energy scale with a complete lack of the momentum transfer resolution. A similar statement applies to the spin-echo neutron spectrometers and speckle interferometry.

2. How to Describe Diffusivity / Relaxation

Diffusivity and relaxation (fluctuation) of the hyperfine interactions are quite similar phenomena as both of them involve time dependent interactions. A diffusivity depends upon the wave-vector transfer to the system as it involves displacement of the interacting particle. On the other hand, pure relaxation is independent of the momentum transfer. Diffusive motion in the condensed matter occurs via atomic jumps or more generally via events of the short duration, e.g., encounters with the vacancy or vacancy complexes [10]. Such a process could be described by a scalar correlation (self-correlation) function as long as the jumps occur within a single Bravais lattice [11]. Encounter diagrams and the Monte-Carlo simulations are powerful methods to evaluate scalar correlation functions, and to compare with the experimental data [12]. One has to note, that for the wave-vector transfer being one of the reciprocal lattice vectors there is no effect on the observables. It is interesting to note, that for a two-dimensional diffusivity such "insensitivity"

TABLE 1. Methods sensitive to atomic diffusive jumps in energy-momentum domain. $\bar{q}$ stands for the wave-vector transfer, ω for frequency, and t for time.

Method	Domain	Sensitivity to relaxation	Comments
QMS	$\bar{q}\omega$	yes	^{57}Fe - absorption $^{57}Co/Fe$ - emission 1 neV / 7.3 $Å^{-1}$ single crystals (SC) incoherent
QNS (incoherent)	$\bar{q}\omega$	no	mainly hydrogen 0.3 μeV / 3 $Å^{-1}$ SC, polycrystalline, amorphous incoherent
QNS (coherent)	$\bar{q}\omega$	no	lack of ω resolution
RSMR	$\bar{q}\omega$	no	any element beyond B (except resonant) 1 μeV / no limit SC, polycrystalline, amorphous coherent
SR (forward) (Bragg nuclear)	$\bar{q}t$	yes	equivalent to QMS except built-in coherency
SR (incoherent)	$\bar{q}t$	yes	unsuitable like QMS in the scattering geometry
SR (RSMR)	$\bar{q}t$	no	like RSMR except 1 neV / 14 $Å^{-1}$

Note that RSMR in the crystalline matter does not see jumps within a particular Bravais sub-lattice.

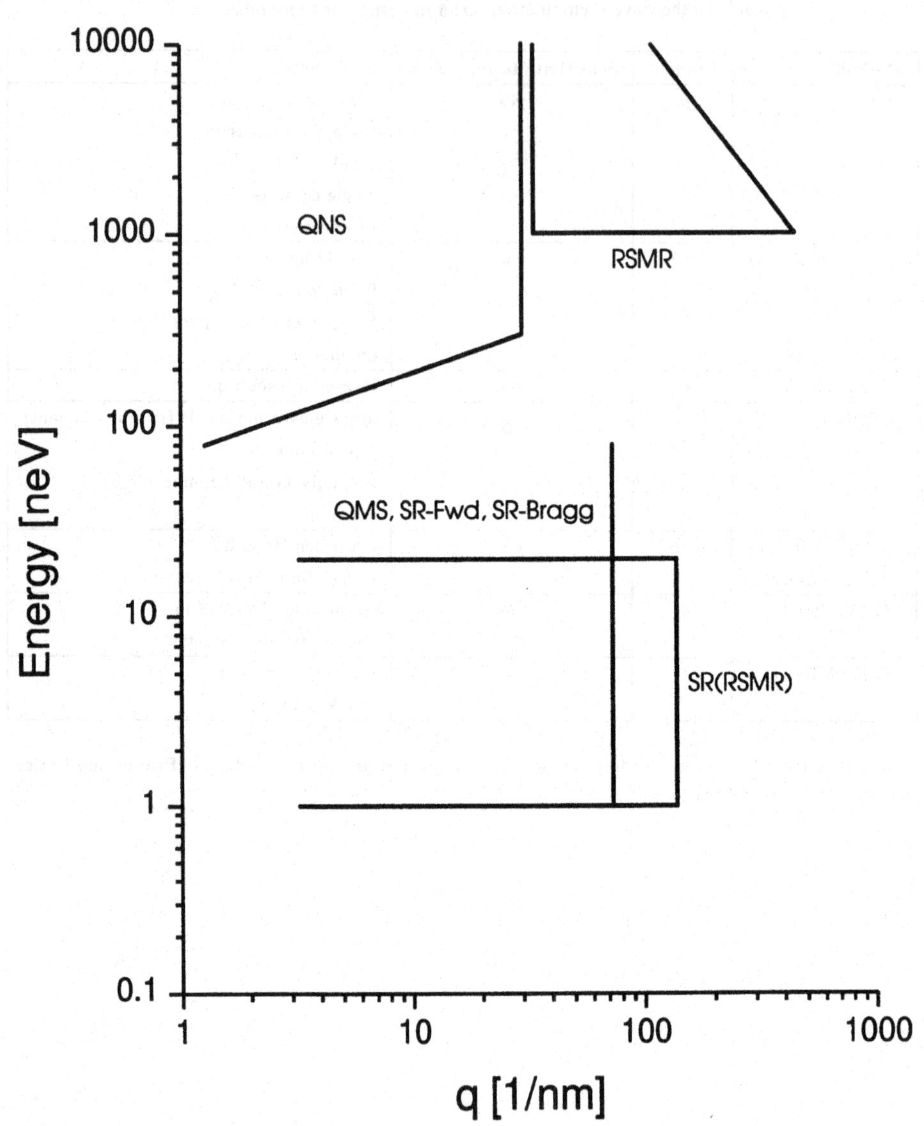

Figure 1. Ranges of different methods in the energy-momentum space.

points transform into lines in the reciprocal space, while for the uni-dimensional diffusivity into surfaces, respectively.

There are many approaches to the relaxation phenomena in the literature, but it seems that the most general is the application of the Liouiville's operators in conjunction with the super-Hamiltonian formalism developed by Blume [13]. The main advantage of the above method is the independence upon the perturbation calculus, and hence, an ability to use a single consistent model all the way from a "static" interaction to the completely averaged case.

Let us consider a very simple example applicable to many real situations, i.e., a spectrum resulting from ^{57}Fe (14.4 keV) embedded in randomly dispersed single domain super-paramagnetic α-Co particles in the zero external magnetic field. Here, a total relaxation operator is a [16 x 16] block diagonal matrix consisting of symmetric [2 x 2] matrices having real off-diagonal elements and complex diagonal elements. The above operator is a complete description of the system as both states (up and down) of the effective field have equal probabilities. Fig. 2 shows eigenvalues of the above [2 x 2] matrices vs reduced relaxation frequency, and the corresponding intensities resulting from the eigenvectors. Resulting Mőssbauer spectra are shown in Fig. 3, while the corresponding SR forward or nuclear Bragg, and incoherent spectra are shown in Fig. 4. One has to note, that the model reproduces underlying physics at all relaxation frequencies, and it gives a motional narrowing in a natural way.

Basically, the same approach works equally well for non-diagonal super-Hamiltonians and/or unequal occupancies of the various relaxation states. One can apply it to the non-equilibrium initial states as well.

2.1. JUMPS BETWEEN VARIOUS BRAVAIS LATTICES

A diffusivity occurring between various Bravais lattices can be described neither via the scalar self-correlation function nor by a more sophisticated multi-particle scalar correlation function. Instead, one has to introduce a diffusion operator being quite similar to the relaxation operator [14]. One can distinguish again symmetrical (equally populated) and asymmetrical cases [15], and/or equilibrium or non-equilibrium initial states [16]. A basic difference between a relaxation and diffusion operator is such, that the latter contains geometrical factors in the off-diagonal elements, and hence, it becomes sensitive to the wave-vector transfer. One has to note as well, that for a long range diffusion there is no motional narrowing. The above approach allows to incorporate a diffusion driven hyperfine relaxation in a straight-forward manner [17]. Namely, the event leading atoms from one to another Bravais lattice might lead from one to another hyperfine Hamiltonian as well. A simple example of such a behavior is a non-spherical molecule containing ^{57}Fe atom, the latter experiencing local electric field gradient (EFG), embedded in a viscous liquid like glycerol. Fig. 5 shows reduced broadenings and reduced splittings vs reduced relaxation frequency for a model with axial EFG fluctuating randomly between three mutually orthogonal axes. One obtains three separate linewidths at sufficiently high frequency, albeit the narrowest one "survives" solely

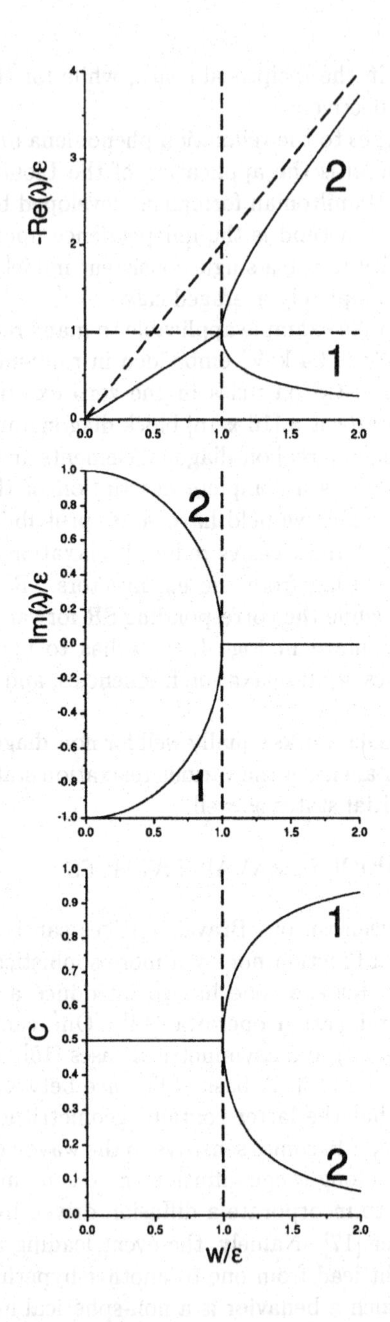

Figure 2. Reduced broadenings *-Re(λ)/ε*, line positions *Im(λ)/ε* and contributions *C* plotted *vs.* reduced relaxation frequency *ω/ε*. A static line separation equals 2*ε*. A simple two state model with equally populated both states is considered.

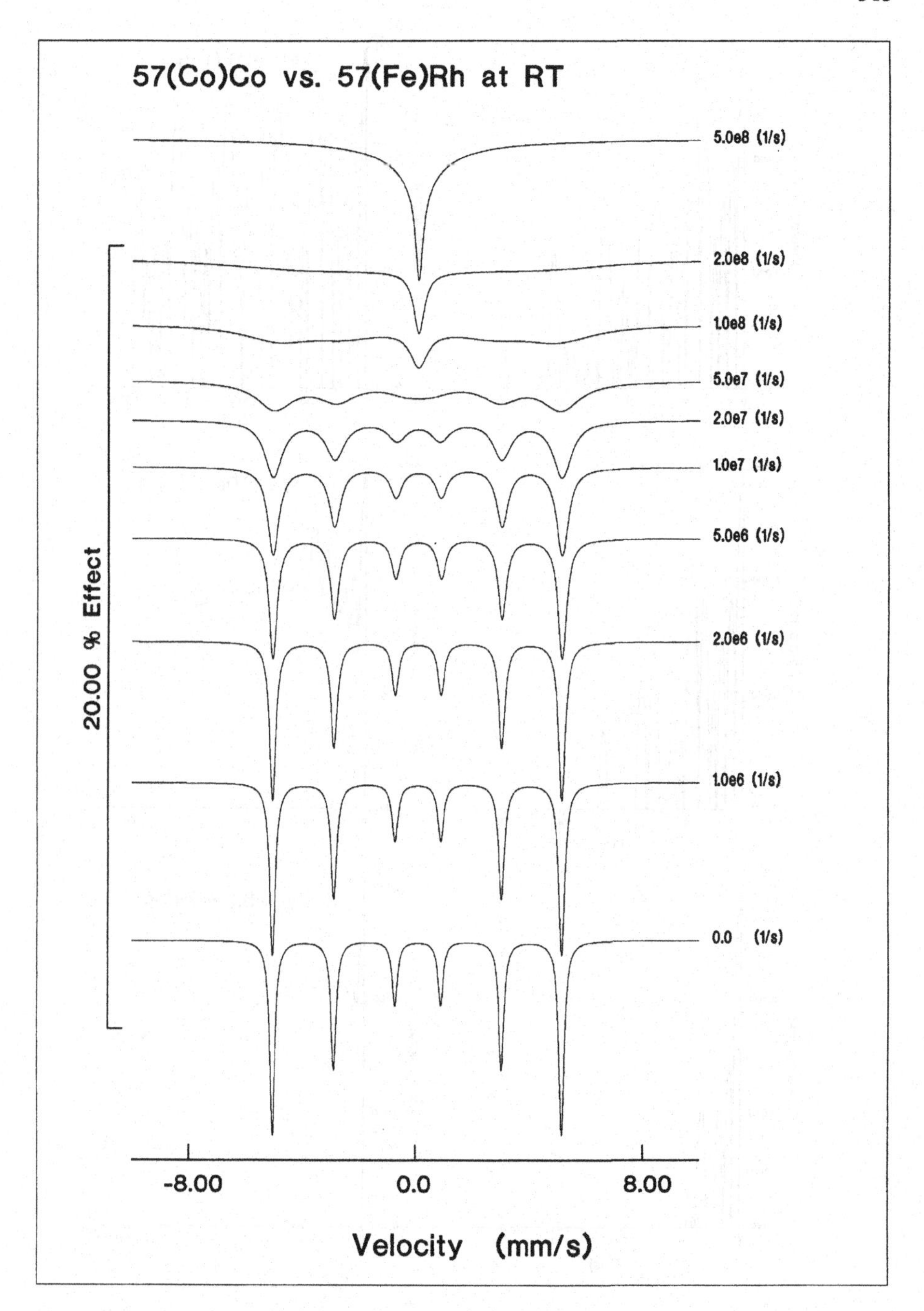

Figure 3. α-Co super-paramagnetic Mössbauer spectra *vs.* relaxation frequency.

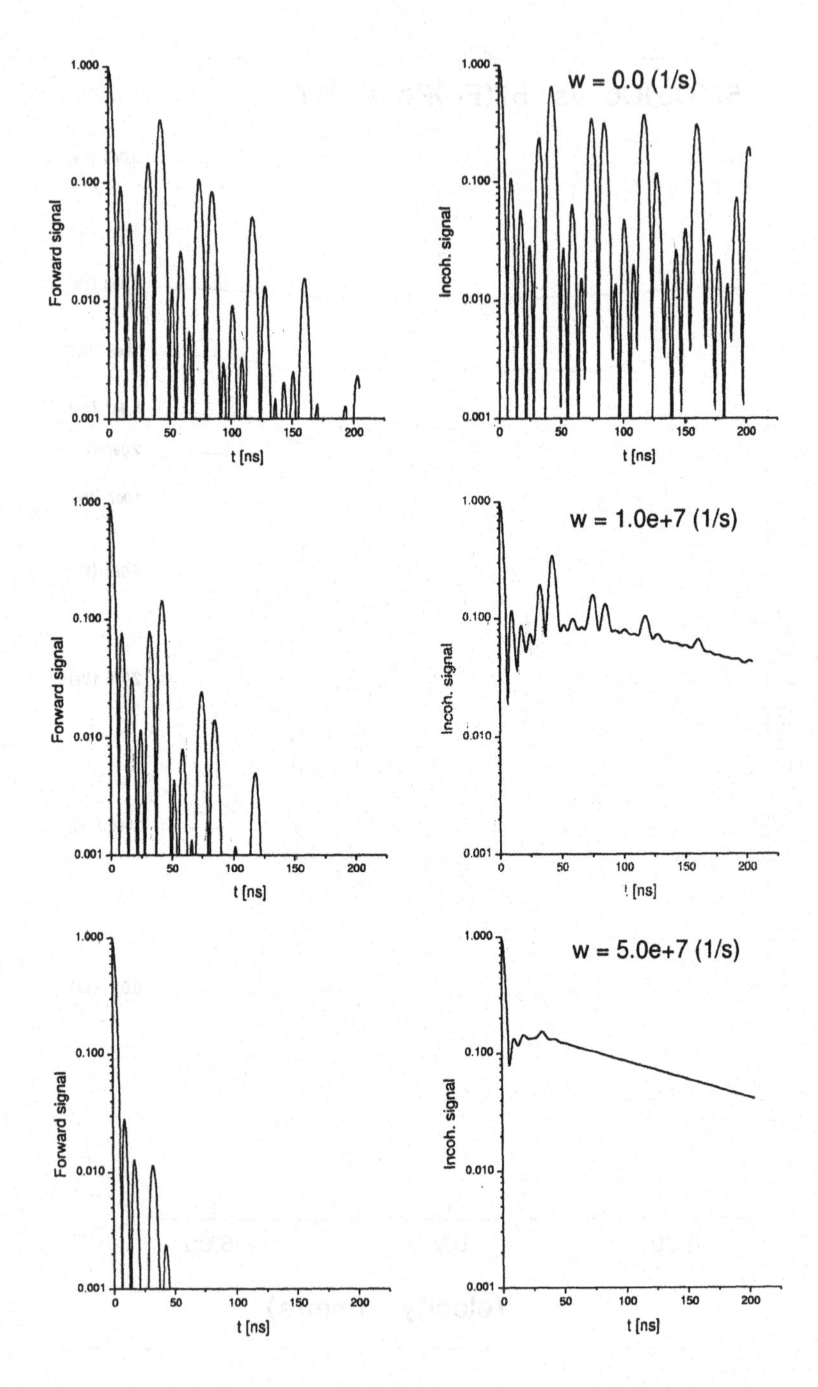

Figure 4. α-Co(Fe) super-paramagnetic forward and incoherent SR signal *vs.* relaxation frequency.

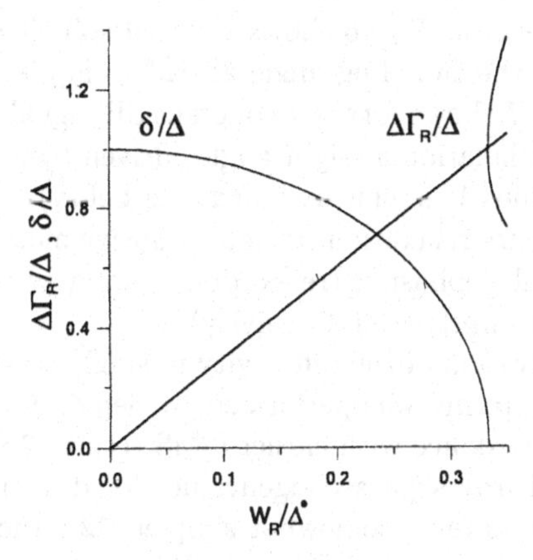

Figure 5. Reduced broadening $\Delta\Gamma_R/\Delta$ and reduced splitting δ/Δ *vs.* reduced relaxation frequency ω_R/Δ for ^{57}Fe quadrupole split spectra. Δ denotes a static splitting due to the axial EFG switching randomly between three mutually orthogonal directions. Note three various linewiths at high frequencies. The narrowest one retains non-zero intensity for the relaxation frequency approaching infinity.

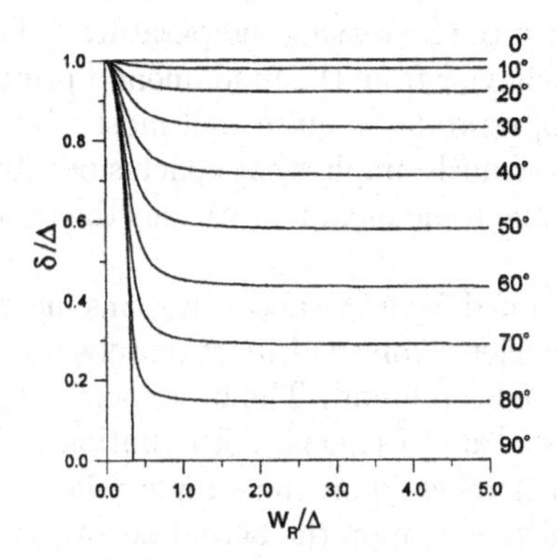

Figure 6. Reduced splitting δ/Δ *vs.* reduced relaxation frequency plotted for various tumbling angles between the original EFG axis and the final EFG axis. Axially symmetric EFG assumed.

at extremely high frequencies. Fig. 6 shows reduced splitting for a slightly more complicated case, where the tumbling angle at each jump is an additional model parameter. Finally, Fig. 7 shows forward coherent SR signal from the above system experiencing EFG relaxation at right angles driven by a basic diffusive jump frequency (a non-Arrhenius behavior assumed, see below). The higher curve at 275 K corresponds to a pure relaxation without diffusive motions. One can clearly distinguish between local dephasing (relaxation leading to motional narrowing) and a global dephasing (unrestricted diffusivity).

The pioneering observation of the diffusivity in single crystals by means of the forward coherent SR scattering was performed by Sepioł *et al.* [6] on the Fe_3Si system in a complete accordance with earlier QMS results [18]. The above results allowed to distinguish clearly separate eigenvalues for direction perpendicular to [113] plane. Similar results for a somewhat simpler B2 structure $FeAl$ were obtained by Vogl and Sepioł [19] using forward SR scattering. All the above data show diffusive speed-up and a reduction of the recoilless fraction with the increasing temperature.

2.2. NON-ARRHENIUS SYSTEMS

Amorphous systems, e.g., fragile glass forming liquids generally do not follow Arrhenius law, and exhibit departures from a Debye behavior as far as the electric susceptibility vs frequency is considered. Biological systems behave in a somewhat similar manner, albeit they usually exhibit some traces of motional narrowing indicative of some restricted motions absent in the former systems. These systems approach Debye liquids with the increasing temperature. There is none consistent theory describing such a behavior from the fundamental principles. A phenomenological Kohlrausch law [20] reproduces quite well most of the observables despite the fact that it leads to obviously unphysical conclusions for short time periods. A more detailed review of the topic including various theoretical approaches could be found in Ref. [4].

Spectroscopic data obtained from the above systems indicate presence of many relaxation/diffusion frequencies involved. Unfortunately, no simple rules governing frequency distributions have been found. The best spectroscopic model being able to cover more than six decades of frequency distribution (still phenomenological, albeit physically consistent) relies upon the simple relationship stating that the rate of fluctuation decay depends upon the actual departure from equilibrium [4]. Such an assumption transforms linear rate equations (diffusion operator) into nonlinear Bernoulli's type differential equations. The model was quite successful in interpreting standard transmission spectra of glycerol doped with ferrous chloride, and RSMR spectra of pure glycerol obtained at the first maximum of the spatial scattering function all the way from a "static" limit to the enormously broadened spectrum [4].

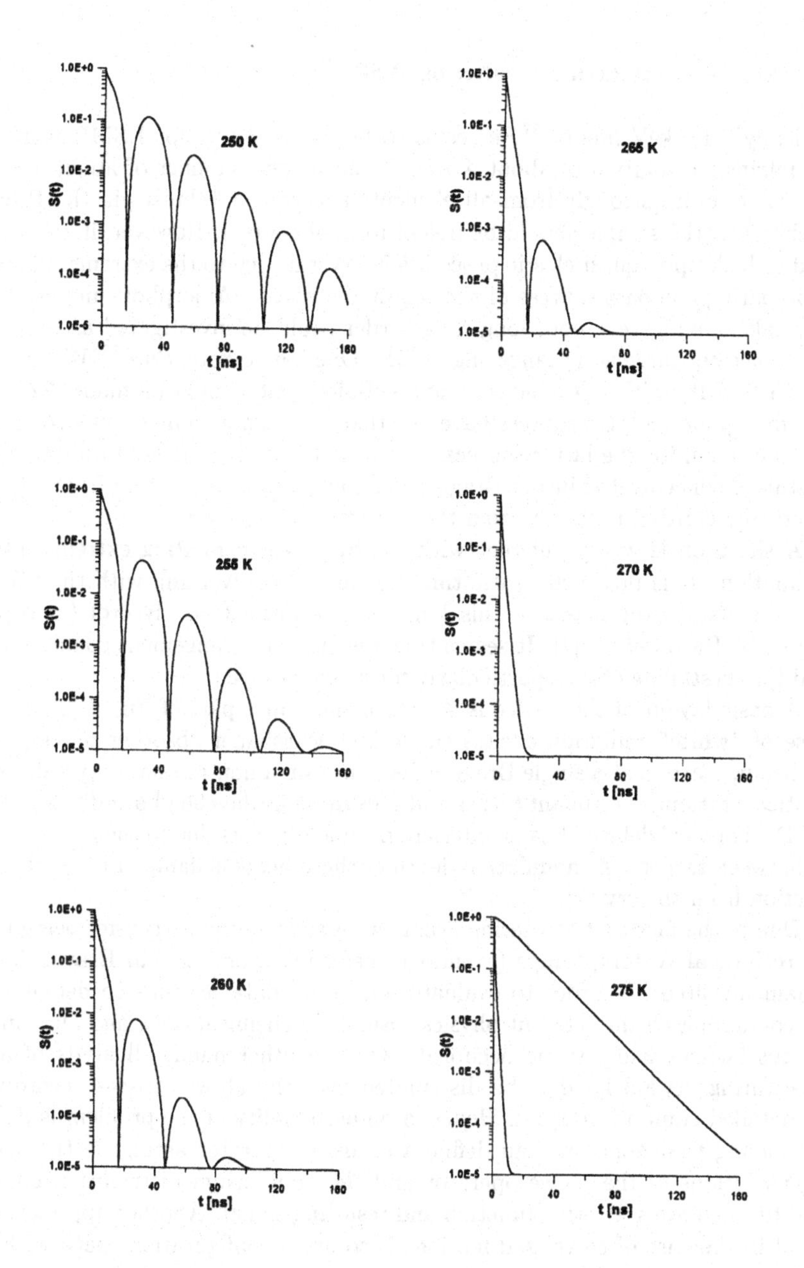

Figure 7. Forward SR signal for axial EFG randomly fluctuating between three mutually orthogonal axes due to the molecule diffusive motion. Upper curve at 275 K shows pure relaxation without diffusive jumps. Non-Arrhenius diffusivity.

3. Non-resonant Rayleigh scattering of SR

Basically, 14.4 keV line of ^{57}Fe seems to be an ideal tool for RSMR work with the intrinsic resolution of about 4.5 neV, and a wave-number of about 7.3 $\AA^{-1}$. It scatters quite strongly from all elements except the lightest via the Rayleigh mechanism (the sample should be free of iron, of course). However, it was seldom used to look upon quasi-elastic processes in crystals due to the extreme difficulties in preparing compact sources of the required activity. Amorphous materials and "crystals" having very poor long range order could be investigated as they allow and "tolerate" large acceptance angles [21]. One has to note, that ^{183}W sources of 100 Ci activity with tolerable resonant self-absorption could be made [22], while the corresponding ^{57}Co sources have less than 0.5 Ci not to mention strong internal conversion for the latter sources. A quasi-elastically scattered radiation from crystals is concentrated in the Bragg reflections, and hence, a tiny fraction of the spherically emitted radiation from the source could be used.

A situation is vastly different with the SR beam exhibiting extreme internal collimation. It is practical to use time domain, while working with the SR. The first successful experiment of this kind was performed on glycerol (amorphous target) by Baron *et al.* [7]. It seems that the similar arrangement could be easily used for crystalline cases - particularly for single crystals.

A basic layout of the device is shown in the upper part of Fig. 8 [8]. A short pulse of "white" radiation excites single line Mőssbauer absorber moving along the beam, scatters at a single Bragg reflection from a non-resonant crystal, excites identical stationary resonant target and falls upon avalanche photo-diode detector (APD). Forward delayed beams interfere producing beats due to the relative velocity between targets. A modulation depth of these beats is dampen by a scattering function from the crystal.

Due to the fact that scattering occurs with a wave-vector transfer being one of the reciprocal vectors, jumps (events) transferring atoms within Bravais lattices remain invisible. In order to evaluate scattering data one has to define a unit cell containing all the relevant physics (usually, a chemical cell), and all primitive Bravais lattices likely to be occupied. On the other hand, all kinds of atoms constituting crystal have to be distributed over the above lattices according to the detailed equilibrium rules. Hence, a dimensionality of the problem is defined, and having that done one can define a diffusion operator acting on the system. Eigensolutions of the above operator and the equilibrium distribution could be used to calculate scattering function and resulting signal. Another approach quite useful in the case of correlated motions is to use a configuration space with each basis vector associated with a particular configuration of the cell, and to define a diffusion operator in the above space together with probabilities of particular configurations following detailed equilibrium [8].

Lower part of Fig. 8 shows a unit cell of the B2 structure (CsCl cell) being composed of two simple cubic lattices interpenetrating each other. Note, that these two primitive lattices converge into a single non-primitive BCC lattice in the case both sites are occupied by the same atoms or there is no energy involved

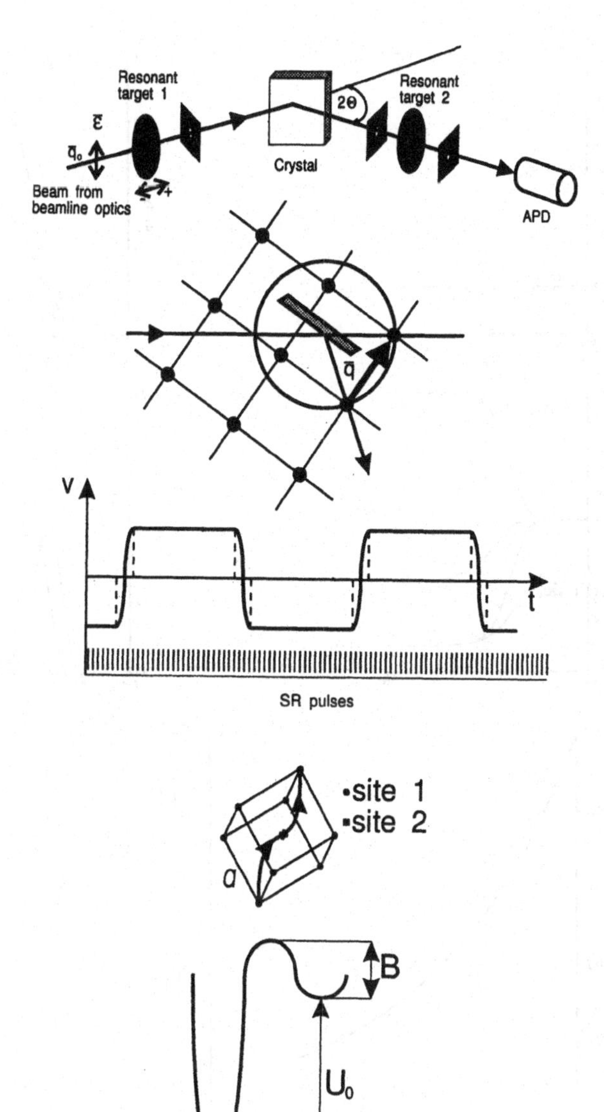

Figure 8. A device to look upon diffusivity in single crystals by means of non-resonant Bragg scattering of the SR generated Mössbauer beams, and a unit cell with corresponding energy barriers for the B2 structure. U_0 stands for the energy of the anti-structural atomic pair.

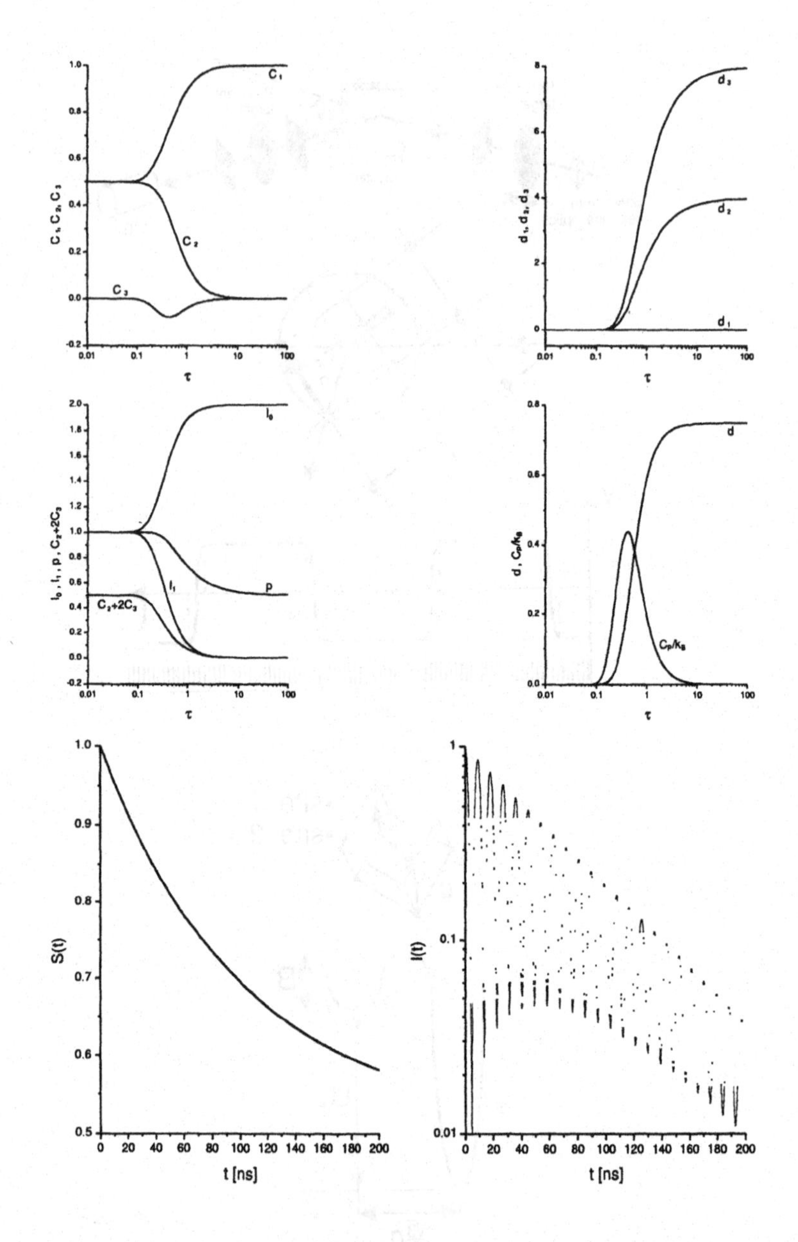

Figure 9. Relative intensities C_1, C_2, C_3 of the exponents constituting B2 structure scattering function *vs.* reduced temperature $\tau = T/U_0$, corresponding reduced arguments of the exponents d_1, d_2, d_3, reduced allowed and forbidden intensities I_0 and I_1, respectively; probability to find anti-structural pair p, a measure of the global monotonicity of the scattering function $C_2 + 2C_3$, reduced diffusion coefficient d, and a reduced heat capacity C_p/k_B. Lower part shows a scattering function $S(t)$ and the corresponding detector signal $I(t)$ *vs.* time.

in segregation of atoms among sites (see, Fig. 8 for "segregation" energy U_0 and a jump barrier B for inversion of the atomic pair into anti-structural pair). The above system could be solved analytically in the configuration space under the assumption that jumps occur to the nearest neighbor sites.

Results are gathered in Fig. 9. Namely, a scattering function is a linear combination of three decaying with time exponents. Contributions C_1, C_2, C_3 and decrements d_1, d_2, d_3 are plotted vs reduced temperature $\tau = T/U_0$ with T being temperature. Additionally, a probability p to find the anti-structural pair, reduced allowed I_0, and reduced forbidden I_1 intensities, respectively, and a measure of the scattering function monotonicity with time $C_2 + 2C_3$ are plotted in the same manner. Some auxiliary results like a reduced diffusion coefficient d due to the above mechanism, and a reduced contribution to the heat capacity under the constant pressure C_p/k_B are shown vs reduced temperature as well. Finally, a typical scattering function $S(t)$ and a resulting signal $I(t)$ are shown. One can note, that a scattering function does not fall here to zero, but to a constant positive value, i.e., a hard-core purely elastic scattering always remains. For very high temperatures some kind of motional narrowing occurs (no decay of scattering function) due to the simple fact that high temperatures restore broken symmetry leading to a BCC like lattice, i.e., a single Bravais lattice.

4. Conclusions

The paper is aimed at a brief review of the recent developments in the methods enabling to look upon diffusivity in the complete energy-momentum space. In our opinion one can expect a rapid development and understanding of the RSMR method in the momentum-time domain. Application of the SR Mőssbauer beams for the purpose is likely to open a new branch of physics, where the scattering functions could be obtained with a very high energy resolution at high momentum transfers. These regions were hardly accessible in the past, if at all.

On the other hand, an understanding of the glassy systems still remains a challenge.

Acknowledgements

Dr. Urszula D. Wdowik, Institute of Physics and Computer Science, P.U., Cracow, Poland is thanked for many comments.

References

1. Petry,W. and Vogl, G. (1987), Mat.Sci.Forum **15-18**, 323.
2. Vogl, G. (1990), Hyperfine Interact. **53**, 197.
3. Blank, H. and Maier, B. (eds.) (ILL, Grenoble, 1988), Guide to Neutron Research Facilities at the ILL - the Yellow Book.
4. Ruebenbauer, K., Mullen, J.G., Nienhaus, G.U., and Schupp, G. (1994), Phys.Rev.B **49**, 15607.
5. Gerdau, E., Rűffer, R., Winkler, H., Tolksdorf, W., Klages, C.P., and Hannon, J.P. (1985), Phys.Rev.Lett. **54**, 835.

6. Sepioł, B., Meyer, A., Vogl, G., Rüffer, R., Chumakov, A.I., and Baron, A.Q.R. (1996), Phys.Rev.Lett. **76**, 3220.
7. Baron, A.Q.R., Franz, H., Meyer, A., Rüffer, R., Chumakov, A.I., Burkel, E., and Petry, W. (1997), Phys.Rev.Lett. **79**, 2823.
8. Ruebenbauer, K. and Wdowik, U.D. Phys.Rev.B - in press.
9. Kwater, M., Ruebenbauer, K., and Wdowik, U.D. (1993), PhysicaB **190**, 199; Kwater, M., Ruebenbauer, K., and Wdowik, U.D. (1993), *ibid.* **190**, 209.
10. Ruebenbauer, K. Sepioł, B. (1986), Hyperfine Interact. **30**, 121.
11. Singwi, K.S. and Sjölander, A. (1960), Phys.Rev. **120**, 1093.
12. Ruebenbauer, K., Sepioł, B., and Miczko, B. (1991), PhysicaB **168**, 80.
13. Blume, M. (1968), Phys.Rev. **174**, 351.
14. Chudley, C.T. and Elliott, R.J. (1961), Proc.Phys.Soc. London **77**, 353.
15. Kutner, R. and Sosnowska, I. (1977), J.Phys.Chem.Solids **38**, 741.
16. Ruebenbauer, K., Wdowik, U.D., Kwater, M., and Kowalik, J.T. (1996), Phys.Rev.B **54**, 12880.
17. Ruebenbauer, K., Wdowik, U.D., and Kwater, M. (1996), Phys.Rev.B **54**, 4096.
18. Sepioł, B. and Vogl, G. (1993), Phys.Rev.Lett. **71**, 731.
19. Vogl, G. and Sepioł, B. - to be published.
20. Kohlrausch, R. (1854), Ann.Phys. (Leipzig) **91**, 56.
21. Champeney, D.C. (1979), Rep.Prog.Phys. **42**, 1017.
22. Schupp, G., Hammouda, B., and Hsueh, C.M. (1990), Phys.Rev.A **41**, 5610.

SYNCHROTRON MÖSSBAUER REFLECTOMETRY IN MATERIALS SCIENCE

Magnetic structure and phase analysis in thin films

D.L. NAGY[1], L. BOTTYÁN[1], L. DEÁK[1], J. DEKOSTER[2], G. LANGOUCHE[2], V.G. SEMENOV[3], H. SPIERING[4] AND E. SZILÁGYI[1]

[1)] *KFKI Research Institute for Particle and Nuclear Physics, Budapest, Hungary*

[2)] *Instituut voor Kern- en Stralingsfysica, University of Leuven, Leuven, Belgium*

[3)] *Chemical Institute, Saint Petersburg University, Saint Petersburg, Russia*

[4)] *Inst. für Anorganische und Analytische Chemie, Universität Mainz, Mainz, Germany*

Abstract

^{57}Fe nuclear resonant scattering experiments are reported on iron-containing thin films using 14.41 keV synchrotron radiation at angles of grazing incidence around and slightly above the critical angle of the electronic total reflection. In partially oxidised α–Fe films of 20 nm original thickness various oxide and oxihydroxide phases are identified at different depth. In a $[Fe/FeSi]_{10}$ multilayer grown on Zerodur® substrate the Fe–Fe interlayer coupling varies with the distance from the substrate. The antiferromagnetic order of the top layers of this multilayer can be suppressed by external magnetic field. These examples demonstrate the efficiency of *synchrotron Mössbauer reflectometry (SMR)*, a new method capable of depth profiling the hyperfine interactions on a nm scale.

1. Introduction

Specular reflection of x-rays and neutrons is being widely used in materials science. The corresponding methods, viz. *x-ray* and *neutron reflectometry* have become established techniques of studying the structure of thin films, implanted semiconductors and metals, multilayers and superlattices. Both x-ray and neutron reflectometry are able to map the depth profile of chemical composition on a nm scale. Besides, neutron reflectometry bears information on the depth profiles of different isotopes of the same element as well as on the magnetic structure of thin films and multilayers. The latter feature makes

M. Miglierini and D. Petridis (eds.), Mössbauer Spectroscopy in Materials Science, 323–336.

neutron reflectometry an efficient technique in studying coupling mechanisms in magnetic multilayers. Since photons have no magnetic moment, x-ray reflectometry is only marginally sensitive to the magnetic structure (by higher order effects). In the vicinity of low-lying nuclear (Mössbauer transition) energies, however, via the hyperfine interactions, the nuclear resonant x-ray reflectivity is also informative of the magnetic structure of the thin film sample. This latter method, often called Synchrotron Mössbauer Reflectometry, or simply SMR, is shortly introduced here with applications in materials science of surfaces and thin films.

1.1. THE PRINCIPLES OF REFLECTOMETRY

Reflectometry uses the fact that the real part of the index of refraction n of most materials for thermal neutrons and of all materials for x-rays is by about 10^{-5} less then unity. At low enough angles Θ of grazing incidence (for n real, i.e. for negligible absorption $\Theta < \Theta_c = \sqrt{2(1-n)}$ the waves are, therefore, totally reflected from a flat surface. The intensity of the reflected specular beam for $\Theta > \Theta_c$ rapidly decreases with increasing wave vector transfer $Q = 2k \cdot \sin\Theta$ where k is the wave vector of the incident radiation. In a stratified medium, reflected and refracted beams appear at each interface. The interference of the reflected beams leads to characteristic patterns in $R(Q)$, the reflectivity vs. wave vector transfer curve, that bear information on the depth profile of the index of refraction, $n(z)$, the argument z being the co-ordinate perpendicular to the sample surface.

The aim of a reflectometric measurement is to reconstruct $n(z)$ from $R(Q)$ since, as it will be shown below, the depth profile of the index of refraction contains information characteristic of the chemical and magnetic structure of the thin film. Various methods exist for calculation of $R(Q) = |r(Q)|^2$ from $n(z)$ one of them being the *method of characteristic matrices* [1] ($r(Q)$ is the reflectivity amplitude). The inverse problem, however, viz. reconstructing $n(z)$ from $R(Q)$ can in most cases only be solved in frames of a given model for the stratified system. Additional information like Rutherford backscattering (RBS) experiments, therefore, turns out to considerably enhance the reliability of the evaluation of reflectometric data [2].

1.2. THE COHERENT FORWARD-SCATTERING AMPLITUDE

The coherent forward scattering of a scalar wave of momentum much higher than that of the scatterers can be described [3] by the index of refraction close to unity

$$n = 1 + \frac{2\pi N}{k^2} f \qquad (1)$$

where N is the density of scatterers and f is the scattering amplitude. The information for materials science applications is implied in the latter quantity. For neutrons, f is the sum of the nuclear and the magnetic scattering lengths b and p, respectively, while for x-rays (far from absorption edges and from nuclear resonances, i.e. Mössbauer transitions) f is determined by the Rayleigh scattering and photoabsorption i.e. by the electron density. Close to a Mössbauer resonance E_0, however, the photon's coherent scattering amplitude f in Eq. (1) is a sum of the electronic and the nuclear coherent scattering amplitudes:

$$f(E) = f_e + f_n(E) \tag{2}$$

the latter being a rapidly varying function of the energy around E_0. The nuclear scattering amplitude $f_n(E)$ can be expressed in terms of the matrix elements $a_{\alpha\beta}$ of the hyperfine Hamiltonian [4]:

$$f_n = \frac{kV}{hc} f_{LM} \frac{1}{2I_g + 1} \sum_{\alpha,\beta} \frac{|a_{\alpha\beta}|^2}{E - (E_\alpha - E_\beta) + i\frac{\Gamma}{2}} \tag{3}$$

where k is the wave number, V the normalisation volume, f_{LM} the recoilless (Lamb–Mössbauer) fraction, I_g the ground state nuclear spin, E_α and E_β are the nuclear excited and ground state energies, respectively ($E_\alpha - E_\beta = E_0$) and Γ is the natural linewidth. For nuclear resonant scattering, f and n are 2×2 matrices rather than scalars since the two polarisation states of photons should be accounted for [4]. The optical approach based on Eq. (1) is exactly valid for the coherent forward scattering of resonant photons and it is also a good approximation for specularly reflected grazing incident photons [5].

1.3. MÖSSBAUER REFLECTOMETRY

X-ray and neutron reflectometry can be used for mapping the isotopic and the electron density structure of thin films. Besides, neutron reflectometry is an efficient method to study thin film magnetism, as well. A drawback of both methods is, however, their insensitivity for local (hyperfine) interactions.

A common feature of methods utilising hyperfine interaction such as nuclear magnetic resonance, Mössbauer spectroscopy and perturbed γ–γ angular correlation (PAC) is the lack of or limited depth selectivity. Conversion electron Mössbauer spectroscopy (CEMS) can be used for depth selective measurements (Depth Selective Conversion Electron Mössbauer Spectroscopy, DCEMS) by analysing the energy of the conversion electrons generated in the resonant process [6]. The information collection speed of DCEMS is very low. Furthermore, the typical depth resolution of this method is about 20–50 nm (may be somewhat improved in the near-surface region).

Close to Θ_c, the penetration depth of the γ–radiation is rapidly decreasing with decreasing Θ down to a few nanometers. This fact is the key to grazing incidence Mössbauer spectroscopy (GIMS) a method capable of depth profiling hyperfine interactions. In most GIMS experiments conversion electrons (Total External Reflection Conversion Electron Mössbauer Spectroscopy, TERCEMS) [7,8] are detected. By changing Θ around Θ_c (for the γ–ray energy $E = 14.41$ keV and an α–Fe sample $\Theta_c = 3.8$ mrad) and taking at each angle a CEMS spectrum one can, roughly speaking, adjust the depth at which the thin film is 'sampled'. The other channel that is often used is that of specularly reflected photons. The approach utilising specularly reflected photons may be called Mössbauer reflectometry (MR) [8,9,10]. MR is based on the fact that, according to Eqs. (1–3), the refraction index $n(E,z)$, at a certain depth, z, contains information on the hyperfine interactions, i.e. on the chemistry, crystal and magnetic structure of the film. Consequently, by measuring $R(E,\Theta)$ in an MR experiment one can reconstruct the depth profile of chemical compounds containing the resonant nucleus along with their hyperfine electric and magnetic structure. (Since k is practically constant in a Mössbauer experiment, henceforth we shall use Θ rather than Q in the argument of R and r.)

The realisation of MR (like of TERCEMS) is quite time-consuming. Even using a strong Mössbauer source of about 5 GBq activity, the typical time of a measurement at a single Θ value is one week or more. Consequently, data collection time of $R(E,\Theta)$ is several months. The reason for this fact is that the radiation of a radioactive source is unavoidably emitted in the full 4π solid angle a tiny fraction of which (about 5×10^{-6} sterad) really hits the sample. MR should be realised, therefore, utilising a source that generates a well-collimated radiation. This is why synchrotrons are best suited for grazing incidence nuclear resonant scattering experiment. This is the approach that we shall call, henceforth, synchrotron Mössbauer reflectometry (SMR).*

In the present paper we shall first outline the basic principles of SRM followed by a brief overview of the SMR experiments published so far. The potentialities of SMR in depth selective phase analysis will be demonstrated on experiments performed on an ^{57}Fe film of 20 nm original thickness exposed to controlled corrosion. It will turn out that the hyperfine interaction in the upper region is different from that in the deeper regions of the film. In a following example we shall show that, using SMR, not only the chemical but also the magnetic structure of a magnetic multilayer can be determined with a high accuracy. In fact, an experiment performed in various external magnetic fields on an artificial $[Fe/FeSi]_{10}$ multilayer grown on Zerodur® substrate shows that the direction of the hyperfine field is different in different external fields and at different distances from the substrate.

2. Synchrotron Mössbauer Reflectometry

Due to its unique characteristics, such as high brilliance, high degree of polarisation, high collimation, small beam size, pulsed time structure, extremely high longitudinal and finite transversal coherence length, tunability, synchrotron radiation (hereafter SR) has rapidly become a powerful tool in solid state physics and materials sciences. Since 1985, the first successful demonstration of nuclear resonant scattering (NRS) [11], the field of hyperfine interactions has also become accessible to SR experiments. In contrast to conventional Mössbauer spectroscopy sensitive to the (energy domain) hyperfine shift and splitting of nuclear transitions NRS of SR shows the effect of hyperfine interactions in form of (time domain) quantum beats.

SMR means measuring the time-dependent reflectivity $R_t(t,\Theta)$, i.e. taking several time-domain grazing incidence spectra after the coherent excitation of the hyperfine split nuclear sublevels by the synchrotron pulse (at $t = 0$) at various angles of grazing incidence Θ and evaluating all these spectra simultaneously in order to determine the

* The same method is often referred to as „grazing incidence nuclear resonant scattering of SR". We prefer, however, SMR, since it is shorter and emphasises the close relation of this technique to x-ray and neutron reflectometry.

depth profile of hyperfine interactions. (Subscript t was introduced only to make a formal difference between the energy and time domain reflectivities). In contrast to x-ray and neutron reflectometry, the depth selectivity of energy domain MR mainly comes from the Θ-dependence of the penetration depth around Θ_c. In the following we shall show that a high number of delayed photons is always available around Θ_c, a fact of crucial importance for SMR.

2.1. THE TOTAL REFLECTION PEAK

As compared to the nuclear lifetime of the ^{57}Fe Mössbauer level of about 10^{-7} s, the electronic scattering is prompt. In a synchrotron time domain experiment, due to the dead time of the detector and electronics, the delayed photons are detected beginning about 10 ns after the exciting pulse. One would expect that all delayed photons stem from the nuclear scattering and will, therefore, not necessarily have significant contribution around Θ_c. Since, however, the photon is coherently scattered on the whole system of nuclei and electrons the electronic scattering amplitude also contributes to the reflectivity amplitude $r(E,\Theta)$ and therefore to the number of delayed photons detected between t_1 and t_2 which is proportional to the integral reflectivity

$$D(\Theta) = \int_{t_1}^{t_2} R_t(t,\Theta)\,dt = \int_{t_1}^{t_2} |r_t(t,\Theta)|^2 dt \tag{3}$$

where

$$r_t(t,\Theta) = \frac{1}{2\pi} \int_{-\infty}^{+\infty} r(E,\Theta) \exp(\frac{-i}{\hbar} Et) dE \,. \tag{4}$$

As a consequence, as first predicted by model calculations [12], not only the reflectivity of the *prompt* photons but, surprisingly, also the integral reflectivity of the *delayed* photons has a maximum at Θ_c (somewhat more precisely at high enough negative derivative of the electronic reflectivity [13]). This "total reflection peak" has also experimentally been observed [13] and, since then, widely used in all synchrotron NRS experiments at grazing incidence. In spite of the fact that various qualitative explanations have been published [13,14,15] admittedly no clear physical picture describes the existence of the total reflection peak so far. The existence of this peak ensures that a time domain MR, i.e. SMR experiment is always feasible around Θ_c.

2.2. LOW ANGLE BRAGG REFLECTIONS, TIME INTEGRAL SMR

Like in neutron- and x-ray reflectometry, $D(\Theta)$ has local maximum if the Bragg condition for an artificial periodic multilayer is fulfilled. An important feature of SMR is its sensitivity to alternating hyperfine interactions in consecutive sublayers. Akin to neutron reflectometry, also in SMR superreflections appear on $D(\Theta)$ due to period-doubling in antiferromagnetically coupled multilayers. From that point of view, the information yield by $D(\Theta)$ or, in other words, by *time integral SMR* (TISMR) is fairly comparable with that of unpolarised neutron reflectometry. The data collection time is, however, in most cases shorter. At the nuclear resonance scattering beamline ID18 of the European Synchrotron Radiation Facility, Grenoble (ESRF) in 16 bunch mode the typical time of a TISMR scan on a 2 cm long sample is 20 minutes.

Apart from the trivial method-dependent correction, Q_c (corresponding to the change of the optical path in a refractive medium) the low angle Bragg reflections will appear at nearly the same value of wave vector transfer

$$Q_B' = \sqrt{Q_B^2 + Q_c^2} \tag{5}$$

for x-ray, neutron as well as TISMR scans. Here Q_B' is the actual wave vector transfer of the low angle Bragg peak, fulfilling the Bragg condition $Q_B = 2\pi/d$, Q_c is the critical wave vector transfer of the total reflection and d is the structural or magnetic period length of the multilayer.

2.3. TIME DIFFERENTIAL SMR

The time response $R_t(t,\Theta)$ of the film at a fixed value of Θ shows quantum beats due, very roughly speaking, to hyperfine interactions in layers of the film from the top down to the penetration depth of photons at the angle of grazing incidence Θ. These are the *time differential SMR* (TDSMR) spectra. The time necessary for such a measurement very much depends on Θ and the structure of the film. Typical data collection times at ID18 range from 10 to 100 minutes.

The full depth selective information in an SMR experiment is contained in the whole two-dimensional spectrum $R_t(t,\Theta)$, i.e. the reflectivity as a function of the angle of grazing incidence and the time elapsed since the synchrotron pulse. Due to the new *continuous* variable t, the two-dimensional $R_t(t,\Theta)$ curves may contain more information than neutron reflectograms yielding one, two or three scans for unpolarised neutron source, polarised neutron source, and polarised neutron source with polarisation analysis, respectively. Neutron reflectometry and SMR, however, are complementary rather than exclusive methods.

2.4. DATA EVALUATION

The evaluation of $R_t(t,\Theta)$ is a more sophisticated task than that of neutron and x-ray reflectivities since not only the interference of the radiation from different depths but also the absorption should be accounted for. Moreover, close to the nuclear resonance, both the real and the imaginary part of the index of refraction are strongly energy-dependent. It has been shown [5] that MR spectra just in the (1 mrad $\leq \Theta \leq$ 10 mrad) region of practical interest for the ^{57}Fe resonance can be expressed in terms of the coherent forward-scattering amplitude. Consequently, a fast numerical algorithm [5] exists which allows for fitting the whole two-dimensional spectrum $R_t(t,\Theta)$ adjusting a few parameters of a model set for the layered structure.

An important feature of the evaluation of SMR data is the necessity of the simultaneous fitting of the quantum beat spectra for different Θ values. Besides, the parameters of hyperfine interactions should agree with those taken from CEMS data measured on the same sample. Also the prompt x-ray reflectivity and the TISMR data as well as, if available, the neutron reflectivities and TERCEMS data should be evaluated simultaneously.

A program allowing for a unified treatment of all these features EFFINO [16] has recently been developed and is available on request from one of the authors (H. S., e-mail: spiering@iacgu7.chemie.uni-mainz.de) or can be downloaded from the MIX website (http://www.kfki.hu/~mixhp/).

2.5. METHODOLOGICAL ASPECTS

SMR experiments can be performed at any nuclear resonant scattering SR beamline since, as compared with forward-scattering experiments, the only necessary additional instrument is a goniometer installed between the high resolution monochromator and the movable detector ensuring the Θ–2Θ geometry. The proper adjustment of the sample goniometer is extremely important in both kinds of SMR experiments: the Θ = 0 position should be established with an accuracy of at least 0.05 mrad.

Like in x-ray and neutron reflectometry, the quality of the substrate is a most important issue. In order not to smear out the Bragg peaks, strict requirements towards surface and interface roughness (not exceeding about 0.5 nm) and substrate flatness (local curvature radius not less than about 100 m) should be fulfilled. If SMR is only used for depth selective phase analysis the roughness can be increased to maximum 1–2 nm, nevertheless, the high flatness should be retained in this case, as well. X-ray mirror materials like float glass or Zerodur® fulfil these criteria. Such substrates are, however, not always applicable. First, being amorphous, no superlattices can be grown on these substrates and, second, their maximum annealing temperature may be limited. If using single crystal substrates like MgO, Si, GaAs or sapphire, the roughness and flatness should be characterised in advance by STM, AFM, x-ray reflectometry and profilometry.

In a TISMR measurement the delayed specularly scattered photons from t_1 to t_2 are collected and stored as a function of Θ in a Θ–2Θ scan. Usually the prompt scattered photons are also measured by another detector downstream in the beam and stored in another memory sector along with the TISMR data; in other words an x-ray reflectometry scan is simultaneously obtained. Since, however, the count rate of the prompt scattering by several orders of magnitude exceeds that of the delayed photons special care should be taken to avoid overloading the detector and electronics in the total reflection region.

In a TDSMR measurement both the sample goniometer (Θ) and the detector (2Θ) are fixed. Delayed counts are collected as a function of time after the last synchrotron pulse. If the distance between two subsequent synchrotron pulses is not much higher than the nuclear lifetime the evaluation of TDSMR spectra should account for the fact that a small part of the delayed photons is a response not to the last but to one of the preceding synchrotron pulses.

Besides the higher performance in data collection speed and depth resolution SMR has a further important advantage as compared to DCEMS and TERCEMS. In fact, in an SMR experiment there is no detector-implied limitation regarding temperature and external magnetic field at sample position.

2.6. PREVIOUS SMR STUDIES

SMR experiments started 1993 with the first successful observation of a pure nuclear reflection of SR from an isotopically periodic $^{57}Fe/Sc/^{56}Fe/Sc$ multilayer by Chumakov et al. [17]. The aim of this work was to contribute to the development of nuclear monochromators of μeV bandwidth rather than to study the structure of the multilayer. Deák et al. [18] have recently shown that such monochromators can be built from electronically homogeneous isotopically periodic $^{56}Fe/^{57}Fe$ multilayers, as well. In an unpublished study [19] the TDSMR spectra of a $^{nat}Fe/^{57}Fe$ multilayer ("nat" for natural isotopic abundance) have been measured.

Applications of SMR to materials science are still scarce. Studies can be divided into two groups, viz. *a)* depth selective phase analysis of thin films and *b)* magnetic structure analysis of thin magnetic films and of magnetic multilayers.

The way to depth selective phase analysis has been open by the demonstration of the existence of the total reflection peak in TISMR by Baron et al. [13]. The feasibility of depth selective phase analysis by TDSMR was first shown at the Mössbauer beamline F4 of HASYLAB, DESY, Hamburg on the example of oxidised ^{57}Fe films and of an $Al/^{57}Fe$ bilayer before and after ion beam mixing with 120 keV Xe ions [20]. In Chapter 3 we briefly present new, high quality TDSMR measurements on the same oxidised iron films taken at ESRF Grenoble. A detailed quantitative analysis of these data will be published elsewhere [21]. In a very recent TDSMR study by Niesen et al. [22] the existence of a nonmagnetic Ge/Fe interlayer and the sensitivity of SMR down to one monolayer ^{57}Fe was demonstrated in $Au/^{57}Fe$ films on Ge substrate. The ultimate depth selectivity of one monolayer reported in the latter work admittedly stems from the fact that the films contained monolayers of the resonant ^{57}Fe probe rather than from the inherent depth selectivity of SMR which is, however, never better than about 10 monolayers. Andreeva et al. [23] studied the depth profile of the electron density and of the hyperfine field in a $Zr(10\ nm)/[^{57}Fe(1.6\ nm)/Cr(1.7\ nm)]_{26}/Cr(50\ nm)$ multilayer by x-ray reflectometry and TDSMR, respectively.

The feasibility of thin film magnetic structure analysis was shown by Toellner et al. [24]. These authors demonstrated the existence of pure nuclear reflections due to alternating hyperfine magnetic fields from an Fe/Cr multilayer with a Cr thickness of 1.0 nm mediating an antiferromagnetic (AF) interaction between the Fe layers in a TISMR experiment. In the same contribution [24], TDSMR spectra at the total reflection peak, the AF Bragg peak and the structural Bragg peak were presented and interpreted in terms of a collinear AF structure of the multilayer and an appropriate distribution of the hyperfine fields taken from conventional CEMS spectra.

In Chapter 4 we present some recent studies on AF coupled Fe/FeSi multilayers. Evaluation [25] of these results shows the extreme sensitivity of TDSMR to the direction of sublayer magnetisation. A detailed analysis of the same data will be published elsewhere [26]. TISRM in a field-cooled Fe/FeSi multilayer revealed an anomalous, apparently ferromagnetic “frozen-in” state [27]. The analysis of the corresponding TDSMR data shows, however, that this state is not pure ferromagnetic but the sublayer magnetisations open to about 7° [28].

3. Depth selective phase analysis of oxidised iron films

A series of 20 nm thick ^{57}Fe films grown onto float glass substrate was exposed to annealing in air up to at 285 °C for 4 hours. X-ray reflectometry, TISMR and TDSMR measurements were performed on all samples at the ID18 nuclear resonance beamline of ESRF (E = 14.413 keV, λ = 0.0860 nm) in an external magnetic field B_{ext} = 0.37 T perpendicular to the scattering plane. Fig. 1 shows spectra at various angles (1.6 mrad ≤

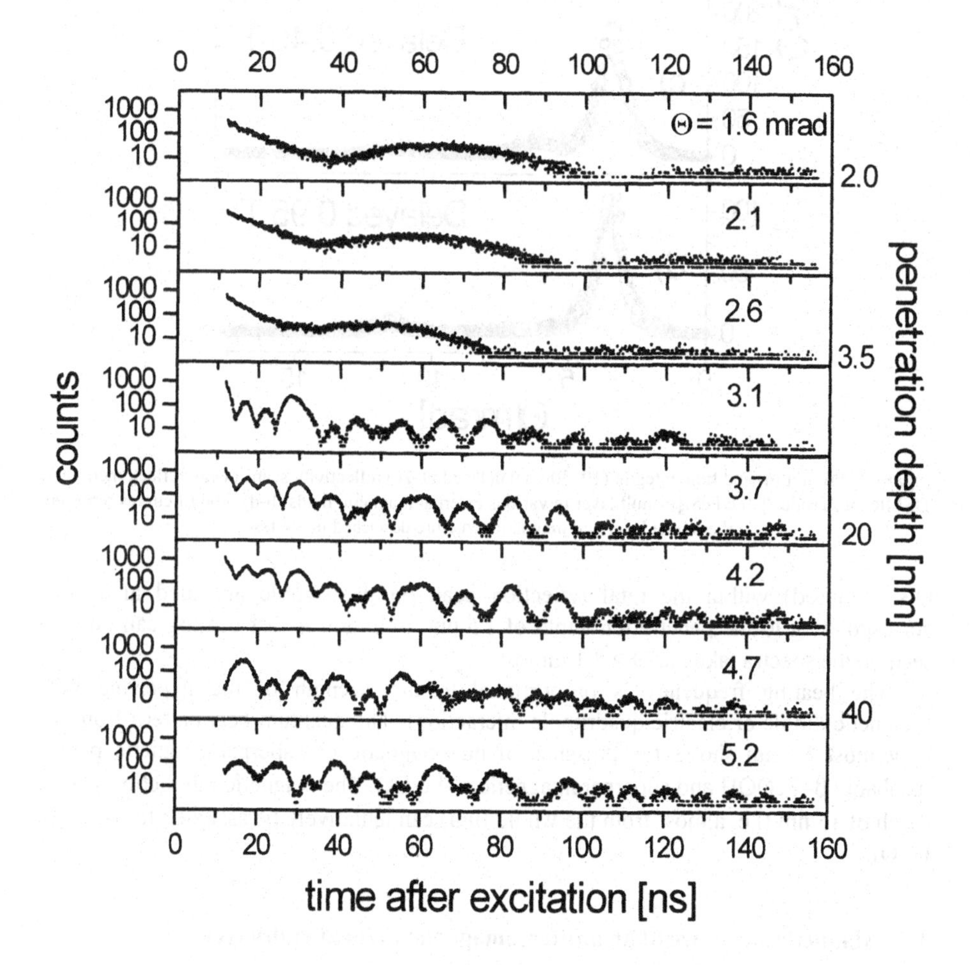

Figure 1. TDSMR spectra measured at different angles of grazing incidence Θ on an originally 20 nm thick ^{57}Fe film grown onto float glass substrate and exposed to annealing in air at 170 °C for 4 hours (E = 14.413 keV). The off-resonance penetration depth of photons is shown on the right side.

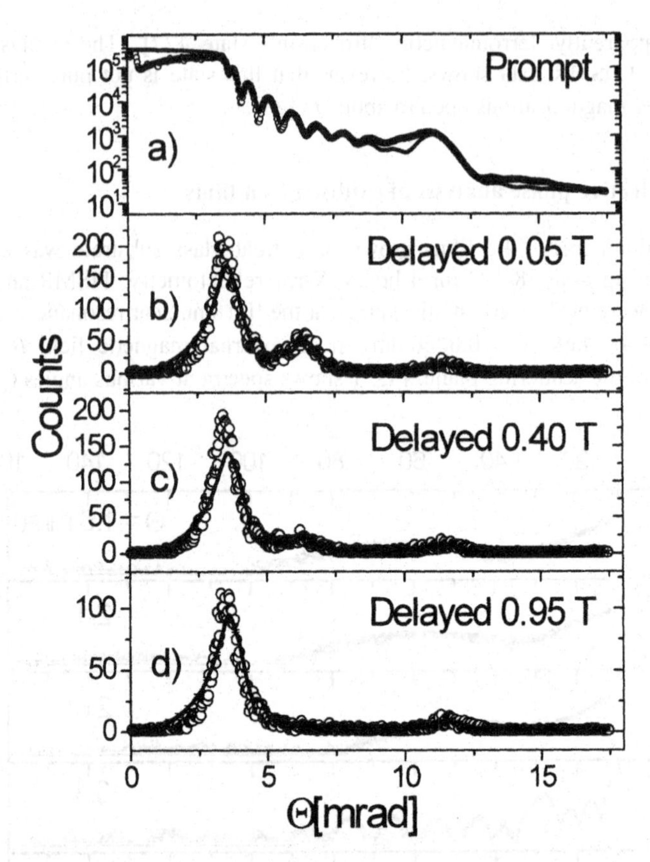

Figure 2. Prompt (a) and time integral (10–300 ns) delayed Θ–2Θ reflectivity scans (x-ray reflectometry and TISMR) of Zerodur/[^{57}Fe/FeSi]$_{10}$ multilayer in various external magnetic fields (b-d). The fit curves represent the model layer and magnetic structure described in the text.

$\Theta \leq$ 5.2 mrad) within the total reflection peak on the sample annealed at 170 °C corresponding to the penetration depth of 1.5 nm to 40 nm. A fast beating can only be seen on the spectra taken at $\Theta \geq 3.1$ mrad.

The beating frequency is proportional to the strength of the hyperfine field (magnetic dipole or electric quadrupole interaction). The quantum beat pattern from the uppermost 1.5 nm shows the presence of non-magnetic or superparamagnetic phases (probably β–FeOOH and superparamagnetic α–Fe_2O_3). The magnetic interaction from a depth of 15 nm (i.e. almost from the whole film) can tentatively be assigned to α-Fe and Fe_3O_4.

4. Magnetic structure of an antiferromagnetic Fe/FeSi multilayer

A Zerodur/[^{57}Fe(2.55 nm)/^{nat}FeSi(1.5 nm)]$_{10}$ multilayer was grown by molecular beam epitaxy. On [Fe/^{57}FeSi]$_n$ multilayers, conversion electron Mössbauer spectroscopy

(CEMS) showed a characteristic resonance of the metastable FeSi phase of CsCl (B2) structure that is known to mediate an AF coupling between the Fe layers [29,30].

Grazing incidence prompt and delayed time integral (10–300 ns) Θ–2Θ scans (x-ray reflectometry and TISMR, c.f. Fig. 2) as well as TDSMR spectra at selected angles were recorded at room temperature at ID18 in various external magnetic fields ($0 < B_{ext} < 0.95$ T) perpendicular to the scattering plane. The prompt x-ray reflectivity curve (Fig. 2, top) shows the structural multilayer Bragg peak at 0.67° and damped Kiessig oscillations corresponding to the total film thickness of 41.2 nm. The delayed nuclear scattering, in contrast, reveals the apparent magnetic multilayer period doubling (AF multilayer Bragg peak at 0.36°) which gradually disappears with increasing transversal magnetic field. This qualitatively shows that the AF order of the Fe sublayer magnetisations is increasingly canted by increasing magnetic field and is fully suppressed in $B_{ext} = 0.95$ T.

Under favourable conditions, the direction of the individual sublayer magnetisations can be determined from the Θ–2Θ scans with an accuracy of ± 5° [25]. A detailed analysis of the Θ–2Θ scans and of the time spectra [25,26] reveals subtle details. It turns out that, in accordance with surface magneto-optic Kerr results [31], the top Fe layers

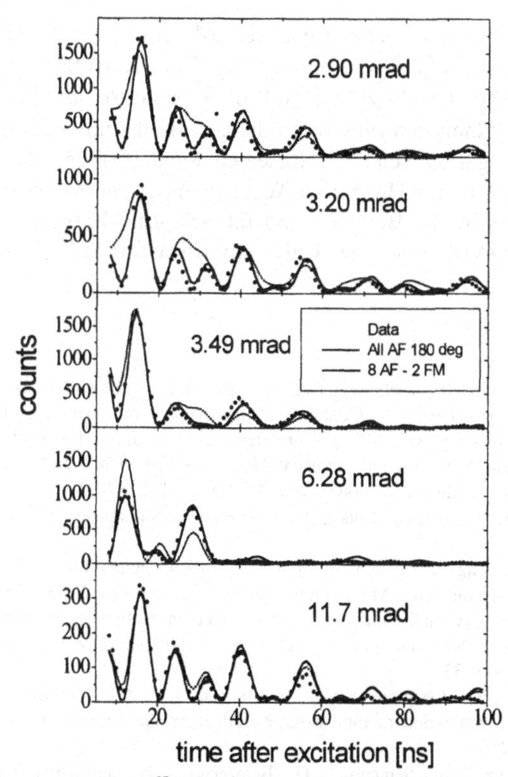

Figure 3. TDSMR spectra of Zerodur/[^{57}Fe/FeSi]$_{10}$ multilayer in an external field of 0.05 T perpendicular to the scattering plane at different angles of grazing incidence. Thin solid fit curve for an 'all AF' (0–180°) sublayer magnetisation alignment relative to scattering plane, thick solid fit curve for the 8AF–2FM model.

are AF-coupled, while those close to the substrate are ferromagnetically (FM)-coupled or they are not coupled to each other at all. This latter leads to a massive misalignment of the top AF layer magnetisations (c.f. "8AF–2FM" in Fig. 3.) as compared to the "naïve" perpendicular alignment of the AF sublayer magnetisations to the small external magnetic field ("All AF 180 deg" in Fig. 3.).

5. Conclusions

In conclusion we have demonstrated the feasibility of SMR, a method yielding both structural (chemical and magnetic) and depth information with a resolution of a few nanometers. The method becomes quantitative if a number of time spectra at different angles of grazing incidence are measured with suitable statistics and simultaneously evaluated in terms of a model for the depth profile of hyperfine interactions. Such measurements can be performed within a reasonable measuring time at dedicated beamlines of third generation SR sources.

Acknowledgement

The beam time for the measurements was generously granted by ESRF Grenoble. It is a pleasure to thank for helpful discussions to Drs. R. Rüffer, A.Q.R. Baron and A.I. Chumakov of ESRF. Thanks are due to Prof. V.N. Gittsovich, Drs. O. Leupold, and N.N. Salashchenko for their contribution to the corrosion studies. This work was partly supported by the Hungarian Scientific Research Fund (OTKA) under Contract Nos. T016667 and F022150, by the Hungarian Academy of Sciences (Contract No. AKP 97–104 2,2/17) as well as by the Belgian Fund for Scientific Research, Flanders (FWO), Concerted Action (GOA) and the Inter-University Attraction Pole (IUAP P4/10).

References

1. Born, M. and Wolf, E. (1970) *Principles of Optics*, Pergamon Press, Oxford, p. 51.
2. Szilágyi, E., Bottyán, L., Deák, L., Gerdau, E., Gittsovich, V.N., Gróf, A., Kótai, E., Leupold, O., Nagy, D.L., and Semenov, V.G. (1997) Corrosion depth profiles by Rutherford backscattering spectrometry and synchrotron x-ray reflectometry, *Mater. Sci. Forum* **248–249,** 365–368.
3. Lax, M. (1951) Multiple scattering of Waves, *Rev. Mod. Phys.* **23,** 287–310.
4. Spiering, H. (1985) Recent developments in the evaluation of Mössbauer line intensities, *Hyp. Int.* **24–26,** 737–768.
5. Deák, L., Bottyán, L., Nagy, D.L., and Spiering, H. (1996) Coherent forward-scattering amplitude in transmission and grazing incidence Mössbauer spectroscopy, *Phys. Rev. B.* **53,** 6158–6164.
6. Bonchev, Zw., Jordanov, A., and Minkova A. (1971) Method of analysis of thin surface layers by the Mössbauer Effect, in I. Dézsi (ed.), *Proc. Conf. Appl. Mössbauer Effect (Tihany, 1969)*, Akadémiai Kiadó, Budapest, pp. 333–337.
7. Frost, J.C., Cowie, B.C.C., Chapman, S.N., and Marshall, J. F. (1985) Surface sensitive Mössbauer spectroscopy by the combination of total external reflection and conversion electron detection, *Appl. Phys. Lett.* **47,** 581–583.
8. Irkaev, S.M., Andreeva, M.A., Semenov, V.G., Belozerskii, G.N., and Grishin, O.V. (1993) Grazing incidence Mössbauer spectroscopy: new method for surface layers analysis — Part I. Instrumentation *Nucl. Instrum. Methods* **B74,** 545–553.

9. Irkaev, S.M., Andreeva, M.A., Semenov, V.G., Belozerskii, G.N., and Grishin, O.V. (1993) Grazing incidence Mössbauer spectroscopy: new method for surface layers analysis — Part II. Theory of grazing incidence Mössbauer spectra *Nucl. Instrum. Methods* **B74,** 554–564.
10. Nagy D.L. and Pasyuk, V.V. (1992) Calculation of Mössbauer reflectometry spectra, *Hyp. Int.* **71,** 1349–1352.
11. Gerdau, E., Rüffer, R., Winkler, H., Tolksdorf, W., Klages, C.P., and Hannon, J.P. (1985) Nuclear Bragg diffraction of synchrotron radiation in yttrium iron garnet, *Phys. Rev. Lett.* **54,** 835–838.
12. Deák, L., Bottyán, L., and Nagy, D.L. (1994) Calculation of nuclear resonant scattering spectra of magnetic multilayers, *Hyp. Int.* **92,** 1083–1088.
13. Baron, A.Q.R., Arthur, J., Ruby, S.L., Chumakov, A.I., Smirnov, G.V., and Brown, G.S. (1994) Angular dependence of specular resonant nuclear scattering of x rays, *Phys. Rev. B* **50,** 10354–10357.
14. Andreeva, M.A. (1996) Time-differential Mössbauer total external reflection of synchrotron radiation, *Phys. Lett.* **A210,** 359–363.
15. Nagy, D.L., Baron, A.Q.R., Bottyán, L., Deák, L., Dekoster, J., Korecki, J., Langouche, G., Rüffer, R., Semenov, V.G., and Szilágyi, E.; Synchrotron Mössbauer reflectometry: general aspects and applications to multilayer magnetism, *J. Alloys and Compounds*, submitted for publication.
16. Spiering, H., Deák, L., and Bottyán, L. (1999) EFFINO, in E. Gerdau and H. de Waard (eds.), *Nuclear Resonant Scattering of Synchrotron Radiation*, Baltzer Science Publishers — special volume of *Hyp. Int.*, in press.
17. Chumakov, A.I., Smirnov, G.V., Baron, A.Q.R., Arthur, J., Brown, D.E., Ruby, S.L., Brown, G.S., and Salashchenko, N.N., (1993) Resonant diffraction of synchrotron radiation by a nuclear multilayer, *Phys. Rev. Lett.* **71,** 2489–2491.
18. Deák, L., Bayreuther, G., Bottyán, L., Gerdau, E., Korecki, J., Kornilov, E.I., Lauter, H.J., Leupold, O., Nagy, D.L., Petrenko, A.V., Pasyuk-Lauter, V.V., Reuther, H., Richter, E., Röhlsberger, R., and Szilágyi: E. (1998 or 1999) Pure nuclear Bragg reflection of a periodic $^{56}Fe/^{57}Fe$ multilayer, *J. Appl. Phys.*, in press.
19. Nagy, D.L., Baron, A.Q.R., Bottyán, L., Degroote, S., Dekoster, J., Korecki, J., Lauter, H.J., Petrenko, A.V., Pasyuk-Lauter, V.V., and Szilágyi, E.: (1997) Time response of synchrotron radiation from a periodic $^{nat}Fe/^{57}Fe$ multilayer; International Conference on the Applications of the Mössbauer Effect ICAME'97, Rio de Janeiro, 14–20 September 1997, Poster TU.T9.P05.
20. Nagy, D.L., Bottyán, L., Deák, L., Gerdau, E., Gittsovich, V.N., Korecki, J., Leupold, O. Reuther, H., Semenov, V.G., and E. Szilágyi (1997) Synchrotron Mössbauer reflectometry: feasibility of depth selective phase analysis of thin films and multilayers, in E.A. Görlich and K. Latka (eds.), *Condensed Matter Studies by Nuclear Methods (Proc. XXXII. Zakopane School of Physics, Zakopane, 10–17 May 1997)*, Institute of Physics, Jagellonian University and H. Niewodniczanski Institute of Nuclear Physics, Cracow, pp. 17–25.
21. Nagy, D.L., Bottyán, L., Deák, L., Gittsovich, V.N., Leupold, O., Rüffer, R., Semenov, V.G., Spiering, H., and Szilágyi, E.; Depth selective phase analysis of 20 nm oxidised ^{57}Fe films by synchrotron Mössbauer reflectometry, to be published.
22. Niesen, L., Mugarza, A., Rosu, M.F., Coehoorn, R., Jungblut, R., Roozeboom, M.F., Baron, A.Q.R., Chumakov, A.I., and Rüffer, R. (1998) Magnetic behaviour of probe layers of ^{57}Fe in thin Fe films observed by means of nuclear resonant scattering of synchrotron radiation, *Phys. Rev. B.*, scheduled for 1 October 1998.
23. Andreeva, M.A, Irkaev, S.M., Semenov, V.G., Prokhorov, K.A., Salaschenko, N.N., Chumakov, A.I., and Rüffer, R.; Mössbauer reflectometry of ultrathin multilayer film Zr(10 nm)/[^{57}Fe(1.6 nm)/Cr(1.7 nm)×26]/Cr(50 nm) using synchrotron radiation, *J. Alloys and Compounds*, in press.
24. Toellner, T.L., Sturhahn, W., Röhlsberger, R., Alp, E.E., Sowers, C.H., and Fullerton, E.E., (1995) Observation of pure nuclear diffraction from a Fe/Cr antiferromagnetic multilayer, *Phys. Rev. Lett.* **74,** 3475–3478.
25. Bottyán, L., Dekoster, J., Deák, L., Baron, A.Q.R., Degroote, S., Moons, R., Nagy, D.L., and Langouche, G. (1998) Layer magnetisation canting in ^{57}Fe/FeSi multilayer observed by synchrotron Mössbauer reflectometry, *Hyp. Int.*, in press.
26. Bottyán, L., Dekoster, J., Deák, L., Baron, A.Q.R., Degroote, S., Moons, R., Nagy, D.L., Vértesy, G., Langouche, G., and Spiering, H.; Magnetic structure of Fe/FeSi multilayers, *Phys. Rev. B.*, submitted for publication.

27. Dekoster, J., Bottyán, L., Moons, R., Degroote, S., Baron, A.Q.R., Pasyuk-Lauter, V.V., Vértesy, G., Langouche, G., and Nagy, D.L. (1997) Synchrotron Mössbauer reflectometric evidence of antiferromagnetic to ferromagnetic transformation in Fe/FeSi multilayers, 15th International Colloquium on Magnetic Films and Surfaces ICMFS 97, 4–8 Aug. 1997, Sunshine Coast Queensland, Australia, Book of Digests.
28. Dekoster, J., Bottyán, L., Moons, R., Degroote, S., Baron, A.Q.R., Pasyuk-Lauter, V.V., Vértesy, G., Langouche, G., and Nagy, D.L.; On the nature of the "ferromagnetic" frozen-in state of Fe/FeSi multilayers, to be published.
29. Chaiken, A., Michel, R.P., and Wall, M.A. (1996) Structure and magnetism of Fe/Si multilayers grown by ion-beam sputtering, *Phys. Rev. B* **53,** 5518–5529.
30. Dekoster, J., Moons, R., Degroote, S., Vantomme, A., and Langouche, G. (1995) Magnetic phase transition in the CsCl FeSi spacer in Fe/FeSi multilayers *Mat. Res. Soc. Symp. Proc.* **382**, 253–258.
31. Kohlhepp, J., Valkier, M., van der Graaf, A., and den Broeder, F.J.A. (1997) Mimicking of a strong biquadratic interlayer exchange coupling in Fe/Si multilayers, *Phys. Rev. B* **55**, R696–R699.

SITE PREFERENCE OF ^{57}Co IN Fe-Si ALLOYS AFTER GRAIN BOUNDARY DIFFUSION

O. SCHNEEWEISS, S. HAVLÍČEK AND T. ŽÁK
Institute of Physics of Materials, Academy of Sciences of the Czech Republic, Žižkova 22, CZ--61662 Brno, Czech Republic

1. Introduction

Properties of crystalline materials are strongly influenced by structure and chemical composition of internal interfaces, namely grain boundaries (GBs). They constitute a transition area between two adjacent crystal lattices and therefore topology of the atomic arrangement distinguishes from that in crystal bulks. This difference in crystallography is accomplished in real materials by different chemical compositions, usually by impurity segregation [1]. There is mutual connection between structure and chemical composition of GBs. Impurities on GBs may cause a structural transformation in comparison with that of pure material [2-5]. Chemical composition of GBs is related to (i) bulk composition of materials [6], (ii) heat and mechanical treatment history [7], and (iii) structure of individual boundaries [8].

Due to substantial influence of grain boundary properties on engineering materials important research activities are devoted to both theoretical and experimental research of GBs [9]. In the field of theoretical studies the most important trends are focused on complex boundary structures and their relations to the behavior of polycrystals. In the experimental research application of high resolution electron microscopy dominates.

Fe-Si alloys belong to the important engineering materials as binary alloys, e.g., the electrical steels, or as components of more complex alloys, e.g., soft magnetic nanocrystalline materials. Fe-Si phase diagram [10] shows that the alloys with less than 25 at.% Si crystalline in α phase (bcc-solid solution) and above approx. 10 at.% the ordered structure B2 and $D0_3$ are formed. The Fe-Si alloys were often investigated because of well established preparation of single crystals and bicrystals [11-13]. These studies showed that at GBs chemical composition differs from that of grain bulk, but the chemical quality of the grain interface depends on chemical composition of bulk. For example, changes in silicon content were observed in dependence on the nitrogen impurity atoms while in pure Fe-Si silicon depletion the grain boundary segregation [13].

Experimental investigations of GBs is usually performed using different electron spectroscopy techniques on thinned samples for the transmission or on surfaces obtained by intercrystalline fracture of a bulk sample. Recently it was shown that emission Mössbauer spectroscopy can yield new information about the structure and chemical quality of grain boundary. Recently this method was applied for investigations of GBs in

M. Miglierini and D. Petridis (eds.), Mössbauer Spectroscopy in Materials Science, 337–348.

the pure α-iron samples [14]. It was shown that about 5% of the emitting atoms at the {112} grain boundary of iron are located at the positions either having impurity atoms in the nearest neighbourhood or characterised by a larger atomic spacing in comparison with the lattice in the bulk.

The aim of this study is to investigate grain boundary in bcc Fe-Si alloys from the point of view of atomic structure and chemical composition.

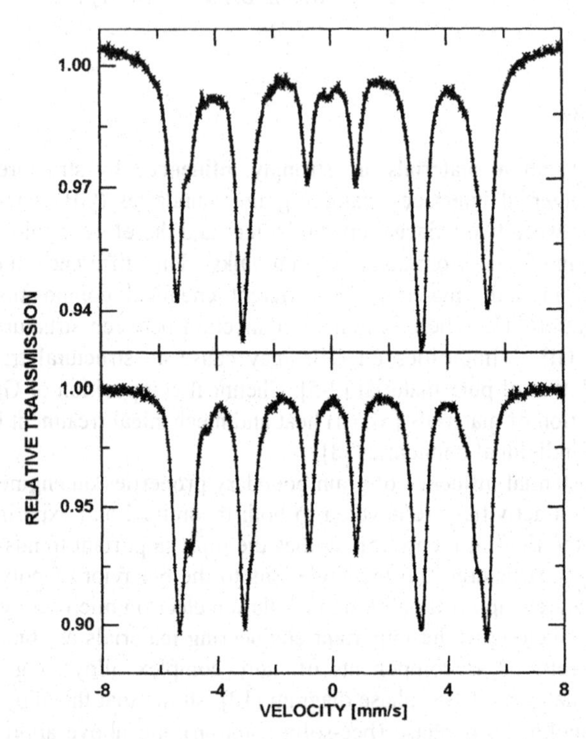

Figure 1. The transmission Mössbauer spectrum of the Fe-4at.% Si sample using ^{57}Co in Cr source (above) and emission ^{57}Fe Mössbauer spectrum of $Na_4Fe(CN)_6$ 10 H_2O taken using ^{57}Co in Fe-4at.% Si source (below).

2. Experimental

The samples of Fe - 4, 12, and 17 at.% Si were prepared in the form of 1mm thick plates cut using spark erosion from fine grain polycrystals of the alloys made of pure (3.5N) elements. Foils <50μm thick were prepared for transmission experiments. The sample surfaces of the samples were carefully ground, polished and cleaned, and annealed in a vacuum furnace in vacuum about 10^{-3} Pa at 973 K for 1 hour with subsequent cooling rate 120 K/h down to room temperature. After the annealing, cleaning by Ar ions was carried out. The aim of the procedure of sample and surface treatments was to remove all traces of plastic deformation produced during cutting and grinding of the plates. The structure of the surfaces of both samples was investigated by optical metallography and ^{57}Fe CEMS. Content of Si was determined according to EDX analysis and chemical wet analysis. These samples were used as matrices for the preparation of ^{57}Co sources. A thin layer of ^{57}Co was deposited on their surfaces in a few drops of cobalt(II) chloride in 0.1M HCl that were dried using an infrared radiation. Diffusion anneals were carried out in vacuum at 873K for 24, 48, and 120hrs. The temperature for diffusion annealing was chosen so that the diffusion along GBs is fast enough and the bulk diffusion is negligible [15,16].

The emission ^{57}Fe Mössbauer spectra were taken at room temperature using $Na_4Fe(CN)_6 \cdot 10\,H_2O$ as absorber with iron enriched to 90% of ^{57}Fe. The standard transmission ^{57}Fe spectra of the polycrystalline foils and CEMS spectra were taken using ^{57}Co in Cr source. Isomer shifts (δ) are reported against pure α-iron foil. Quadrupole splittings (σ) and hyperfine inductions (B_{hf}) are given with respect to B_{hf} of α-iron (3.9156 mm/s corresponding to 33T).

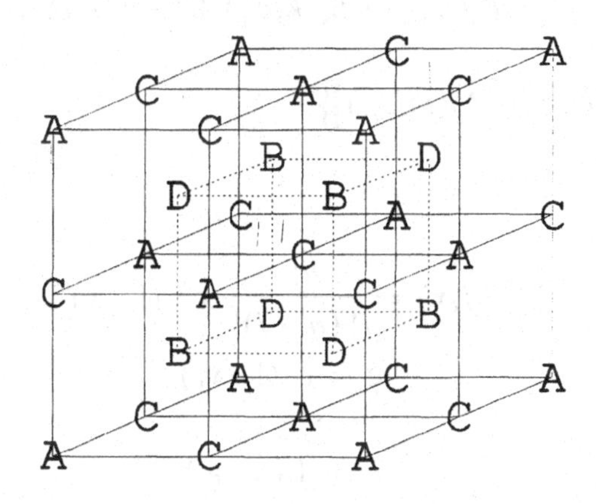

Figure 2. Description of the sublattices in the $D0_3$ superstructure.

3. Results and Discussion

Spectrum analysis were carried out using standard procedure by fitting a set of independent six line patterns which represent iron atoms with different chemical constitution of the first three nearest neighbour shells. The atomic arrangements will be described using (klm) indexes, where k, l, and m represent number of silicon and or impurity atoms in the first, second and third neighbour shells, respectively. From numerous experiments as well as theoretical calculations [e.g., 17-19] it is well known that silicon and some other atoms cause a decrease in B_{hf} if they are placed in the first and second nearest neighbourhood, and an increase from the third nearest neighbourhood. The third nearest neighbour shell was taken into account due to a component which represents iron atoms having iron atoms only in the nearest and second nearest shells and at least one silicon atom in the third nearest shell. These iron atoms denoted further as (00m) are represented by a component having a higher B_{hf} than the (000) component. An example of the spectra is shown in the Fig. 1.

For interpretation of experimental results theoretical models of different iron atom surroundings were taken into account. For a low Si content in Fe-Si alloys a solid solution of Si in bcc structure is expected. Probability of the iron atoms surroundings can be described by a binomial function. An ordered structure must be taken into account in the alloys with Si contents above 10 at.%. Therefore function which yields probabilities of individual configurations in an ordered structure or solid solution in the bcc lattice was prepared. It can be written in following way:

$$P_c(k,l,m) = \frac{1}{4\cdot(1-c)}\{2\cdot(1-c_A)\cdot B_6(l,c_A)\cdot B_{12}(m,c_A)\cdot$$

$$\sum_{i=\max(0,k-4)}^{\min(k,4)} B_4(k-i,c_D)\cdot B_4(i,c_B) + B_8(k,c_A)\cdot\left[(1-c_B)\cdot B_6(l,c_D)\cdot B_{12}(m,c_B) + \right.$$

$$\left. + (1-c_D)\cdot B_6(l,c_B)\cdot B_{12}(m,c_D)\right]\} \tag{1}$$

where

$$B_n(j,x) = \frac{n!}{j!\cdot(n-j)!}\cdot x^j\cdot(1-x)^{n-j} \tag{2}$$

$$c_A = c\cdot(1-\nu_1) \tag{3}$$

$$c_A = c_C \tag{4}$$

$$c_B = c\cdot(1+\nu_1+2\nu_2) \tag{5}$$

$$c_D = c\cdot(1+\nu_1-2\nu_2) \tag{6}$$

$$c = (2c_A + c_B + c_D) \tag{7}$$

The lower indexes A, B, C, and D correspond to the description of sublattices in the DO_3 superstructure is given in Fig. 2. This function may be applied for an alloying element concentration c (e.g. for Si, $0 \le c \le 0.25$) and the order parameters ν_1 and $\nu_2 \in \langle 0,1 \rangle$. For $\nu_1 = \nu_2 = 0$ the function gives probabilities for the solid solution, for $\nu_1 = 1$ and $\nu_2 = 0$ for the superstructure B2, and $\nu_1 = 1$ and $\nu_2 = 1$ for superstructure DO_3. It should be noted that this model can be varied by different choice of the partial concentrations c_A, c_B, c_C, and c_D of the sublattices. Their sum must be equal to the total concentration. A change of that would cause slight changes in the probabilities.

For an inhomogeneous distribution of alloying element atoms a model based on the computer simulation of occupation of bcc lattice positions was applied. The lattice positions were filled by Fe and Si atoms using random numbers generator with constrains for individual sublattices and/or mutual distances between the alloying element atoms. This model allows to simulate clustering or segregation of Si atoms and also a homogeneous solid solution or an ordered structure. The probabilities P(k,l,m) were derived for Si concentrations c_{Si} of the samples studied and for concentration $c=c_{Si} \pm 5$ at.% with step 0.5 at%.

TABLE 1. The components and their hyperfine parameters of the transmission spectrum of the Fe-4at.% Si sample using ^{57}Co in Cr source and emission spectrum of $Na_4Fe(CN)_6 \cdot 10 H_2O$ absorber using ^{57}Co in Fe-4at.% Si alloy as a source. Description of the components is given in the last column.

Component	Transmission spectrum			Emission spectrum			Description
#	δ [mm/s]	σ [mm/s]	B_{hf} [T]	δ [mm/s]	σ [mm/s]	B_{hf} [T]	
1				0.004 ±0.002	-0.02 ± 0.01	35.73 ± 0.05	Co on surface
2	0.003	-0.00	33.76	0.005	-0.00	33.76	(0 0 m), m =1÷12
3	0.005	0.00	33.00	0.006	0.00	33.00	(0 0 0)
4	-0.010	0.01	31.74	-0.005	0.01	31.74	(0 l), l = 1÷6
5	0.054	0.05	30.39	0.062	0.05	30.39	(1)
6	0.076	0.03	27.47	0.081	0.03	27.47	(2)
7	0.095	-0.05	21.85	0.099	-0.05	21.85	(3)

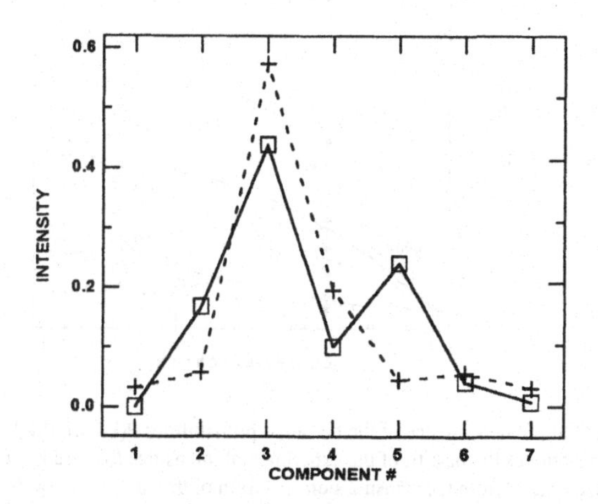

Figure 3. The intensities of components determined in the transmission spectrum of the Fe-4at.% Si sample using ^{57}Co in Cr source (the squares connected by the solid line) in comparison with the intensities of components derived from the emission spectrum of $Na_4Fe(CN)_6 \cdot 10 H_2O$ absorber using ^{57}Co in Fe-4at.% Si source (the crosses connected by dashed line). The components are described in Table 1.

The components and their hyperfine parameters derived form the spectra of the Fe-4 at.% Si samples are shown in Table 1. Fig. 3 demonstrates differences in the intensities of the components of the spectrum representing grain bulks (the transmission spectrum) and the spectrum representing GBs (the emission spectrum) in this alloy. A good agreement of the intensities of the bulk spectrum with the calculated probabilities P(k,l,m) for solid solution for concentration 3.7±0.5 at.% Si was found (Fig. 4). The intensities of GBs spectrum do not fit any model for homogeneous arrangement of Si atoms satisfactory. The most important differences can be observed for the intensity of the component 3 which is interpreted as iron with none silicon atom in the first, second and third nearest neighbourhood and for the intensity of the component 5 representing iron with one silicon atom in the nearest neighbourhood. These differences cannot be explained by an increase in Si content or by higher concentration of impurities at GBs only. Therefore probabilities for a structure with a segregation were calculated. In this model formation of Si and impurity atoms chains was simulated, where Si and impurity atoms are in the nearest neighborhoods and occupy regular positions in bcc structure. The probability derived from this model was in substantially better agreement with the experimental intensities derived from the GBs spectrum than the probabilities calculated for solid solution. This indicates that similar clusters are probably formed in the sample at the GBs.

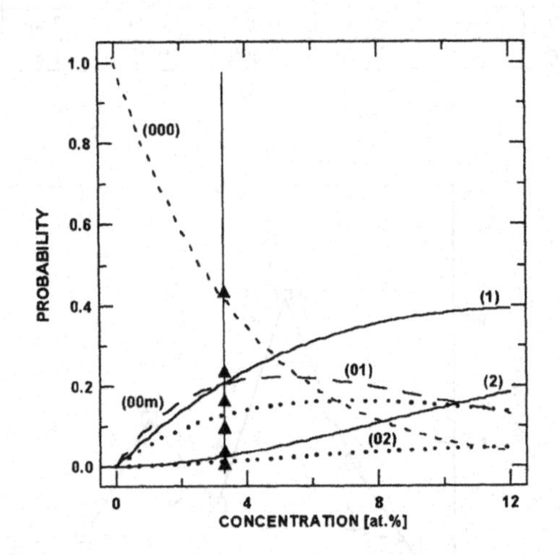

Figure 4. The concentration dependence of the binomial probabilities P(k,l,m), P(k,l) and P(k). The vertical straight line denotes the best fit of the experimental intensities (labeled by full triangles) determined from the transmission spectrum of the sample Fe-4at.%Si .

The spectra of the Fe-12 at.% Si alloy are shown in Fig. 5 and the parameters of components identified in the spectra are summarized in the Table 2. In this case four components are mentioned only. The sextets representing (00m), (000), and (01) are given as their sum because their hyperfine field parameters do not allow to distinguish them in this case unambiguously. The fine differences in hyperfine induction are mixed by changes in isomer shifts and quadrupole splittings. Similar complications appear in resolution of the components representing iron with 1 and 2 silicon atoms in the nearest neighborhood. An increase in the quadrupole splitting for the component (1) causes its

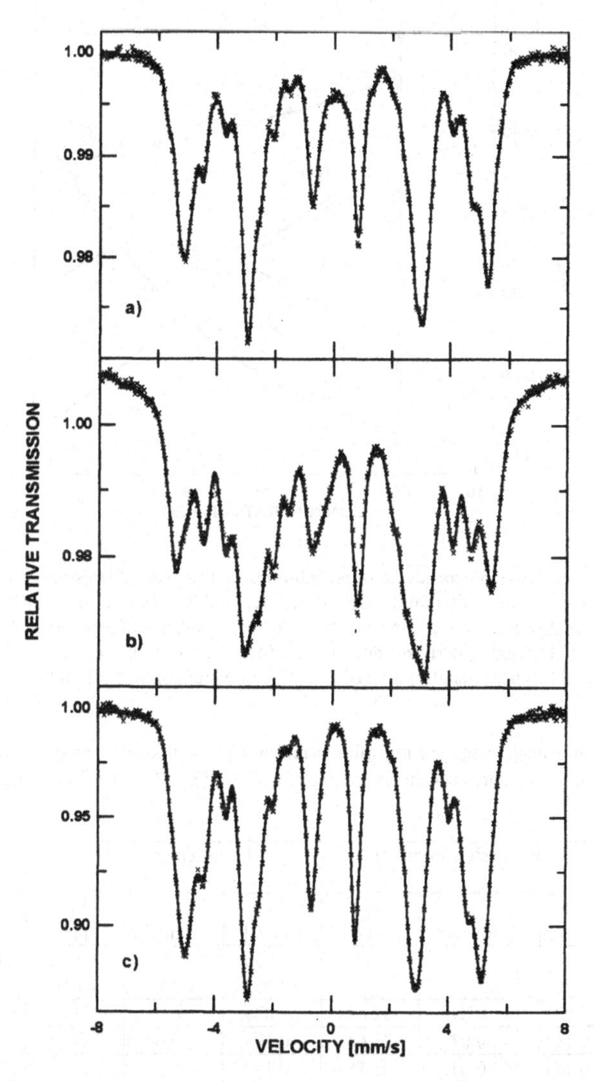

Figure 5. The transmission Mössbauer spectrum of the Fe-12at.%Si- 5at.%Co sample (a) and Fe-12at.% Si sample (c) using ^{57}Co in Cr source, and the emission ^{57}Fe Mössbauer spectrum of $Na_4Fe(CN)_6 \cdot 10\ H_2O$ using ^{57}Co in Fe-12at.% Si source (b).

overlapping with the component (2). Therefore their integral intensity (1)+(2) is used for comparison with the calculated model probabilities. According to the phase diagram $D0_3$ long range order (superstructure) can be expected. Therefore the probabilities were calculated for $D0_3$ and B2 ordered structures. The results of these calculations are drawn in Fig 6.

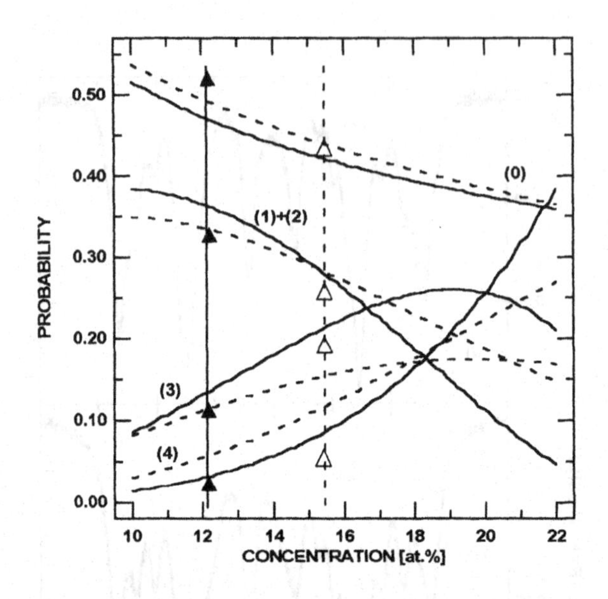

Figure 6. The concentration dependence of probabilities for the selected (most intensive) iron nearest neighbourhood configuration for the $D0_3$ (solid lines) and B2 (dashed lines) superstructures. The solid vertical straight line denotes best correspondence of experimental data (labeled by full triangles) derived from bulk spectrum and the dashed one experimental data (labeled by open triangles) derived from GBs spectrum of the Fe-12at.%Si alloy.

TABLE 2. The components and their hyperfine parameters of the transmission spectrum of the Fe-12at.% Si sample using ^{57}Co in Cr source and emission spectrum of $Na_4Fe(CN)_6 \cdot 10\ H_2O$ using ^{57}Co in Fe-12at.% Si alloy source.

Component	Transmission spectrum			Emission spectrum			Description
#	δ [mm/s]	σ [mm/s]	B_{hf} [T]	δ [mm/s]	σ [mm/s]	B_{hf} [T]	
1				0.030 ±0.002	-0.02 ± 0.01	35.28 ± 0.05	Co on surface
2	0.035	0.01	32.62	0.029	0.01	33.30	(0)
3	0.111	0.01	28.64	0.108	0.02	28.71	(1)+(2)
4	0.181	0.01	23.78	0.190	0.01	24.13	(3)
5	0.320	0.09	18.78	0.315	0.05	19.25	(4)

From the comparison of the experimental values derived from the bulk and GBs spectra and comparison with the model results the following conclusion can be derived: (i) intensities of the components in the transmission spectra are in good agreement with $D0_3$ superstructure, (ii) the emission data indicate that higher content of Si (and impurities) appears at GBs. The increase in Si content can be estimated roughly at 3 at.%.

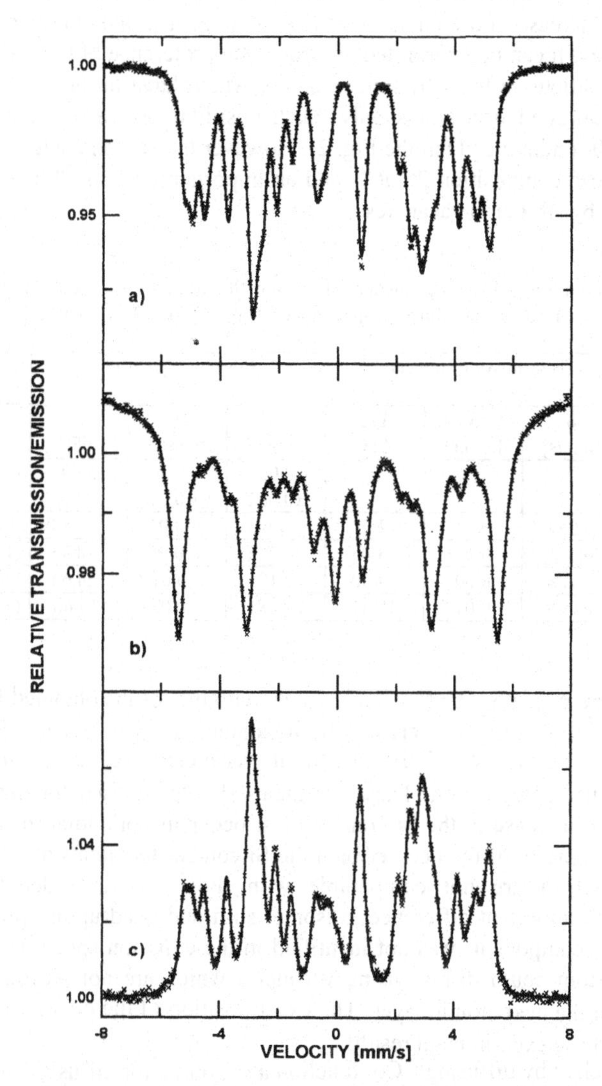

Figure 7. The transmission (a) and conversion electron (c) Mössbauer spectra of the Fe-17at.% Si sample using ^{57}Co in Cr source, and emission spectrum of $Na_4Fe(CN)_6 \cdot 10\ H_2O$ using ^{57}Co in Fe-17at.% Si source (b).

The data derived from spectra of Fe-17 are given in Table 3. A high degree of $D0_3$ order must be expected in this alloys and so the model probabilities for this atomic arrangement were calculated. The data from the emission spectrum showed an important disagreement with the transmission one (Fig. 7). To insure that the surface of the sample for the ^{57}Co in Fe-17at.% Si source was not affected by preparation procedure the CEMS spectrum was taken on the sample prepared in the same way. It was proved that the surface and the bulk give the same spectra as documented in Fig. 7. The important increase in the intensity of (0) component in the emission spectrum was observed (Fig. 8). It can be interpreted as strong site preference of the diffusing ^{57}Co for those lattice positions in the $D0_3$ superstructure which have in the nearest and second nearest neighbourhood iron atoms only. Such positions are regularly occupied by Si atoms in the $D0_3$ (sublattice D in the Fig. 2). However the silicon content does not reach the stoichiometric composition 25 at.% and an important part could be occupied either by Fe atoms or by structural vacancies.

TABLE 3. The component hypefine. parameters from the transmission spectrum of the Fe-17at.% Si sample using ^{57}Co in Cr source and emission spectrum of $Na_4Fe(CN)_6 \cdot 10\ H_2O$ absorber using ^{57}Co in Fe-17at.% Si alloy source.

Component	Transmission spectrum			Emission spectrum			Description
	δ [mm/s]	σ [mm/s]	B_{hf} [T]	δ [mm/s]	σ [mm/s]	B_{hf} [T]	
1				0.004 ±0.002	-0.02 ± 0.01	35.97 ± 0.05	Co on surface
2	0.049	0.02	32.02	0.043	0.01	32.88	(0)
3	0.071	0.03	28.05	0.092	0.05	28.75	(1) + (2)
4	0.188	0.00	24.30	0.195	0.01	24.64	(3)
5	0.266	0.01	19.37	0.281	0.02	19.66	(4)

It could be expected, that in accordance with the results obtained from previous alloys, also in this case an increase in Si content at GBs occurs. Simultaneously, the probability occurence of the structural vacancies increases with approaching the stoichiometric composition. These vacancies should be used for diffusion of ^{57}Co along GBs. An increase in the Si content must occur in both adjacent grains. If there would be an important difference between the Si content the atoms at GBs would be in surroundings which are not comparable with those in the ordered structure. A component with important difference in isomer shift and quadrupole splitting would be expected. Such component was not identified in the emission spectrum. On the other hand the Co atom could diffuse using vacancies which are not placed exactly at the interface but in the next atomic layer. The exact position of the diffusion path cannot be decided from these experimental results.

GBs, enriched by diffusing ^{57}Co, function as a source for diffusing these atoms into volumes of grains [20,21]. Therefore the emission spectra were measured using the sources prepared with three diffusion annealing times (24, 48 and 120 hrs). Only slight changes in the spectra were observed due to the increase in the annealing time. These differences can be explained as an increase in the part of spectrum representing grain bulks in agreement with expected diffusion of ^{57}Co from GBs into grain bulks.

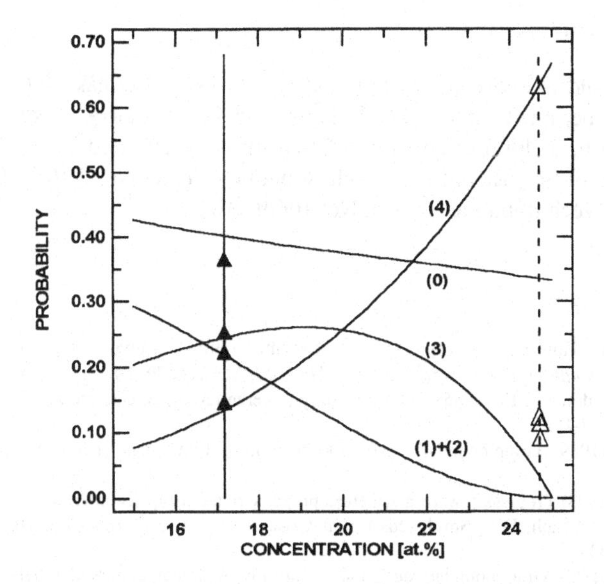

Figure 8. The concentration dependence of probabilities for selected (the most intensive) iron nearest neighbourhood configuration for the $D0_3$ superstructures. The solid vertical straight line denotes the best correspondence of experimental data (labeled by full triangles) derived from bulk spectrum and the dashed one indicates the experimental data (labeled by open triangles) derived from the emission spectrum of the Fe-17at.%Si alloy.

4. Conclusions

The comparison of the results derived from the transmission and emission spectra with the model of atomic arrangements in the studied alloys shows that the GBs contain higher concentration of Si and impurities. The model of clustering of Si and impurity atoms at grain boundary allows to explain experimental data satisfactorily in the samples Fe-4 at.% Si. In the alloy with 12 at.% Si, the enrichment of the GBs by Si can be estimated at 3 at.%. The important site preference of diffusing atoms in the sample of Fe-17at.%Si alloy can be explained by an increase of Si concentration at GBs as well. The preferred sites were identified to be positions in the $D0_3$ superstructure having iron atoms in the nearest neighbourhood shell. These positions are known by the formation of structural vacancies in the alloys and so the found site preference is in agreement with the accepted diffusion mechanism along grain boundaries [15].

Acknowledgement

The authors would like to express their gratitude to Dr. J. Čermák for carrying out the diffusion experiments and to Dr. I. Turek and Dr. P. Lejček (Institute of Physics ASCR, Prague) for helpful discussions. This work was supported by the Grant Agency of the Academy of Sciences of the Czech Republic (project No. A1010708) and Grant Agency of the Czech Republic (project No. 106/97/1044).

References

1. Lejček, P. and Hofmann, S. (1995) Thermodynamics and structural aspects of grain boundary segregation, *Critical Review in Solid State and Materials Sciences* **20**, 1-85.
2. Clarke, D.R. and Wolf , D. (1986) Grain boundary in ceramics and at ceramic-metal interfaces, *Mater. Sci. Eng.* **83**, 197-204.
3. Watanabe, T. (1989) Grain boundary design for the control of intergranular fracture, *Mater. Sci. Forum* **46**, 25-48.
4. Smith, D.A. (1993) Progress toward a structure property relationship for interfaces, in S. Ranganathan, C.S. Pande, R.B. Rath, D.A. Smith (eds.), *Interfaces: Structure and Properties*, Oxford & IBH, New Delhi, pp. 87-111.
5. Watanabe, T. (1993) Grain boundary design and control for high temperature materials, *Mater. Sci. Eng.* **A166**, 11-28.
6. Briant, C.L. (1988) Competitive grain boundary segregation in Fe-P-S and Fe-P-Sb alloys , *Acta Metall.* **36**, 1805-1813.
7. Valiev, R.Z., Mulyukov, R.R., and Ovchinikov, V. V. (1990) Direction of a grain-boundary phase in submicrometre-grained iron, *Phil. Mag. Lett.* **62**, 253-256.
8. Lejček, P., Adámek, J., Hofmann, S. (1992) Anisotropy of grain boundary segregation in Σ=5 bicrystals of α-iron, *Surface Sci.* **264**, 449-454.
9. Paidar,V., Gemperlová, J., Lejček, P., Vitek, V. (1992) A report on current research on grain-boundary structure and chemistry, *Mater. Sci. Eng.* **A154**, 113-123.
10. Massalski, T.B., Murray, J.L., Bennett, L.H., Baker, H. (eds.) (1986) *Binary alloy phase diagrams*, American Society for Metals, Metals Park, Ohio, p. 1108.
11. Hofmann, S. and Lejček, P. (1990) Segregation at special grain boundaries in Fe-Si alloy bicrystals, *J. de Physique, Colloque de Physique, Colloq. C1, suppl. 1,* **51,** C1-179 - C1-184.
12. Hofmann, S., Lejček, P., and Adámek, J. (1992) Grain boundary segregation in [100] symmetrical tilt bicrystals of an Fe-Si alloy, *Surface and Interafce Analysis* **19**, 601-606.
13. Lejček, P., Krajnikov, A.V., Ivashchenko, Yu.N., Militzer, M., and Adámek, J. (1993) Solute segregation to grain boundaries and free surfaces in an Fe-Si multicomponent alloy, *Surface Sci.* **280,** 325-334.
14. Schneeweiss, O., Turek, I., Čermák, J., and Lejček, P. (1998) Properties of iron atoms at grain boundaries in Fe and $Fe_{72}Al_{28}$, Materials Research Society Symposium Proceedings Series., Philladelphia, Vol. 527, in print.
15. Kaur, I., Gust, W., and Kozma, L. (1989) Handbook of grain and Interphase boundary diffusion data, Vol. 2, Ziegler Press, Stuttgart, pp. 837-847; Vol. 1, pp. 12 - 48.
16. Million, B. (1977) Diffusion of Fe-59 in α-Fe-Si alloys, Czech. J. Phys B **27**, 928-934.
17. Stearns, M.B. (1971) Measurement of conduction-electron spin-density oscillations in ordered FeSi alloys, *Phys. Rev. B* **4**, 4069-4080.
18. Haggström, L., Granäs, L., Wäppling, R., and Devanarayanan, S. (1973) Mössbauer study of ordering in FeSi alloys, *Physica Scripta* 7, 125-131.
19. Van der Woude, F., and Sawatzky, G.A. (1974) Mössbauer effect in iron and dilute iron based alloys, *Physics Reports* **12C**, 335-374.
20. Girifalco L.A. (1964) *Atomic Migration in Crystals*, Blaisdell Pub. Company, New York, p. 117-123.
21. Adda, Y. and Philibert, J. (1966) *La diffusion dans les solides*, Inst. Nat. Sci., Techn. Nucleaires, P. Univ. France, Saclay, p. 667.

HIGH PRESSURE MÖSSBAUER SPECTROSCOPY

MOSHE P. PASTERNAK[1] AND R. DEAN TAYLOR[2]
[1)] *School of Physics and Astronomy, Tel Aviv University, 69978 Tel Aviv, ISRAEL*
[2)] *MS-K764, MST-10, Los Alamos National Laboratory, Los Alamos, NM 87545, USA*

1. Introduction

Pressure is the most effective method available to solid-state scientists to alter many properties of matter. Since its implementation by Bridgman in the beginning of this century, high pressure (HP) research has provided substantial information, in all aspects, on properties of matter. Today, as a result of the development of static high pressure devices based on the *diamond-anvil cell* (DAC) experimenters can reach pressures in the megabar region corresponding to generating *energy densities* in matter of the order of eV/Å^3. With such energy densities insulators with gap energies in the eV regions become metals, new structural and electronic phases become stable and new aspects of magnetism are being revealed. In other words the DAC has become a most powerful ultra-high-pressure device, helping scientists discover new states of matter. Some of the modern DAC's for generating pressures into the megabar region can fit into the palm of the hand (see Fig. 1) and allow a variety of sophisticated measurements to be performed even though samples are of almost *microscopic* dimensions. The principle underlying

Figure 1. A photograph of a miniature piston/cylinder DAC contrasting its miniature size with a US 25¢ coin and capable of reaching static pressures of 140 GPa [Ref. 1].

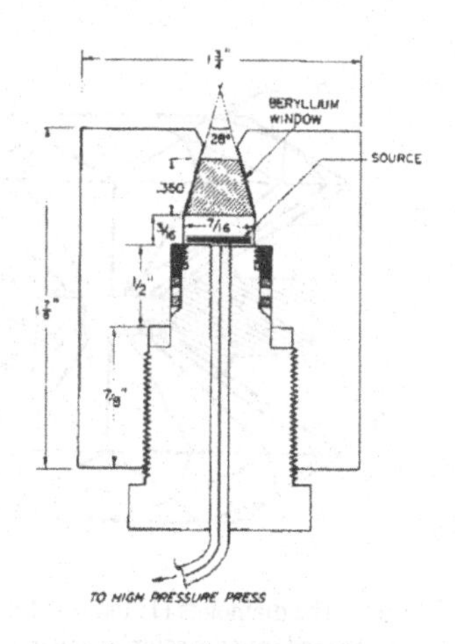

Figure 2. The first pressure cell used for Mössbauer spectroscopy.

M. Miglierini and D. Petridis (eds.), Mössbauer Spectroscopy in Materials Science, 349–358.

the DAC, its potential applications, and pressure calibrations are thoroughly described in the review paper by Jayaraman [2].

About three years after the discovery of the Mössbauer effect, R.V. Hanks published a short paper [4] entitled *Pressure Dependence of the Mössbauer Effect* suggesting a new application of Mössbauer spectroscopy (MS) for high pressure studies of the state of matter. Six months later, Pound, Benedek, and Drever reported [5] the first high pressure MS experiment. Citing these authors: "The source, about 2 mC of Co^{57} diffused into iron, was enclosed in a beryllium-copper pressure bomb equipped with one-half-inch thick Be window. The bomb was cemented to the ferroelectric transducer and connected to a Bridgman press by a stainless steel tube". The pressure bomb used for this experiment is shown in Fig. 2. The highest pressure achieved was 3000 Kg/cm^2 (~3 Kbar). Pioneering studies in MS high pressure experiments of the *first generation* using Bridgman's-type cells and starting in early 60's and into the mid 70's can be attributed to H.G. Drickamer and collaborators. An excellent review on the *first generation* HP methodology Mössbauer studies to 1975 was written by W.B. Holzapfel [6]. The present paper reviews the most recent ^{57}Fe HP-MS using DACs and conventional ^{57}Co(Rh) sources. A general detailed review on HP-MS of the *second generation,* using DACs, has been recently published by Pasternak and Taylor [7].

2. Experimental

2.1. DIAMOND ANVIL CELLS FOR HIGH PRESSURE MÖSSBAUER SPECTROSCOPY

DAC's can be classified into two main classes: *opposing-plates* (o/p) and *piston-cylinder* (p/c) cells. Whereas o/p cells are relatively simple, inexpensive, easy to handle,

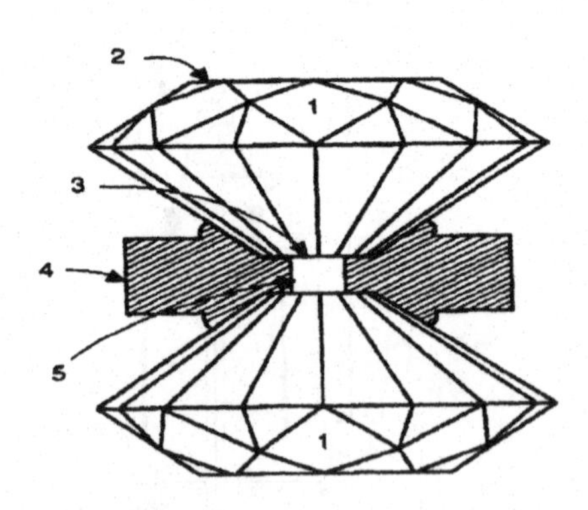

Figure 3. The diamonds (1) modified to have flat culets (3) are pressed into the sample, which is confined in the sample chamber (5) formed in a preindented metal gasket (4). Force is applied to the tables (2) *via* the backing plates.

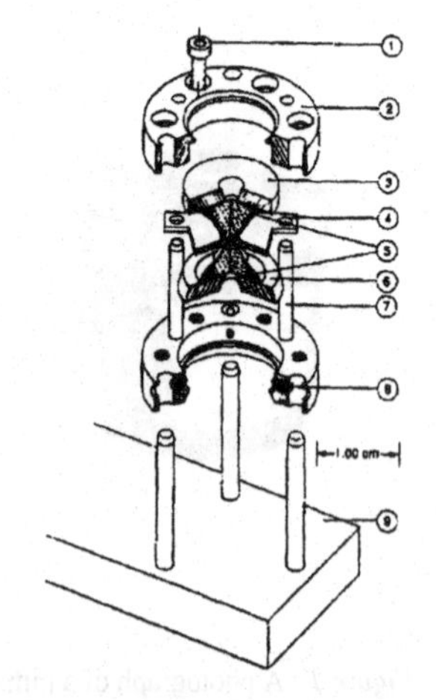

Figure 4. A miniature DAC of the opposing plates type built at Tel Aviv University. For details and operation of cell see Ref. 3.

and suitable for pressures up to 30 GPa with ~ 600 μm culets [8], p/c cells being much more precise in its closure are suitable for pressures beyond 100 GPa with culet sizes of less than 300 μm. A generic anvils configuration is shown in Fig. 3. A sketch of a o/p miniature DAC is shown in Fig. 4.

2.2. ^{57}Fe MÖSSBAUER SPECTROSCOPY WITH DIAMOND ANVIL CELLS

Common problems specific to all HP-MS studies employing DACs are: (*a*) the small size of absorbers and consequently the need for (*b*) high specific activity source, (*c*) reasonable transmission through the diamond anvils of the pertinent γ-rays, and (*d*) collimation of unwanted radiation. The typical height of a pair of 0.30-carat anvils is 3.5 mm. Thus, for ^{119}Sn or ^{151}Eu MS the anvils allow 65% transmission, whereas for ^{57}Fe MS only 35% of the 14.4-keV γ-rays is transmitted (see Fig. 5). The situation is even worse in case one needs to use the K-x-rays of Fe, say, for inelastic scattering using synchrotron radiation. Also the selective absorption by the diamonds of the 14 keV adds considerably to the background. However recent testing with *perforated* diamond anvils resulted in considerable improvement in the S/N ratio (see Figs 6 and 7).

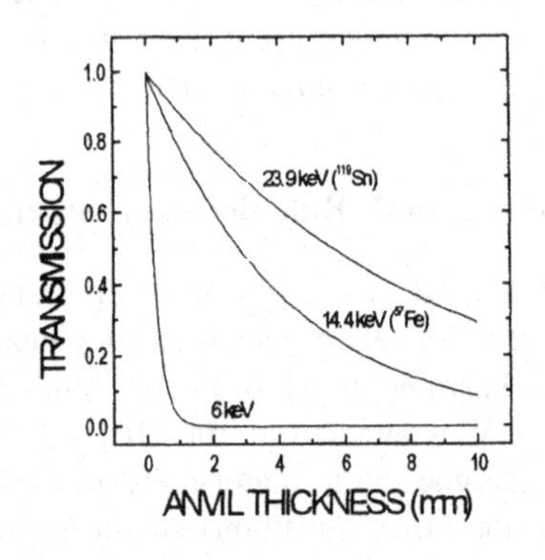

Figure 5. Transmission curves of diamond for ^{119}Sn, ^{57}Fe, and Fe K-x-rays.

Sources. Due to the minuscule size (75 - 150 μm diameter) of typical absorbers used for

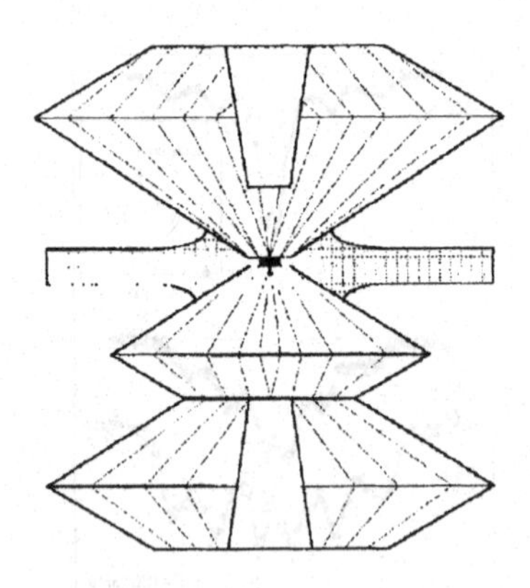

Figure 6. DAC setup of perforated anvils for improved S/N in ^{57}Fe MS. The upper anvil has been partially drilled to ¾ of its height, the lower one is fully drilled, and a 0.1C anvil (window) was mounted as shown for good optical viewing.

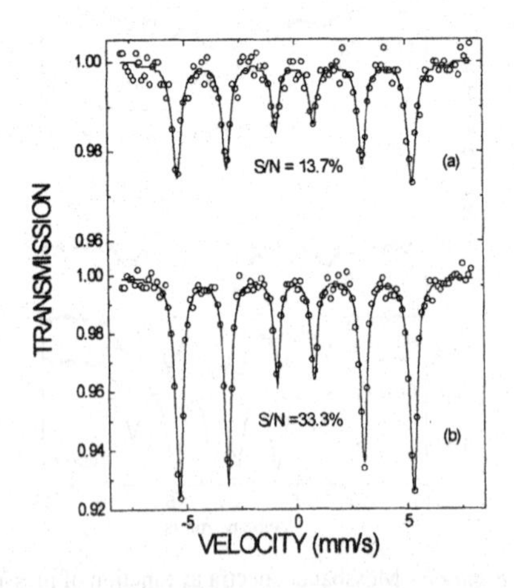

Figure 7. A total effect of 33% (b) was obtained with **perforated** anvils (see Fig. 6) compared to 13.7% with **filled** anvils (a). The same sample was used in both cases.

pressures in the vicinity of 100 GPa, sources need to have a high specific activity concentrated in the possible smallest practical area. Today one can purchase ~ 10 mCi ^{57}Co(Rh) sources with 0.5 x 0.5 x 0.006 mm dimensions. Due to this high concentration, a magnetic splitting often develops below 25 K within a source life-time. Therefore, for experiments using LHe, it becomes necessary to keep the source temperature at $T \geq 30$ K [9].

Absorber thickness in DAC's is optimized for maximum total effect and is normally enriched in ^{57}Fe to about 15 - 25%. For $P < 40$ GPa we use gaskets of $Ta_{0.9}W_{0.1}$ whereas for $P > 40$ GPa we use Re. Both materials are excellent as gaskets and as collimators for the 14.4 keV γ-rays. To fill up the sample volume and to assure hydrostatic conditions, we use liquid Ar.

We now present recent results with ^{57}Fe Mössbauer spectroscopy using DAC's to pressures beyond 100 GPa.

3. Hund's Rule Becomes Dispensable at Very High Pressures; the Case of FeO

The need to space apart the (parallel) spins in order to *lower* the Coulomb repulsion becomes redundant when the density of the applied mechanical energy becomes comparable to that of the Coulomb energy. Mechanical energy density of the order of 1 eV/Å^3 is attained at 100 GPa. At this high pressure an antiferromagnetic insulator (*i*) undergoes an insulator-metal transition due to loss of electron correlation within the *d*-bands (*Mott* transition) and/or (*ii*) undergoes a spin-crossover, from *high* to *low* spin (HS → LS). Whereas the onset of a pressure-induced *Mott* transition has been unequivocally demonstrated [10] in the case of NiI_2 by using ^{129}I MS, recent ^{57}Fe MS

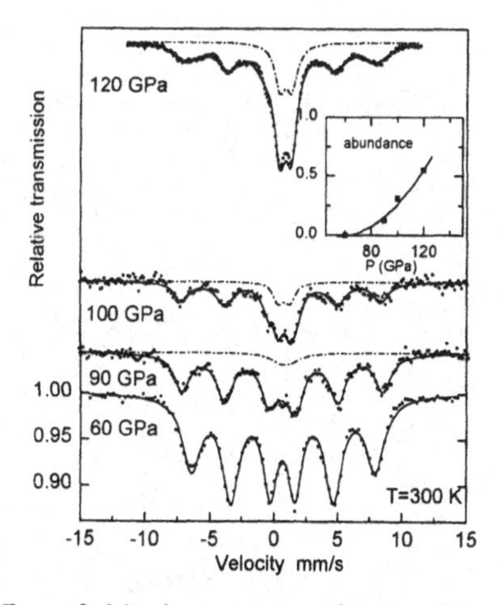

Figure 8. Mössbauer spectra as function of pressure at 300 K. A non-magnetic component appears at 90 GPa, and its abundance increases to 60% of the total at 120 GPa. From the inset one extrapolates to full conversion at 140 GPa.

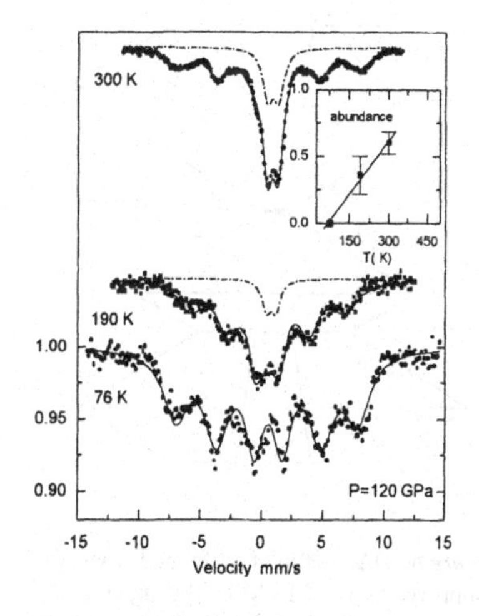

Figure 9. Temperature dependence of the HS→LS at 120 GPa. The strong T-dependence is a manifestation of a small gap separating the magnetic HS and diamagnetic LS states.

studies [11] in wustite (FeO) showed a clear case of HS → LS induced by pressure.

Mössbauer studies in $Fe_{0.94}O$ were performed as a function of pressure and temperature in a piston/cylinder Mao-Bell cell using a 20% enriched ^{57}Fe sample of 75 μm diameter. The dependencies of pressure (T = 300 K) and of temperature (P = 120 GPa) of the Mössbauer spectra are shown in Figs. 8 and 9, respectively.

The HS state configuration of the d^6 Fe(II) is a $^5T_{2g}$. With a decrease of interatomic distances the crystal field is drastically enhanced due to the strong -*r*- dependence (10*D*q ~ $1/r^5$) overcoming the spin-spin exchange energy and leading to a diamagnetic $^1A_{1g}$ ground state. Based on previous studies of the equation of state to 200 GPa, where it was shown that no structural phase transition occurred in the 50 - 200 GPa range, we concluded that the spin crossover is a second order phase transition.

A pressure-induced HS → LS did not take place in another divalent-iron compound, namely, the antiferromagnetic insulator FeI_2 where application of high pressure resulted in a transition to a *Mott-metal.*

4. The Breakdown of *Electron-electron* Correlation in the Mott-insulator FeI_2

The antiferromagnet FeI2 (T_N = 12 K) crystallizes in the CdI_2 structure where the Fe layers are separated by two iodine layers. In the magnetic state the Fe^{2+} moments are ferromagnetically coupled within the layer, forming at 55° to the *c*-axis. With increasing pressure T_N increases reaching 120 K at ~ 20 GPa. During this process the resistivity decreases, and for P > 20 GPa its temperature dependence shows a positive slope *e.g.*, an insulator-metal transition occurs. No structural transition has been observed. Mössbauer spectra recorded in the 20 - 30 GPa range show a relaxation-like spectra and at P > 30 GPa a non-magnetic spectrum is observed to the lowest temperature.

The magnetic phase diagram of FeI_2 is shown in Fig. 10 (upper curve). A model for a 3-D Mott-Hubbard phase diagram as proposed by DeMarco *et al.*[12] is shown in the lower part. The variables are the normalized temperature versus w/U where w is the bandwidth and U is the Hubbard energy. The resemblance between the relatively simple model proposed by DeMarco and the experimental results is quite striking. It is quite clear that HP-MS is the only experimental method for obtaining these kinds of magnetic phase diagrams where w/U changes with decreasing inter-atomic distances. Another interesting result concerning the evolution of the magnetic state with increasing pressure is the case of the orthoferrite $PrFeO_3$.

5. Pressure-induced Collapse of the Magnetic State of $PrFeO_3$

The $RFeO_3$ (R = rare earth) orthoferrites are well known antiferromagnetic insulators having a very large T_N. The orthorhombic structure is a distorted version of the cubic perovskite; the distortion is caused by the smaller size of the rare earth ion. The FeO_6 octahedral units tilt and rotate in order to fill the extra space otherwise present around the rare-earth iron. The presence of this extra space is critical when pressure is increased, forcing the FeO_6 units to tilt to compensate for the reduced volume. In fact,

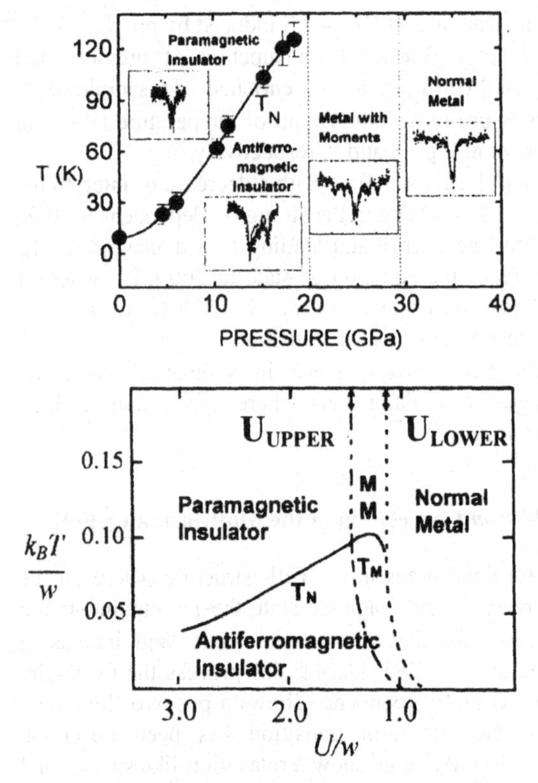

Figure 10. Magnetic phase diagram of FeI_2 (upper part). The icons display the characteristic Mössbauer spectra for different phases. The lower part figure was drawn on the basis of DeMarco's model (see text).

the volume collapse has been recently [13] discovered by us during HP studies of $PrFeO_3$ using XRD and MS.

The pressure evolution of the Mössbauer spectra at 300 K is shown in Fig. 11. At ~38 GPa one can see a nonmagnetic, quadrupole-split component appearing whose abundance increases with P, reaching unity at 50 GPa. Fig. 12 shows the *equation of state* (*a*), the IS(P) (*b*), and the HP-phase abundance (*c*) deduced from the Mössbauer data. As is clearly seen, a volume discontinuity is observed at 42(8) GPa corresponding to a first order phase transition, and a sharp decrease of the IS corresponding to an **increase** in electron density. No hysteresis is found during decompression (*c*).

6. Experimental Confirmation of a *p-p* Intra-band Gap in Sr_2FeO_4

A large proportion of the heavy transition metal (TM) oxides within the *iron-group* family are supposed to have an insulating Mott-Hubbard (M-H) *d-d* type intra-band gap U or a charge-transfer (C-T) energy gap Δ, where metal-to-metal or ligand-to-metal $d^n + d^n \rightarrow d^{n+1} + d^{n-1}$ and $d^n \rightarrow d^{n+1}\,\underline{L}$ ($\underline{L}$ is a hole in the ligand band) charge fluctuations, respectively, are expected to occur upon approaching an insulator-metal transition [14]. Recently it has been suggested [15] that a third mechanism may be involved in the opening of a band gap in correlated systems. This may occur in the case of a strong *p-d* hybridized TM-compound with a high valence TM-ion. Based on X-ray photo-spectroscopy by Mizokawa *et al.* [15], it has been suggested that the gap opening will occur within the ligand *p*-band; thus creating a *p-p* gap. If indeed it is a ligand intraband gap, the pressure-induced metallization at P_m should not affect the properties of the TM-*d* bands; the magnetic moment will persist at $P \geq P_m$. The motivation behind this work was to elucidate the mechanism responsible for the presence of the band-gap.

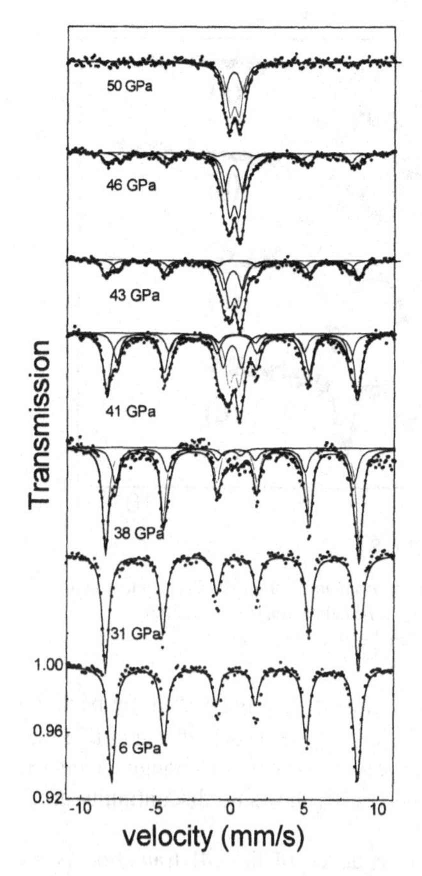

Figure 11. Spectra of $Pr^{57}FeO_3$ as a function of pressure at 300K. Note the appearance of a non-magnetic component at 38 GPa and the final onset of a pure non-magnetic phase at 50 GPa.

Figure 12. (*a*) The equation of state of $Pr^{57}FeO_3$. Note the volume discontinuity at 42(4) GPa. (*b*) Pressure dependence of the IS. The sharp decrease in IS corresponds to an increase in density. (*c*) The high pressure phase abundance. Data were taken at 300 K .

In recent years a few of the Fe-based perovskite-type compounds have been synthesized with the Fe ion in the unusually high oxidation of +4. Of particular interest is Sr_2FeO_4, an antiferromagnetic semiconductor with T_N = 60 K, that crystallizes in the K_2NiF_4 structure with tetragonal space group *I 4/mmm*. Electrical-transport, magnetic and structural properties of Sr_2FeO_4 (Fe^{4+}, d^4) were investigated by resistance, Mössbauer spectroscopy, and X-ray diffraction to ~30 GPa in a DAC [16]. At T < T_N and ambient pressure the Mössbauer spectrum consists of four magnetic components with identical IS suggesting the possibility of a helical magnetic structure [17] (Fig. 6a and b). With increasing pressure the spectra become simpler, and at P > 19 GPa a single-component spectrum is observed (see Fig 13*c* and *d*).

A sharp *insulator-metal* transition occurs at ~ 18 GPa. The R(T) dependence for 16 and 20 GPa is shown in Fig. 14. It is clearly seen that R(T) at 16 GPa is characteristic of a *gapped* state, whereas at 20 GPa dR/dT is positive, typical of a

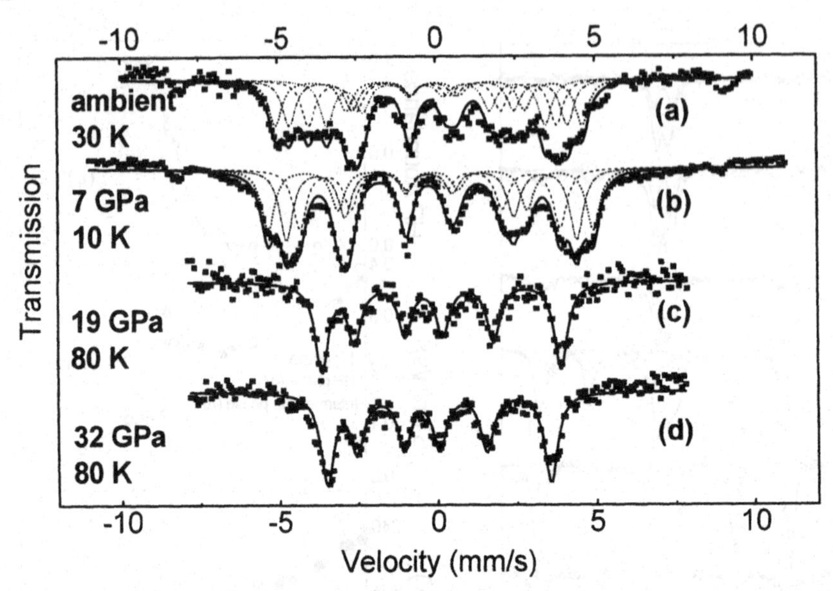

Figure 13. Spectra of Sr_2FeO_4 at $T < T_N$ for pressures ranging from ambient to 32 GPa. Up to 18 GPa one finds three different magnetic sites due to a possible helical magnetic structure. At $P > 19$ GPa the data could be fitted with a single site.

metallic state. The transition into a metallic state is further accentuated in the behavior of σ(P) where it discontinuously increases at ~ 18 GPa by an order of magnitude (see inset). Evidence that this transition does not involve a structural change is clearly indicated in the equation of state (see Fig. 15) where V(P) shows no discontinuity at the 16 - 20 GPa range.

Proof that the *d*-band is unaltered as a consequence of the MI transition is the persistency of the magnetic hyperfine interaction in ^{57}Fe (see Fig 13). Thus, it can be concluded that in Sr_2FeO_4 the gap-closure does not involve fluctuations of the $d-d$ or $d-\underline{L}$ types, and its band description should account for a p - p gap as suggested by Mizokawa *et al.* (Ref. 16).

7. Conclusions

Recent years have witnessed a constant growth in the number of high pressure studies involving Mössbauer spectroscopy in diamond anvil cells. During this period, cells originally designed and used for other spectroscopic methods have been adapted, and new cells more appropriate for MS studies have been developed. To help overcome the stringent requirement for minute absorber sizes (that are trending ever smaller in the quest for even higher pressures), some success has been achieved in producing compact, high-areal-density sources. The culmination of such efforts allow DAC-based MS with absorbers at hydrostatic pressures in the 100 GPa regime. The cells are small enough to allow temperature and/or applied magnetic field studies at pressure. Preliminary tests

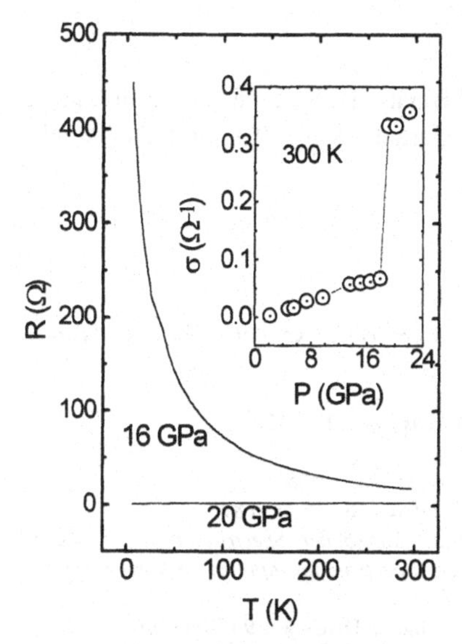

Figure 14. Resistance as a function of temperature at two selected pressures. Semiconducting behavior is evident at 16 GPa whereas the sample is metallic at 20 GPa. The inset shows conductance at 300 K as a function of pressure, revealing the onset of metallization at 18(1) GPa.

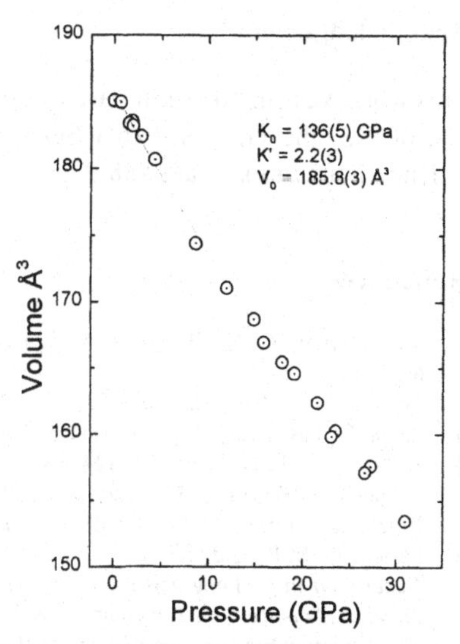

Figure 15. The room-temperature isotherm of Sr_2FeO_4. The solid line is a fit using the Birch-Murnaghan third-order finite-strain equation of state. Parameters derived from this fit are tabulated in the figure.

with *perforated* anvils are promising which will allow studies to higher pressures with ^{57}Fe enrichment close to natural abundance.

Presently the stimulus for reaching ever-increasing pressures in MS is the desire for studies of *metal-iron*-oxides which (1) play an extremely important role in **earth-sciences** and (2) provide further cases of **insulator-metal** transitions in correlated systems. High pressure ^{57}Fe Mössbauer spectroscopy is presently the only adequate tool for determining multiple phases and for investigating the intricate influence of reduced interatomic distances upon the electronic structure and the magnetic state of iron based compounds. Pressures needed for such studies are in the 100 GPa range and beyond.

We are recently witnessing the full implementation of MS with Synchrotron Radiation. One of the most interesting, unique, and recently developed subjects is the study of the *phonon density of states* under pressure using the inelastic channel of the nuclear resonant scattering. The phonon density of states is obtained by monitoring the K-fluorescence radiation of the de-exciting nuclei while the energy of the incident synchrotron radiation is tuned with a meV high-resolution monochromator across the 14.4-keV resonance. With *perforated* anvils one should be able to perform inelastic high pressure studies in ^{57}Fe compounds using the 6 keV x-rays to study the phonon density of states at very high pressures.

Acknowledgement

This work was supported in part by the USA-Israel Binational Science Foundation grant #95-00012, the German-Israel Science Foundation grant #I-086.401, and the Israeli-Science-Foundation grant #88/97.

References

1. Machavariani, G.Yu., Pasternak, M.P., Hearne, G.R., and Rozenberg, G.Kh. (1998) *Rev. Sci. Instrum.* **69**, 1423.
2. Jayaraman, A. (1983) *Rev. Modern. Phys.* **55**, 65.
3. Sterer, E., Pasternak, M.P., and Taylor, R.D. (1990) *Rev. Sci. Instrum.* **61**, 1117.
4. Hanks, R.V. (1961) *Phys. Rev.* **124**, 1319.
5. Pound, R.V., Benedek, G.B., and Drever, R. (1961) *Phys. Rev. Letters* **7**, 405.
6. Holzapfel, Wilfried B. (1975) in *CRC Critical Reviews in Solid State Sciences*, p 89.
7. Pasternak, M.P. and Taylor, R.D. (1996) *High Pressure Mössbauer Spectroscopy: The Second Generation* in G. Long and F. Grandjean (eds)., *Mössbauer Spectroscopy Applied to Magnetism and Materials Science, Vol. 2*, Plenum Press, New York, p167.
8. The maximum pressure reported with a 300 μm culet anvil using o/p DACs was 90 GPa (Ref. 11).
9. Hearne, G.R., Pasternak, M.P., and Taylor, R.D. (1994) Rev. Sci. Inst. **65**, 3787.
10. Pasternak, M.P., Taylor, R.D., Chen, A., Meade, C.,. Falicov, L.M, Giesekus, A., Jeanloz, R., and Yu, P.Y. (1990) *Phys. Rev. Letters* **65**, 790 and references therein.
11. Pasternak, M.P., Taylor, R.D., Jeanloz, R., Li, X., Nguyen, J.H., and McCammon, C.A. (1997), *Phys.Rev.Letters* **79**, 5046.
12. DeMarco, J. *et al.* (1978), *Phys. Rev.* **B18**, 3968.
13. Xu, W., Naaman, O., Rozenberg, G.Kh., Machavariani, G.Yu, Pasternak, M.P., and Taylor, R.D. (1998), to be published
14. Bocquet, A.E., Saitoh, T., Mizokawa, T., and Fujimori, A. (1992), *Solid St. Comm.* **83**, 11.
15. Mizokawa, T., Namatame, H., Fujimori, A., Akeyama, K., Kondoh, H., Kuroda, H., and Kosugi, N.,(1991), *Phys. Rev. Lett* **67**, 1638.
16. Rozenberg, G. Kh., Milner, A. P., Pasternak, M. P., Hearne, G. R., and Taylor, R. D. (1998), accepted for publication in *Phys. Rev.* B.
17. Adler, P. (1994), *J. Sol. State Chem.* **108**, 275.

THE CHALLENGE OF AN AUTOMATIC MÖSSBAUER ANALYSIS

PAULO A. DE SOUZA JR.[1] AND VIJAYENDRA K. GARG[2]
[1)]Departamento de Física, Centro de Ciências Exatas, Universidade Federal do Espírito Santo, 29060-900 Vitória, E.S., Brazil
[2)]Instituto de Física, Universidade de Brasília, 9000-970 Brasília, DF, Brazil

Abstract

The present paper reports on our efforts to obtain an intelligent and automatic Mössbauer analysis. For this propose, a program was implemented capable to analyse and fit a Mössbauer spectrum, identify the substance under study, and give information of the micro-environment. The implemented program uses expert systems, genetic algorithm, fuzzy logic and artificial neural networks to process the present information based on previous knowledge of the substance. The program, also, is able to process additional information from other analytical techniques such as XRD and chemical analysis. It is possible to update the program´s knowledge.

1. Introduction

The Mössbauer spectroscopy is useful to identify the resonant atom-bearing compounds present in the sample. In spite of its potential industrial applications, this resonance effect is not popular as an analytical technique mainly because the Mössbauer analysis is complex, human dependent and time-consuming. Recently, application of genetic algorithms in the analysis of Mössbauer effect has been introduced, and later using fuzzy logic some improvements were proposed. Simultaneously we used artificial neural network in Mössbauer pattern analysis and identification of substances and micro-environment properties. A Mössbauer spectrum could contain a number of components, which may involve physical, mineralogical and chemical principles. The analyzed parameters could alter with the changes in temperature, pressure, external magnetic field and other experimental variables and numerical procedures (fitting of the spectrum) allow a better spectral interpretation [1]. A genetic algorithm was implemented as a first approach to the identification of resonant-phases present in the sample.

M. Miglierini and D. Petridis (eds.), Mössbauer Spectroscopy in Materials Science, 359–372.

2. Analytical Information from Mössbauer Spectra

The basic task of the spectral analysis is to identify the individual Mössbauer species from the corresponding patterns in the spectrum. Some of these Mössbauer patterns can vary according to their complexity (e.g. elementary or superimposed patterns) and experimental conditions (being standard or induced patterns). Most experiments are performed at room temperature, under atmospheric pressure and without an external magnetic field. Induced patterns are obtained under non-standard conditions. In this case the differences between the induced and standard patterns can give important contributions to the analysis [2]. Generally, the Mössbauer spectrum of a multi-compound/multi-phase material is a superposition of patterns. Mössbauer spectra encountered in the practice can be quite complex. In order to obtain some analytical information from such a spectrum it has to be decomposed to elementary patterns (singlets, doublets and sextets). The Mössbauer parameters are determined by least square fitting programs. The Mössbauer parameters in the case of ^{57}Fe are derived from the line parameters (line position, line width and intensity).

2.1. CLASSICAL APPROACH

The conventional analysis of a Mössbauer spectrum starts with the choice of a spectral model and initial values to the spectral parameters. An experienced analyst does this by an "eye fit" and use of a fitting program to improve the initial parameters [3]. The main characteristics of the classical approach of a Mössbauer fitting procedure is the large time consumed, the excessive human effort required and the dependence on the specialist experience.

Investigation on precision and interlaboratory reproducibility of measurements of the Mössbauer effect in minerals [4 - 6] indicated that the possible causes for imprecision in Mössbauer data are both the experimental setup and the applied analysis. From the experimental point of view the imprecision comes from the counting statistics which can be minimized by through testing of the spectrometer, the long term drift of the equipment's driver, sample concentration, electronic relaxation which can cause differences in peak amplitudes, preferred orientation in the samples during the sample preparation and by calibration of the equipment. From the analytical point of view, imprecision comes from the overlap of one or more peaks of different subspectrum and errors in the guessed model for fitting, specially how many peaks can be fit. Experimental standards are still an open discussion in Mössbauer spectroscopy and some solutions have been proposed [2 - 7].

3. Preliminary Automated Mössbauer spectral Analysis

3.1. TOPOLOGICAL ANALYSIS

For the identification of phases in a Mössbauer spectrum it is necessary to make a topological analysis. Important hints are given for the typical doublet positions which appear always at low velocity because, in fact, in ^{57}Fe a doublet is a quadrupole interaction and a sextet is a result of a bigger interaction, the magnetic one. No one is expecting to find a singlet out of the middle of the velocity scale. The line width at half maxima (Γ, mm/s) can be considered, in a first analysis, as constant for both lines of a doublet or for all six lines of a sextet. Dollase [8] showed that the peak overlap is the source of large uncertainty in Mössbauer analysis where peak separation is less than 0.6 Γ. If there is a superposition of sextets, one of these lines will have a large Γ. This can be understood as two or more superimposed lines of sextets. From these general assumptions, a program capable to perform the topological analysis of the Mössbauer spectra was implemented. The topological analysis consists of a list of possible collection of singlets, doublets and sextets that fits better the Mössbauer spectrum. The peak position of a set of singlets, doublets and sextets are made from the results obtained from the topological analysis and from the application of a second derivative method [9]. There is important physical information for determining the reliability of some models (set of subspectra). The solution with large number of lines may be quite difficult to interpret. There is still an open discussion concerning the criteria for determining the reliability of a given spectral model. The χ^2 parameter, which should be nearly unit was used. It was assumed that the physical aspects that keeps a fitting model for a spectra are correct. The use of expert systems to solve this problem automatically is in advanced development [10].

TABLE 1. Mössbauer parameters for the nine main phases on atmospheric corrosion products of steel.

Iron phase Name	Chemical Formulae	Temperature (K)	IS (mm/s)	QS (mm/s)	B_{eff} (Tesla)
goethite	α-FeOOH	300	0.35	-0.24	36.7
		77	0.47	-0.24	49.2
akaganeite	β-FeOOH	300	0.37	0.96	
		77	0.51	-0.15	47.4
lepidocrocite	γ-FeOOH	300	0.37	0.94	
		77	0.48	0.92	
feroxyhite	δ-FeOOH	300	0.36	0.10	45.1
		77	0.51	0.20	53.3
hematite	α-Fe_2O_3	300	0.37	-0.20	51.7
		77	0.48	0.40	54.3
	γ-Fe_2O_3	300	0.30		50.3
maghemite		77	0.53		52.6
	γ-Fe_2O_3	300	0.36	0.94	
	(paramagnetic)	77	0.47	-0.07	52.0
		300	0.28		48.7
magnetite	Fe_3O_4		0.66	-0.16	45.7
		77	0.47	-0.05	52.2
			0.81	-0.42	49.9
matalic iron	α-Fe	300			33.0

For small group of substances, such as corrosion products, the iron-bearing substances are well known (and also its Mössbauer parameters) [11]. Tests were carried out with this restricted group for finding the peak position and intensity by using of a genetic algorithm. In this application, the model is the possible combination of nine typical atmospheric corrosion products formed (Table 1).

Each solution is a string of the peak intensity for the nine possible substances. When the relative intensity is smaller than 5%, this phase is excluded from the solution (chromosome). This way, the program will find the main phases present at the Mössbauer spectra with more than 5% in relative peak intensity. This procedure works with a restricted group of known substances in a concentration superior to 5%. This may be interpreted as a limitation, but is reasonable for applications in material industry.

4. The Genetic Algorithm Fitting Analysis

4.1. GENETIC ALGORITHM

A genetic algorithm [3, 12-15] is an iterative algorithm based on the idea of letting several solutions (individuals) compete with each other for the opportunity of being selected to create new solutions (reproduction). Each solution is represented as a string of binary digits (chromosome). New representations are created by pairwise exchanging bits in the binary representations of two parent solutions. Some of the bits in the new solutions may be switched randomly from zero to one or conversely (mutation). The algorithm starts with a randomly determined set of individuals (initial population) and iteratively generates new populations through the application of genetic operators (selection, cross-in-over, mutation and elitism or clone).

The quality of each parameter combination, i.e., the fitness of each solution, is defined as the inverse value of the calculated χ^2 value. The ratio of the fitness and the sum of all fitness values of the individuals in the population is used as probability for an individual to be selected to the reproduction. Thus, individuals with larger fitness will be chosen with greater probability to the reproduction process.The iteration is usually stopped when the average fitness of the population starts to be saturated. This is due to the fact that the genetic operators tend to favor those chromosome structures for which the corresponding fitness values are large. The convergence of the algorithm is usually improved by exploiting the so called "elitism" or "clone", in which the best individual in the old population replaces the worst in the new one.

4.2. THE IMPLEMENTED GENETIC ALGORITHM

The implementation of the genetic algorithm includes the following steps:
(1) *The input* - The user provides the model. In addition some experimental conditions and calibration will be informed. The model file contains the maximum velocity and the positions of peaks, given as a separate list for each spectral component (i.e., one peak position for each singlet, two for each doublet and six for each sextet).
(2) *The background* - The absorption level and standard deviation of the baseline is evaluated on the basis of a given number of initial and final points in the spectrum.

(3) *The parameter intervals* - The parameter intervals are determined on the basis of the given model, the computed background level and the maximum spectral velocity used.

(4) *Initial population* - A given number (e.g., 100) of solutions is created by a random generation of binary strings of length b.

(5) *Fitness calculation* - The fitness of each solution is calculated as the inverse value of χ^2 computed on the basis of the measured spectrum and the linear combination of Lorentzian functions corresponding to the parameters picked from the binary string. The sum and average of all fitness values are computed.

(6) *Stopping* - The stopping conditions are tested (e.g., whether a given number of generations has been created or whether a good solution quality has been achieved without any improvement in the average fitness during the last generations). The use of Fuzzy logic as a dynamic stopping condition criterion will be discussed. If the conditions hold, then the iterations stop and the program jumps to step 9.

(7) *Elitism* or *clone* - The worst solution in the new population is replaced with the best solution of the previous population.

(8) *Reproduction: cross-in-over and mutation* - Each individual is assigned a selection probability proportional to its share of the total fitness, and individuals are pairwise selected for reproduction by a "roulette wheel" with sector sizes proportional to their selection probabilities. A new population is formed with the help of the operations of cross-in-over and mutation, which are performed with probabilities indicated by the parameters cross-in-over and mutation probability, respectively.

(9) *Final output* - The program computes the final parameters on the basis of the decoded parameter values corresponding to the best solution, writes the results into a file and reports to the user.

The quality of the fitted Mössbauer spectra were as good as that obtained by NORMOS [16], a comercial Mössbauer fitting program. The advantage that the input parameters (peak positions) were given just only once or obtained from the topological analysis to assess the final fitted spectrum. The genetic algorithm is faster, and, is less tedious to use, requiring minimal user interaction [3, 13].

4.3. THE ALGORITHM SENSIBILITY

The results of the genetic algorithm were obtained for different mutation probabilities and population sizes (number of individuals). There are small differences in the obtained parameters, with distinct input parameters for the genetic fitting program, which can be considered the same in the experimental error. The major difference is obtained in area ratio. This is a typical result of delay of convergence of this parameter (peak intensity and Γ) in comparison with the hyperfine parameters (IS, QS and B_{eff}). A gradient search for peak intensity was included and this delay in convergence was minimized. Table 2 depicts some results obtained by changing the genetic parameters.

TABLE 2. Sensibility of the genetic algorithm to the mutation probability.

Probability rate (%)	IS (mm/s)	QS (mm/s)	Relative area (%)	χ^2	Additional elapsed time *
0.005	- 0.03	1.39	64.7	1.80	39.7 %
	1.80	3.72	35.3		
0.004	- 0,03	1.41	61.6	1.77	4.7 %
	1.80	3.72	38.4		
0.003	- 0.03	1.38	65.6	1.81	36.8 %
	1.80	3.72	34.4		
0.002	- 0.03	1.40	64.4	1.77	optimal
	1.80	3.72	35.6		
0.001	- 0.03	1.38	65.8	1.81	36.3 %
	1.08	3.72	34.2		

(population size: 100 individuals; number of bits: 8 bits;
* percentual elapsed time in addition to the fastier solution (optimal).
The fitted spectrum is a superposition of two doublets)

5. Fuzzy Logic

Fuzzy sets theory has been extensively used in applications where due to imprecise information or the empirical nature of the problem the solution is highly dependent an human experience [17, 18]. A fuzzy system incorporates fuzzy heuristics and knowledge that defines the terms being used in the former level. Indeed, it is possible to encode linguistic rules and heuristics directly reducing the solution time since the expert's knowledge can be built in directly. In addition, its qualitative representation form makes fuzzy interpretations of data very natural and an intuitively plausible way to formulate and solve several problems. The fuzzy system was used to control and accelerate the least-squares fitting of a Mössbauer spectrum.

5.1. THE FUZZY IMPLEMENTATION

The relevant variables to the Fuzzy implantation are the Mössbauer parameters B_{eff}, Γ, relative area of each subspectra and the difference (D_{fi}) between final and initial values of IS and QS after a simple least-squares fitting - D_{fi}IS and D_{fi}QS respectively. A full description of these relationship was published in [3, 18].

It is not hard to judge the efficiency of the fuzzy control. All the actions of the fuzzy controller can be reported, analyzed and improved. By this way, a significative improvement in the inference rules is realized. The fitting of Mössbauer spectra from genetic algorithm and fuzzy logic is fast and produces parameters similar to those from human analysis. Generally, the χ^2 value is a measure of quality of the spectra and it has been used as a stopping condition. Although it is relatively much simpler to fit a good quality spectrum, the Mössbauer parameters are almost independent of the statistical quality of the spectra [18, 19]. However, the gradient of χ^2 value indicates how far the program is from the optimal result. Some improvement might be achieved using the fitting program extensively. For the substance and the crystalline identification, small differences are filtered by an artificial neural identification process. It is also possible that experimental noise appears for the spectra which can not be detected or controlled.

A cost-benefit ratio is strongly decreased with the full automation procedure in the spectral analysis.

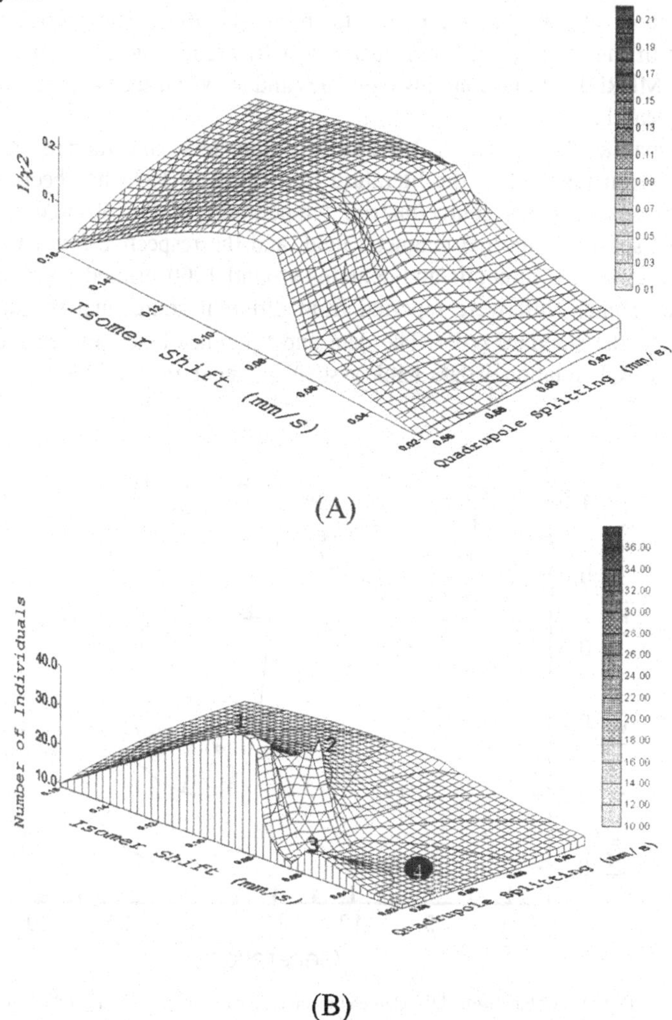

Figure 1. (A) The three dimensional representation of the convergence of the genetic algorithm in δ and Δ plane with the fitness function $(1/\chi^2)$; (B) number of individuals in each with a given δ and Δ. The point 1 is the optimal solution, 2 is a local maximum (or minimal χ^2), 3 was a weak solution (dead individuals) and in 4 region no solution was indicated.

6. Data bank

There are useful data collections that can be found in the literature for Mössbauer reference data of a given compound. The most extensive source is the Mössbauer Information System (MIS) of the Mössbauer Effect Data Center (MEDC) at the

University of North Carolina at Ashville, USA. The MEDC is a scientific center collecting Mössbauer data since the discovery of the Mössbauer effect. The MIS contains about 30,000 bibliographic references, and its update entries are regularly published in the volumes of Mössbauer Effect Reference and Data Journal (MERDJ) [20]. One MERDJ volume contains ten issues and a special index, covering one year of literature search.

A separate data bank of Mössbauer parameters and references of minerals containing iron reported in the literature from 1958 to 1996 has been created. This computer bank contains source/matrix, temperature of the absorber, isomer shift, quadrupole splitting, internal magnetic field, and the respective reference. The stored data cover some 500 minerals and contains around 1000 printed pages of a standard book. The isomer shift were reported from different source/matrix, and have been converted to the standard α-Fe. The data and reference bank was developed under a WATCOM SQL relational data bank system manager in a Windows™ environment (Figure 3).

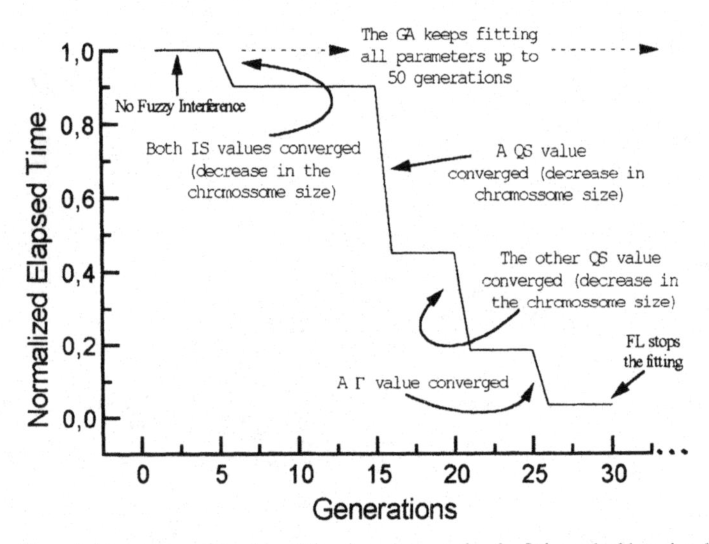

Figure 2. Comparison of the elapsed time in each generation for fitting a doublet using the fuzzy logic controlling the genetic algorithm and using only the genetic algorithm.

7. Artificial Neural Networks

There are two main reasons that exclude the possibility of a general substance or property identification system based only on a Mössbauer set of data, namely:

- Mössbauer spectroscopy is a sensitive technique, and, as discussed, it is possible to find some small difference in interlaboratory measurements of the same sample [4 - 6];

- It is possible to find some mistakes in the published parameters from spectral misinterpretation or experimental imprecision.

The search for exact information using a conventional data bank is of limited help. Data bank search for exact information became incapable to differentiate the values of Mössbauer parameters within the experimental errors (e.g., IS = 0.22 mm/s from IS = 0.23 mm/s, but physically or depending of the experimental error both values may be considered the same). We must therefore select a different approach and try to identify the substance and its properties.

Artificial neurons learn from experience, generalize from previous examples to new ones, and extract essential characteristics from inputs containing irrelevant data. The ability to see through noise and distortion to the pattern that lies within is vital for the pattern recognition in a real world environment. An artificial neural network was trained with the Mössbauer parameters published in the literature and stored in our data bank (Figure 3). After adequate training it could successfully identify several substances and their crystalline structures from new values of isomer shift, quadrupole splitting and internal magnetic field.

7.1. APPLIED ARTIFICIAL NEURAL NETWORK

Recently, the use of artificial neural network in Mössbauer analysis of corrosion products from Mössbauer spectra [21] and in identification of a substance from Mössbauer parameters of iron-bearing minerals has shown success [3, 22-25]. Identification of non-magnetic phases [22, 23] and magnetic phases of iron-bearing minerals [24, 25], and their crystalline structure [26] from Mössbauer parameters has also been reported.

7.2. IDENTIFICATION OF SUBSTANCES

This program is formed by a hybrid network associated with a computer data bank. Input layer is followed by a Kohonen's layer [24] with 1428 units that realize an unsupervised learning. The fine-tuning of the first mapping is done by a Learning Vector Quantization (LVQ) network (supervised and competitive learning) using the same 1428 neurons to identify 478 substances [23-26]. The source program [27] of the ANN was written using Pascal language. A Delphi™ for window's program has been used to develop the user's friendly interface. The data bank contains the Mössbauer parameters of 476 substances from the literature since 1958 to 1996 and they were used to train and test the ANN. Three trainings were carried out according to the measured temperature (room, liquid nitrogen and liquid helium temperature). To use the ANN the program restores the experimental Mössbauer parameters of δ, Δ and B_{eff} obtained by the genetic algorithm and Fuzzy sets. The ANN output will give the name of the corresponding substance, or up to three other options to choose. The options are given to eliminate possible errors in the ANN identification and let the operator choose the right option. The correct option can be chosen from other physical or chemical data and/or by the researcher's experience. Sometimes it may happen that the ANN gives the same name to all four options. Once the ANN classifies the substance, the user may start the data bank mode to obtain the references of these substances that were published. The neural identification of substances is reliable. One of the great advantages is that the LVQ network can classify the substance from the Mössbauer parameter with almost a hundred percent of accuracy. The major limitation of this approach is that it doesn't work for materials for which data are not available in the data bank used.

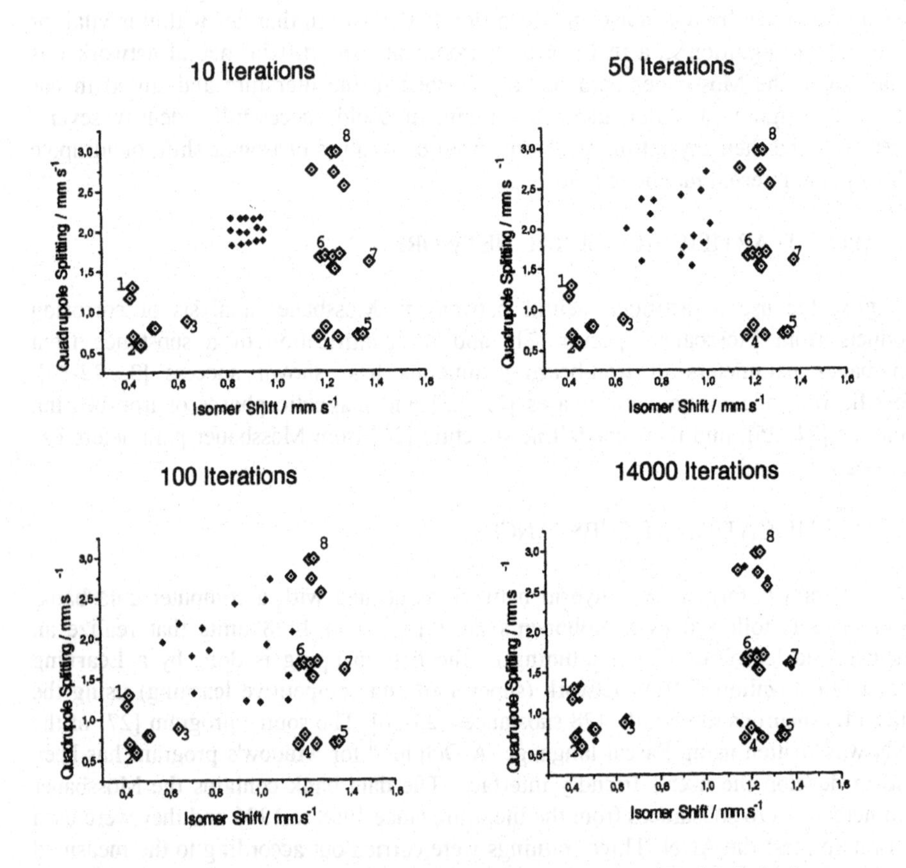

Figure 3. **The convergence of the LVQ for a two dimensional space (IS, QS) for 8 substances after 10, 50, 100, and 14000 iterations.**

7.3. IDENTIFICATION OF CRYSTALLINE STRUCTURE

An ANN was applied in the identification of different crystalline structures and, also, its degree of distortion (Table 3). Although an appropriate training with published data could not be performed, the ANN could successfully identify, from new values of Mössbauer parameters, the crystalline structure of the substance under study.

Burns [28] tried to relate different crystalline structures graphically using Mössbauer parameters of IS and QS and correlating these parameters with various bonds length-related parameters as mean metal-oxygen distance of a coordination site, and the volume per oxygen in a unit cell and polyhedral volume. But this method has had limited success in several cases to delimit δ ranges for Fe^{2+} and Fe^{3+} ions in silicates with tetrahedral, octahedral and five-fold coordination sites. For this purpose Burns reported these parameters graphically in pairs making it in two dimensional graphics. In some cases (orthopyroxene IS =1.12 mm/s and QS = 1.90 mm/s, octahedral; and pigeonite IS =1.12 mm/s and QS =1.96 mm/s, 6-7 coordination) there is superposition of clusters that make it impossible to determine precisely their crystalline structures using only IS and QS. For a correct identification it would be necessary to make various two dimensional graphics. The program performed the identification of different degrees of distortion in octahedral crystalline structure of silicates with Fe^{2+} valences, the identification of different crystalline structures with IS and QS Mössbauer parameters of silicates with Fe^{2+} valences and the identification of different crystalline structures with IS and QS Mössbauer parameters and polyhedral volume of a coordination site (Vp) to silicates with Fe^{3+} valences. The values of Vp used here are as reported by Burns [29 - 30] for the respective minerals. Now the ANN could classify any crystalline structure from new values of IS, QS and Vp.

8. Conclusion

The Mössbauer spectroscopy is a powerful technique in several fields of the human activities. The analysis of a Mössbauer spectrum is a time-consuming task, require excessive human effort and depends on the specialist experience. The use of genetic algorithm and fuzzy logic were proposed to obtain the Mössbauer parameters (Figure 4). An artificial neural network trained with a published Mössbauer parameters data bank is capable to identify, from new experimental Mössbauer parameters (IS, QS, and B_{eff}), the substance under study and its crystalline structure (Figure 5). The implemented program is fast and presents a high correct rate. Table 4 depicts the Mössbauer parameters obtained in from the spectrum of Figure 4. The ANN identification was correct (magnetite) for both GA and standard fitted Mössbauer parameters. This illustrates the robustness of this patter recognition identification.

TABLE 3. Results of ANN identification: A general example.

Mineral	IS	QS	ANN Classif	Mineral	IS	QS	Vp	ANN Classif.
Biotite *	1.560	2.620	Octa.	Epidote	0.360	2.010	-	Octa.
	1.115	2.570	Octa.		0.400	2.020	-	Octa.
	1.102	2.240	Octa.		0.338	2.020	-	Octa.
	1.280	2.600	Octa.		0.196	2.020	-	Octa.
	1.240	2.230	Octa.		0.378	2.150	-	Octa.
	1.135	2.605	Octa.		0.344	2.075	-	Octa.
	1.105	2.190	Octa.		1.052	1.990	-	Octa.
	1.182	2.580	Octa.		0.350	2.056	-	Octa.
	1.010	2.223	Octa.		0.354	2.085	-	Octa.
	1.040	2.600	Octa.		0.450	2.020	-	Octa.
	1.013	2.550	Octa.	Glaucophane	0.360	0.480	-	Octa.
	1.007	2.107	Octa.		0.358	0.480	-	Octa.
Diopside	1.380	2.440	Octa.		0.378	0.420	-	Octa.
	1.205	2.320	Octa.		0.340	0.480	-	Octa.
	1.295	2.130	Octa.		0.380	0.460	-	Octa.
	1.372	2.480	Octa.		0.360	0.470	-	Octa.
	0.921	2.214	Octa.		0.380	0.470	-	Octa.
	1.177	2.160	Octa.	Grandidierite	0.330	1.200	-	5-CN
Olivine	1.158	2.930	OD		0.365	1.252	-	5-CN
	1.128	2.930	OD		0.334	1.133	-	5-CN
	1.183	2.968	OD		0.333	1.201	-	5-CN
	1.168	2.930	OD	Kyanite	0.380	0.990	-	Octa.
	1.187	3.018	OD	Muscovite	0.370	0.860	-	Octa.
	1.062	2.990	OD		0.640	0.700	-	Octa.
	1.208	3.030	OD		0.390	0.990	-	Octa.
Orthopyroxene	1.123	2.015	OVD		0.480	0.720	-	Octa.
	1.120	1.960	OVD	Osumilite	0.250	1.710	-	Tetra.
Pigeonite	1.173	2.430	Octa.	Phlogopite	0.170	0.500	-	Tetra.
Aegerine**	0.380	0.300	Octa.		0.490	0.390	-	Tetra.
Andalusite	0.220	2.650	5-CN		0.308	0.360	-	Tetra.
	0.350	1.830	Octa.		0.240	0.470	-	Tetra.
	0.360	1.810	Octa.	Sanidine	0.210	0.480	-	Tetra.
Aemigmatite	0.430	0.990	Octa.		0.220	0.610	-	Tetra.
	0.390	0.310	Octa.	Vesuvianite	0.330	0.560	-	5-CN
Andradite	0.390	0.550	Octa.		0.290	0.560	-	5-CN
	0.380	0.570	Octa.		0.390	0.460	-	5-CN
	0.430	0.530	Octa.		0.550	0.420	-	5-CN
	0.420	0.560	Octa.	Orthopyroxene **	1.120	1.900	-	Octa.
	0.390	0.640	Octa.	Pigeonite	1.120	1.960	-	6-7 CN
	0.390	0.560	Octa.	Fe Sapphirine ***	0.330	0.850	10.78	Octa.
Actinolite	0.380	0.890	Octa.		0.350	1.490	12.72	Octa.
Babingtonite	0.380	0.890	Octa.	Mg Sapphirine	0.290	1.230	11.63	Octa.
	0.380	0.880	Octa.		0.300	0.760	9.47	Octa.
	0.400	0.870	Octa.		0.290	1.230	2.75	Tetra.
Chloritoidite	0.330	0.950	Octa.		0.300	0.760	2.78	Tetra.
Clintonite	0.500	1.100	Octa.	Yoderite	0360	1.000	9.82	Octa.
	0.270	0.680	Tetra.		0.360	1.000	6.07	5-CN

(IS = isomer shift relative to α-Fe, mm/s., at room temperature, QS = quadrupole splitting, mm/s, Vp = Polyhedral volume in a coordination site, $Å^3$, Octa. = octahedral, OVD = octahedral very distorted, OD = octahedral distorted, Tetra. = tetrahedral; *. first test, **. second test, ***. third test).

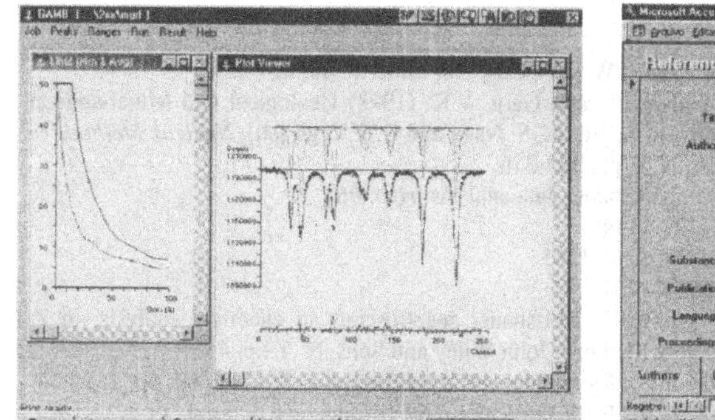

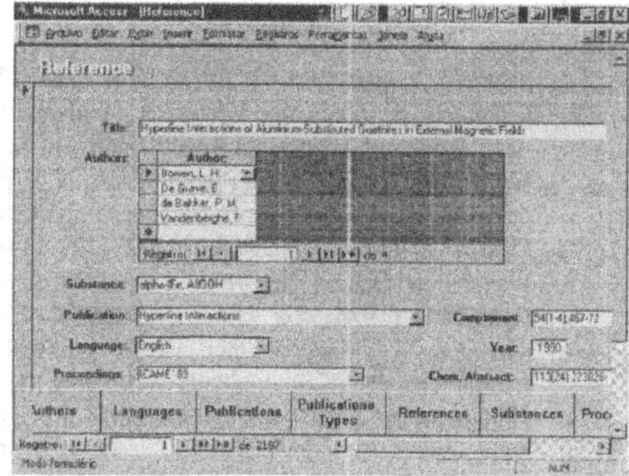

Figure 4. Fitting program's (left) and databank's interface (right).

TABLE 4. Comparison of output parameter from genetic algorithm and a reference program (like-Normos).

I.S. (mm/s)	Q.S. (mm/s)	Hn (Tesla)	R.A. (%)	χ^2
genetic algorithm				
0.21	-0.02	48.95	33.8	3.35
0.54	0.01	45.90	66.2	
reference				
0.14	-0.02	48.97	33.7	2.16
0.54	-0.01	45.64	66.3	

R.A. = Relative Area

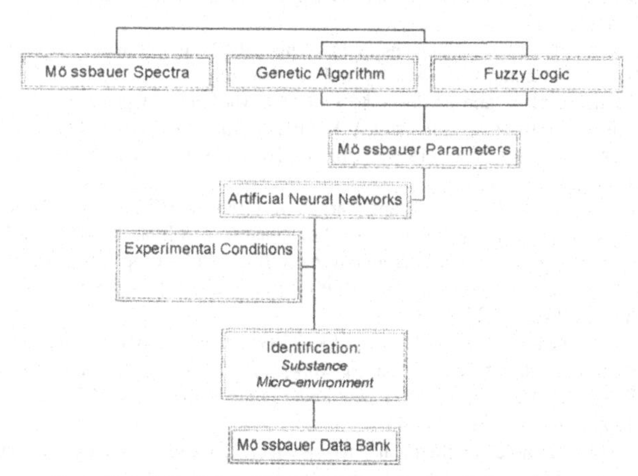

Figure 5. Automation in Mössbauer spectroscopy fitting process

Acknowledgements

Financial support from Companhia Siderúrgica de Tubarão (grant project #17 UFES/CST - 1995/98) is thankfully acknoledge. We also thank Dr. Richard E. Brand for a copy of Normos fitting program used in the text.

References

1. Fabris, J.D., Coey, J.M.D., Qi, Q., and Mussel, W.N. (1995), *Am. Mineral.* **80**, 664-669.
2. Kuzmann, E., Nagy, S., Vértes, A., Weiszburg, and Garg, V.K. (1998) Geological and Mineralogical Applications of Mössbauer Spectroscopy, in A. Vértes, S. Nagy and K.S. Vegh (ed), *Nuclear Methods in Mineralogy and Geology*, Plenum Press, N.Y. pp. 285-376.
3. de Souza Jr., P.A. (in press) Advances in Mössbauer data analysis, *Hyp. Int.*
4. Dyar, M.D. (1984), *Am. Mineral.* **69**, 1127-1144.
5. Wayshunas, G.A. (1986), *Am. Mineral.* **71**, 1261-1265.
6. Dyar, M.D. (1986), *Am. Mineral.* **71**, 1266-1267.
7. Kuzmann, E., Nagy, S., and Vértes, A. (1994) Mössbauer spectroscopy in chemical analysis, in Z. Alfassi (ed), Chemical Analysis by Nuclear Methods, John Wiley and Sons, N. Y., p. 455.
8. Dollase, W.A., (1975) Statistical limitations of Mössbauer spectral fitting, *Am. Mineral.* **60**, pp. 257-264.
9. Afanas'ev, A.M. and Chuev, M.A. (1995), JETP **80**(3), pp. 560-567.
10. de Souza Jr., P.A. (1997) Industrial impact in the air quality in metropolitan Vitória region (ES): Technical Report to CST steel industry, personal communication.
11. Cook, D.C. (1998), *Hyp. Int.* **111**, pp. 71-82.
12. Goldberg, D.E. (1989) *Genetic Algorithms in Search, Optimization, and Machine Learning*, Addison-Wesley Publishing Company, Reading, MA.
13. Ahonen, H., de Souza Jr., P.A., and Garg, V.K. (1997), *Nucl. Instr. Meth. Phys. Res. B* **124**, pp. 633-638.
14. Klencsár, Z. (1997), *Nucl. Instr. Meth. Phys. Res. B* **129**, pp. 527.
15. Kléncsár, Z., Kuzmann, E., and Vértes, A. (1998), *Hyp. Int.* **112**, pp. 269-273.
16. Brant, R.A. (1994) *Normos Mössbauer Fitting Program: Demo Version*, Wissel gmbH, Stanberg, Germany.
17. Marks II, R.J. (Ed), (1994) *Fuzzy Logic Technology and Applications*, IEEE Technology Update Series.
18. de Souza Jr., P.A., Lamego, M.M., and Garg, V.K. (1997) Applications of Fuzzy Logic in the Analysis of ^{57}Fe Mössbauer Spectra, in G. Cameron, M. Hausson, A. Jerdee and C. Melvin (eds.), *Proceedings of the 39th Midwest Symposium on Circuits and Systems*, IEEE press, New York, pp. 569-571.
19. Mitra, S. (1992) *Applied Mössbauer Spectroscopy: Theory and Practice for Geochemistry and Archeologists*, Pergamon Press, New York.
20. Stevens, J.G., Pollak, H., Zhe, L., Stevens, V.E., White, R.M., and Gibson, J.L. (1992) *Mineral: Data*, Mössbauer Effect Data Center, Univ. North Caroline, Asheville, NC.
21. Souza, M.N., Figueira, M.A., and da Costa, M.S. (1993), *Nucl. Instr. Meth. Phys. Res. B* **73**, p. 95.
22. Salles, E.O.T., de Souza Jr., P.A., and Garg, V.K. (1994), *Nucl. Instr. Meth. Phys. Res. B* **94**, p. 499.
23. de Souza Jr., P.A. and Garg, V.K. (1997) Artificial Neural Networks in Mössbauer Material Science, *Czech. J. Phys.* **47**, pp. 513-516.
24. de Souza Jr., P.A, Salles, Salles, E.O.T., and Garg, V.K. (1996) Artificial Neural Networks in Mössbauer Mineralogy, in L.P. Calôba, P.S.R. Diniz, A.C.M.de Queiroz, and E.H. Watanabe (Eds.), *Proceedings of the 38th Midwest Symposium on Circuits and Systems*, IEEE press, **1**, pp. 558-561.
25. de Souza Jr, P.A., Garg, R., and Garg, V.K. (1998) Automation of the analysis of Mössbauer spectra and parameters, *Hyp. Int.* **112**, pp. 275-278.
26. Salles, E.O.T, de Souza Jr., P.A. and Garg, V.K. (1995) Identification of Crystalline Structure using Mössbauer Parameters and Artificial Neural Network, *J. Radioanal. Nucl. Chem.* **190**, 439 (1995).
27. de Souza Jr., P.A. and Garg, V.K. (1995) Patent under number: INPI 95000457.
28. Burns, R.G. (1994), *Hyp. Int.* **91**, 739-745.
29. Burns, R.G., (1993) Remote geochemical analysis elemental and mineralogical composition, in C.M. Pieters, P.A.J. Englerd (eds.), Cambridge University Press, Cambridge.
30. Burns, R.G. and Solberg, T.C. (1990) Spectroscopy characterization of minerals and their surfaces, in L.M. Coyne, S.W.S. Mckeever, D.F. Blake (eds.), Am. Chem. Soc., Washington.

EVALUATION OF EXPERIMENTAL DATA : LINESHAPE AND GOODNESS OF FIT

G. PEDRAZZI, S.Z. CAI AND I. ORTALLI
Istituto di Scienze Fisiche ed Unità INFM
Plesso Biotecnologico Integrato - Università di Parma
Via Volturno, 39 - 43100 Parma, Italy

1. Introduction

All the fitting procedures, irrespective of the chosen optimization algorithm and of all the intriguing details involved [1, 2, 3, 4], have in common two steps : 1) the initial choice of a proper fitting model, and 2) the final evaluation of the agreement between the model and the experimental data, i.e. the goodness of the fit.

The present work deals in some way with both the above steps. In a first section we propose new goodness of fit parameters that we believe are particularly useful in Mössbauer spectroscopy, and in a following section we propose the use of a peak profile different from the conventional Lorentzian when treating crystalline sites and thickness effects.

2. Statistics and Goodness of Fit (GOF)

The evaluation of the agreement of a model with the experimental data is of major importance for any data analysis. If the model, in fact, does not describe the data adequately any derived conclusion may be quite misleading.

The judgment of the adequacy of the fit relies on the statistical properties of some mathematical distribution like, for instance, F or χ^2 [5].

In Mössbauer spectroscopy the variance of any experimental point is in principle known since the underlying counting statistics should follow a Poisson distribution, that is approximated by the Normal distribution for large number of counts [6, 7].

Assuming, therefore, that the stochastic errors are Normal the sum of the weighted squared deviations

$$\chi^2 = \sum_i \frac{(y_i - yc_i)^2}{\sigma_i^2} \tag{1}$$

follows a χ^2– distribution with $n - m$ degrees of freedom [5] ; y_i represents the i-th experimental point, yc_i is the corresponding calculated value from the model, σ_i^2 is the

M. Miglierini and D. Petridis (eds.), Mössbauer Spectroscopy in Materials Science, 373–384.

variance associated to the i-th experimental point ($\sigma_i^2 = y_i$ in the present case), n is the number of experimental points and m is the number of parameters to be fitted.

The mean value (expected value) of a χ^2–distribution is equal to the number of degrees of freedom, i.e. to $n-m$, while the variance is twice as large. Therefore, in case of a perfect fit we expect to find a value of χ^2 very close to $n-m$.

A very used small modification of χ^2 is the so called *reduced* χ^2, or χ^2_r, that is equal to χ^2 divided by $n-m$. The advantage of such formulation is that for a perfect fit $\chi^2_r = 1$ irrespective of the degrees of freedom. Indeed this is the main GOF parameter used in Mössbauer spectroscopy.

Consulting the statistical tables of the χ^2–distribution we can easily find the 95% (or 99%) confidence limits, i.e. those values of χ^2 that we use to discriminate (statistically) an acceptable fit from a non acceptable one. As an example, using 256 points, 4 parameters to fit, and assuming a probability significance level of 5% (≡ 95% confidence limits) a fully satisfactory fit should have a value of χ^2_r in the range ~ [0.82 – 1.18]. Outside this interval the fitting is not acceptable at the 5% probability level.

It might appear that using χ^2 we have all what we need to jugde a fit. This is not completely true as already stated by other authors [8, 9].

Let consider for example figure 1 and figure 2.

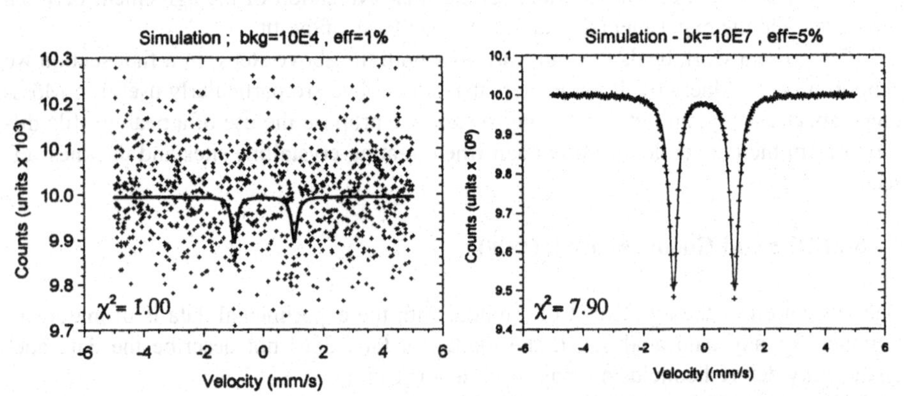

Figure 1. Simulated spectrum with low S/N ratio.

Figure 2. Simulated spectrum with high S/N ratio fitted with an incorrect model.

At a first look we could think that the χ^2_r values reported in the pictures have been exchanged since the first spectrum is a complete mess while the second seems acceptable. Indeed the reported χ^2_r values are the correct ones. We should keep in mind what is the question to which any statistical test gives an answer. The right question to pose is not "*how good is the fitting*" but rather "*how much the chosen model is consistent with that particular set of data*", in a probabilistic sense.

It is therefore clear why the first spectrum provides a $\chi^2_r = 1.0$ and the second 7.9. For the first spectrum we can say that the data are so spread (large variance) that almost any model could stay in between. About the second, although acceptable at a first look, the answer is that the model is inconsistent with the data and should be rejected. In fact there is another doublet buried under the main peaks.

In a statistical sense χ^2 works fine and is invaluable for the information it provides. From the point of view of an experimentalist χ^2 has some limitations. For instance, it is rather insensitive to the quality of the data and if we have two spectra that provide an equal value of χ^2_r, we can not say (without seeing the spectra) for which of them the signal to noise ratio (S/N) is higher.

Because of these and other reasons people have tried to introduce additional goodness of fit criteria to cover the apparent deficiencies of χ^2 [see for example 8, 10, 11].

All these methods provide useful information but they lack of an underlying statistical distribution, in other words it is not possible to define a confidence limit, like in χ^2, that states the range of acceptability of a fit.

2.1 DERIVATION OF NEW GOF PARAMETERS : χ_s, CD_χ and R_χ

We will now show that χ^2 contains much more information of what is normally used and will derive from it three new goodness of fit parameters.

Let's first calculate the the mean value of the experimental data, $\bar{y}$

$$\bar{y} = \frac{\sum_i y_i}{n} \tag{2}$$

We now add and subtract $\bar{y}$ inside the numerator of χ^2 (eq. 1). Of course, adding and subtracting a constant does not change the value of the function. So

$$\chi^2 = \sum_i \frac{(y_i - \bar{y} + \bar{y} - yc_i)^2}{y_i} \tag{3}$$

Separating the terms we obtain

$$\chi^2 = \sum_i \frac{(y_i - \bar{y})^2}{y_i} + \sum_i \frac{(\bar{y} - yc_i)^2}{y_i} + 2\sum_i \frac{(y_i - \bar{y})(\bar{y} - yc_i)}{y_i} \tag{4}$$

Let's now define :

$$\chi_1 = \sum_i \frac{(y_i - \bar{y})^2}{y_i} \;;\quad \chi_2 = \sum_i \frac{(\bar{y} - yc_i)^2}{y_i} \;;\quad \chi_3 = 2\sum_i \frac{(y_i - \bar{y})(\bar{y} - yc_i)}{y_i} \tag{5}$$

so that

$$\chi^2 = \chi_1 + \chi_2 + \chi_3 \tag{7}$$

Let's have a look to the three quantities :

χ_1 represents the weighted sum of the squared deviations of the experimental data from their mean value. If we indicate with w_i the weight for the *i*-th square we can also write :

$$\chi_1 = \sum_i \frac{(y_i - \bar{y})^2}{y_i} = \sum_i w_i (y_i - \bar{y})^2 \tag{8}$$

The latter term can be viewed as a measure of the dispersion of the experimental data around their mean. χ_1 can therefore be seen as a parameter related to the variance of the experimental data. Most important, as it will become clear going on, χ_1 depends *only* on the experimental data, and does *not* depend on the model used for fitting or on any other value determined from the fit. χ_1 is just a *constant* value for the set of experimental data. This latter property is the base of the definition of our first goodness-of-fit parameter.

Let's consider the ratio :

$$\chi_s = \frac{\chi^2}{\chi_1} \tag{9}$$

The name «χ_s» has been chosen by the fact that it actually represents a *scaled* χ^2 (scaled by the data). Since χ_1 is a constant, it follows immediately that χ_s has exactly the same properties of χ^2. Therefore, as for χ^2, we can define confidence limits (ex: 95% or 99%) for χ_s that can be used in a real statistical way to accept or reject the fitting model. But unlike χ^2, whose optimal values are independent on data quality, χ_s is scaled by χ_1 and, thanks to its properties, the better are the model and the data the smaller is χ_s. In the limit of infinite counts (and for a correct model) $\chi_s \to 0$.

This behaviour of χ_s resembles that of MISFIT [8]. We can say that χ_s takes the good properties of both χ^2 and MISFIT but without their limitations. Moreover, we are not forced to define an equivalent of ΔMISFIT because what we need to test the model is already included in the statistical properties of χ_s.

The second parameter of equation 7 , χ_2, is related to the amount of variance of the data that is described by the model.

We can now define the second GOF parameter :

$$CD_x = \frac{\chi_2}{\chi_1} = \frac{\sum_i \frac{(yc_i - \bar{y})^2}{y_i}}{\sum_i \frac{(y_i - \bar{y})^2}{y_i}} = \frac{\sum_i w_i (yc_i - \bar{y})^2}{\sum_i w_i (y_i - \bar{y})^2} \tag{10}$$

Since χ_1 is a measure of the total variance of the data then CD_x is a measure of the fraction of the variance of the data that is described by the model. It resembles what in statistics is known as *Coefficient of Determination* [12]. As a goodness of fit $CD_x \to 1$ for a satisfactory model with data of high quality, while $CD_x \to 0$ for extremely dispersed data.

The third GOF parameters is defined by

$$R_x = -\frac{1}{2}\frac{\chi_3}{\sqrt{\chi_1}\sqrt{\chi_2}} = \frac{\sum_i w_i(y_i - \bar{y})(yc_i - \bar{y})}{\sqrt{\sum_i w_i(y_i - \bar{y})^2}\sqrt{\sum_i w_i(y_i - \bar{y})^2}} \tag{11}$$

R_x is a measure of the correlation that exists between y_i and yc_i. It is an indication of how changes in y_i are correlated to changes in yc_i. For an appropriate model $R_x \to 1$.

2.2 APPLICATIONS OF χ_s, CD_χ and R_χ

To characterized the performance of the GOF parameter defined above we have generated 80 synthetic spectra of varing S/N. The spectra consist of one lorentzian doublet superimposed to a flat background plus gaussian noise. We have allowed the background to vary from 10^4 to 10^7 counts per channel and the maximum absorption to vary from 0.1% to 40% (it is unrealistic but useful for the test). The equations used to generate the data were

$$D_i = bkg - \varepsilon\left(\frac{(\Gamma/2)^2}{(v_i - v_1)^2 + (\Gamma/2)^2} + \frac{(\Gamma/2)^2}{(v_i - v_2)^2 + (\Gamma/2)^2}\right) \tag{12}$$

$$y_i = D_i + N(0,1)\sqrt{D_i} \qquad i = 1,\ldots,256 \tag{13}$$

where D_i is a lorentzian doublet, $N(0,1)$ is random variable normally distributed about zero with unit variance and y_i are the final synthetic data.

We have then fitted the spectra using 2 different models, the correct one, i.e. the same used for generating the data, and an incorrect one based on a pseudovoigt profile (70% Gaussian , 30% Lorentzian). A total of 160 fits were obtained.

Figure 3 shows the trend of χ_s after fitting with the correct model. Table 1 and Table 2 reports 22 of the 80 estimations of the GOF parameters obtained for the correct and incorrect model, respectively.

As it can be seen in figure 3, χ_s has the desired property of being sensitive to the quality of the data and depends on both counting statistics and % absorption. Therefore, χ_s can provide a useful criterium to jugde the quality of a fit even without knowing the details of the spectrum. We can also use χ_s as a measure of iso-quality for spectra of different absorption. For example, from figure 3 we can see that a spectrum with 5x10^6 counts per channel and 0.5% absorption is equivalent to a spectrum of 5x10^4 counts per channel but with an absorption of 5%. We could have derived these conclusion also on the base of S/N considerations but χ_s is a more direct indication.

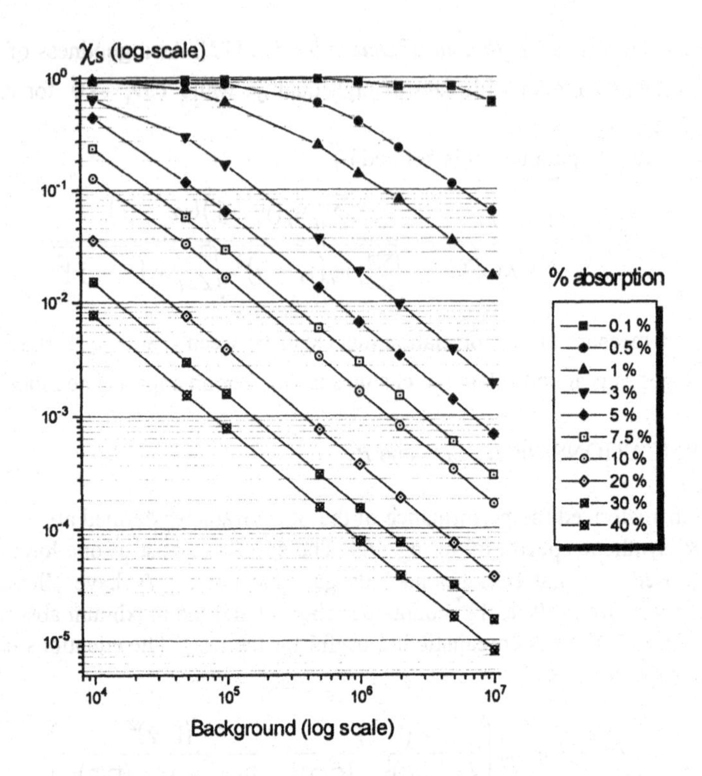

Figure 3. Trend of the GOF parameter χ_s in dependence of the counts per channel and absorption (correct model).

From Table 1 we can now see how CD_χ and R_χ complement the information of χ_s. For example, a spectrum like the one of figure 1 is easily recognized using our GOF. A so spread spectrum corresponds to the third line of table 1 (10^4 counts and 1 % absorption). In this case χ^2 just say that the fitting is consistent with the model ($\chi^2 \cong 1$). If we look at χ_s we have the same information (we are in the acceptable range) but the high value of χ_s suggests that the quality of the data is not good. If we pass to CD_χ and R_χ we can see that 5% only of the total variance is described by the model with a correlation of 0.22, very low. The conclusion is evident, a consistent model that describes only 5% of the variance means that the data are of very poor quality i.e. extremely dispersed.

In conclusion the use of χ_s, CD_χ and R_χ can provide many useful information on both the quality of the fitting and on the data. Moreover, comparison of spectra of different S/N and from different groups is easily achievable. A more detailed description of the method and extensive report of the results will be found in a forthcoming paper [13].

TABLE 1. GOF parameters with different absorptions and counts
(22 of 80 ; correct model, Lorentzian profile)

Abs %	Bkg	χ^2	χ^2_r	χ_s	CDχ	Rχ	Acceptable range for χ_s
0.1	1.00E+06	241.0	0.960	9.15E-01	0.085	0.291	[0.788 - 1.129]
0.1	1.00E+07	244.6	0.975	6.06E-01	0.394	0.628	[0.514 - 0.736]
1.0	1.00E+04	245.4	0.978	9.50E-01	0.050	0.224	[0.803 - 1.150]
1.0	1.00E+05	246.8	0.983	5.99E-01	0.401	0.633	[0.504 - 0.721]
1.0	1.00E+06	240.3	0.958	1.37E-01	0.863	0.929	[0.118 - 0.169]
1.0	1.00E+07	240.2	0.957	1.67E-02	0.983	0.992	[0.014 - 0.020]
5.0	1.00E+04	246.5	0.982	4.43E-01	0.557	0.747	[0.373 - 0.533]
5.0	1.00E+05	249.2	0.993	6.40E-02	0.936	0.967	[0.053 - 0.076]
5.0	1.00E+06	244.0	0.972	6.57E-03	0.993	0.997	[0.0056 - 0.0080]
5.0	1.00E+07	246.8	0.983	6.78E-04	0.999	1.000	[0.00057 - 0.00081]
7.5	1.00E+04	244.9	0.976	2.33E-01	0.767	0.876	[0.197 - 0.283]
7.5	1.00E+05	248.9	0.992	2.89E-02	0.971	0.985	[0.024 - 0.035]
7.5	1.00E+06	247.6	0.987	2.93E-03	0.997	0.999	[0.0024 - 0.0035]
7.5	1.00E+07	244.8	0.975	2.92E-04	1.000	1.000	[0.00024 - 0.00035]
10.0	1.00E+04	239.7	0.955	1.26E-01	0.874	0.935	[0.109 - 0.156]
10.0	1.00E+05	250.3	0.997	1.65E-02	0.984	0.992	[0.013 - 0.019]
10.0	1.00E+06	244.1	0.972	1.62E-03	0.998	0.999	[0.0013 - 0.0019]
10.0	1.00E+07	248.6	0.991	1.64E-04	1.000	1.000	[0.00013 - 0.00019]
40.0	1.00E+04	248.9	0.992	7.71E-03	0.992	0.996	[0.0064 - 0.0092]
40.0	1.00E+05	248.0	0.988	7.74E-04	0.999	1.000	[0.00064 - 0.00092]
40.0	1.00E+06	245.8	0.979	7.70E-05	1.000	1.000	[0.000065 - 0.000093]
40.0	1.00E+07	248.7	0.991	8.00E-06	1.000	1.000	[0.0000065 - 0.0000093]

TABLE 2. GOF parameters with different absorptions and counts
(22 of 80 ; incorrect model, pseudo-Voigt profile)

Abs %	Bkg	χ^2	χ^2_r	χ_s	CDχ	Rχ	Acceptable range for χ_s
0.1	1.00E+06	243.0	0.968	9.23E-01	0.077	0.278	[0.788 - 1.129]
0.1	1.00E+07	238.4	0.950	5.90E-01	0.410	0.640	[0.514 - 0.736]
1.0	1.00E+04	244.6	0.974	9.46E-01	0.054	0.232	[0.803 - 1.150]
1.0	1.00E+05	250.2	0.997	6.07E-01	0.393	0.627	[0.504 - 0.721]
1.0	1.00E+06	263.8	1.051	1.51E-01	0.849	0.922	[0.118 - 0.169]
1.0	1.00E+07	527.5	2.102	3.67E-02	0.963	0.981	[0.014 - 0.020]
5.0	1.00E+04	240.9	0.960	4.33E-01	0.567	0.753	[0.373 - 0.533]
5.0	1.00E+05	305.7	1.218	7.85E-02	0.921	0.960	[0.053 - 0.076]
5.0	1.00E+06	819.6	3.265	2.21E-02	0.978	0.989	[0.0056 - 0.0080]
5.0	1.00E+07	7103.5	28.301	1.95E-02	0.980	0.990	[0.00057 - 0.00081]
7.5	1.00E+04	251.3	1.001	2.39E-01	0.761	0.872	[0.197 - 0.283]
7.5	1.00E+05	431.3	1.718	5.01E-02	0.950	0.975	[0.024 - 0.035]
7.5	1.00E+06	1896.8	7.557	2.25E-02	0.978	0.989	[0.0024 - 0.0035]
7.5	1.00E+07	15480.9	61.677	1.85E-02	0.982	0.991	[0.00024 - 0.00035]
10.0	1.00E+04	250.4	0.997	1.32E-01	0.868	0.932	[0.109 - 0.156]
10.0	1.00E+05	487.5	1.942	3.21E-02	0.968	0.984	[0.013 - 0.019]
10.0	1.00E+06	3192.5	12.719	2.11E-02	0.979	0.989	[0.0013 - 0.0019]
10.0	1.00E+07	27558.7	109.796	1.82E-02	0.982	0.991	[0.00013 - 0.00019]
40.0	1.00E+04	737.6	2.939	2.28E-02	0.977	0.989	[0.0064 - 0.0092]
40.0	1.00E+05	5078.1	20.231	1.58E-02	0.984	0.992	[0.00064 - 0.00092]
40.0	1.00E+06	49867.9	198.677	1.55E-02	0.984	0.992	[0.000065 - 0.000093]
40.0	1.00E+07	493576.5	1966.440	1.54E-02	0.985	0.992	[0.0000065 - 0.0000093]

3. Expo-lorentzian lineshape : an alternative to Lorentzians

The ability of Mössbauer spectroscopy to determine the presence of different oxidation states and their relative amounts is one of the most appreciated feature of the technique. Much of the accuracy in quantitative estimations, however, is strictly dependent on the use of correct shape functions to fit the experimental data, and indeed much efforts have been put on this problem [14-21].

As known, a measured Mössbauer spectrum is given by the so called transmission integral, which is the convolution of the gamma source shape and the transmission function. The mathematical complexities of the transmission integral prevent an exact analytical solution and up to now this has limited its applications in practice. Notwithstanding that nowadays personal computer can fit an experimental spectrum in a minute (even using the transmission integral) most of the people working in the Mössbauer field still show a marked preference to use simple Lorentzians to fit their data, the so called *thin sample* approximation. But if this limit is not applicable, as it is often the case, Lorentzian fittings are inappropriate and may provide quite misleading results.

Several techniques have been developed to overcome the direct calculation of the integral, among them, for example, we can cite the Fourier convolution and deconvolution methods [16, 17, 18 and references inside]. In present work, we avoid all the above mathematical complexities by constructing a simple analytical function to represent the transmission integral. Computer simulations have been performed to test the behavior of the function. The results show that it is able to reproduce the main spectral features and to produce values very closed to the transmission integral. The advantage, of course, is that it is much faster and easier to implement.

3.1 THE CONSTRUCTED FUNCTION

Assuming one single absorption line and a Lorentzian cross section we can express a Mössbauer transmission spectrum by the transmission integral [14, 18] :

$$N(v) = B_0 + B_m(1 - f_s) + \\ + B_m f_s \frac{2}{\pi\Gamma_s} \int_{-\infty}^{+\infty} \frac{(\Gamma_s/2)^2}{[E - E_s(1 + v/c)]^2 + (\Gamma_s/2)^2} e^{-\frac{1}{4} t_a \frac{\Gamma_{nat}\Gamma_a}{(E-\delta)^2 + (\Gamma_a/2)^2}} dE \tag{14}$$

where $N(v)$ is the number of counts per unit time in the channel centered at v (in units of velocity), B_0 is the count rate per channel from non Mössbauer γ - rays, B_m is the count rate per channel from Mössbauer γ - rays including both recoiless and nonrecoiless events, f_s is the Mössbauer recoiless fraction of the source, f_a is the Mössbauer recoiless fraction of the absorber, Γ_s is the linewidth of the source, Γ_a is the linewidth of the absorber, $t_a = f_a n_a \sigma_0$ is the usual dimensionless effective thickness parameter.

Let's now introduce the following «Expo-Lorentzian» function (generalization to more than one component is straightforward):

$$M(\nu) = A + Be^{-T_a \frac{\Gamma^2/4}{(\nu-\delta)^2+(\Gamma^2/4)}} \tag{15}$$

where $M(\nu)$ has the same meaning as $N(\nu)$ above, and A and B are parameters related to the spectral background. T_a is the counter part of the thickness parameter t_a, and Γ is related to the line width.

Since this function resembles the transmission function in the transmission integral there is a closed relation between them. We will show now that this function has also the same main properties of the transmission integral :

(a) Symmetry : since the absorption cross section enters in the same way in both the transmission integral and in the present function it is straightforward that the spectral symmetry is maintained.

(b) Spectral Area : the spectral area from the transmission integral, S_N, is, following Lang [15]

$$S_N = \int_{-\infty}^{\infty} N(\nu)d\nu = B_m f_s \frac{\pi\Gamma_a}{2} \sum_{p=1}^{\infty} \frac{(-1)^{p+1}}{p!} \frac{(2p-3)!!}{(2p-2)!!} t_a^p = B_m f_s \frac{\pi\Gamma_a}{2} L(t_a) \tag{16}$$

while the spectral area of the function $M(\nu)$ can be demonstrated to be:

$$S_M = \int_{-\infty}^{\infty} M(\nu)d\nu = B\frac{\pi\Gamma}{2} \sum_{p=1}^{\infty} \frac{(-1)^{p+1}}{p!} \frac{(2p-3)!!}{(2p-2)!!} T_a^p = B\frac{\pi\Gamma}{2} L(T_a) \tag{17}$$

If we define the parameter A and B as : $A = B_0 + B_m(1-f)$, $B = B_m f$, where f is an appropriate factor (different from f_s and f_a) we have that eq. 17 takes exactly the same form of eq. 16.

(c) Thin Sample Approximation : In the thin limit approximation ($t_a \ll 1$ and $T_a \ll 1$), the above expressions become :

$$N(\nu) \approx B_0 + B_m - B_m f_s \frac{t_a}{2} \frac{(\Gamma_s+\Gamma_a)^2/4}{(\nu-\delta)^2+(\Gamma_s+\Gamma_a)^2/4} \tag{18}$$

$$M(\nu) \approx A + B - BT_a \frac{\Gamma^2/4}{(\nu-\delta)^2+\Gamma^2/4} \tag{19}$$

Again we see that they have the same form, the frequently used Lorentzian shape. If we impose $N(\nu) = M(\nu)$ we obtain : $A + B = B_0 + B_m$, $BT_a = \frac{B_m f_s}{2} t_a$, $\Gamma = \Gamma_a + \Gamma_s$.

In the following we present the fitting results obtained by the "expo-lor" shape function on simulated spectra obtained by the direct calculation of the transmission integral. Comparison with Lorentzian fittings is also shown.

3.2 SIMULATION PROCEDURE

We have performed various simulation of Mössbauer spectra at different t_a. The transmission integral was calculated by the Cranshaw's method as described in [17].

In order to show the line shape clearly we have at first generated many spectra without noise (1773 spectra, ideal case), with the values of t_a spanning a huge scale, from 1 to 99. Then we have performed the same procedures including Gaussian noise to simulate practical case. All the simulations and fitting routines were executed both by MATLAB® and by the equivalent procedures in FORTRAN. We have also tested the case of broader intrinsic linewidth allowing the width in the absorber to vary from $\Gamma_a = \Gamma_s = \Gamma_{nat}$ up to $\Gamma_a = 5\Gamma_s$.

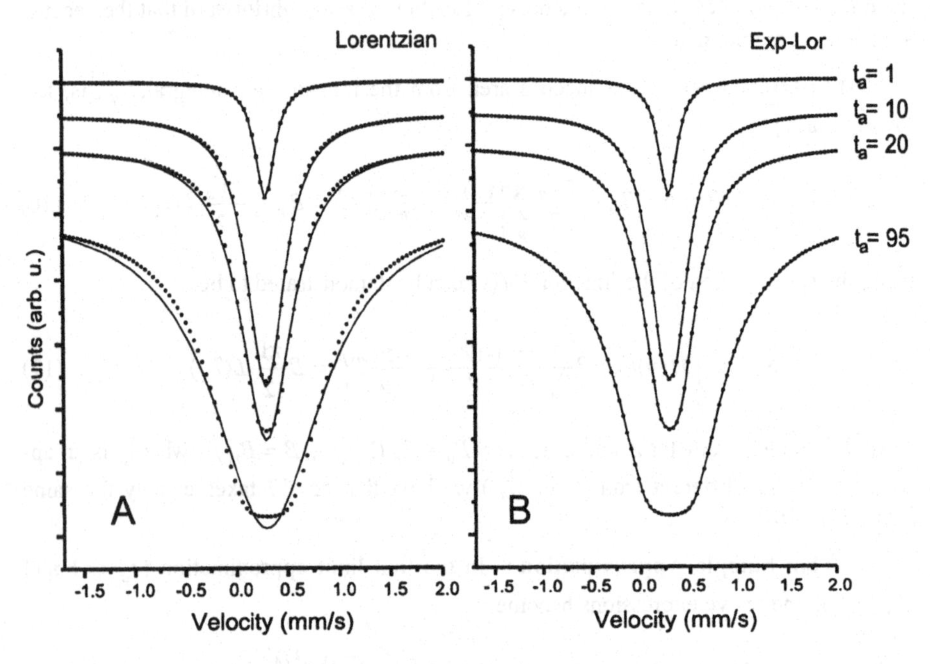

Figure 4. A) Transmission Integral calculation (t_a 1-95) and fitting by Lorentzians ; B) Transmission Integral calculation (t_a 1-95) and fitting by Expo-Lorentzians. Source and absorber linewidth are set to 0.1 mm/s.

Lorentzian fittings show a consistent deviation from the correct shape already at moderately low value of t_a ($t_a < 10$). As it can be seen in figure 4 A) at $t_a = 10$ the difference is well marked and at $t_a = 20$ becomes completely inadequate. Just to stress this behaviour a simulation at $t_a = 95$ is also shown. From figure 4 B) it can be seen, on the contrary, that the expo-lorentzian profile follows very well the transmission integral shape. Even at $t_a = 95$ the two shapes are practically identical.

Table 3 reports the comparison of the areas obtained by the transmission integral and by the expo-lor function in the range $t_a = 1 \cdots 20$.

TABLE 3. Spectral areas (mm/s) for various values of the absorber thickness t_a. Area_F are the results obtained by fitting with the expo-lor function and Area_T are those calculated from the transmission integral. Source and absorber linewidth were set at 0.1 mm/s.

t_a	Area_F	Area_T	t_a	Area_F	Area_T	t_a	Area_F	Area_T
1.0	0.1258	0.1259	7.5	0.4682	0.4682	14.0	0.6518	0.6510
1.5	0.1721	0.1723	8.0	0.4848	0.4848	14.5	0.6638	0.6630
2.0	0.2115	0.2116	8.5	0.5008	0.5007	15.0	0.6756	0.6747
2.5	0.2457	0.2459	9.0	0.5163	0.5162	15.5	0.6873	0.6863
3.0	0.2762	0.2764	9.5	0.5314	0.5313	16.0	0.6987	0.6976
3.5	0.3037	0.3039	10.0	0.5461	0.5459	16.5	0.7100	0.7088
4.0	0.3289	0.3291	10.5	0.5604	0.5601	17.0	0.7210	0.7198
4.5	0.3523	0.3525	11.0	0.5743	0.5740	17.5	0.7319	0.7306
5.0	0.3742	0.3743	11.5	0.5879	0.5875	18.0	0.7427	0.7413
5.5	0.3948	0.3950	12.0	0.6012	0.6008	18.5	0.7533	0.7518
6.0	0.4144	0.4145	12.5	0.6142	0.6137	19.0	0.7638	0.7622
6.5	0.4331	0.4332	13.0	0.6270	0.6264	19.5	0.7741	0.7725
7.0	0.4510	0.4510	13.5	0.6395	0.6388	20.0	0.7843	0.7826

A complete report of all the results and relations between the parameters (T_a vs t_a, Γ vs Γ_a and Γ_s, as well as fittings with intrinsic broad lines will be found in [22].

One of the possible applications of the exp-lor fitting can be in the determination of quantitative site polulations, as already suggested in [15]. If, for example, we collect two spectra of the same specimen, one β times thicker than the other, and consider an absorption peak, the normalized area in the two cases will be :

$$A_1 = f_s \frac{\pi\Gamma_a}{2} L(t_a) \;\; ; \;\; A_2 = f_s \frac{\pi\Gamma_a}{2} L(\beta t_a) \tag{20}$$

if we take their ratio A_1/A_2 we obtain :

$$\frac{A_1}{A_2} = \frac{f_s \frac{\pi\Gamma_a}{2} L(t_a)}{f_s \frac{\pi\Gamma_a}{2} L(\beta t_a)} = \frac{L(t_a)}{L(\beta t_a)} \tag{21}$$

By a simple iterative procedure is then easy to find the requested t_a value for any peak and consequently the iron population. It's also possible to determine Γ_a.

In conclusion it's our befief that expo-lor fittings are well suited for quatitative estimations. In case the thin sample approximation holds the expo-lor functions provide results equivalent to lorentzians and as t_a increases, as in practical cases, expo-lor fittings can be a reliable alternative to the transmission integral.

References

1. Press W.H., Teukolsky S.A., Vetterling W.T., Flannery B.P. (1992) *Numerical Recipes in Fortran. The Art of Scientific Computing*, 2nd edition, Cambridge University Press.
2. Walsh G.R. (1977) *Methods of Optimization*, John Wiley & Sons.

3. Nash J.C. (1990) *Compact Numerical Methods for Computers*, 2nd edition, Adam Hilger, Bristol and New York.
4. Box G.E.P. and Tiao G.C. (1973) *Bayesian Inference in Statistical Analysis*, Addison-Wesley.
5. Brandt S. (1983) Statistical and Computational Methods in Data Analysis, North Holland Publishing Company.
6. Knoll G.F. (1979) *Radiation Detection and Measurement*, John Wiley & Sons.
7. Taylor J.R. (1982) An introduction to Error analysis. The Study of Uncertainties in Physical Measurements, University Science Books, Mill Valley, California.
8. Ruby S.L. (1973) Why Misfit if you already have χ^2, in I.J. Gruverman, C.W. Seidel and D.K. Dieterly (eds.), *Mössbauer Effect Methodology*, Plenum Press, New York, Vol. 8, pp. 263-276.
9. Waychunas G.A. (1986) Performance and use of Mössbauer goodness-of-fit parameters : Response to spectra of varying signal/noise ratio and possible misinterpretation, *Am. Mineral.* **71**, 1261-1265.
10. Dollase W.A. (1975) Statistical limitation of Mössbauer spectral fitting, *American Mineralogist* **60**, 257-264
11. Miglierini M. (1993) Dispersion versus Absorption (DIPSA) : a criterion for validity of fitting models in Mössbauer spectroscopy. I theory and simulations, *Acta Phys. Slov.* **3**, 207-220.
12. Spiegel M. R. (1994) *Statistics*, 2nd edition, McGraw-Hill.
13. Pedrazzi G. , manuscript in preparation.
14. Margulies S. and Ehrman J.R. (1961) Transmission and line broadening of resonance radiation incident on a resonant absorber, *Nucl. Instr. and Meth.* **12,** 131-137.
15. Lang G. (1963) Interpretation of experimental Mössbauer spectral Areas, *Nucl. Instr. and Meth.* **24,** 425-428.
16. Ure M. C. D. and Flinn P.A. (1971), in I.J. Gruverman, C.W. Seidel and D.K. Dieterly (eds.), *Mössbauer Effect Methodology*, Plenum Press, New York, Vol. 7, pp. 245-262
17. Lin T.M. and Preston R.S. (1974) Comparison of techniques for folding and unfolding Mössbauer spectra for data analysis in I.J. Gruverman, C.W. Seidel and D.K. Dieterly (eds.), *Mössbauer Effect Methodology*, Plenum Press, New York, Vol. 9, pp. 205-223.
18. Rancourt D.G. (1989), Accurate site population from Mössbauer spectroscopy, *Nucl. Instr. and Meth.* **B44,** 199-210.
19. Rancourt D.G., McDonald A.M., Lalonde A.E., and Ping J.Y. (1993) Mössbauer absorber thicknesses for accurate site population in Fe-bearing minerals, *American Mineralogist* **78,** 1-7.
20. Ping Y.J. and Rancourt D.G. (1992) Thickness effects with intrinsically broad absorption lines, *Hyperfine Interactions* **71**, 1433-1436.
21. Mørup S. and Both E. (1975) Interpretation of spectra with broadened lines, *Nucl. Instr. and Meth.* **124**, 445-448.
22. Cai S.Z., Pedrazzi G., Ortalli I., manuscript in preparation.

CONFIT FOR WINDOWS® 95

T. ŽÁK
Institute of Physics of Materials, AS CR
Žižkova 22, CZ-616 62 Brno, Czech Republic

A new coat was given to the well-tried software for the fitting of Mössbauer spectra. The program is based on the Veslý's idea of convolution [1-3] and on the optimization package of Kučera et al. [4]. It enables the use of transmission integral and of distribution of magnetic splitting and/or isomer shift allowing to fit precisely overlapped spectra of amorphous, microcrystalline or multiphase materials. The process of fitting can be checked and stopped or continued after the parameter change using on-screen controls. The newly implemented dialogues simplify the man-machine interface of this 32 bit Windows® 95 application. Internally, the computation was speed-up by omitting all obsolete parts and program desegmentation, the numerical stability was improved by the 8 byte representation for the substantial part of variables. Finally, some pieces of information concerning current and future extension of the described software package are added.

1. Introduction

Materials studied by Mössbauer spectroscopy are not pure compounds or well defined alloys any more. Among them, complicated alloys, intermetallics, and more or less crystallised amorphous structures are not of an exception. As a consequence there are complicated spectra having many components overlapping each other, or lines distributed within a certain, usually Gaussian, profile. Final sample thickness results in the necessity of application of transmission integral. To reach the best physical interpretation of such spectra, a mighty and quick enough software package is necessary for measurement pre-processing and evaluating, nowadays on the 32-bit platform of Windows® 95.

2. Historical background

The recent version of software used in our Mössbauer laboratory is based on two fundamental papers [1, 2]. Let us to review briefly basic features of these two articles.

The first one [1] introduces the idea of convolution being involved in exact analytical models of distributed spectra. As a suitable numerical procedure for finding least-squares estimates of unknown parameters the one based on the fast Fourier transform algorithm is presented. As every convolution is evaluated here in Fourier transform domain as simple multiplication, problems with numerical approximation of convolution are overcome. Here also numerical tests have been confronted with theoretical error analysis.

The second paper [2] describes a FORTRAN IV implementation of such a processing. It comprises, among others, a computer-aided matrix model design. This matrix serves for transposition between vector of "physical" Mössbauer parameters (as mag-

M. Miglierini and D. Petridis (eds.), Mössbauer Spectroscopy in Materials Science, 385–390.

netic or quadrupole splitting, isomer shift etc.) and "mathematical" parameters characterising the individual lines (intensity, width, position etc.).

The implementation, a computer system called CONFIT (Convolution Fit [3]) accomplishes the least-squares minimization by the modified Newton-Marquardt gradient method (in the optimization package [4] denoted as OPNMC) and was originally developed on the ICL 2950/10 mainframe computer to combine fast processing of experimental spectra with a maximum rate of flexibility in the choice of the appropriate mathematical model.

In that ancient time of mainframe computers there were serious memory limits resulting in program segmentation and in not too large (however sufficient for common purposes) space for spectra components. In case of its exceeding the virtual memory on disk was used, substantially slowing down the processing.

In the early PC age the program was simply transferred with only minor changes (usually because of compiler differences), not taking advantage of all new PC abilities, such as graphic screen presentation. Later on, the evolution of the fitting process could be observed on the screen allowing to stop it and change parameters accordingly. However, it was possible to run the program in the DOS environment only. Most of limits including the program segmentation remained. Therefore it was decided to adapt the program to the 32-bit platform of Windows® 95 as much as possible and put it as a keystone of a larger package for Mösbauer spectra processing.

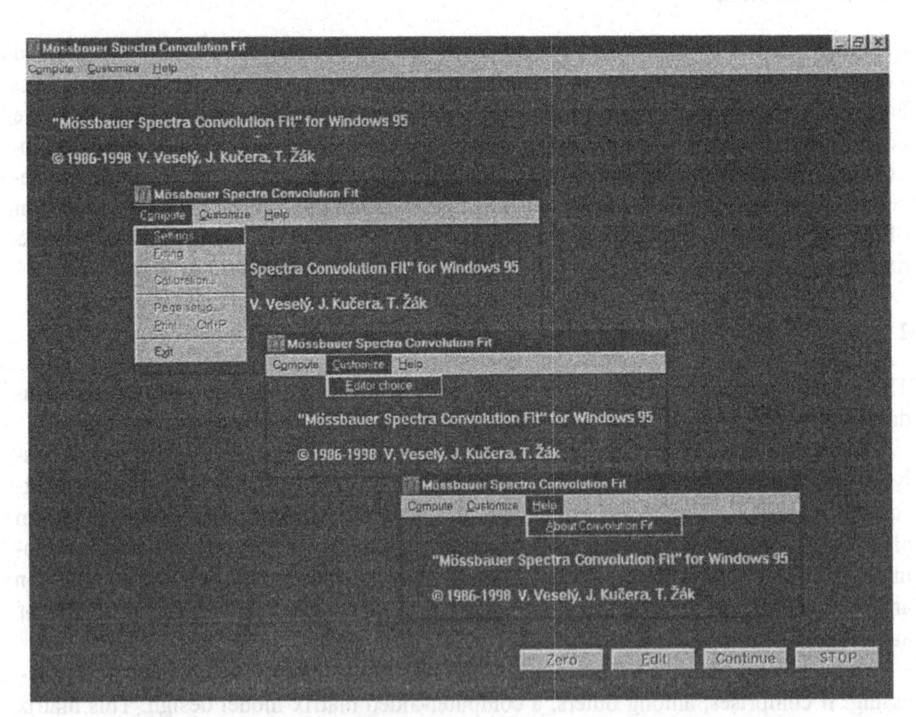

Figure 1. CONFIT main window; in addition, menu examples are inserted

3. New features

The program was transformed into the Windows® 95 application with its typical appearance including its own icon, main and sub menus, and common or custom dialogues (see Figure 1). The main window area is used for every communication with the user, showing besides the name of the measurement all changes in the spectrum fit: value of the least-squares residual, graphical representation of experimental points, of curve of the fitted function and of the differences between them for each point. It also contains buttons allowing the start/stop progress of fitting and parameter editing. As a default editor, the Windows' Notepad is implemented, however, another suitable one can be chosen (see below). Nowadays, the application is localised both as a Czech and English version. However, strings in any other language can be introduced any time.

Internally, the fitting method, including the convolution with forward and backward FFT, was more or less left in its original form. Minor changes were done because of compiler and FORTRAN 90 specifics. Computation was speed-up predominantly by omitting all obsolete parts, further, program segments were brought together making all internal data transfer via disk redundant. The numerical stability was improved by the 8 byte representation for all variables used in FFT and convolution.

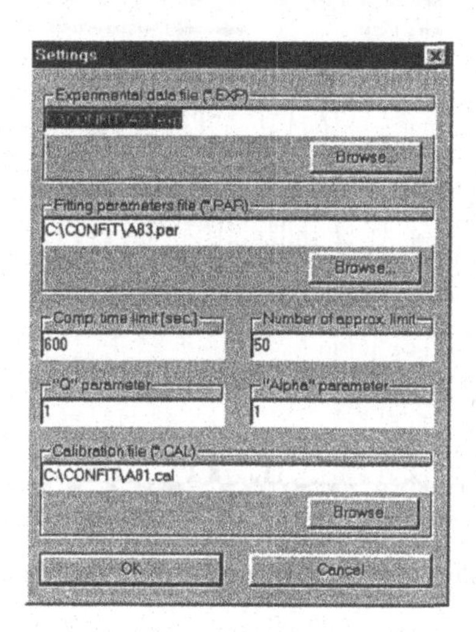

Figure 2. Dialogue box for settings of fitting procedure

Before the user can start the fitting process, the files of experimental data (extension "EXP") and of fitted parameters (extension "PAR") must be specified. To this purpose serves the first option of the "Compute" menu, opening the "Settings" dialogue box (see Figure 2). Choosing the name of file of experimental data in the first field, the name of file of parameters is automatically completed including appropriate extension in the second one. However, it could be changed manually as well. In the next two fields limits for computational time and number of approximations can be set. Just below you can set values of two parameters influencing the accuracy of the FFT method. For most purposes the basic value of 1 is enough. They both can be increased up to 16.

In the last field the name of file with calibration coefficients can be specified eventually. All three file names can be either specified directly or found browsing in the available disk space using the common file-open dialogue. After closing the "Settings" dialogue box with the OK button the existence of all files and plausibility of all other parameters is checked. In case of success the second line in the menu is enabled allowing to start the processing. In it, as a first task the consistency of the selected model with its parameters is checked. The occurrence of errors prevents to start the fitting. If some warnings or comments are signalised, the processing can be still triggered after the user approval. Details concerning the individual problems are written in the output file, that can be reviewed. In absence of any errors, warnings or comments the computation starts immediately.

During the fitting procedure the user can influence its evolution by buttons in the lower right corner of the main window. The most left one serves to displaying the *status quo* without any subsequent fitting. The next one (from left to right) triggers the editor, the last two buttons serve for continuing and stopping of the processing at arbitrary approximation step, respectively.

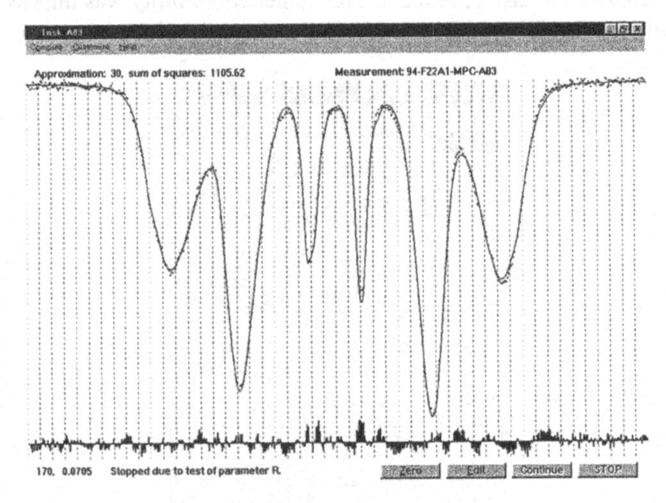

Figure 3. Main window with the fitting finished

After the fitting process finished, reason of its stopping is signalised; e.g. reciprocal value of relative changes in least-squares residual (represented by the parameter R) is above some limit or number of approximation is exceeded. At the same time, the position of the cross-hair cursor starts to be signalised digitally yielding the position and depth (or height) of individual parts of the spectrum (Figure 3). The final values of the parameters are written in the output file together with their errors gained from covariance matrix. If the calibration file was specified, the all possible calibrated values are added together with mean magnetic splitting and area ratios of individual subspectra. Now also menu options concerning the printer are enabled and the graphic representation of both measured and fitted spectra and their differences can be printed out.

In case of a six-line iron spectrum an extra menu option is enabled. Using it, the velocity calibration coefficients are estimated from basic physical constants and stored in a calibration file for later use by other fits.

The second item of the main menu (from the left), "Customize", has in its menu only one option: to choose and set an appropriate editor application for editing and viewing of common text files, if the default "Notepad" is not suitable or is missing (see Figure 4).

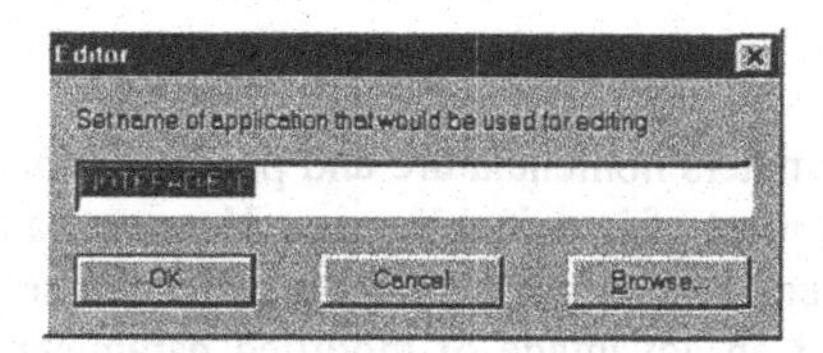

Figure 4. Dialogue box for editor choice

The "Help" item (the most right one), or better its single option, gives the basic information about the application up to now only. However, in the future a short user's guide will be available here.

4. Auxiliary software

The fitting of spectra represents only a part of the whole processing. The data as gained from the Ortec multiscaling PC card is in the binary form and unfolded. Therefore a simple Windows® 95 application MOSSPROC was written bringing together all steps necessary to prepare experimental data in a form suitable for the input of the CONFIT application (Figure 5).

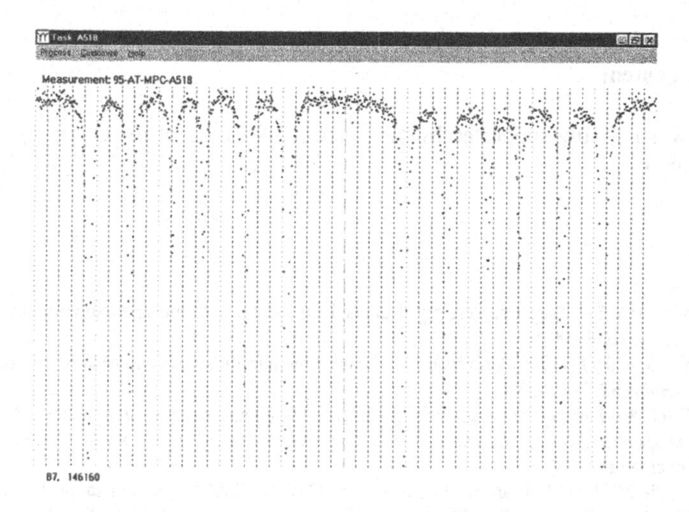

Figure 5. Example of a raw unfolded spectrum

It introduces a user-friendly man-machine interface for binary to text data transformation, for a folding procedure (with either rigid or fitted folding point) and for reversion of channels order. In addition, the text format of data files can be edited by an external editor in a way similar to that of CONFIT. All steps and data changes can be immediately checked watching the graphical representation of the spectrum. Coordinates of individual experimental points are read by the cross-hair cursor.

5. Future development

The used system of parameters nomenclature and parameter matrix construction is very sophisticated and covers most of imaginable cases. However, although some mnemonic is introduced, it is not very easy to put together a parameter set for a qualitatively new spectrum (without taking an advantage of modified parameters of some older similar one).

This is a reason why a next complementary part of this package is thought. It should be able to guide the user through the construction of fitted function including all constrains and interrelations and to be helpful by setting a parameter estimates. An environment of dialogue windows would be very suitable for such a purpose.

6. Conclusions

This new version of CONFIT for Windows® 95 not only simplifies and speeds up the process of fitting of Mössbauer spectra and eliminates the necessity to switch to DOS environment but also, conserving all the quality of implemented mathematical model, represents a qualitatively new way to a friendly interaction between the user and this application.

Acknowledgement

This work was partially supported by the grant 102/95/1251 of the Grant Agency of the Czech Republic.

References

1. Veselý, V. (1986) FFT based processing of unresolved spectra with multiple convolutions, *Nucl. Instr. Meth. Phys. Res.* **B18**, 88–100.
2. Veselý, V., Zemčík, T. (1990) Universal FFT / matrix method for processing Mössbauer spectra, *Hyperfine Interactions* **58**, 2679–2686.
3. Veselý, V. (1986) Fast computer analysis of complex Mössbauer spectra, *Proc. of the "Applications of Mössbauer spectroscopy in physics of metals and physical metallurgy"*, 12–16 May 1986, Valtice, Czechoslovakia (abstract).
4. Kučera, J., Hřebíček, J., Lukšan, L., Kopeček, I. (1985) OPTIPACK – Optimization Program Package, User Guide Modif. 2.2, Institute of Physical Metallurgy, Czechoslovak Acad. of Sci., Brno, *Research report VZ 609/730*, in Czech.

YAP:Ce SCINTILLATION DETECTOR FOR TRANSMISSION MÖSSBAUER SPECTROSCOPY

M. MASHLAN [1], D. JANCIK [1] AND A.L. KHOLMETSKII[2]
[1] Department of Experimental Physics, Palacky University, Svobody 26, 771 46 Olomouc, Czech Republic.
[2] Department of Physics, Belarus State University, Skorina 4, 220 080 Minsk, Belarus.

Abstract

A YAP:Ce scintillation crystal has a decay time that is one order of magnitude shorter than that of NaI:Tl, and therefore this allows its use for rather high counting rates. The optimal thickness of a YAP:Ce crystal for ^{57}Fe and ^{119}Sn Mössbauer spectroscopy is 0.35 mm. The photomultiplier tube R6095 (Hamamatsu) is used in the scintillation detector. The signal is amplified by means of a fast preamplifier constructed on the base of a NE 5539 operational amplifier. Standard pulse-height discriminators, used with the NaI:Tl scintillation detector and proportional counter, respectively, cannot be used with a YAP:Ce scintillation detector, because the duration of the pulses on the preamplifier output is less than 200 ns. A fast pulse-height discriminator was built using two fast MAC 160 comparators.

1. Introduction

The physical and chemical properties, the relatively low cost of the components and well-developed technology make yttrium aluminum perovskite crystal doped by cerium (YAP:Ce) rather promising as a detector of ionizing radiation [1,2]. The high luminescence efficiency and chemical inertness enable its application in aggressive media without any shortening of its high lifetime or any deterioration of the scintillation properties in contradiction to plastic or glass scintillators. Thanks to its high radiation resistance, the YAP:Ce detector is suitable for a long-term use under high radiation doses. The YAP:Ce detector is suitable for the detection of scattered and secondary electrons in scanning electron microscopy, the detection of ^{14}C and ^{3}H in liquid chromatography, fast coincidence measurement, the detection of radioactive rays under aggressive chemical conditions, high temperature or high pressure and screens for electron or X-ray imaging, respectively. We ourselves were using this scintillation detector for ^{57}Fe transmission Mössbauer spectroscopy [3,4,5] during the last six years.

M. Miglierini and D. Petridis (eds.), Mössbauer Spectroscopy in Materials Science, 391–398.

The features and use of the YAP:Ce detector in relation to Mössbauer spectroscopy are presented and discussed in this paper.

2. Physical properties of YAP:Ce

Yttrium aluminum perovskite with $YAlO_3$ formula, activated by cerium, is a fast, mechanical and chemical resistant scintillation material. Its mechanical properties enable precise machining. YAP:Ce crystals are grown by Chochralsky [1] or horizontal oriented crystallization [2] methods.

The physical and scintillation characteristics of YAP:Ce crystal in relation to the most extended inorganic scintillation crystals are listed in Table 1. Absorption and luminescence spectra of YAP:Ce crystals manufactured by the Chochralsky method in PRECIOSA Turnov (Czech Republic) and by the horizontal oriented crystallization method in RINC Minsk (Belarus) are given in Fig.1 and Fig.2, respectively. The luminescence band agrees well with the maximum sensitivity spectral range of the most widely used photomultipliers.

TABLE 1. The physical and scintillation characteristics of the most extended inorganic scintillators

	YAP:Ce	NaI:Tl	YAG:Ce	BGO	CaF_2:Eu
Density [g/cm^3]	5.37	3.67	4.57	7.13	3.18
Hardness [Mho]	8.6	2	8.5	5	4
Index of refraction	1.95	1.85	1.82	2.15	1.44
Crystal structure	Rhombic	Cubic	Cubic	Cubic	Cubic
Melting point [°C]	1875	651	1970	1050	1360
Hygroscopic	No	Yes	No	No	No
[1]Therm. Exp. [10^{-5}/K]	4-11	4.75	8-9	7	1.95
Cleavage	No	Yes	No	No	Yes
Chemical formula	$YAlO_3$	NaI	$Y_3Al_5O_{12}$	$Bi_4(GeO_4)_3$	CaF_2
Integrated light output [%NaI:Tl]	40	100	15	15-20	50
Max. of emission [nm]	350-360	415	550	480	435
Decay constant [ns]	28	230	70	300	940
[2]Absorbtion length [cm]	2.22	-	2.61	-	-
Photon yield at 300 K [10^3ph/MeV]	10	38	8	2-3	23
[3]Relative efficiency [%]	150	-	125	-	-

[1]Linear coefficient of thermal expansion
[2]Length of absorbtion to the 1/e value for 511 keV of gamma ray
[3]Relative luminescence efficiency for beta rays compared with a plastic scintillator

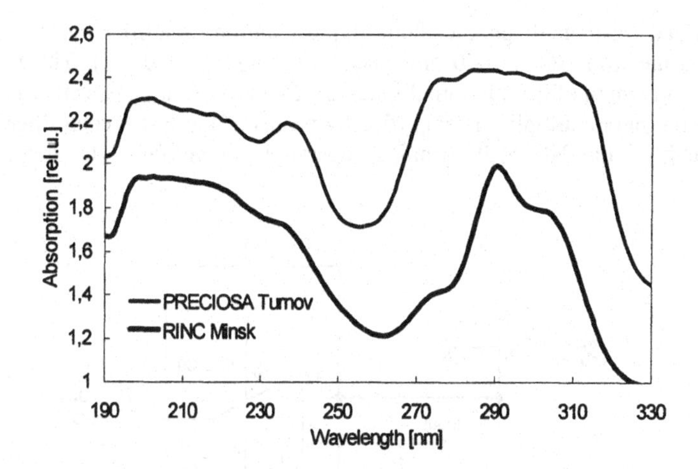

Figure 1. Absorption of the two types of $YAlO_3$:Ce crystals prepared by the Chochralsky and horizontal oriented crystallization methods. Sample thickness 0.35 mm, T = 300 K (Spectrophotometer SPECORD).

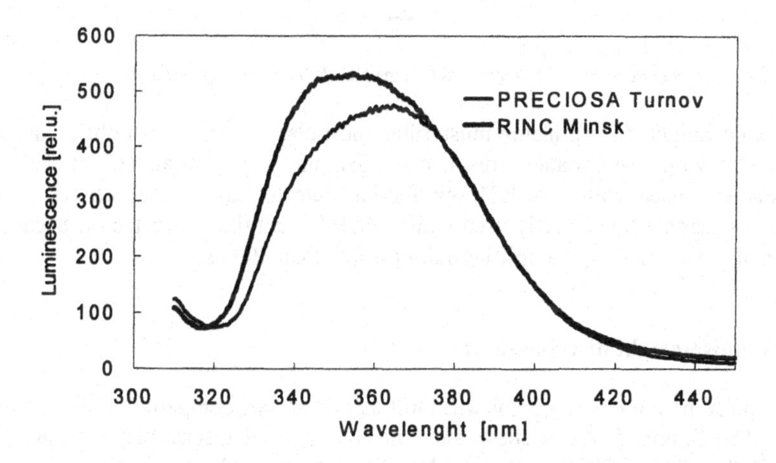

Figure 2. Luminescence of the two types of $YAlO_3$:Ce crystals prepared by the Chochralsky and horizontal oriented crystallization methods. Sample thickness 0.35 mm, T=300 K.
Wavelength of excitation light is 300 nm. (Hitachi Fluorescence Spectrophotometer F-4500).

3. Scintillation detector

The thickness of YAP:Ce scintillation crystal was optimised for ^{57}Fe (14.4 keV) and ^{119}Sn (23.8 keV) Mössbauer spectroscopy. The optimal thickness of $YAlO_3$:Ce crystal is 0.35 mm and its absorption coefficient is nearly 100 percent for both isotopes. The scintillation detector uses the photomultiplier tube R6095 (Hamamatsu) [6] that is characterized by bialkali photocathode, typical current amplification of $2.1 \cdot 10^6$, and spectral range from 300 to 650 nm (peak wavelength is 420 nm). The photocathode diameter and length of this photomultiplier are 28 and 127 mm, respectively. The output signal of the photomultiplier is amplified by means of the fast preamplifier (see Fig.3) constructed on the NE 5539 (Philips) operational amplifier. The output signal is negative.

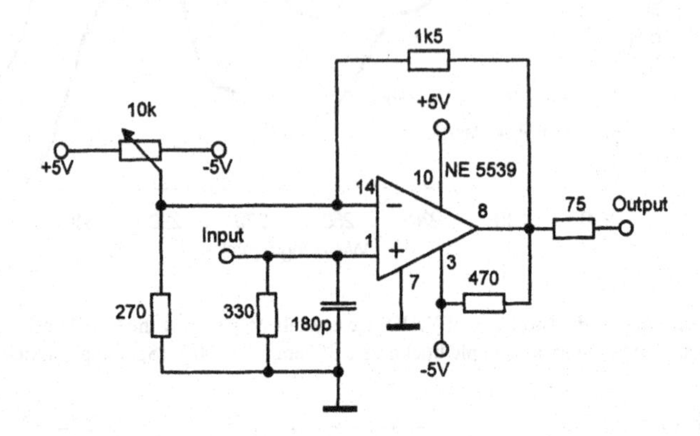

Figure 3. The fast preamplifier on the NE 5539 operational amplifier.

A pulse-height discriminator must filter the pulses of the preamplifier output in accordance with Mössbauer resonance gamma rays. Standard pulse-height discriminators, used with a NaI:Tl scintillation detector and a proportional counter, respectively, cannot be correctly used with a YAP:Ce scintillation detector, because the duration of pulses on the preamplifier output is less than 200 ns.

4. Fast pulse-height discriminator

The fast pulse-height discriminator was built using two fast comparators MAC 160 (see Fig. 4). The figures 5 and 6 show the time diagrams of discriminator output signal (**imp**), if the output detector signal (**det**) fits into the selected interval (**V**) of the amplitude value and out of this interval, respectively. The input **C** of the 74ALS74 circuit triggers from the high logic state to the low logic state, when the value of an amplitude of the detector output signal is more than the low voltage level (**LL**) of the pulse-height discriminator. The 74ALS123 circuit is ready to generate the output pulse (**imp**), because its input **R** is in the high logic stage. The input **B** of the 74ALS123 circuit changes over from the low logic state to the high logic state only after 30 ns delay. This delay is determined by the front edge of the detector output signal (**det**) and it prevents the generating of an output pulse (**imp**) by the 74ALS123 circuit if the value of the amplitude of the detector output signal (**det**) is more than the high voltage level (**HL**) of the pulse-height discriminator. The input **S** of the 74ALS74 circuit triggers the changeover from the high logic state to the low logic state, when the value of an amplitude of the detector output signal is more than the high voltage level (**HL**) of the pulse-height discriminator. The input **R** of the 74ALS123 circuit changes over from the high logic state to low logic stage and it prevents the generating of an output pulse (**imp**).

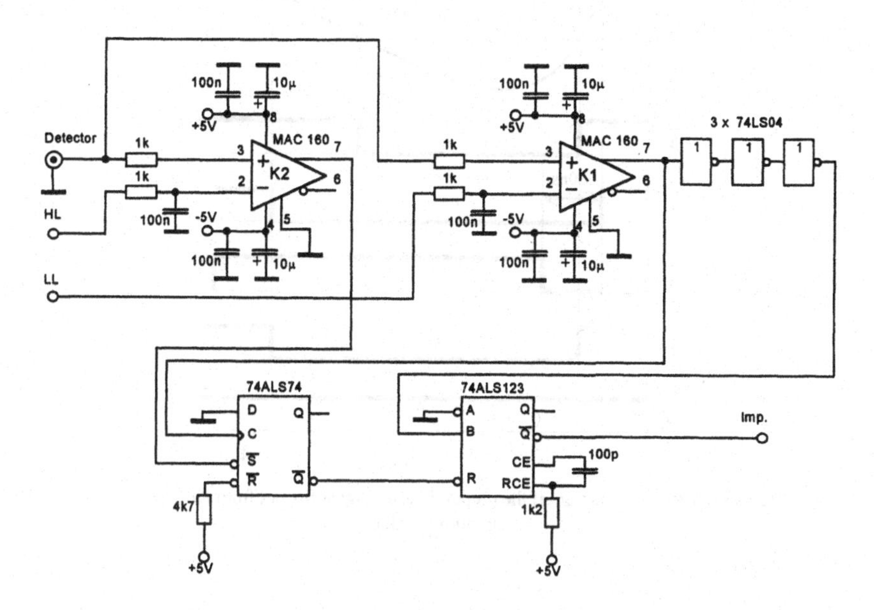

Figure 4. The fast pulse-height discriminator.

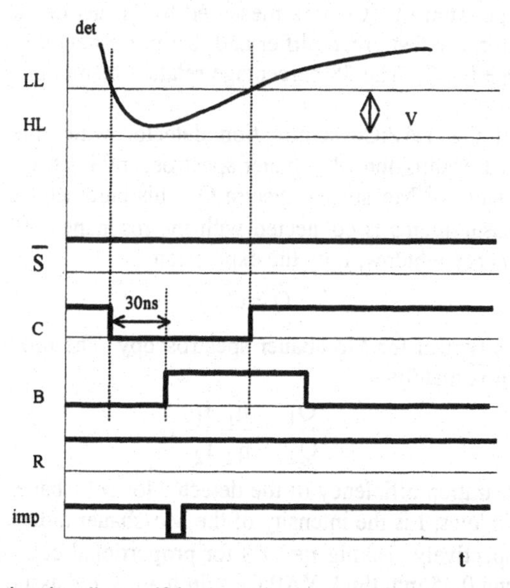

Figure 5. The appropriate time diagram if the output detector signal fits into the selected interval of the amplitude value.

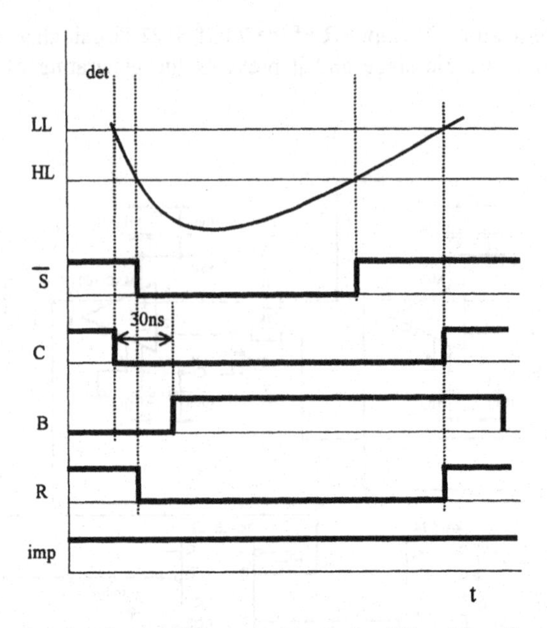

Figure 6. The appropriate time diagram if the output detector signal fits out from the selected interval of the amplitude value.

5. Results

The pulse-height spectrum of ^{57}Co was measured by means of the YAP:Ce (0.35 mm thickness) scintillator, the fast preamplifier and fast pulse-height discriminator that was described above (see Fig.7). The 38 percentage relative resolution was obtained for the 14.4 keV energy.

Comparison of the YAP:Ce scintillation detector with other detectors that are commonly used for transmission Mössbauer spectroscopy [7-9] can be made by means of the statistical quality of Mössbauer spectra Q. This parameter that characterizes the productivity of measurements is connected with the resonance effect, ε, and the count within a selected energy window, I, by the expression

$$Q = \varepsilon^2 I \ , \tag{1}$$

if ε is small, that is typical for Mössbauer spectroscopy. The productivity ratio of two detectors [3] is approximately

$$\frac{Q_1}{Q_2} \approx \frac{\eta_1}{\eta_2}\frac{I_1}{I_2} \ , \tag{2}$$

where η is the registration efficiency of the detector for Mössbauer radiation within the selected energy windows, I is the intensity of the Mössbauer radiation in the solid angle of registration, respectively. Taking η=30% for proportional counter, η=80% for a 0.1 mm thick NaI:Tl and 0.35 mm thick YAP:Ce scintillators and assuming a limiting count rate ratio of

$$\frac{I_{YAP}}{I_{Prop.counter}} \approx \frac{I_{YAP}}{I_{NaI}} \approx 8 \tag{3}$$

one gets

$$\frac{Q_{YAP}}{Q_{Prop.counter}} \approx 18, \quad \frac{Q_{YAP}}{Q_{NaI}} \approx 7. \tag{4}$$

These results show a considerable increase in productivity by using the YAP:Ce detector in the case of limiting count rate.

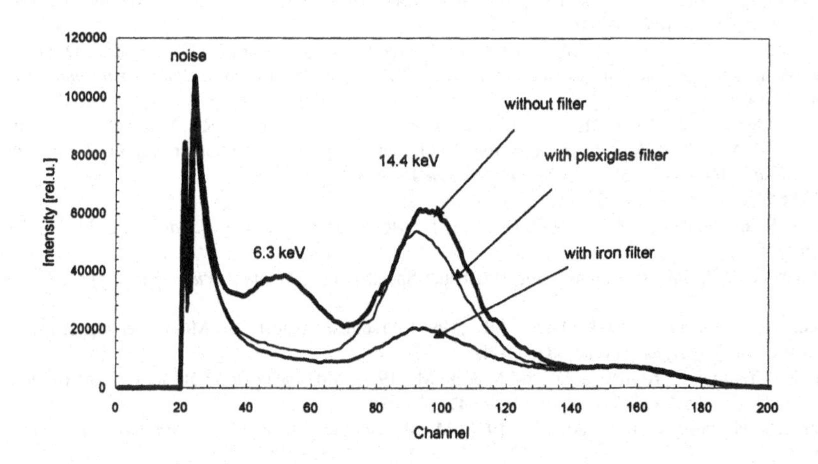

Figure 7. The pulse height spectrum of ^{57}Co, measured using the $YAlO_3$:Ce scintillation crystal, the fast preamplifier and pulse-height discriminator.

6. Conclusion

Comparative analysis of characteristics of the YAP:Ce detector [3,4] and good experiences over a relatively long period of using this detector for other investigation measurements [10-12], allow one to make an unambiguous conclusion regarding the unconditional advantages of this detector unit over all the detectors commonly used for transmission Mössbauer spectroscopy. The considerable improvement in the production of data (short time measurement, wide range of the velocity, use of stronger sources, *etc.*) enhances the capabilities of the Mössbauer spectroscopy as an industrial, research and analytical tool.

Acknowledgement

Financial support from the Ministry of Education of the Czech Republic under Project 23/1998 is gratefully acknowledged.

References

1. Autrata, R., Schauer, P., Kvapil, J., and Kvapil, Jos. (1983) *Scanning* 91.
2. Baryshevsky, V.G., Korzhik, M.V, Moroz, V.I., Pavlenko, V.B., Fyodorov, A.A., Smirnova, S.A., Egorycheva, O.A., and Kachanov, V.A. (1991) $YAlO_3$:Ce-fast-acting for detection of ionizing radiation, *Nucl. Instr. and Meth.* **B58**, 291-293.
3. Fyodorov, A.A., Kholmetskii, A.L., Korzhik, M.V., Lopatik, A.R., Mashlan, M., and Misevich, O.V. (1994) High-performance transmission Mössbauer spectroscopy with $YAlO_3$:Ce scintillation detector, *Nucl. Instr. and Meth.* **B88**, 462-464.
4. Kholmetskii, A.L., Mashlan, M., Misevich, O.V., Chudakov, V.A., Lopatik, A.R., and Zak, D. (1997) Comparison of the productivity of fast detectors for Mössbauer spectroscopy, *Nucl. Instr. and Meth.* **B124**, 143-144.
5. Žák, D., Mašláň, M., Kholmetskii, A.L., Evdokimov, V.A., Lopatik, A.R., Misevich, O.V., and Fyodorov, A.A. (1995) Mössbauer spectrometer based on personal computer equipped with the $YAlO_3$:Ce scintillation crystal, *Acta Physica Slovaca* **45**, 85-88.
6. HAMAMATSU, Technical data sheet.
7. Havlíček, S. and Schneeweiss, O. (1987) Detectors for Mössbauer spectroscopy, *Jad. Energ.* **33**, 287-291 (in czech).
8. Nagarajan, R. (1989) Instrumentation for Mössbauer Spectroscopy, *Indian J. Pure Appl. Phys.* **27**, 393-406.
9. Gancedo, R., Gracia, M., and Marco, J.R. (1994) Practical Aspects of Mössbauer Spectroscopy Instrumentation, *Hyperfine Interact.* **83**, 71-78.
10. Zbořil, R., Grambal, F., Krausová, D., and Mašláň, M. (1997) Mössbauer study of thermal conversion of $FeSO_4 \cdot 7H_2O$, *Czechoslovak Journal of Physics* **47**, 565-570.
11. Mašláň, M., Martinec, P., and Černá, K. (1996) Mössbauer study of tourmaline, *Materials Structure* **3**, 178-179.
12. Zboril, R., Mashlan, M., and Krausova, D. (1999) The mechanism of β-Fe_2O_3 formation by the solis-state reaction between NaCl and $Fe_2(SO_4)_3$, in M. Miglierini and D. Petridis (eds)., *Mössbauer Spectroscopy in Materials Science*, Kluwer Academic Publishers, Dordrecht, pp. 49-56.

THE MÖSSBAUER SPECTROMETER AS A VIRTUAL INSTRUMENT

M. MASHLAN[1], D. JANCIK[1], D. ZAK[1], F. DUFKA[2], V. SNASEL[2], AND A.L. KHOLMETSKII[3]

[1)] *Department of Experimental Physics, Palacky University, Svobody 26, 771 46 Olomouc, Czech Republic.*

[2)] *Department of Computer Science, Palacky University, Tomkova 10, 771 46 Olomouc, Czech Republic.*

[3)] *Department of Physics, Belarus State University, Skorina 4, 220 080 Minsk, Belarus.*

Abstract

New programming methods have been used for building a virtual Mössbauer spectrometer. Data acquisition is realized via the PIGGY 32/154/320 microcontroller. This microcontroller is connected with a computer by RS232 interface. The software communicates with the spectrometer module and provides an intelligent graphical user interface for the control of the spectrometer and for the presentation of the measured results. Mössbauer spectra of 256, 512, 1024, 2048 channels can be accumulated in constant acceleration or constant velocity modes. Pulse height spectra can be collected in an energy window scanning mode. The velocity range is ±100 mm/s and the non-linearity of the velocity scale is lower than 0.1 %, the maximum input count rate is 10 MHz and the channel capacity is 2^{32}-1 counts.

1. Introduction

The computer technique was intensively used in the design of physical measurements and experiments during the last twenty years. Microprocessors and microcomputers as well as the wide base of electronic circuits have radically changed the approach to the building of measurement and experimental devices. Two trends have occurred in Mössbauer methodology. First, the microcomputer and microprocessor are used as replacements for the multichannel analyzer for data acquisition [1-7]. Second, experimental data are accumulated and analyzed using a microcomputer instead of a mainframe computer [8-14]. Some published designs of spectrometers have been in the form of plug-in boards into the slot bus of the computer [5,9,10,14], other spectrometers have used some commercially-available cards [11], others have a built-in CAMAC system with a standard transparent GPIB bus [12] or use the RS232 serial link [4,6,7]. Different programming languages have been used for the software implementation of

M. Miglierini and D. Petridis (eds.), Mössbauer Spectroscopy in Materials Science, 399–406.

spectrometers, for example ASSEMBLER [4,7,10,14], BASIC [6,9,10,11], PASCAL [14], C++ [12].

The spectrometer that is described in this paper uses the RS232 serial interface to connect the specific spectrometer hardware with the main computer. The user interface has a form of a virtual instrument, which is realized on the main computer using a graphical programming language. The virtual instruments are the programs that imitate the actual instruments. Thanks to this software design, users with graphical programming experience can create personal virtual instruments. An example of a virtual instrument for the fast determination of the Fe^{3+}/Fe^{2+} ratio in minerals is presented.

2. Hardware design

The schematic block diagram of the spectrometer is shown in Figure 1. The mini transducer [15] with a radiation source, two discs of collimators, an absorber holder as well as a scintillation detector are assembled on the roller (diameter 65 mm, length 300 mm) that is placed on the mounting bench. The assembly of the spectrometer is very flexible and allows the sample holder to be changed to a cryostat or furnace. The velocity generator with a transducer control unit, two single channel analyzers and the unit accumulating spectra are controlled using the PIGGY 32/154/320 microcontroller unit and are situated together upon the mounting bench. The scintillation detector consists of a YAP:Ce crystal, a photomultiplier tube, a fast preamplifier and a high-voltage supply [16].

3. Software implementation

The software plays an important role in the development of the automated data acquisitions and instrument control systems. The quality and flexibility of the software that is used in the development of the instrumentation systems ultimately determines its overall quality and usefulness.

The software support of the presented spectrometer can be divided into two parts. The first part is the software for the onboard microcontroller unit burned in EPROM memory of PIGGY 32/154/320 microcontroller unit. This one controls all the other hardware units, implements all the measurements, and provides communication with the outer world through a RS232 serial interface. The second part is the software that communicates with the spectrometer module and provides an intelligent graphical user interface for the control of the spectrometer and for the presentation of the measured results. This part is optional, because all the measurements can be controlled directly, through the serial terminal connected to the spectrometer. There are two main software packages for the spectrometer, both for IBM PC compatible running under Microsoft Windows 95, or Windows NT operating systems. The first one is written in LabVIEW programming environment and the second one is written in Microsoft Visual C++ 5.0 and Microsoft Visual Basic.

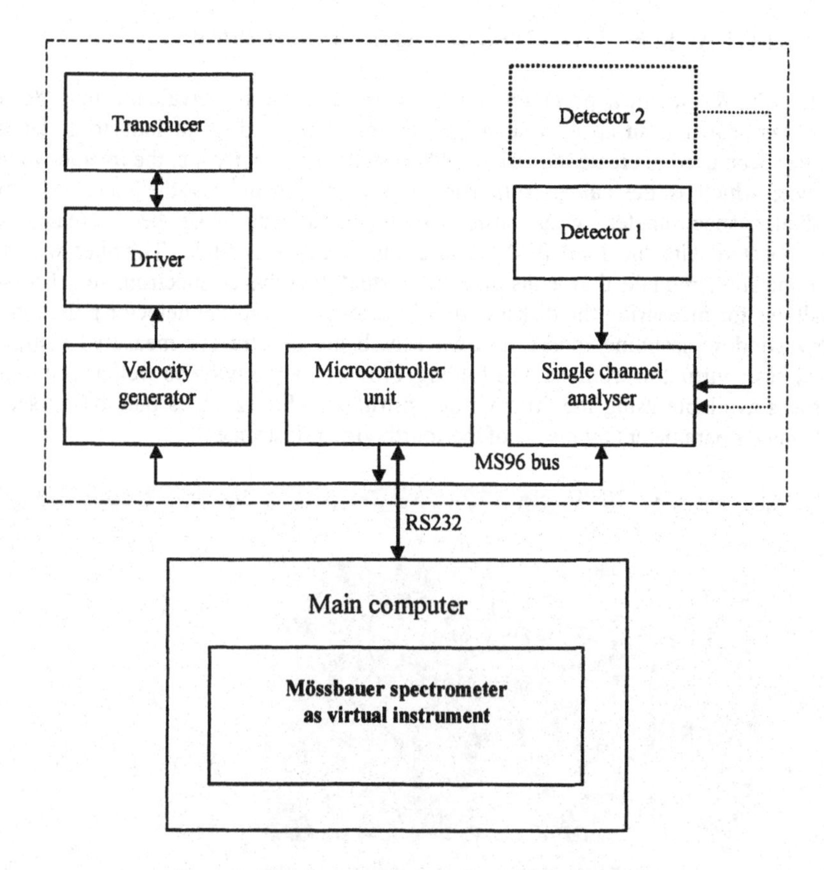

Figure 1. The schematic block diagram of the spectrometer.

3.1. SOFTWARE FOR THE MICROCONTROLLER UNIT

The PIGGY 32/154/320 unit includes the 80C32 microcontroller, a RS232 serial interface, an external bus for communication with other hardware units, an EPROM memory for the program code, a real time clock, a battery backed-up RAM memory for 4 spectra (32 Kb) and other measuring data.

The program is written in ASSEMBLER. If the user turns on the spectrometer, this program is started and provides all core functions of the standard spectrometer device. The program implements algorithms for three main measuring modes (amplitude analysis, constant velocity and constant acceleration measurement) and contains a command line interpreter for communication through the serial interface using simple textual commands.

These commands allow the setting or getting of various measuring parameters, starting and stopping of measurement, and retrieving of the measured results.

3.2. A VIRTUAL SPECTROMETER CREATED BY LABVIEW

The LabVIEW graphical programming language, that became available in 1986 and which has grown from an alternative programming method to an industrial standard [17], has been used to create the virtual Mössbauer spectrometer, i.e. the interactive user interface, which is the called front panel, because it simulates the panel of a real Mössbauer spectrometer. Four basic virtual instruments have been created and interconnected with the LabVIEW serial communication module. Together with this communication module they constitute the virtual Mössbauer spectrometer. The first one allows for measuring the distribution of the amplitude of the detector pulses in the energy window scanning mode. Its own Mössbauer spectra are measured using the second basic virtual instrument. The folding, browsing and saving as well as printing of spectra is available using the third virtual instrument (Fig. 2). It is possible to set the spectrometer parameters by means of the fourth virtual instrument.

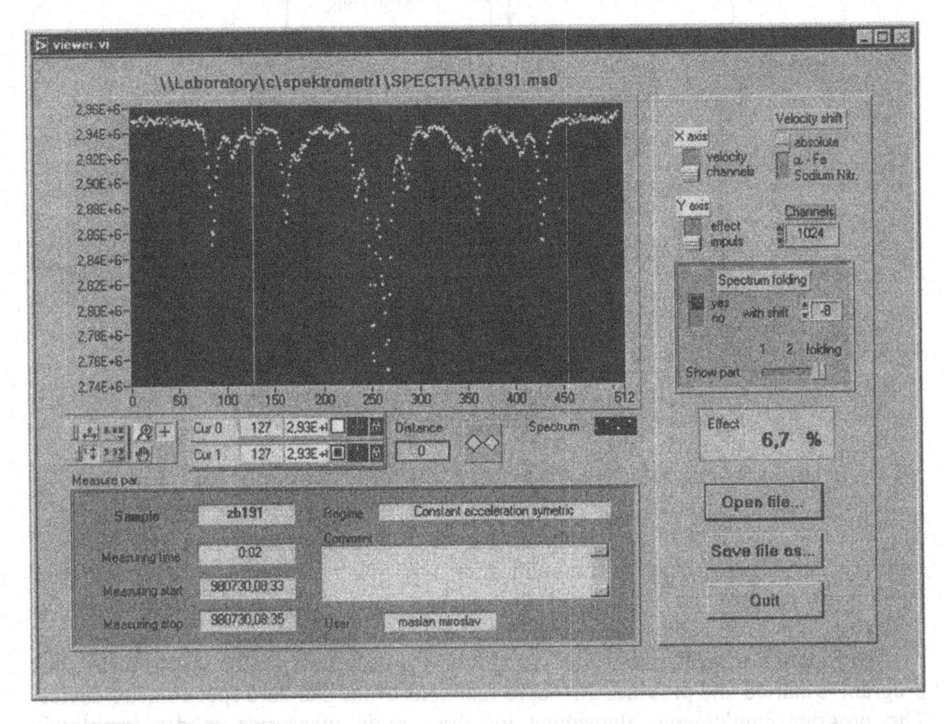

Figure 2. **The front panel of the virtual instrument for viewing, folding, saving and printing of the Mössbauer spectra**

3.3. SOFTWARE FOR NETWORK USE

In addition to the LabVIEW design other software has been developed for the use of the spectrometer across computer networks. It is based on the distributed COM (Component Object Model) technology [18] that is an integral part of Windows 95 and Windows NT operating systems.

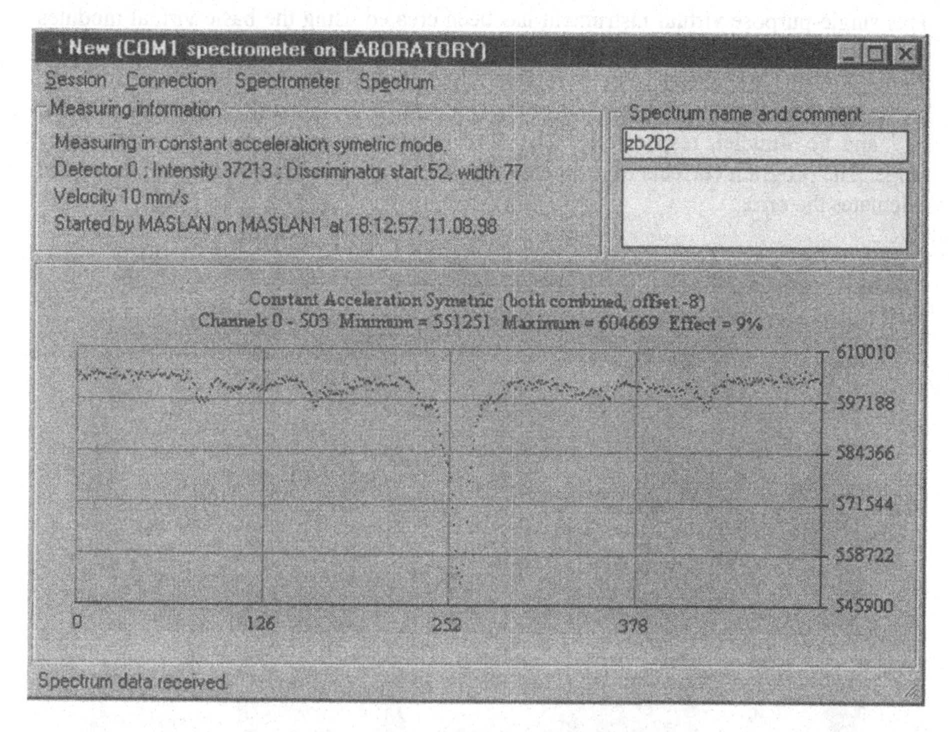

Figure 3. The front panel of Mössbauer spectrometer for the network use

This software solution consists of two main parts. The first part is a framework of COM components that allows the attaching and use of the spectrometer connected to the appropriate computer in a computer network. This component framework is programming language independent and can be used in programming environments, which support the distributed COM and Automation technology (such as Microsoft Visual C++, Borland C++, Borland Delphi and others). This framework can be used for making customized spectrometer applications. All the components were written in Visual C++. The second part consists of two applications written in Visual Basic. These applications use the COM components and actually provide the user interface for the use of the spectrometer in a computer network. The first one allows the spectrometer to be working in connection with the network computer. The spectrometer can be connected to different serial ports on a single computer and can be made public to the computers on the network under different names and can be protected by passwords if desired. The second application allows the use of such a configured spectrometer in a network. It provides a graphic user interface for making various types of measurements similarly to the program written in LabVIEW (Fig. 3). This software can also be used on a single computer, which is not connected to the network with a locally connected spectrometer, if required.

3.4. A VIRTUAL INSTRUMENT FOR THE Fe^{3+}/Fe^{2+} RATIO DETERMINATION

This single-purpose virtual instrument has been created using the basic virtual modules of spectrometer denoted above. The front panel of this virtual instrument is shown in Figure 4. Using a constant velocity regime, three points of the Mössbauer spectrum are measured. The first and the second point conform to the maximum resonance of the Fe^{3+} and Fe^{2+} nuclei, respectively. The third point is measured out of the resonance range. The program corrects experimental Fe^{3+}/Fe^{2+} ratios using a calibration curve and calculates the error.

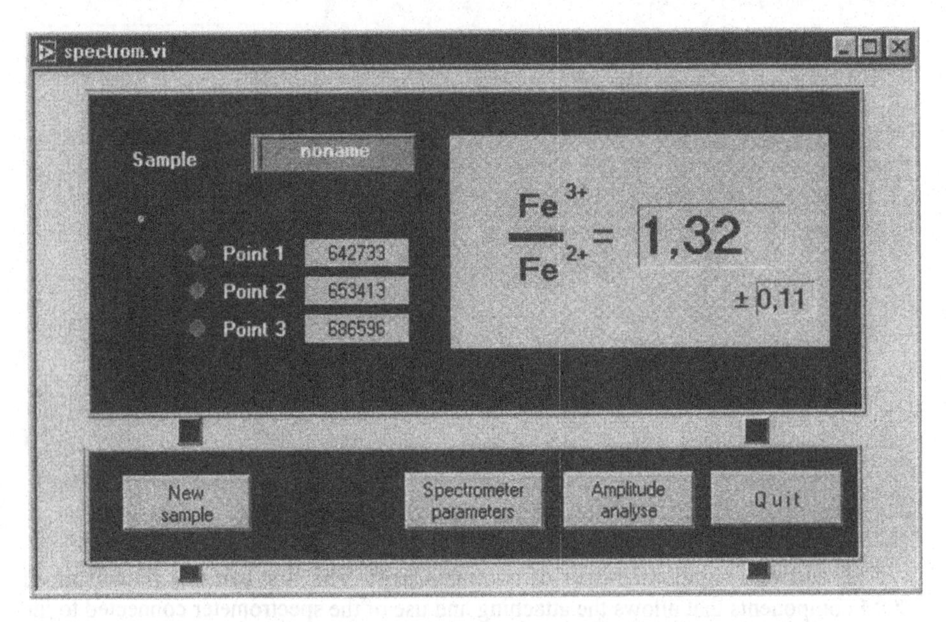

Figure 4. The front panel of the virtual instrument for Fe^{3+}/Fe^{2+} ratio determination

4. Results

4.1. THE NON-LINEARITY OF THE SPECTROMETER

The main parameter that characterize the Mössbauer spectrometer's quality (particularly the driving system) is the non-linearity of the velocity scale. Results of the non-linearity study are shown in Table 1. The six lines of an α-Fe experimental spectrum were approximated by Lorentzian functions and the non-linearity of all lines has been calculated by means of the least squares method fitting the following function

$$non(i) = \frac{x(i) - a \cdot v(i) - b}{v(6) - v(1)},$$

where i (i = 1 to 6), x(i), v(i), a, b are line number, experimental position of the line, theoretical position of the line and parameters from the least square method, respectively.

TABLE 1. The results of non-linearity measurements

Number of line	Constant acceleration mode						Constant velocity mode	
	symmetric velocity signal				asymmetric velocity signal			
	ascending part		aescending part					
	x(i) [mm.s^{-1}]	non(i) [%]	x(i) [mm.s^{-1}]	Non(i) [%]	x(i) [mm.s^{-1}]	non(i) [%]	x(i) [mm.s^{-1}]	non(i) [%]
1.	-5,042	+0,03	-5,048	-0,05	-5,038	+0,07	-5,040	+0,06
2.	-2,811	+0,02	-2,808	+0,03	-2,816	-0,03	-2,820	-0,06
3.	-0,582	-0,03	-0,573	+0,04	-0,585	-0,05	-0,591	-0,10
4.	1,092	-0,05	1,101	+0,01	1,092	-0,06	1,109	+0,09
5.	3,331	-0,02	3,334	-0,01	3,332	-0,01	3,343	+0,07
6.	5,576	+0,05	5,570	-0,02	5,577	+0,06	5,568	-0,05

4.2. THE Fe^{3+}/Fe^{2+} RATIO IN STILPNOMELANE

The capabilities of the Mössbauer spectrometer described above are demonstrated in the example of the determination of the Fe^{3+}/Fe^{2+} ratio in the Stilpnomelane. Stilpnomelane is a silicate mineral with the general formula of $K(Fe^{2+}MgFe^{3+}Al)_8(SiAl)_{12}(O,OH)_{27}\cdot 2H_2O$. Two samples of Stilpnomelane from Horní údolí (Czech Republic) were measured. The measuring time was four minutes per one point of the spectrum. The results of the Mössbauer analysis as well as the results of the conventional chemical method are listed in Table 2.

TABLE 2. The Fe^{3+}/Fe^{2+} ratio measured by different methods for Stilpnomelane samples

Sample	Fe^{3+}/Fe^{2+} ratio		
	Mössbauer spectroscopy		Chemical analysis
	A	B	
HU-1	1.75±0.09	2,26	1.99
HU-2	1.38±0.08	1,85	1.72

A - the fast determination from three points of the spectrum, without any correction
B – the Fe^{3+}/Fe^{2+} ratio determined from the areas of the subspectra in the constant acceleration mode

5. Conclusion

New programming methods have been used for building a virtual Mössbauer spectrometer. Using the graphical programming language, the users can create their own virtual instruments. The example of an analytical instrument for the Fe^{3+}/Fe^{2+} ratio determination has been shown.

Acknowledgement

Financial support from the Ministry of Education of the Czech Republic under Project 23/1998 is gratefully acknowledged.

References

1. Wang, G.W., Chirovski, L.M., Lee, W.P., Becker, A.J., and Groves, J.L. (1978) A simple Mössbauer data acquisition system using a minicomputer, *Nucl. Instr. and Meth.* **155**, 273-277.
2. Sundqvist, T. and Wappling, R. (1983) A microcomputer controlled Mössbauer spectrometer, *Nucl. Instr. and Meth.* **205**, 473-478.
3. Linares, J. and Sundqvist, T. (1984) Mössbauer data collection using an Apple-11 microcomputer, *Rev. Sci. Instrum.* **17**, 350-351.
4. Grogan, S.and Thornill, J. (1986) Remote microprocessor controlled Mössbauer spectrometer, *Nucl. Instr. and Meth.* **A256**, 525-528.
5. Schaaf, P., Wenzel, T., Schemmerling, K., and Lieb, K.P. (1994) A simple six-input multichannel systém for Mössbauer spectroscopy, *Hyperfine Interact.* **92**, 1189-1193.
6. Verrasto, C., Trombetta, G., Pita, A., Saragovi, C., and Duhalde, S. (1991) 5nsec dead time multichannel scaling systém for Mössbauer spectrometer, *Hyperfine Interact.* **67**, 689-694.
7. Jing, J., Campbell, S.J., and Pellegrino, J. (1992) A stand-alone Mössbauer spectrometer based on the MC68008 microprocessor, *Meas. Sci. Technol.* **3**, 80-84.
8. Sisson, K.J. and Boolchand, P. (1982) A microcomputer system for the analysis of Mössbauer spectra, *Nucl. Instr. and Meth.* **198**, 317-332.
9. Zhou, R., Li, F. and Zhang, Z. (1988) A microcomputer system for data-acquisition, fitting and plotting of Mössbauer spectra, *Hyperfine Interact.* **42**, 1181-1184.
10. Bil'dyukevich, E.V., Gurachevskii, V.L., Litvinovich, Yu.A., Mashlan, M., Misevich, O.V., Kholmetskii, A.L., and Chudakov, V.A. (1985) Gamma-resonance complex on line with a microcomputer, *Instrum. Exp. Tech.* **28**, 1303-1304.
11. Faigel, Gy., Haustein, P.E., and Siddons, D.P. (1986) An IBM PC based Mössbauer spectrometer and data analysis system, *Nucl. Instr. and Meth.* **B17**, 363-367.
12. Kwater, M., Pochron, M., Ruebenbauer, K., Terlecki, T., and Wdowik, U.D. (1994) Mössbauer controller and data acquisition unit MsAa-1, *Hyperfine Interact.* **92**,
13. Klingelhöfer, G., Foh, J., Held, P., Jäger, H., Kankeleit, E., and Teucher, R. (1992) Mössbauer backscattering spectrometer for mineralogical analysis of the Mars surface, *Hyp. Int..* **71**, 1449-1452.
14. Mašláň, M., Žák, D., Kholmetskii, A.L., Evdokimov, V.A., Misevich, O.V., Fyodorov, A.A., Lopatik, A.R., and Snášel, V. (1994) PC-AT based Mössbauer spectrometer, *Avta UPO Physica* **116**, 9-20.
15. Evdokimov, V.A., Mahlan, M., Zak, D., Fyodorov, A.A., Kholmetskii, A.L., and Misevich, O.V. (1995) Mini and micro transducers for Mössbauer spectroscopy, *Nucl. Instr. and Meth.* **B95**, 278-280.
16. Mashlan, M., Jancik, D., and Kholmetskii, A.L. (1999) YAP:Ce scintillation detector for transmission Mössbauer spectroscopy, in M. Miglierini and D. Petridis (eds)., *Mössbauer Spectroscopy in Materials Science*, Kluwer Academic Publishers, Dordrecht, pp. 391-398.
17. *LabVIEW – Graphical Programming for Instrumentation. User Manual*, National Instruments Corporation, Austin.
18. Eddon, G. and Eddon, H. (1998) *Inside Distributed COM*, Microsoft Press.

CALIBRATION OF THE MÖSSBAUER SPECTROMETER VELOCITY BY OPTICAL METHODS

Mössbauer Spectrometer Calibration

M. KWATER, K. RUEBENBAUER AND U.D. WDOWIK
Mössbauer Spectroscopy Laboratory
Institute of Physics and Computer Science, Pedagogical University
PL-03-084 Cracow, ul. Podchorazych 2, Poland

1. Introduction

Optical methods could be used for the absolute velocity calibration in almost all channels independently [1-18]. The best, and practically unperturbing method is the use of a laser light source in combination with the interferometer [2].

The electromagnetic unpolarized plane-wave having frequency ν_0 is transformed into plane-wave having frequency $\nu_0 + \Delta\nu$ upon completely elastic back-scattering from a moving reflector, where the frequency shift $\Delta\nu$ can be expressed as follows:

$$\Delta\nu = \nu_0 \left(([(c/n)^2 - v^2]/[(c/n) + v\, cos(\Theta)]^2)^k - 1 \right) \tag{1}$$

with n being the absolute refraction index of the homogeneous, isotropic, stationary, non-absorbing and non-dispersive medium in the vicinity of the moving reflector ($n \geq 1$), $v < c/n$ stands for the absolute value of the reflector velocity in the reference frame of the original wave with c being a velocity of light in vacuum, and Θ is the angle between the incident beam and the reflector velocity. The index $k = 1, 2, 3, \ldots$ enumerates multiplicity of the scattering. The expression (1) can be approximated as follows for $v \ll c/n$:

$$\Delta\nu = -[2nkv\nu_0 cos(\Theta)]/c\,. \tag{2}$$

Hence, it is important to have well defined angle Θ, i.e., to keep a beam parallel to the transducer movable rod, and to have well defined multiplicity of the scattering. One obtains

$$\Delta\nu = \pm 2nv\nu_0/c \tag{3}$$

for $k = 1$, where the sign (+) corresponds to the velocity anti-parallel to the original beam direction, while the sign (-) corresponds to the velocity parallel to the original beam direction. The above formula allows for an absolute velocity calibration in vacuum ($n = 1$) provided the frequency ν_0 is well known or for an almost absolute calibration in air or He, where the corrections to the refraction

M. Miglierini and D. Petridis (eds.), Mössbauer Spectroscopy in Materials Science, 407–412.

index are very small at the typical laser frequencies. The above corrections could be taken into account adopting refraction indices for frequencies close to the laser basic frequency ν_0 and further correcting for the approximate values of the gas pressure and temperature (see, Table 1). Hence, it is important to apply light sources, where the basic frequency is sufficiently well defined (10^{-5} accuracy suffices) from the first principles by the emitting medium. On the other hand, it is important to have as high frequency ν_0 as practical in order to increase sensitivity at low velocities. Therefore, it seems that He-Ne lasers lasing red light in a continuous mode are the best light sources, while the frequency shift $\Delta\nu$ (for $k = 1$) could be measured applying Michelson-Morley interferometer and a fast photo-diode due to the high internal coherency of the above light source. An unpolarizing laser is a better choice as for such a case the laser beam can be easily split into reference and measuring beams with almost equal intensities giving the maximum possible modulation depth. An intensity $I(t)$ of the combined beam obeys the following expression for the above conditions:

$$I(t) = (a_1^2 + a_2^2) + 2a_1a_2 \cos\left(\phi + \frac{4\pi n\nu_0}{c}\int_0^t dt'\, V(t')\right), \tag{4}$$

where t stands for time, a_1 and a_2 stand for the absolute values of the amplitudes of reference and measuring beams, respectively, ϕ is the phase shift between the above beams originating in the unequal optical paths of the above beams, and $V(t)$ stands for the velocity (velocity component along the measuring beam) with positive velocities for a motion anti-parallel to the incoming beam. One has to note, that $V(t)$ is a periodic function of time without constant component. In order to obtain a deep modulation the amplitude of the transducer motion has to be several times bigger than the wave length of the laser light in the medium surrounding movable reflector at least.

2. Design and Performance of the Calibration Device

A basic layout of the device is shown in Fig. 1. An unpolarizing 0.5 mW red light He-Ne laser is rigidly mounted on the axis of the device. The laser has cavity length of about 20 cm (1090 MHz longitudinal mode separation frequency), and it operates on three longitudinal modes transforming one into another at random times having scale of several seconds. Such an arrangement prevents phase locking allowing to measure precisely low velocities at the acceptable broadening of the bandwidth (note that at high velocities and high repetition frequencies of the spectrometer an intermodal modulation assures additional and efficient phase randomizer without excessive increase in bandwidth). A beam expander mounted on the laser axis expands the beam to about 2 mm diameter assuring immunity against transversal motions and misalignments. The beam is subsequently passed through a diaphragm having smooth edges and split into 50-50 proportions by a

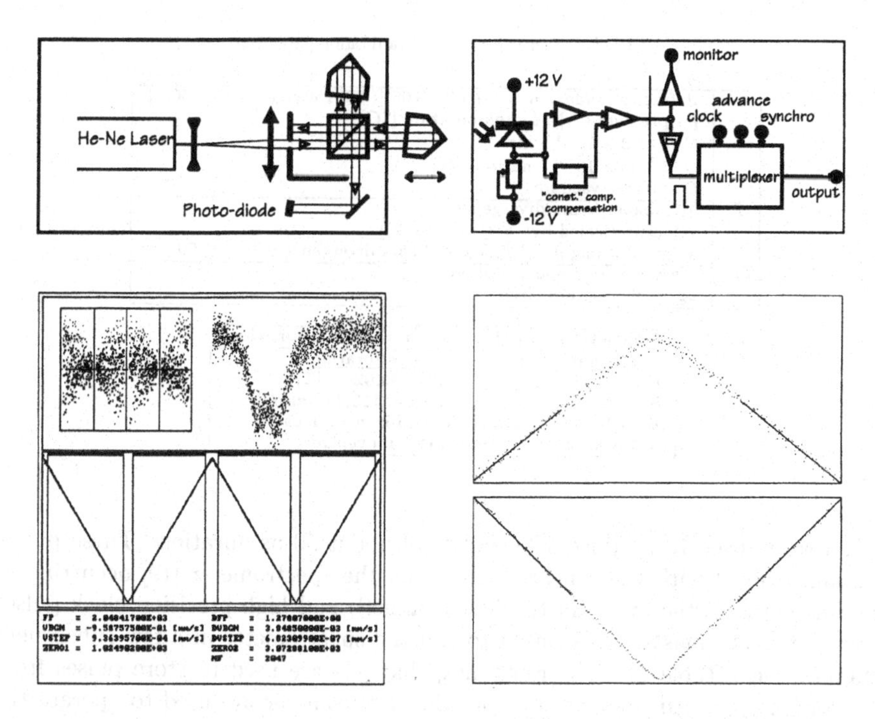

Figure 1. A schematic diagram of the mechanical design and optical system. A simplified diagram of the electronics. Typical results for low velocities (see text for details).

coated beam splitter at right angles. The splitter is offset to produce spatially separated beams in order to assure that $k = 1$ condition is satisfied. Tilted corner prisms are used to back-scatter reference and measuring beams, the latter being obtained by scattering from a movable prism. Such a design prevents multiple scattering. A combined beam passes via widely opened diaphragm, scatters from a flat metallic mirror and it is fully intercepted by a tilted photo-diode. Hence, a photo-diode remains in a relatively dark environment (a whole device is encapsulated in a light tight container), and a multiplicity of the scattering ($k = 1$) is precisely defined.

The electronic system is shown in Fig. 1 as well. The photo-diode is reversely polarized to assure about 13 ns rise time, and the resulting signal is filtered to about 20 MHz bandwidth with the subsequent compensation of the "constant" component. An integration time constant of the "constant" component circuit is set to several seconds in order to allow for the low count rate. The analog signal could be monitored on the oscilloscope. A wide range fast comparator with a small

TABLE 1. Corrections to the calibration parameters.

1.	Refraction index: $n = 1 + C(p/760)[(273.1502)/(t + 273.1502)]$ p - pressure [Tr] ; t - temperature [^oC] $C = 2.92 \times 10^{-4}$ - air $C = 3.50 \times 10^{-5}$ - natural He $C = 0.00$ - vacuum
2.	Mőssbauer beam divergence: velocities scaled by: $\frac{1}{2}[1 + cos(\gamma/2)]$ γ - a total angular Mőssbauer beam divergence $0 \leq \gamma \ll 90^o$
3.	Master clock frequency used: $\nu_c = 3999992.6(8)$ [Hz]

Consistency of the laser and self-calibration	
Laser calibration	Self-calibration
H = 33.03(1) [T]	H = 33.02(1) [T]
S = -0.110(1) [mm/s]	S = -0.113(1) [mm/s]
H - effective magnetic field on iron nucleus	
S - position of the iron spectrum centre	

hysteresis is used to produce TTL pulses of about 50 ns duration. These pulses are multiplexed with the synchro pulse from the spectrometer [19] occurring at the beginning of the cycle (in the first channel), and high precision clock pulses from the clock (master clock) used to run the spectrometer in a few subsequent channels (see, Table 1). The remaining channels are used to store pulses from the comparator. Advance pulses from the spectrometer are used to operate the multiplexer. A combined output is fed to the second multi-scaler bank of the spectrometer, while the first bank is used to collect simultaneously the Mőssbauer spectrum. A movable prism could be attached to the back-end of the transducer rod or just to the back of the Mőssbauer source (sample) for transducers having driving rod replaced by a driving hollow tube [20].

One covers velocity ranges from about 0.5 mm/s at minimum to more than 1 m/s. It is important to correct velocity scale for a Mőssbauer beam divergence in the case of a close geometry (see, Table 1).

Table 1 shows consistency between iron foil self-calibration and a simultaneous laser calibration, while the lower part of Fig. 1 shows a spectrum of partly hydrated ferrous sulphate obtained at low velocity (left part of two doublets) for a round-corner triangular reference function used to run the spectrometer. Upper left corner shows residuals of the laser signal with a proper phase in all four quarters of the cycle, upper right corner shows folded Mőssbauer spectrum vs velocity, while the lower part exhibits processed laser signal. Here, FP stands for the folding point on the 4096 channel scale used to collect the raw Mőssbauer data with DFP being a corresponding error, ZERO1 and ZERO2 denote "left" and "right" zeros of the velocity on the above channel scale, while VBGN stands for the velocity in the first folded channel with DVBGN being a corresponding error. VSTEP and DVSTEP denote velocity increment per folded channel and corresponding error, respectively. Finally, NF stands for the resulting number of folded channels [21].

Finally, the lower right part of Fig. 1 shows a laser signal around the folding point (note a reproduction of the round-corner), and a laser signal in the vicinity

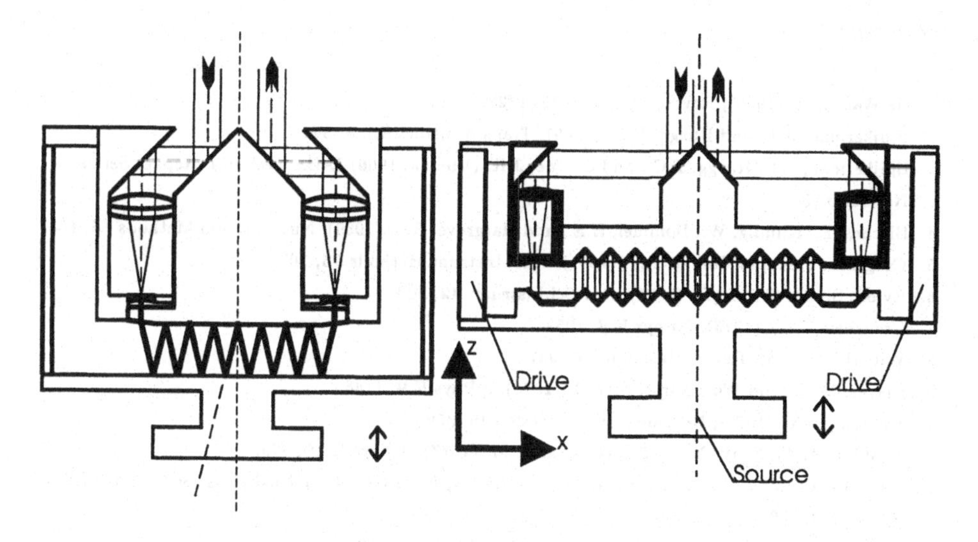

Figure 2. Possible multiple scattering optical systems for very low velocities.

of the "left" velocity zero (note a reproduction of the compressional wave in the driving rod at a turning point despite a low repetition frequency used).

3. Discussion and Conclusions

There is basically no limit at the high velocities as far as the Mőssbauer spectroscopy is considered. Very low velocities require low repetition frequencies of the spectrometer and their calibration suffers from the low count rate of the TTL pulses. Due to the fact that a basic laser frequency cannot be increased significantly one has to use multiple scattering with a well defined index k. However, standard multiple scattering interferometers, e.g., Fabry-Perot or Lummer-Gehrcke devices cannot be used as the multiplicity k is poorly defined for them. Fig. 2 sketches two possible modifications of the Fabry-Perot system to achieve the above goal. The left one is easier to make and it is insensitive to transversal motions, but it has a limited longitudinal range. The right one does not suffer from the limited longitudinal range, but it remains very sensitive to the transversal misalignment. The above devices (particularly the right one) would allow to extend the interferometric calibration to the velocities characteristic for zinc spectroscopy.

In conclusion one could say that a relatively simple, easy to use, and robust laser calibration system based on the Michelson-Morley interferometer has been built and put into operation. A precision better than 10^{-3} is attainable at all accessible velocities being close to 10^{-4} for typical iron spectroscopy velocity ranges.

References

1. de Waard, H. (1965), Rev.Sci.Instrum. **36**, 1728.
2. Spijkerman, J.J., and Ruegg, F.C. (1965), Trans.Am.Nucl.Sci. **8**, 344.
3. Spijkerman, J.J., Ruegg, F.C., and de Voe, J.R. (Vienna, 1966) IAEA Technical Report Series No. 50, p.53.
4. Biscar,J.P., Kündig, W., Bommel, H.E., and Hargrove, R.S. (1969), Nucl.Instrum.Methods **75**, 165.
5. Cosgrove, J.G. and Collins, R.J. (1971), Nucl.Instrum.Methods **95**, 269.
6. Eylon, S. and Treves, D. (1971), Rev.Sci.Instrum. **42**, 504.
7. Cranshaw, T.E. (1973), J.Phys.E **6**, 1053.
8. Wit, H.P. (1975), Rev.Sci.Instrum. **46**, 927.
9. Player, M.A., and Woodhams, F.W.D. (1976), J.Phys.E **9**, 1148.
10. Carson, D.W. (1977), Nucl.Instrum.Methods **145**, 359.
11. Yoshimura, T., Syoji, Y., and Wakabayashi, M. (1977), J.Phys.E **10**, 829.
12. Ali, M., Alimuddin, Kareem, S., and Rama Reddy, K. (1979), Proc.Nucl.Phys. Solid State Phys. Symp. **22C**, 566.
13. Reiman, S.I., and Mitrofanov, K.P. (1980), Prib.Tekh.Eksp. **2**, 66.
14. Gu, Y-J., Yang, X., and Zhao, J-M. (1981), Huadong Shifan Daxue Xuebao, Ziran Kexueban **3**, 59.
15. Otterloo, B.F., Stadnik, Z.M., and Swolfs, A.E.M. (1983), Rev.Sci.Instrum. **54**, 1575.
16. Xia,Y-F., Liu, R., Wang, S., Cao, Z-Q., Chen,Q-T., Shen, L-P., Gan, Z., Mao, W-K., Li, Y-J., and Jiang, S. (1984), Nanjing Daxue Xuebao, Ziran Kexue **20**, 59.
17. Amulyavichyus, A.P., and Davidonis, R.Yu. (1987), Prib.Tekh.Eksp. **4**, 218.
18. Chow, L. and Kimble, T. (1987), Hyperfine Interact. **35**, 1049.
19. Kwater, M., Pochroń, M., Ruebenbauer, K., Terlecki, T., Wdowik, U.D., and Górnicki, R. (1995), Acta Physica Slovaca **45**, 81.
20. Górnicki, R. - unpublished.
21. Ruebenbauer, K., MOSGRAF - Mössbauer Data Processing System, contact: sfrueben@cyf-kr.edu.pl for more information.

THE MINIATURIZED SPECTROMETER MIMOS II

The 2001 and 2003 US Mars Missions and Terrestrial Applications in Materials Science and Industry

G. KLINGELHÖFER
Darmstadt University of Technology, Schlossgartenstr. 9, D-64289 Darmstadt, Germany

Abstract

A miniaturized Mössbauer (MB) spectrometer for extraterrestrial applications has been developed at Darmstadt. The instrument MIMOS-II operates in backscattering geometry. It meets the requirements for space application as low mass (≤500g), small volume, and low power consumption (≤2 W), by using advanced technologies as SMD- and hybrid-technology as well as state of the art electronic components. The instrument has been tested extensively in the laboratory but also in the field mounted on the robotic arm of a prototype Martian Rover Rocky-7 under development at JPL/NASA, United States. The MIMOS II spectrometer has been selected recently as part of the ATHENA Rover payload for the US Mars missions Mars Surveyor 2001 and 2003. By determining the oxidation state of iron and the iron mineralogy on the surface of Mars, MB spectroscopy will contribute to a much deeper understanding of the evolution of the planet Mars, its surface and atmosphere, and the history of water. Because of the small size, the backscattering geometry, and the portability of the instrument there is a lot of possible terrestrial applications, e.g. in materials science and in industry. Examples are quality control in steel industry and iron ore processing plants, monitoring of oxidation processes in power plants, pipes etc., and in-situ investigations of materials processing in materials sciences.

1. Introduction

The element iron is one of the key elements in the evolution of our solar system. Its chemistry is strongly coupled to the chemistry of abundant elements as hydrogen, oxygen and carbon. The identification and determination of the abundance of iron-containing minerals, their weathering products, the measurement of the ferric (Fe^{3+}) to ferrous (Fe^{2+}) ratio, and the determination of the properties of possible magnetic phases will provide significant information about formation processes leading to the present state of the surface of the planet, as well as about the nature and extent of atmosphere – surface interactions of planetary surfaces, the chemistry and physics of weathering

M. Miglierini and D. Petridis (eds.), Mössbauer Spectroscopy in Materials Science, 413–426.

processes involving these Fe-bearing phases [4, 6, 8, 9, 25, 31, 32]. These results will contribute to the understanding of the history and development of the planetary surface as for instance the Martian surface, where evidence has been found that liquid water probably has been present on the surface [3, 10] which might have led in connection with a dense atmosphere to the formation of life similar to the Earth.

At the moment the planet Mars is in the focus of the planetary exploration programs of the US-American space agency NASA, the Japanese space agency, and European space agency ESA. In particular NASA has implemented a long term program for the exploration of Mars called, Mars Surveyor Program [51, 52, 53]. The primary goals are the search for water and its history, the evidence of past or present life, the understanding of the history of climate and its evolution, and the search for resources. The preliminary culmination of this program will be a sample return mission scheduled for the year 2005. Before this there will be missions in 2001 and 2003 respectively to investigate the mineralogy of different landing sites and to collect samples for a possible sample return mission. A set of instruments called the ATHENA instrument package [51, 52] has been selected for these missions including the Mössbauer spectrometer MIMOS II. This miniaturized Mössbauer spectrometer, similar to the one under development in the US [1, 41, 45] has been developed at the Darmstadt University of Technology [13, 21, 27, 30, 33, 34] for the Russian Mars mission Mars96/98 which never happened to go, and was modified to meet the requirements of the Mars Surveyor missions as well as those of the European Mars-Express 2003 mission, currently under serious consideration [54].

Because of its small size, portability, and the backscattering geometry a lot of terrestrial applications especially in Materials Science and in industry open up, which have not been feasible up to now with standard laboratory equipment. This is documented by the growing interest in using this newly developed instrumentation.

2. Mössbauer Spectroscopy on Mars

The scientific basis for landing a Mössbauer spectrometer on Mars is extensively discussed by Knudsen [36, 37, 38]. Briefly, a Mössbauer spectrometer on a Martian lander can identify and measure the relative abundance of iron-bearing minerals (e.g., carbonates, phyllosilicates, hydroxyoxides, phosphates, oxides, silicates, sulfides, sulfates), including those that are magnetically ordered, measurement of the ferric (Fe-3+) to ferrous (Fe-2+) ratio, and the size distribution of magnetically-ordered particles (i.e., nanophase or superparamagnetic particles versus multidomain particles). These data characterize the present state of Martian surface materials and thus provide constraints on weathering processes by which the surface evolved to its present state. Fe-Mössbauer spectroscopy (FeMS) can, for example, identify primary igneous mineralogies such as iron-bearing olivine and pyroxene and weathering products which do (e.g., goethite and jarosite) and do not (e.g., hematite and maghemite) contain volatiles as a part of their structures. Iron-bearing sulfides (e.g., pyrite and phyrrhotite) and carbonates (e.g., siderite) can also be detected. By determining the size distribution of oxide particles, Mössbauer analyses can differentiate between low-temperature (e.g., palagonitization) and high-temperature (e.g., oxidative alteration in an impact melt) hydrothermal processes.

Another question is whether the SNC meteorites can be related to actual materials on Mars. Mössbauer spectroscopy can contribute significantly to answering this question by comparing the laboratory Mössbauer results on SNC meteorites with the data from the in-situ analysis on Mars.

3. MIMOS-II and the US-Mars-Surveyor Project

3.1. THE MARS-SURVEYOR PROJECT

The US-American Space Agency NASA has implemented a 10 year 'plus' program called Mars Surveyor, which is aimed to launch two spacecrafts (an Orbiter and a Lander) at every approximately two years opportunity, starting in 1996 with the MGS-orbiter. Main goals of this program are the search for water, to get a better understanding of the climate as it may have been in the past and now is in the present, and the search for life, which may have evolved in the past, or even still might be present.

The search for water is the main goal of the MS'98 mission named MVACS (Mars Volatiles and Climate Surveyor), which will land at the rim of the Martian south polar cap in 1999.

For the 2001, 2003, and the 2005 opportunities the missions are targeted to either the ancient highlands, and/or the borders between the ancient highlands and e.g. the channel outflows, similar to the Mars-Pathfinder landing site in 1997.

These missions will carry a suite of instruments, called the ATHENA instrument package (in 2003 and 2005) and APEX (ATHENA Precursor Experiment for 2001), optimized for the mineralogical and geological characterization of the landing sites, its rocks and soil. In 2003 and in 2005 samples will be selected by theses instruments and stored in a cache, to be taken back to Earth in the years 2005 and/or later. So real Martian samples will be brought back to Earth in the year 2008 at the earliest.

The instruments selected for these missions are, besides the Mössbauer spectrometer MIMOS II, a microscopic imager for close-up images with a resolution of about 50 micrometers, an alpha-proton-X-ray spectrometer for the elemental analysis of the samples, a miniaturized Thermal Emission Spectrometer (Mini-TES), a Raman spectrometer and a drilling device (mini corer).

The APEX instrument package, which will not include all ATHENA instruments, will be mounted on the 2001-lander deck, with the exception of the APX spectrometer, which will be mounted on the Sojourner-class rover similar to the Mars-Pathfinder mission in 1997. The Mössbauer spectrometer will be attached to the Robotic arm, and therefore will have access to different kind of samples as there are soil, rocks, and the magnet array supplied by Denmark [19,40,49], collecting the magnetic particles of the airblown Martian dust. Furthermore the Robotic arm will dig a trench of a depth of up to 50 cm in the ground. The Mössbauer spectrometer will be used to determine the Fe^{2+}/Fe^{3+} ratio as a function of depth.

The ATHENA payload for the 2003 and 2005 missions, which includes all the selected instruments, will be carried by a rover. This rover is significantly larger than the Pathfinder rover Sojourner or Rocky 7 [2,50]. The Mössbauer spectrometer, the APXS, the Raman spectrometer, and the microscopic imager will be mounted on a so called instrument arm.

3.2. THE MINIATURIZED MÖSSBAUER SPECTROMETER MIMOS II

The instrument MIMOS II is operating in backscattering geometry measuring the scattered 14.41 keV MB radiation and the 6.4 keV Fe x-rays [33]. In backscattering geometry no sample preparation is needed. The instrument is easy portable (e.g. by a rover) and can be taken to the sample. The main parts of the instrument are the gamma- and x-ray detector system, the collimator, the MB drive and its control unit, and the data acquisition and spectrometer control unit. The detector system consists of Si PIN-diodes, charge sensitive pre- and filter amplifiers, and single channel analyzers. The miniaturized MB drive is of the double loudspeaker type and has a weight of about 50 g (see fig. 2). The spectrometer control and data acquisition system generates the velocity reference signal and collects the data of up to 5 individual detector channels. The data are stored in its own on-board memory, but can be transferred at any time via a standard serial interface to a computer. The instrument can be configured in different ways. For the field tests with the Rocky-7 rover the whole instrument, with two detectors (see fig. 1), including all the electronics, was assembled in a single housing with dimensions of about 90 mm x 65 mm x 45 mm and a total mass of less than 500g.

For the Mars Surveyor missions the instrument is split into two parts, an electronics board and a sensor head. The sensor head (see fig. 2) incorporates the drive, 4 detector systems, a reference detector system, the drive control unit, two MB sources, and a collimator. The electronics board, which will be placed in the temperature controlled compartment of the rover or lander, carries the microprocessor unit and data acquisition

Figure 1. The prototype of the Mössbauer spectrometer MIMOS II with two detector channels, and without outer housing.

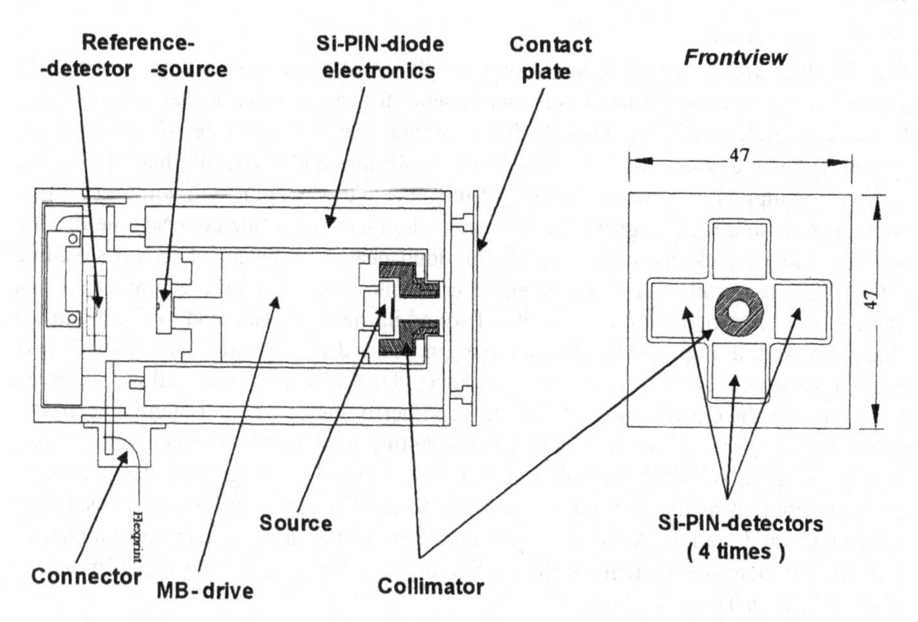

Figure 2. Scheme of the 5-detector-channel instrument sensor head design for the Mars-Surveyor missions; dimensions are in mm.

system, voltage converters, and electrical and data interfaces. Dimensions of the sensor head are 47 mm x 47 mm x 85 mm, having a mass of about 400 g, whereas the electronics board is 100 mm x 160 mm x 20 mm and has a mass of less than about 150 g. The design (scheme) of the instrument sensor head with 5 detector channels, one of them used to measure in transmission mode a calibration spectrum, is shown in fig. 2.

3.2.1. *Mössbauer Source and Shielding*

Especially for a space mission a high source intensity is desirable, with the constraint that the source line width should not increase significantly (a factor of 2-3) over the 1-3 year duration of the mission. The optimum specific activity for ^{57}Co is 1 Ci/cm^2 [12,16]. Sources of 150 to 300 mCi ^{57}Co in Rh with a specific activity close to this value, and with sufficiently narrow source line width (< 0.16 mm/s) have been tested successfully.

Very important is an effective shielding of the detector system, in particular the Si-diode, from direct and cascade radiation from the ^{57}Co source. A graded shield consisting of concentric tubes of brass, uranium, tungsten and another outer brass cylinder has good performances, but also a shield made out of tantalum works sufficiently good. This shielding also acts as a collimator, limiting the maximum emission angle to about 25° and reducing the cosine smearing [35] to a level that still allows a reasonable separation of the outer lines of γ- and α-Fe_2O_3.

3.2.2. *Drive System*
The simplest way to meet the space and weight constraints was to scale down drive systems we have built for laboratory instruments for many years at Darmstadt [7,20]. We constructed a drive system which had about one fifth the size of our standard system. It has a diameter of 22 mm, a length of 40 mm and about 50g mass [47]. The system is equipped with SmCo permanent magnets and was optimized with regard to a homogeneous and high magnetic field in the coil gap. A rigid tube connects the driving and the velocity pick up coils in the double-loudspeaker arrangement. Good shielding between the two coils is needed to avoid crosstalk. The short tube guarantees a fast transfer of information with the velocity of sound in the aluminium and thus a minimum phase lag and a high feedback gain margin. The drive operates at about 25 Hz frequency, which is also its main resonance. This low frequency allows a broad bandwidth for the closed loop system, and good performance with a triangular reference signal, but requires rather soft springs. The resulting nonlinearity between velocity and pickup voltage is still tolerable within its 0.1% accuracy. The design provides limiters to avoid destruction of the soft Kapton springs during the large accelerations associated with launch and landing. Vibration and shock tests of the drive system and the analog part of the detector system, with shocks up to 100 g have been performed at CNES/CNRS in Toulouse, France.

3.2.3. *Detector System and Electronics*
The instrument operates in backscattering geometry, which does not need special sample preparation for the measurement itself, and is therefore the geometry of choice for a spacecraft mission [15]. The main disadvantage of backscattering is the secondary radiation caused by primary 122 keV radiation. For a reduction of the background at the 14.4 keV gamma-ray and the 6.4 keV x-ray line, good energy resolution of the detector system is required. In addition a detector system covering a large solid angle and strong ^{57}Co sources are needed to minimize data acquisition time. For this reason, Si-PIN-diodes were selected as detectors [17,18,48] instead of a set of gas-counters as considered by other authors [1,43]. We have chosen Silicon-PIN-diodes with 10 x 10 mm^2 active area and a thickness of about 400 micrometer. The efficiencies for 6.4 and 14.4 keV radiation are nearly 100 % and about 65 % respectively [17]. The energy resolution of the Si-PIN-diodes improves at lower temperature [17,18]. The leakage current is the dominant source for line broadening at higher temperatures. The 100 V DC voltage for the diodes are generated by a high frequency cascade circuitry with a power consumption of less than 5 mW. Noise contributions have to be minimized by incorporating preamplifier-amplifier-SCA systems for each detector.

For velocity calibration a separate (fifth) channel has been added. The calibration spectrum of a reference sample will be recorded simultaneously with the backscatter spectra using a second source at the other end of the moving tube. A combination of reference absorbers is considered, as for instance α-Fe, γ- or α-Fe_2O_3, both magnetically split, and the quadrupole split SNP ($Na_2[Fe(CN)_5\ NO]2H_2O$).

The operational control of the experiment and the data handling is done by a microprocessor which is suitable for the special applications.

4. Laboratory and Field Tests on a Prototype Rover

The exploration of Mars within the next few years will focus on climate and life and the determination of the presence of water. For this the identification of the mineralogy of rocks and soils is of great importance. A major element of future Mars missions will be a Rover carrying instruments for in-situ investigations. One of these instruments will be the MB spectrometer MIMOS II. To investigate such a possible application and to demonstrate the capabilities of the instrument we have participated in a field test of the prototype Mars-Rover Rocky-7, the next generation of this type of Rover after Sojourner, who landed on Mars in July 1997 aboard NASA spacecraft 'Pathfinder'.

The field experiments with Rocky-7 have been performed at Lavic Lake, a dry lake bed with surrounding lava flows, located in the Mojave desert in California. This area includes sites that are analogs for primary rover targets on Mars, especially sites with

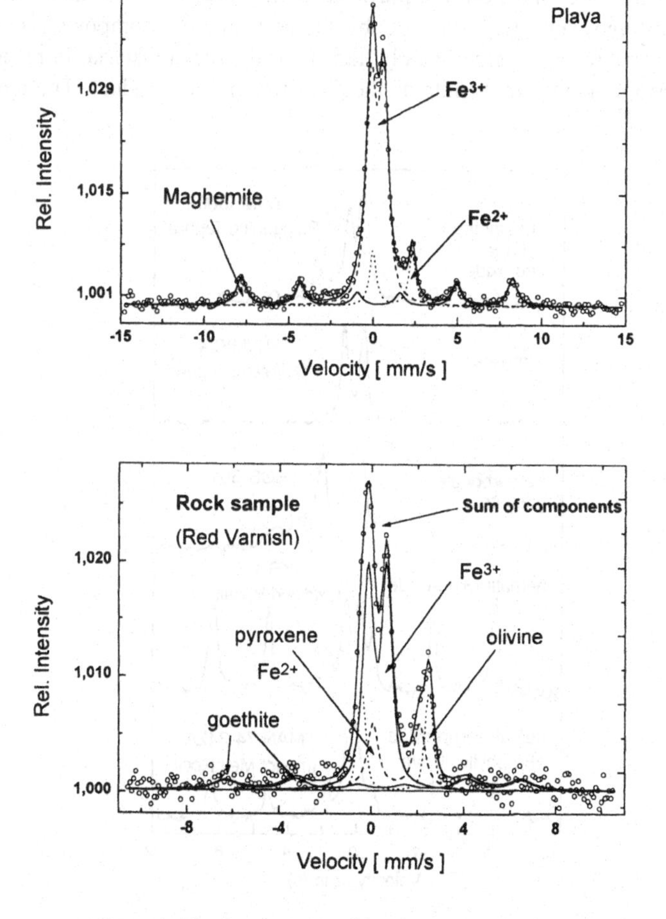

Figure 3. Mössbauer spectra of the playa and one rock sample.

extensive alteration of rocks by hydrothermal processes and landforms indicative of past wet climate. Main goals for the MIMOS II have been to investigate the influence of wind on the quality of the MB spectra, and to demonstrate that MB spectra with sufficient statistical quality can be obtained within a couple of hours (6-16h) depending of course on the iron content of the sample, the source strength and other parameters like the energy resolution of the detector system. To check the influence of wind on the spectra, the instrument, mounted on the robotic arm of the rover, was placed against a chunk of hematite to increase the signal to background ratio. We took data at different environmental conditions and compared the results with data taken in the laboratory. No significant line broadening (much less than the natural line width) could be observed. Furthermore we took data from a number of different kind of rocks found at the test site. Most of them have been analyzed with the instrument not being mounted on the robotic arm but inside a trailer, at temperatures ranging from +20 C to +45 C. Because of the high temperatures the noise level of the detectors increased.

In fig. 3 as an example data are shown obtained from a rock found at one of the basalt flows at the side, and from the playa of the dry lake bed. The data clearly show the presence of pyroxene (two Fe^{2+}-doubletts) and an Fe^{3+} component in the rock sample. The Fe^{3+} intensity is significantly higher for the playa material in respect to the rock, and it also contains maghemite as the final weathering product. The data for the

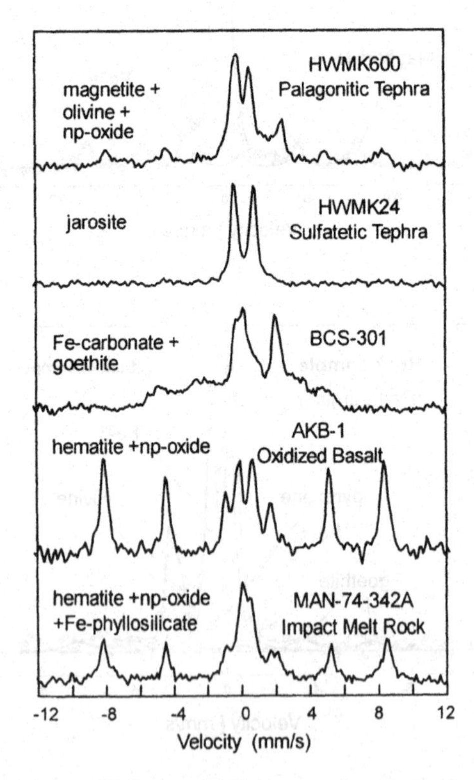

Figure 4. MIMOS II spectra of Martian sample analogues (see [42]).

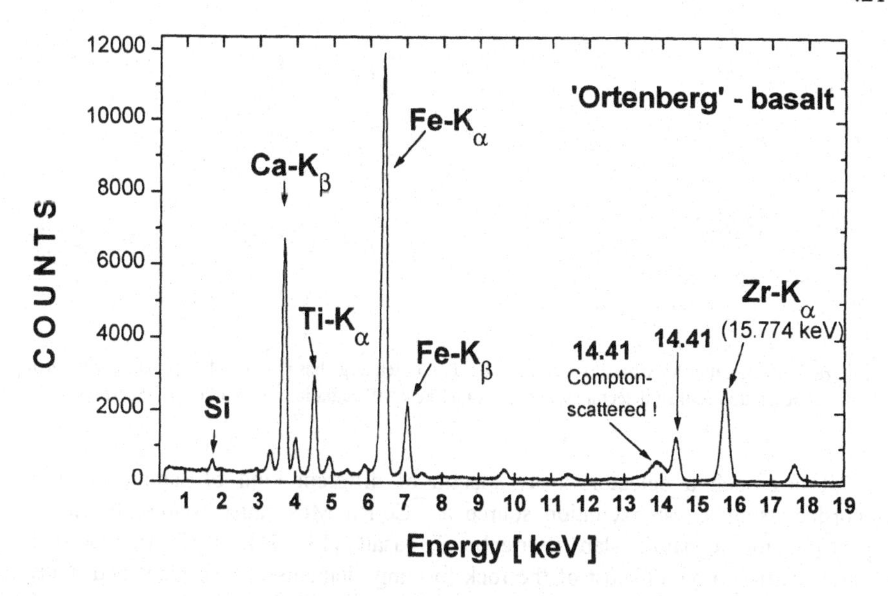

Figure 5. X-ray spectrum of a basalt („Ortenberg“ basalt; see [14,31]), taken with the high resolution Si-drift detector system. Excitation source: ^{57}Co in Rhodium matrix (Mössbauer source). Energy resolution: about 170 eV at +10° C.

rock samples also show the presence of goethite as an intermediate weathering product of the base rock material. Similar results have been found for other rocks. The data shown in fig. 3 have been recorded within less than 20 hours per spectrum using a 2-detector system and a 80 mCi $^{57}Co/Rh$ Mössbauer source.

Laboratory test measurements have been performed at room temperature to demonstrate the sensitivity of the instrument [42]. Several Martian sample analogues have been analyzed with the instrument MIMOS II. Results are shown in fig. 4.

5. Signal to Noise Ratio Considerations; High Resolution Detector Systems

The quality of a Mössbauer spectrum is related directly to the signal to noise ratio, where 'signal' corresponds to the Mössbauer resonance line, and noise corresponds to the non-resonant background in the spectrum. Expressions have been formulated in the literature to describe the quality of a Mössbauer spectrum [5, 7, 22, 23]. They have in common that the quality of the spectra increases with decreasing 'noise' (non-resonant background) much more than with increasing 'signal' intensity. This has been verified in this work using a state of the art high resolution x-ray detector developed recently at the MPI for Extraterrestik in Munich, Germany [39, 46]. These so called Si-drift detectors do have an extremely low capacity because of their special design and therefore an extraordinarily high energy resolution even at room temperature for x-rays in the keV range. With a moderate cooling to about –10°C an energy resolution of less than about 140 eV has been achieved recently [24].

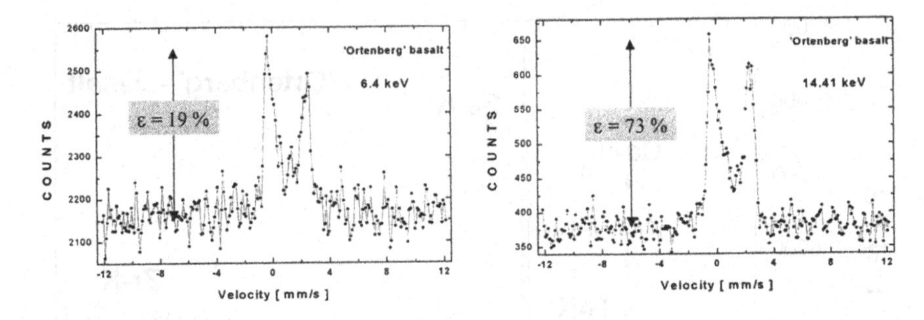

Figure 6. Backscatter Mössbauer spectra of a basalt („Ortenberg" basalt; see [14,31]), taken with the high resolution Si drift detector system: (a) 14.41 keV MB radiation; (b) 6.4 keV K_α-Fe-xrays.

Using such a detector, x-ray spectra of different kind of samples have been recorded using as an excitation source a ^{57}Co/Rh Mössbauer source. In fig. 5 the spectrum for a basalt slab ('Ortenberg' basalt [14, 31] is shown. Clearly the characteristic x-rays of most of the rock-forming elements can be identified. Only Al, Na, and Mg are difficult to see, which is due to the low excitation cross sections using x-ray sources instead of α-particle sources [44]. Of particular interest is the achieved separation of the 14.41 keV Mössbauer radiation from the Compton scattered 14.41 keV radiation (fig. 5). Backscatter Mössbauer spectra have been taken for this sample at the 14.41 keV Mössbauer line as well as at the Fe-K_α (6.403 keV) and Fe-K_β (7.057 keV) x-ray lines. The resonance effect ε (defined as the ratio (peak intensity – background) / background) for the 14.41 keV and the 6.4 keV data is 73% and 19%, respectively (fig. 6). This has to be compared to $\varepsilon \cong 4\%$ obtained for this sample with our standard Si-PIN diode at 14.41 keV with an energy resolution of about 2 keV. This clearly shows that the quality of the MB data can be improved significantly by using high resolution detectors. Unfortunately currently the sensitive area of these newly developed detectors is only about 7 mm^2 in comparison to 100 mm^2 of the Si-PIN diodes used in MIMOS II. Developments are underway to built arrays of those detectors with total sensitive areas in the order of 50 to 100 mm^2. First prototypes have been tested recently [39].

Besides the improvement in the quality of the data, with these high resolution detectors simultaneously the Fe mineralogy (via Mössbauer data) and the elemental composition (via the x-ray fluorescence spectrum) of the sample will be determined.

6. Terrestrial Applications in Industry, Materials and Fundamental Science

The small dimensions of our instrument in combination with the backscattering mode open up a variety of new terrestrial applications. Here we can give only very few examples. As described elsewhere (see [28, 29]), in iron ore processing plants the composition of the product depends strongly on the processing conditions. Therefore industry is interested in monitoring in-situ the composition of the output to be able to optimize the operation conditions. For instance the iron carbide processing of iron ores typically involves the phases α-Fe_2O_3, Fe_3O_4, $Fe_{1-x}O$, Fe_3C and Fe. We have used standard transmission MB spectroscopy during a one week long operation of an iron ore processing pilot plant to determine sufficiently fast quantitatively the amount of the different phases to control the plant and optimize the iron carbide output [28]. The use of the above described miniaturized instrument offers now both the prospect of 'real-time' measurements as well as scope of for on site evaluations of mineral findings.

Other possible applications are the inspection of pipes from the inside or outside to look after corrosion products. This can be of importance for industry dealing e.g. with corrosive products. The monitoring of the steel vessel of nuclear power plants can be of importance for the safety of nuclear power plants. Investigation of interfaces and coated materials are of great interest in different areas of fundamental research and in industry. The quality control of galvanized steel is of great commercial interest for industry, and Mössbauer backscattering spectroscopy is probably the only method to identify in short time online in a steel plant the buried iron-zinc intermetallics in galvannealed steel coatings [11]. In materials research the miniaturized instrument MIMOS II can be used to monitor in-situ the modification of materials. For instance the modification of thin surface layers of metals as well as of insulators (especially of current interest are ceramics) by ion implantation can be monitored directly. An experimental setup for such investigations is under construction at the GSI (Gesellschaft für Schwerionenforschung) near Darmstadt.

Recently we have used a prototype of the instrument to investigate the iron oxide composition of rock paintings in the field [26]. The archaeological place is located close to Santana do Riacho, about 100 km north-east of Belo Horizonte in Brasil. The instrument was mounted on a tripod, and the power was supplied by battery. Another application in the field is the monitoring of the Fe oxidation state and the iron mineralogy in the field over time as function of environmental conditions. We have installed an instrument in a two meter long tube with X-ray windows, which is buried in the soil of Brittany in France and will operate for a couple of months continuously [Klingelhöfer et al. in preparation]. First data have been obtained recently and are very promising.

Acknowledgements

The development of the instrument is funded by the German Space Agency DLR. Substantial additional funding is supplied by the Russian Space Research Institute IKI for the Mössbauer sources. The following people and / or institutions have been / are involved in the development of the instrument: B. Bernhardt, P. Held, J. Foh, E. Kankeleit, U. Bonnes, R. Teucher, F. Schlichting, O. Priloutskii, E. Evlanov, B. Zubkov, C. d'Uston, J.M. Knudsen, M.B. Madsen, B. Fegley Jr., R.V. Morris, and the teams of the mechanics and electronics workshops at the institute in Darmstadt. The Si drift detector was kindly supplied for tests by the company RÖNTEC, Berlin, Germany.

References

1. Agresti, D.G., Wills, E.L., Shelfer, T.D., Iwanczyk, J.S., Dorri, N., and Morris, R.V. (1990) Development of a solid-state Mössbauer spectrometer for planetary missions, *Lunar and Planetary Science* **XXI**, 5-6.
2. Arvidson, R.E., Acton, C., Blaney, D., Bowman, J., Kim, S., Klingelhoefer, G., Marshall, J., Niebuhr, C., Plescia, J., Saunders, S., and Ulmer, C.T. (1998) Rocky 7 prototype Mars rover field geology experiments 1. Lavic Lake and Sunshine Volcanic Field, California, *Journal of Geophysical Research*, **103**, 22671-22688.
3. Banin, A., Clark, B.C., and Wänke, H. (1992) Surface Chemistry and Mineralogy, in: H.H. Kieffer, B.M. Jakosky, C.W. Snyder, and M.S. Matthew (eds.) *Mars*, The University of Arizona Press, Tucson and London.
4. Bell III, J.F., McCord, T.B., and Owensby, P.D. (1990) Observational Evidence of Crystalline Iron Oxides on Mars, *J.Geophys. Res.* **95**, 14447-14461.
5. Bernhardt, B. (1997) Dipl. Güteuntersuchungen am Mössbauerspektrometer MIMOS. Dipl.Arbeit TU Darmstadt, Inst. für Kernphysik.
6. Bishop, J.L., Pieters, C.M., and Burns, R.G. (1993) Reflectance Spectra of Sulfate- and Carbonate-bearing Fe3+- doped Montmorillonites as Mars soil Analogs, *Lunar and Planetary Science* **XXIV**, 115-116.
7. Bokemeyer, H., Wohlfahrt, K., Kankeleit, E., and Eckardt, D. (1975) Mössbauer Conversion Spectroscopy: Measurements on the first excited states of $^{180,182}W$ and ^{145}Pm, *Z. Phys.* **A274**, 305-318.
8. Burns, R. and Banin, A. (eds.) (1992) Workshop on Chemical Weathering on Mars. *LPI Tech.Rpt.* 92-04, Part 1, Lunar and Planetary Institute, Houston.
9. Burns, R. and Banin, A. (eds.) (1993) Workshop on Chemical Weathering on Mars. *LPI Tech.Rpt.* 92-04, Part 2, Lunar and Planetary Institute, Houston
10. Carr, M.H. (1981) *The Surface of Mars*, Yale University Press, New Haven and London.
11. Cook, D.C. (1998) In-situ identification of iron-zinc intermetallics in galvannealed steel coatings and iron oxides on exposed steel, *Hyp. Int.* **111**, 71-82.
12. Evlanov, E.N., Frolov, V.A., Prilutski, O.F., Rodin, A.M., Veselova, G.V., and Klingelhöfer, G. (1993) Mössbauer Spectrometer for Mineralogical Analysis of the Mars Surface: Mössbauer Source considerations, *Lunar and Planetary Science* **XXIV**, 459-460.
13. Evlanov, E.N., Mukhin, L.M., Prilutski, O.F., Smirnov, G.V., Juchniewicz, J., Kankeleit, E., Klingelhöfer, G., Knudsen, J.M., and d'Uston, C. (1991) Mössbauer Backscatter Spectrometer for Mineralogical Analysis of the Mars Surface for Mars-94 Mission, *Lunar and Planetary Science* **XXII**, 361-362.
14. Fegley, Jr., B., Klingelhöfer, G., Lodders, K., and Wiedemann, T. (1997) Geochemistry of Surface-Atmosphere Interactions on Venus. in: S.W. Bougher, D.M. Hunten, and R.J. Phillips (eds.) *Venus II - Geology, Geophysics, Atmosphere, and Solar Wind Environment*, University of Arizona Press, 1997, pp. 591-636.
15. Galazkha-Friedman, J. and Juchniewicz, J. (1989) Martian Mössbauer Spectrometer MarMös, *Project Proposal, Space Research Center*, Polish Academy of Sciences, February 1989.
16. Gummer, A.W. (1988) Effect of accumulated Decay Product on the Mössbauer Emission Spectrum, *Nucl. Inst. Meth.* **B34**, 224-227.
17. Held, P. (1993) PIN-Photodioden als Detektoren für das Mössbauerspektrometer MIMOS zur Untersuchung der Marsoberfläche, Diploma Thesis, TH Darmstadt, Inst. f. Nuclear Physics.
18. Held, P., Teucher, R., Klingelhöfer, G., Foh, J., Jäger, H., and Kankeleit, E. (1993) Mössbauer Spectrometer for Mineralogical Analysis of the Mars Surface: First temperature dependent tests of the detector and drive system, *Lunar and Planetary Science* **XXIV**, 633-634.
19. Hviid, S.F., Madsen, M.B., Gunnlaugsson, H.P., Goetz, W., Knudsen, J.M., Hargraves, R.B., Smith, P., Britt, D., Dinesen, A.R., Mogensen, C.T., Olsen, M., Pedersen, C.T., and Vistisen, L. (1997) Magnetic Properties Experiments on the Mars Pathfinder Lander: Preliminary Results, *Science* **278**, 1768-1770.
20. Kankeleit, E.(1964) Velocity Spectrometer for Mössbauer Experiments, *Rev. Sci. Instr.* **35**, 194-197.
21. Kankeleit, E., Foh, J., Held, P., Klingelhöfer, G., and Teucher, R. (1994) A Mössbauer Experiment on Mars, *Hyperfine Interactions* **90**, 107-120.
22. Kajcsos, Zs., Sauer, Ch., and Zinn, W. (1990) Criteria for optimizing DCEMS, *Hyp. Int.* **57**, 1889-1900.
23. Kajcsos, Zs., Sauer, Ch., Zinn, W., Kurz, R., Meyer, W., and Ligtenberg, M.A.C. (1990) On the performance of UHV-ICEMS experiments, *Hyp. Int.* **58**, 2519-2524.

24. Kemmer, J. and Lechner, P. (1998) private communication.
25. Kieffer, H. H., Jakosky, B.M., Snyder, C.W., and Matthew, M.S. (1992) *Mars,* The University of Arizona Press, Tucson.
26. Klingelhöfer, G., da Costa, G.M., Prous, A., and Bernhardt, B. (1998) In-situ Mössbauer spectroscopy of Rock Paintings from Minas Gerais (Brazil), in preparation.
27. Klingelhöfer, G. (1998) In-Situ analysis of planetary surfaces by Mössbauer spectroscopy, *Hyp. Int.* **113**, 369-374.
28. Klingelhöfer, G., Campbell, S.J., Wang, G.M., Held, P., Stahl, B., and Kankeleit, E. (1998) Iron ore processing – in-situ monitoring, *Hyp. Int.* **111**, 335-339.
29. Klingelhöfer, G., Held, P., Bernhardt, B., Foh, J., Teucher, R., and Kankeleit, E. (1998) In-situ phase analysis by a versatile miniaturized Mössbauer spectrometer, *Hyp. Int.* **111**, 331-334.
30. Klingelhöfer , G., Fegley Jr., B., Morris, R.V., Kankeleit, E., Held, P., Evlanov, E., and Priloutskii, O. (1996) Mineralogical analysis of Martian soil and rock by a miniaturized backscattering MB spectrometer, *Planet. Space Sci.* **44**, 1277.
31. Klingelhöfer, G., Fegley,Jr., B., Held, P., and Osborne, R. (1996) Mössbauer Studies of Basalt Oxidation in Simulated Martian Atmospheres, *Meteoritics* **31**, A71.
32. Klingelhöfer, G., Fegley Jr., B., Morris, R.V., Kankeleit, E., Evlanov, E., Priloutskii, O., Knudsen, J.M., and Madsen, M.B. (1996) Mineralogy of the Martian surface analyzed in-situ by Mössbauer spectroscopy, and implications for volatile evolution on Mars, in: Jakosky B. and Treiman A. (eds.) *Workshop on Evolution of Martian Volatiles,* LPI Tech.Rpt. 96-01, Part 1, Lunar and Planetary Institute, Houston.
33. Klingelhöfer, G., Held, P., Teucher, R., Schlichting, F., Foh, J., and Kankeleit, E. (1995) Mössbauer Spectroscopy in Space, *Hyp. Int.* **95**, 305-339.
34. Klingelhöfer, G., Foh, J., Held, P., Jäger, H., Kankeleit, E., and Teucher, R. (1992) Mössbauer Backscattering Spectrometer for Mineralogical Analysis of the Mars Surface, *Hyp. Int.* **71**, 1449-1452.
35. Klingelhöfer, G., Imkeller, U., Kankeleit, E., and Stahl, B. (1992) Remarks on depth selective CEMS - backscattering measurements, *Hyp. Int.* **71**, 1445-1448.
36. Knudsen, J. M. (1989) Mössbauer Spectroscopy of ^{57}Fe and the Evolution of the Solar System, *Hyp. Int.* **47**, 3-31.
37. Knudsen, J.M., Moerup, S., and Galazkha-Friedman, J. (1990) Mössbauer Spectroscopy and the Iron on Mars, *Hyp.Int.* **57**, 2231-2234.
38. Knudsen, J.M., Madsen, M.B., Olsen, M., Vistisen, L., Koch, C.B., Moerup, S., Kankeleit, E., Klingelhöfer, G., Evlanov, E.N., Khromov, V.N., Mukhin, L.M., Prilutskii, O.F., Zubkov, B., Smirnov, G.V., and Juchniewicz, J. (1992) Mössbauer spectroscopy on the surface of the planet Mars. Why?, *Hyperfine Interactions* **68**, 83-94.
39. Lechner, P., Eckbauer, S., Hartmann, R., Krisch, S., Hauff, D., Richter, R., Soltau, H., Strüder, L., Fiorini, C., Gatti, E., Longoni, A., and Sampietro, M. (1996) Silicon drift detectors for high resolution room temperature X-ray spectroscopy, *Nuclear Instruments and Methods* **A 377**, 346-351.
40. Madsen, M.B., Knudsen, J.M., Vistisen, L., and Hargraves, R.B. (1993) Suggestion for extended Viking Magnetic Properties Experiment on future Mars Missions, *Lunar and Planetary Science* **XXIV**, 917.
41. Morris, R.V., Agresti, D.G., Shelfer, T.D., and Wdowiak, T.J. (1989) Mössbauer Backscatter Spectrometer: a new approach for Mineralogical Analysis on Planetary Surfaces, *Lunar and Planetary Science* **XX**, 721-722.
42. Morris, R.V., Squyres, S.W., Bell III, J.F., Christensen, P.H., Economou, T., Klingelhöfer, G., Held, P., Haskin, L.A., Wang, A., Jolliff, B.L., and Rieder, R. (1998) Analyses of Martian surface materials during the Mars Surveyor 2001 mission by the ATHENA instrument payload, *Lunar and Planetary Science* **XXIX**.
43. Prilutskii, O. (1990) Space Research Institute (IKI), Moscow, internal report from Minsk.
44. Rieder, R., Economou, T., Wänke, H., Turkevich, A., Crisp, J., Brückner, J., Dreibus, G., and McSween Jr., H.Y. (1997) The Chemical Composition of Martian Soil and Rocks Returned by the Mobile Alpha Proton X-ray Spectrometer: Preliminary Results from the X-ray Mode, *Science* **278**, 1771-1774.
45. Shelver, T. D., Morris, R.V., Nguyen, T.Q., Agresti, D.G., and Wills, E.L. (1995) Backscatter Mössbauer spectrometer (BaMS) for solid-surface extraterrestrial mineralogical analysis, *Lunar and Planetary Science* **XXVI**, 1279-1280.
46. Strüder, L. and Kemmer, J. (1996) Neuartige Röntgendetektoren für die Astrophysik, *Phys. Blätter* **52**, 21-26.

47. Teucher, R. (1994) Miniaturisierter Mössbauerantrieb, *Diploma Thesis*, TH Darmstadt, Inst. f. Nuclear Physics.
48. Weinheimer, Ch., Schrader, M., Bonn, J., Loeken, Th., and Backe, H. (1992) Measurement of energy resolution and dead layer thickness of LN2-cooled PIN photodiodes, *Nucl. Inst. Meth.* **A311**, 273-279.
49. Knudsen, J. M., University of Copenhagen, Denmark; private communication, 1995.
50. 'Rocky-7 field test' web page at Washington University, St.Louis, Missouri, USA: http:// wundow.wustl.edu/rocky7/
51. ATHENA mission web page at Cornell University, USA: http://astrosun.tn.cornell.edu/athena/index.html
52. Jet Propulsion Laboratory (JPL/NASA) home page (Pasadena, California, USA): http://www.jpl.nasa.gov
53. US-American space agency NASA home page: http://www.nasa.gov
54. European space agency ESA home page: http://www.esa.int/esa/new.html

Author Index

SUBJECT INDEX